Springer Series on
Wave Phenomena 7

Springer Series on

Wave Phenomena

D. F. Parker G. A. Maugin (Eds.)

Recent Developments in Surface Acoustic Waves

Proceedings of
European Mechanics Colloquium 226,
University of Nottingham, U.K., September 2–5, 1987

With 139 Figures

Springer-Verlag Berlin Heidelberg New York
London Paris Tokyo

Dr. David F. Parker

University of Nottingham, Department of Theoretical Mechanics, Nottingham, NG7 2RD, U.K.

Professor Gérard A. Maugin

Laboratoire de Modélisation en Mécanique, Université Pierre et Marie Curie, 4 Place Jussieu, F-75252 Paris Cédex 05, France

Series Editors:

Professor Leonic M. Brekhovskikh, Academician

P. P. Shirsov Institute of Oceanology, Academy of Sciences USSR, SU-117218 Moscow, USSR

Professor Leopold B. Felsen, Ph. D.

Polytechnic University, Route 110, Farmingdale, NY 11735, USA

Professor Hermann A. Haus

Department of Electrical Engineering & Computer Sciences, MIT, Cambridge, MA 02139, USA

Managing Editor: Helmut K. V. Lotsch

Springer-Verlag, Tiergartenstraße 17,
D-6900 Heidelberg, Fed. Rep. of Germany

ISBN-13:978-3-642-83510-0 e-ISBN-13:978-3-642-83508-7
DOI: 10.1007/978-3-642-83508-7

Library of Congress Cataloging-in-Publication Data. European Mechanics Colloquium (226th : 1987 : University of Nottingham) Recent developments in surface acoustic waves : proceedings of European Mechanics Colloquium 226, University of Nottingham, U.K., September 2–5, 1987 / editors, D.F. Parker and G.A. Maugin. – (Springer series on wave phenomena ; 7) Includes index. 1. Acoustic surface waves–Congresses. I. Parker, D.F. (David F.), 1940-. II. Maugin, G.A. (Gérard A.), 1944-. III. Title. IV. Series. QC176.8.A3E97 1987 530.4'1-dc 19 88-16049

Softcover reprint of the hardcover 1st edition 1988

2154/3150-543210

Preface

The topic of surface waves lies at the interface between a number of disciplines – physics, theoretical and applied mechanics, electroacoustics, applied mathematics, surface science and seismology. This volume, based on papers delivered at European Mechanics Colloquium 226, reflects this diversity in approach and background, while showing strong links between phenomena arising from different fields. The emphasis is on recent developments such as nonlinear and other nonclassical effects,which have great importance for both pure science and for applications such as signal processing, nondestructive evaluation and seismic studies.

In recent years there has been considerable progress in the mathematical treatment of nonlinear effects, of viscoelastic and of more novel constitutive effects which modify the predictions of linear elastic and piezoelectric theory for surface acoustic wave (SAW) propagation. A number of these themes serve to group the contents of this volume. Part I contains recent advances in the rigorous mathematical treatment of nonlinearity, together with a paper giving experimental results showing the need for further theoretical development. Part II deals with anisotropic elasticity, showing that even the linear theory presents many possible behaviours, which are still not fully categorized.

Parts III and V concern two themes which emerged strongly from the papers presented. The first is the analysis of waves travelling across rough or corrugated surfaces. The papers show that these (in practice unavoidable) irregularities can produce a variety of effects such as wave slowing, stop bands and retroreflection. The second theme is the interest in waves on composites, particularly periodically layered laminates. As shown in the survey which opens Part V, application is not just to waves in structural composites but also to phenomena on the microscopic scale of superlattices. The accompanying papers emphasize the need for approximations. They well represent the two main approaches, either simplification of the modal analysis within the layers of a composite, or else idealization to "strictly inextensible" materials.

In Part IV, a number of electromechanical interaction processes are described. These couplings are basic to the design of all SAW devices such as are used in signal processing. One paper discusses current procedures for transducer design, while others describe a number of features of piezoelectric waves and of surface waves in materials with more novel interaction mechanisms. The final part, Part VI, includes analysis of waves guided along cylindrical cavities, surface waves used both in flaw detection (material evaluation) and in instruments for evaluating rheological properties of a surrounding fluid. It also includes treatment of surface waves in viscoelastic and thermoelastic materials.

It is hoped that the reader will find, in this collection, an up-to-date account of theoretical methods for analysis of surface waves, treatment of a number of novel SAW phenomena and suggestions for future research. For this the editors have to thank the many authors for their enthusiasm at the Colloquium and their readiness to prepare written accounts.

We wish to put on record our gratitude to the European Research Office of the U.S. Army for financial support and encouragement in running the Colloquium. Also, we wish to thank the University of Nottingham for providing a wide range of help and facilities, and in particular Florence Boot Hall for providing a pleasant ambience for the meeting.

Nottingham, Paris, February 1988

D.F.Parker
G.A.Maugin

Contents

Part II Surface Waves on Anisotropic Substrates

Part III Surface Waves on Rough Surfaces

Part IV Surface Waves on Electromagnetic Bodies

Part I

Nonlinear Surface Acoustic Waves and Applications

State-of-the-Art Elastic SAW Convolvers

H.P. Grassl, L. Reindl, and H.-W. Wörz

Siemens AG, Zentrale Forschung und Entwicklung,
Otto-Hahn-Ring 6, D-8000 München 83, Fed. Rep. of Germany

0. Summary

Elastic convolvers are the only devices in the broad spectrum of modern SAW-components which use a nonlinear effect. With a well-founded design scheme based on reproducible, all-planar technologies, flexible and efficient small-aperture focussing chirp transducers, and with accurate CAD-tools to simulate and optimize these planar structures, devices with excellent performance suited for mass production can be made. This article gives a short review of convolver design principles and an outline of design procedures for state-of-the-art devices.

1. General introduction on SAW convolvers

Surface acoustic wave filters as they are widely used in many kinds of communication and signal processing equipment are normally assumed to be linear time-invariant filters. In fact, any nonlinearity or time dependence of such devices is very undesirable, because it would lead to intermodulation, phase noise or ageing effects. The elastic convolver is an exception, because it cultivates a nonlinear effect for its normal function. The principle of the surface wave convolver is depicted in Fig. 1. Broadband interdigital transducers convert two independent signals (the "signal" s(t) and a "reference" r(t)) to acoustic surface waves which counterpropagate at a velocity v on a common path. Thus they form time-dependent wave fields

$$a_s(t-(x+L/2)/v) \quad \text{and} \quad a_r(t+(x-L/2)/v). \qquad (1)$$

Between the transducers a spatially extended detector senses the polarisation produced by the wave field. When the piezoelectric response of the substrate is essentially linear, it may be expanded into a power series and the output signal at the output plate assumes the form

$$c(t) = \int_{-L_i/2}^{L_i/2} \{\, c_L\cdot(a_s(x,t)+a_r(x,t)) + c_{NL}\cdot(a_s(x,t)+a_r(x,t))^2 + \ldots\} \, dx . \qquad (2)$$

It can be shown that integration over the linear terms and those quadratic in a_s and a_r leads to vanishing contributions in the output signal, because their phase varies rapidly with space. (The output plate is several thousand wavelengths long typically.) Only the bi-linear term remains and can be written as

$$c(t/2+T_d/2) = C_{NL} \cdot \int_{(t-T_i)/2}^{(t+T_i)/2} s(t-\tau)\cdot r(\tau)\, d\tau, \qquad (3)$$

which is virtually the convolution of the two input signals. This expression deviates from a true convolution due to the finiteness of the integration boundaries and to a factor of 2 by which the time scale in the output signal is compressed. Accordingly all frequencies of the output spectrum are doubled.

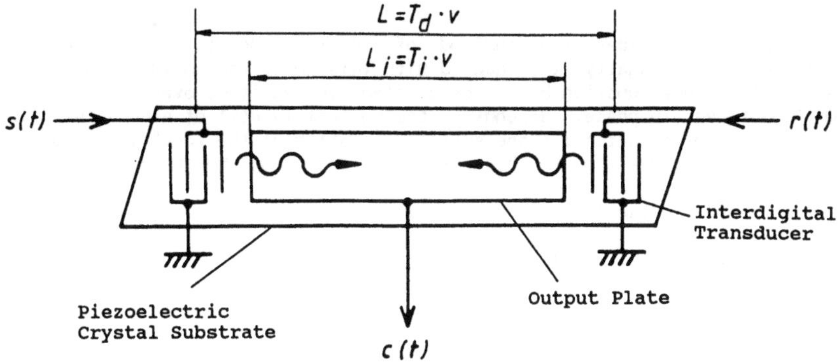

Fig. 1. Principle of SAW convolver

In some SAW convolvers the nonlinearity is produced via the interaction of the surface wave piezoelectric fields with free charge carriers in semiconductor substrates or overlays ("acousto-electric convolvers"), whereas the "elastic" convolver uses the very weak inherent nonlinearity of the piezoelectric insulating substrate, usually YZ-$LiNbO_3$, involving third-order elastic, dielectric, piezoelectric, and electrostrictive material constants as well as geometric nonlinearities /1,2/. In this case the spatially integrating and sensing element for the output signal is simply a metal film electrode along the interaction region. This form of monolithic convolver can be fabricated with the same highly developed technology which is used to produce other SAW filters very reproducibly at low cost and in large quantities. It has today reached the highest degree of maturity, with both large bandwidth and integration time, and is ready to function reliably in such systems as spread spectrum communications equipment /3/, radar signal processors /4/, image processors /5/, speech recognition, etc., where, its signal processing capability equivalent to more than 10^{11} multiplications per second (with about 8 bit resolution of the output) is utilized. A comprehensive review article on the theory and practice of nonlinear Rayleigh-wave components has already been written by MAERFELD /1/ and on acousto-electric convolvers and other devices e.g. by STERN /6/.

2. Analysis of elastic convolvers

A device which produces an output signal according to (3) would be considered an ideal convolver. For practical applications the constant C_{NL} should be as large as possible to provide enough dynamic range with a minimum requirement for input driving power. The ideal convolver delivers, for signals which completely overlap within the device, exactly the convolution in the time domain, which corresponds to the product of the two input signal spectra in the frequency domain. A real convolver will introduce some amount of linear and nonlinear distortions and perhaps add other spurious output signal components. The design goal for a real convolver is therefore to maximize the output signal while minimizing distortions and spuria. In addition, the design has to be such that the device can be manufactured reproducibly at a low cost and meet the requirements of system designers regarding ruggedness, hermeticity, temperature stability, compactness, and weight.

The quality of a convolver is judged primarily by its efficiency E(f), which is defined as the ratio of output signal amplitude to the product of input signal amplitudes at corresponding frequencies 2f and f respectively:

$$C(2f) = S(f) \cdot R(f) \cdot E(f) \,. \qquad (4)$$

Since on a logarithmic scale the magnitude of the efficiency is a summation of values measured in dBm usually it is quoted with this unit. For an ideal convolver the complex efficiency function has to be constant in magnitude over the required bandwidth and linear in phase. Secondly, the uniformity $u(\tau)$ of a convolver is regarded. This term characterizes the spatial variation of the local contribution to the convolution integral due to physical variations along the integration path or due to distortions of the electrical output signal components on their way from where they originate to the output port of the device:

$$c(t) = \int_{T_i} s(t-\tau) \cdot r(t+\tau) \cdot u(\tau)\, d\tau \,. \qquad (5)$$

The ideal convolver has constant uniformity along the whole interaction path.

For a general convolver, which is a device with two inputs and one output involving some nonlinearity, the question arises how such a device can be appropriately characterized by measurements in a manner similar to the measurements made to characterize linear filters. Analysis of such a device yields that in addition to regular transfer functions, which characterize the linear (spurious) signal transmission from the inputs to the output port, three two-dimensional transfer functions are required to fully characterize the nonlinearly (bilinearly) transmitted signals /7/. The convolution response is characterized in the frequency domain by complex functions $H_{sr}(f_1,f_2)$, $H_{ss}(f_1,f_2)$, and $H_{rr}(f_1,f_2)$:

$$\begin{aligned} C(2f) = \int_{-\infty}^{\infty} \{ & H_{sr}[(f+\delta),(f-\delta)] \cdot S[(f+\delta)] \cdot R[(f-\delta)] + \\ & H_{ss}[(f+\delta),(f-\delta)] \cdot S[(f+\delta)] \cdot S[(f-\delta)] + \\ & H_{rr}[(f+\delta),(f-\delta)] \cdot R[(f+\delta)] \cdot R[(f-\delta)] \}\, d\delta \,. \end{aligned} \qquad (6)$$

$H_{sr}(f_1,f_2)$ contains the information on how the two input signals mix to produce the (desired) bi-linear output signal, while $H_{ss}(f_1,f_2)$ and $H_{rr}(f_1,f_2)$ contain information on the most critical spurious signal in the convolver, called self-convolution. It occurs when the input signals convolve with portions of themselves due to reflections of acoustic waves somewhere along the propagation path, most likely at the opposite transducer. With a computer-controlled network analyser the two-dimensional transfer functions can be measured very accurately. The CW output at the sum of frequencies f_1 and f_2 is recorded, when these frequencies are applied to the inputs. Figure 2 shows an example of one-dimensional scans measured through $H_{sr}(f_1,f_2)$ of a real convolver. The scan along the main diagonal of the

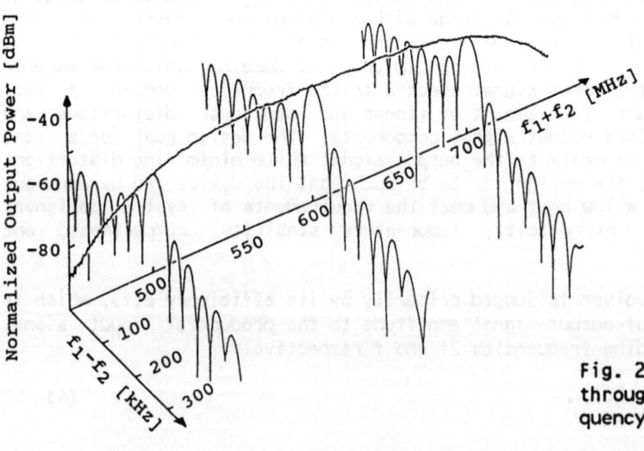

Fig. 2. Measured scans through two-dimensional frequency response $H_{sr}(f_1,f_2)$

(f_1,f_2)-plane is the efficiency function $E(f_1+f_2)$. It can be shown that the Fourier transform of a scan perpendicular to the main diagonal represents the uniformity $u(\tau)$. In /8/ it is demonstrated that also the off-diagonal areas of $H_{sr}(f_1,f_2)$ may contain specific information on spurious signals. If the uniformity is independent of frequency, then $H_{sr}(f_1,f_2)$ can be written as a product of the efficiency and the Fourier transform of the uniformity, and (4) applies. In that case any distortion occurring in the efficiency function could be corrected with a linear time-invariant filter cascaded with the output of the convolver.

The nonlinearity used in an elastic convolver is very weak and great care has to be taken to maximize the useable output signal. Figure 3 shows schematically the elements needed to make an efficient elastic convolver. The input transducers have to be electrically matched to the signal sources. The acoustic output of the transducers has to be concentrated to an output plate as narrow as possible. The narrow acoustic beam needs to be confined by a waveguiding structure to prevent diffraction. Fortunately the metal strip which serves as the output integrating electrode also guides the acoustic waves. Propagation of surface waves is in distinct waveguide modes then, which means nonuniform transverse amplitude distributions and frequency dispersion. The output signal occurs distributed along the waveguides. The thin metal strip on the high-ϵ dielectric behaves also as an electromagnetic waveguide of considerable electrical length and loss. Therefore the output signal has to be taken from multiple, evenly distributed taps to ensure good signal yield and uniformity.

In Fig. 3 two convolver paths are seen to be connected in parallel. The reason for this is to suppress unwanted reflections at the transducers. On one end of the device the transducers are connected acoustically in phase while on the other end they are connected with opposite acoustic phase. Thereby the acoustic waves arrive at each end with such a relative phase, that their electrical responses in the connected transducers cancel; otherwise the voltages induced in the transducers would lead to re-emission of "regenerated" waves, the result of which would be a high level of self-convolution. The self-convolution effect is the most serious spurious effect in convolvers for spread spectrum jam resistant systems, because any signal which exhibits a symmetry with respect to time inversion will lead to a high self-convolution output. Thus a simple CW signal can be used to jam such a system where self-convolution is not sufficiently suppressed. Other applications may be much more insensitive with respect to self-convolution, since self-convolution signals have twice the group delay of the ordinary convolution, and can be removed by gating in certain cases.

There are a number of other spurious effects which may occur in convolvers. Direct feedthrough between input and output ports of the device is to be watched.

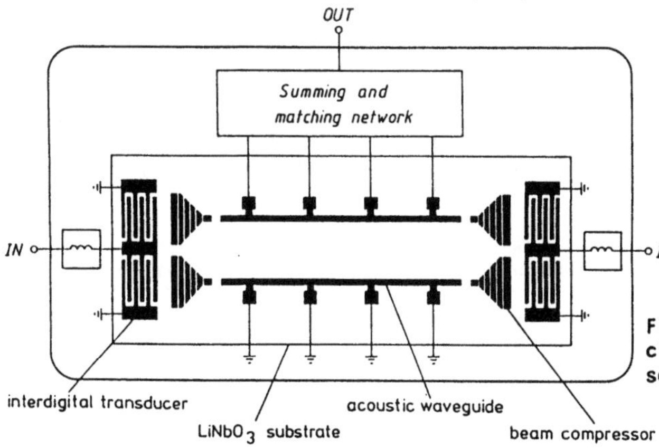

Fig. 3. Schematic of efficient elastic convolver with self-convolution suppression

However, since input and output signals lie in different frequency bands, ordinary bandlimiting filters may be used to suppress the crosstalk between input and output. Good isolation in the convolver reduces or eliminates the requirement for such filters. There are two possible kinds of crosstalk. One is purely electromagnetic and may be removed by shielding measures. The other is acoustic and may be minimized by proper shaping of metal edges of the output plate structure, where some signal may be picked up. Echoes in the convolved signal may occur if multiple waveguide modes are excited, which propagate at different velocities. Multimoding can be avoided by using monomode waveguides or by forming the acoustic beam which is fed to the waveguide so that only a single waveguide mode is excited. Another effect which may degrade the convolver performance is the dispersion introduced by the acoustic waveguide. A narrow metal strip waveguide leads to dispersion by two different causes. One is the geometrical confinement and the other is the finite thickness of the metal layer, which varies in relation to the wavelength over frequency. It is possible, however, to find a combination of metal thickness and strip width where dispersion is minimized over a fairly large bandwidth.

In addition to the above mentioned linear spurious effects there are possible nonlinear parasitic effects. One effect which is often observed is due to the fact that simultaneously with the nonlinear polarization, which is responsible for the electrical output signal, a nonlinear strain occurs which leads to the immediate emission of acoustic waves perpendicular to the surface. They travel through the substrate and bounce back to the output plate, where they are detected and appear as a delayed convolution signal. Roughening the back of the substrate and use of an adhesive which is acoustically matched to the substrate reduces the effect by absorption and diffuse scattering. The power density in the compressed acoustic field may become so large that one would expect noticeable second and higher harmonic generation /9/ and possibly coherent third order intermodulation. Fortunately the dispersion in the waveguide in practice is sufficient to reduce the coherence length for these effects enough that they do not become deleterious or even measurable.

3. Convolver configurations

The nonlinear signal yield is expressed by the figure of merit M, which is defined by

$$V_{OC} = M \cdot \sqrt{(P_1 \cdot P_2 / W^2)}, \qquad (7)$$

where V_{OC} is the open circuit output voltage on the output plate and P_1/W and P_2/W are the acoustic power densities. There are two factors by which the device designer can influence V_{OC}: the choice of substrate material and crystal orientation, and the geometry of the output detection electrode(s), which affect M and the power densities. The maximum available output power P_{max} does, however, not only depend on V_{OC} but also on the capacitance of the output electrode which decides how well the output port can be matched to a real load impedance over a given bandwidth. Therefore, compressing the acoustic beam into a narrow output plate, either by separate focussing structures or focussing transducers, will enhance both V_{OC} by increasing the power densities and P_{max} by reducing the strip capacitance.

The smallness of the nonlinearity made the elastic convolver initially inferior to acoustoelectric convolvers. However, its ruggedness and simplicity in fabrication made it very attractive, so means were devised to overcome the disadvantage of the weak nonlinearity. GANGULY and DAVIS /2/ published a theoretical analysis of the nonlinear signal generation in an elastic convolver. They calculated a theoretical figure of merit $M=2.6 \cdot 10^{-4}$ Vm/W on YZ-$LiNbO_3$. Their theory also predicted the emission of nonlinearly excited bulk waves in agreement with the observation. PLANAT et al. /10/ conducted a finite element analysis of the nonlinear field, taking into account the actual geometry of the output electrode and the wave field, and arrived at $M=0.5 \cdot 10^{-4}$ Vm/W. LIM et al. /11/ made an experimental survey of different materials with regard to their suitability as substrates for elastic convolvers. A suitable material for convolvers has to have a large electromechanical coupling constant for two reasons. Firstly, the insertion loss of the input transducers goes proportionally into the convolver efficiency, and, secondly, the nonlinearity is also related to the electromechanical coupling constant. However, a material must not only have appropriate values of its constants, but also be available in large enough substrate sizes and with reproducibly homogeneous quality. In view of all these requirements, to date no material has appeared which is better suited for elastic convolvers than YZ-$LiNbO_3$. SHTYKOV et al. /12/ verified by experiment that the convolution figure of merit on the Y-cut of $LiNbO_3$ is in fact the highest in the Z-direction of propagation. As no more efficient substrates for elastic convolvers were found, some workers explored possibilities to find other geometrical configurations which allow higher efficiency convolvers. MONKS et al. /13/ described convolvers using a transverse bilinear field configuration (see Fig. 4) on various substrates. NAKAGAWA et al. /14/ tried a convolver configuration where bilinearly excited surface waves propagating normally to a waveguide consisting of multiple parallel strips are detected by very large aperture interdigital transducers placed along the waveguide structure on both sides. Similarly one could think of using the nonlinearly excited bulk waves under the strip waveguide, which are normally regarded as spurious waves, as was suggested in /18/.

After all, the best practical results were obtained with the conventional configuration, combined with diligent designs for interdigital transducers, beam compression, strip waveguide and output circuitry. Nondispersive interdigital transducers with a given bandwith need a certain aperture of about 60 wavelengths to obtain a match to a real 50 Ω source with simple series inductance tuning. When a much smaller aperture is used, the matching network becomes complicated and lossy. Therefore it is more efficient to use the optimum aperture for the transducer and to compress the acoustic wavefield in order to achieve a high power density. DEFRANOULD and MAERFELD /15/ introduced a multistrip beam compressor in the elastic convolver which boosted its efficiency to levels comparable to that of acoustoelectric convolvers. Figure 5 compares different beam compressing schemes. RØNNEKLEIV /16/ designed a diffraction-compensated multistrip beam compressor consisting of smoothly curved strips which transform an incoming plane wave with rectangular profile into a beam whose profile matches exactly that of the fundamental waveguide mode. Figure 6 demonstrates the excellent performance of this beam compressor as seen in the voltage contrast mode of a SEM. Parabolic horn compressors adopted from integrated optics have been used successfully in /17,18/. Other ways of coupling into the narrow waveguide are provided with the prism coupler shown in Fig. 5c /19/ and focussing transducers as described in /20/ and /21/.

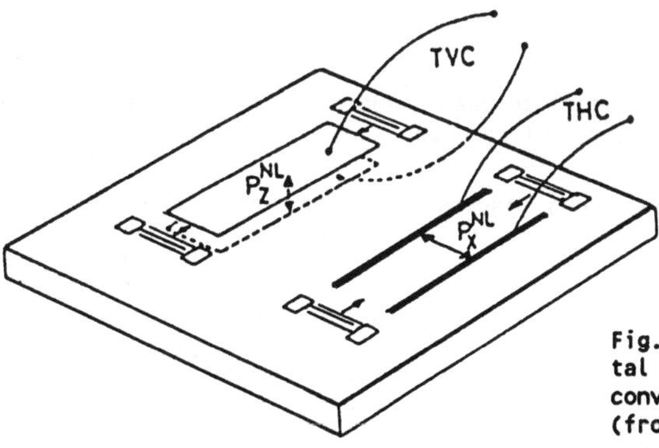

Fig. 4. Transverse horizontal vs. transverse vertical convolver (from /13/)

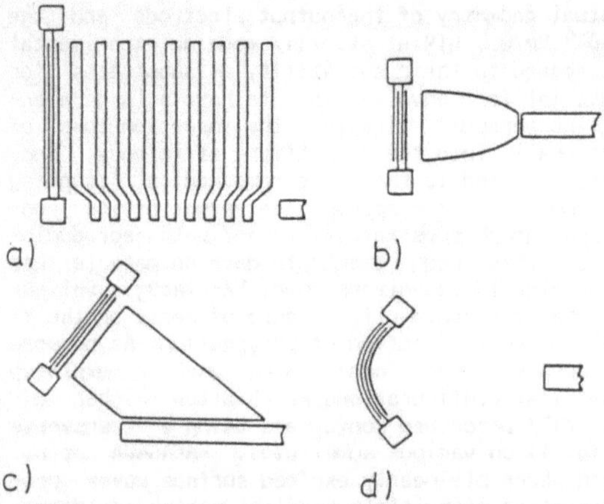

Fig. 5. Beam compressor types
a) Multistrip beamcompressor
b) Parabolic horn compressor
c) Planar prism coupler
d) Focussing transducer

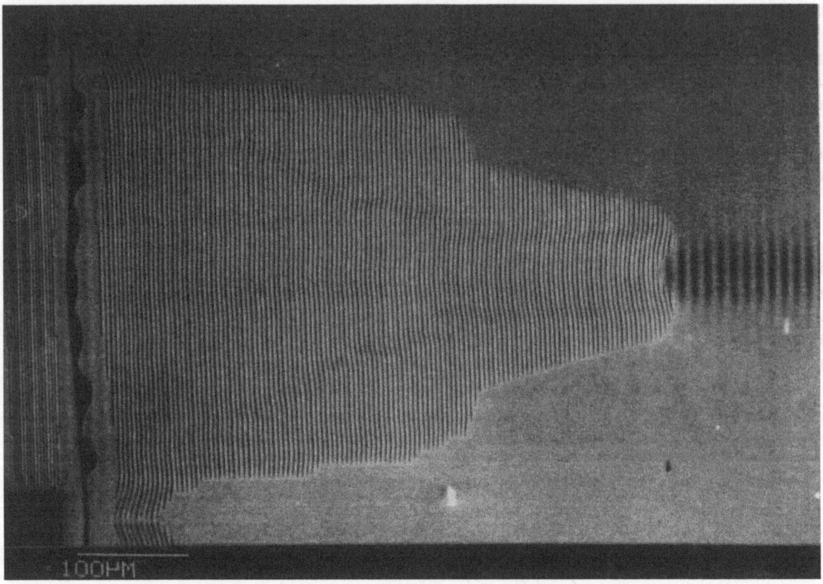

Fig. 6. Diffraction compensated multistrip beam compressor seen with voltage contrast in scanning electron microscope

SMITH /22/ proposed using nonlinearly dispersive transducers as a pair of up-chirp and down-chirp as a means to precisely control the magnitude as well as the phase of the transfer function of a convolver. Thereby a wide bandwidth flat frequency response with a phase correction to compensate for waveguide dispersion could be realized. MORGAN /23/ first demonstrated that such transducers could be made with a very small aperture and the same impedance as non-dispersive transducers with a much wider aperture, so that separate beam compression structures would become obsolete. ENGAN /24/ finally invented the small-aperture focussing chirp transducer which avoids the coupling losses between transducer and waveguide mode due to diffraction effects. ENGAN also demonstrated a simple design procedure by which one finds the optimum aperture and curvature of fingers in the transducer.

4. Design of efficient and precise convolvers with small aperture focussing chirp transducers

Up to the introduction of the small-aperture chirp transducer, broadband convolvers had to be manufactured with two different metal layers - a thin one for the acoustic waveguide to minimize waveguide dispersion and and a thicker one for transducers and beam-compressors to minimize ohmic losses. The chirp transducers have enough degrees of freedom to shape their transfer function so as to incorporate compensation of distortions due to the acoustic waveguide as well as matching networks. This, however, is only possible if the matching networks are made with a technology which is as reproducible and predictable as the surface wave technolgy itself. The authors therefore use thin film microstrip matching networks on alumina substrates which are integrated in the convolver device. Thus, with a standard single layer SAW device technology, using small-aperture focussing chirp transducers, efficient broadband convolvers with a well-behaved transfer function have been made, which are well suited for mass production. Figure 7 shows a photograph of such a device, in this case consisting of two complete convolvers for simultaneous convolution of one signal with two different references together in one package. The component has outer dimensions of 77 x 29 mm^2. The $LiNbO_3$ substrate (65 x 2 mm^2) in the center is mounted over an alumina substrate which contains both the input and the output matching networks.

Figure 8 visualizes the layout of a small-aperture focussing chirp transducer and its radiation pattern. The apodizaton of the finger overlap region, which defines the active aperture, is designed such that it matches the pattern which is formed by the fundamental waveguide mode as it propagates beyond the waveguide end under the influence of diffraction. The ends of the apodization are tapered to smooth out Fresnel-ripple in the frequency characteristic. The fingers are curved to conform to the phase fronts of the diffraction pattern at each distance. By virtue of reciprocity the emission pattern of the waveguide end is equal to its receiving pattern. Therefore the coupling loss between the transducer and the fundamental waveguide mode is minimized by this design technique.

While the apodization of the transducers is determined by the requirement of optimum coupling to the waveguide, the transducer length and the finger positions determine the frequency characteristic of the transducers. The basic finger arrangement in a chirp transducer is such that the finger pitch varies linearly (or monotonically) in the direction of propagation. Accordingly, fingers in various

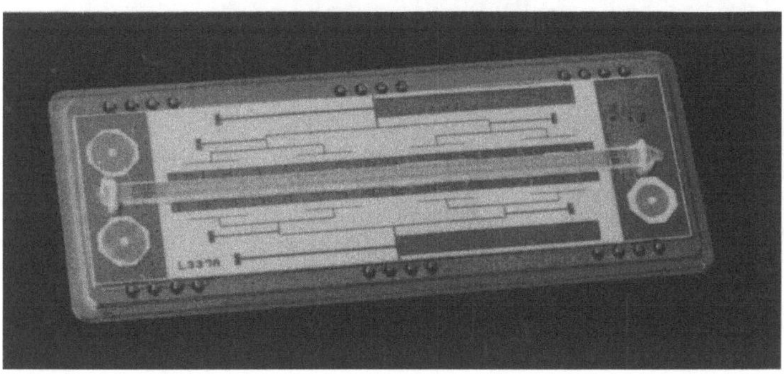

Fig. 7. Photograph of dual channel SAW convolver with integrated matching networks

regions of the transducers are synchronous with different frequencies. The excitation strength of a frequency is inversely dependent on the steepness of the gradient of the finger pitch, i.e. there are many fingers per frequency interval where the pitch varies slowly and vice versa. If the frequency dispersion of the transducers on opposite ends of the convolver has the opposite slope, then the overall phase of the device can be made linear. With an iterative process of adjusting the finger positions within the envelope, both the magnitude and the phase of the device characteristic can be optimized. This design process is described in the following.

First, the output electrode configuration has to be determined. Paramaters to choose are waveguide width, metal thickness, and distance between output signal taps. The number of output taps should be a power of 2 in order to use a binary branching output summing network. The waveguide width is typically three wavelengths at the center frequency, so that only the lowest symmetrical waveguide mode is supported. The optimum choice of metal thickness is more complicated to obtain. It is influenced by the implied sheet resistance and mechanical attenuation A_{wg} of surface waves. Wave attenuation on long strip waveguides can vary considerably depending on the techniques of thin film deposition. It is usually not limited by the ohmic loss but by mechanical friction or scattering effects. The interface between substrate and layer may play an important role /9,25/. For minimum acoustic attenuation the layer thickness should be chosen as small as possible. However, the resistance of the layer increases with decreasing metal thickness. The lower the resistance the higher becomes the electrical Q of the tap impedance and the less is the achievable maximum value of the transfer function $|S_{21}|$ of a matching network for a given bandwidth. The higher becomes, on the other hand, the available power at the output, as less energy is absorbed by ohmic loss on the output electrodes. These effects are still dependent on the number of taps used. In principle, the number of taps should be as large as possible. A minimum can be determined in view of a tolerance for the ripple in the magnitude

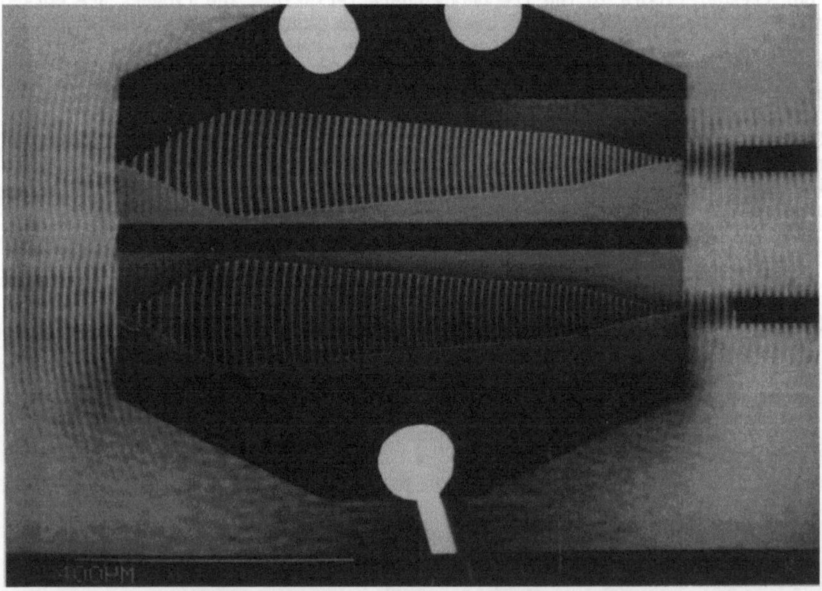

Fig. 8. Small-aperture focussing chirp transducers feeding three wavelengths wide strip waveguides

or phase of the uniformity. A maximum is imposed by practical considerations such as summing network complexity or wire bonding expense. Once the waveguide dimensions are chosen, an output network can be designed. Solving the equations for a pair of coupled lossy microstrip lines, the impedance of one tap connected to an open ended section of output plate is obtained (extension of method in /26/). With the program SUPER COMPACT a microstrip summing and matching network containing shorted and open stubs for impedance transformations is optimized for minimum insertion loss, regardless of flatness /27/. Both magnitude and phase distortions are obtained and will be corrected in the transducer design.

Two transducers with a linear frequency modulation are taken as a starting solution for the iteration. The length of the transducers has to be chosen such that the desired impedance level is obtained. A detailed simulation of the acoustic transfer function of the acoustic delay line is done with an equivalent circuit type analysis which includes all relevant effects such as electrostatic charge distribution on transducer fingers /28/, reflections at finger edges, and coupling between the waveguide mode and the exciting fingers. The transfer function of the acoustic waveguide is computed using an extended model after COLDREN and SCHMIDT /29/, which takes into account the exact anisotropy data of YZ-$LiNbO_3$ (the phase nonlinearity of the convolver is critically dependent on these data) /30/ and the wave attenuation in the metalized and free regions /31/. The effect of spiral matching inductors and parasitic transducer impedances such as capacitance and resistance of bond pads and the response of the output network are cascaded with the acoustic delay line. This gives the simulated overall efficiency function. The difference between this function and the desired flat amplitude and linear phase is then computed and used to correct the dispersion function of the transducers. The finger positions are derived from a function t(f) which describes the relation of the instantaneous frequency $f=v/p$ to the position $x=v\cdot t$ in the transducer, where p is the local interdigital period. According to the principle of stationary phase the magnitude A(f) and the phase $\phi(f)$ of the transfer function are given by

$$A(f) \propto \pm dt(f)/df \qquad \text{and} \qquad \phi(f) = 2\pi f\cdot t(f). \tag{8}$$

Therefore a deviation in magnitude can be corrected by a change

$$\delta t(f) \propto \pm\int \delta A(f)\, df, \tag{9}$$

where the plus-sign applies to the down-chirp and the minus-sign to the up-chirp, and a deviation in phase by a change

$$\delta t(f) \propto \delta\phi(f). \tag{10}$$

When the corrections for amplitude and phase are divided evenly between up- and down-chirp transducers then they are mutually independent. Since the formulas in (8) are rough approximations anyway, we use the linearized correction terms (9) and (10). The compensation procedure still converges within a few iterations.

As an example of results of such a design procedure Fig. 9 shows the efficiency of a convolver with 160 MHz input signal bandwidth and a time-bandwidth product of 2600 with excellent flatness and phase linearity /32/.

5. Conclusion

History of convolver development has shown so far that optimization of the design in the linear SAW device aspects of the elastic convolver aiming at a relatively simple but standard SAW technology can lead to mass producible high performance devices. Efforts to improve the efficiency by engineering the nonlinear signal generation have remained relatively unsuccessful (including the authors' own

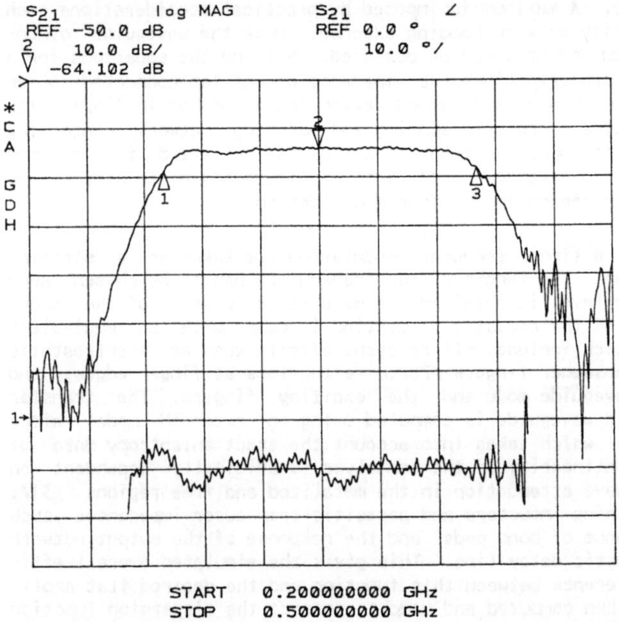

Fig. 9 . **Measured efficiency of wide bandwidth convolver with time-bandwidth product 2600.**

casual efforts to manipulate the nonlinearity of the elastic $LiNbO_3$ convolver by such processes as ion implantation or proton exchange, which are known to influence the linear elastic properties).

6. References

1 C.Maerfeld: "Rayleigh-Wave Non-Linear Components", Rayleigh-Wave Theory and Application, ed. by E.A.Ash & E.G.S.Paige, (Springer, Berlin,Heidelberg, 1985), p.191-218

2 A.K. Ganguly, K.L.Davis: "Nonlinear Interactions in Degenerate Surface Acoustic Wave Elastic Convolvers", J.Appl.Phys. 51, 1980, p. 920-926

3 J.H. Goll, D.C. Malocha: "An Application of SAW Convolvers to High Bandwidth Spread Sppectrum Communications", IEEE Trans. on Microwave Theory and Techniques, MTT-29, 1981, p.473-483

4 I. Yao, E.M. Hauser, C.A. Bouman, G.T. Flynn, J.H. Cafarella: "Wideband Radar Signal Processor based on SAW Convolvers", Proc. 1984 IEEE Ultrasonics Sympos., p.132-137

5 J. Becker, H.P. Grassl: "Application of a SAW-Convolver as a Fast Special Processor in a Multiprocessor-System", Proc. 1986 IEEE Ultrasonics Sympos., p. 169-172

6 E. Stern: "Acoustoelectric Rayleigh Wave Devices", Rayleigh-Wave Theory and Application, ed. by E.A.Ash & E.G.S.Paige, (Springer, Berlin,Heidelberg, 1985), p. 219-237

7 H.P. Grassl, R. Veith: "High Performance SAW Convolvers Characterized by Two-dimensional Frequency Domain Measurements", Proc. 1982 IEEE Ultrasonics Sympos. p. 149-154

8 H.P. Grassl, R. Veith: "Cross Convolution of Acoustic Waveguide Modes in Elastic Convolvers Measured in the Frequency Domain", IEEE Trans. on Sonics and Ultrason. SU-30, 1983, p. 100-104

9 K.L. Davis, J.F.Weller: "SAW Attenuation in Metal Film Coated Delay Lines", Proc. 1979 IEEE Ultrasonics Sympos., p. 659-662
10 M. Planat, H. Gautier, C. Maerfeld: "Calculation of the Figure of Merit of Piezoelectric Waveguide Convolvers", Proc. 1983 IEEE Ultrasonics Sympos., p. 406-409
11 T.C. Lim, E.A.Kraut, R.B. Thompson: "Nonlinear Materials for Acoustic-Surface-Wave Convolver", Appl. Phys. Lett. 20, 1972, p.127-129
12 V.V. Shtykov, I.M Mason, M.D. Motz: "On the Anisotropy of Guided Acoustic Surface Wave Degenerate Convolution", IEEE Trans. on Sonics and Ultrason., SU-22, 1975, p. 131-136
13 T.J. Monks, E.G.S. Paige, R.C. Woods: "SAW and SSBW Convolvers Utilizing Bilinear Field in Various Directions", Proc. 1983 IEEE Ultrasonics Sympos., p. 402-405
14 Y. Nakagawa, S. Makio: "SAW Convolver Using Multi-Channel Waveguide", Jap. J. Appl. Phys. 24, Supplement 24-1), p. 154-156
15 P. Defranould, C. Maerfeld: "Acoustic Convolver Using Multistrip Beamwidth Compressors", Proc. 1974 IEEE Ultrasonics Sympos., p. 224-227
16 A. Rønnekleiv: "Diffraction Compensated SAW Multistrip Beam Compressors", Proc. 1983 IEEE Ultrasonics Sympos., p. 410-414
17 I. Yao: "High Performance Elastic Convolver with Parabolic Horns", Proc. 1980 IEEE Ultrasonics Sympos., p. 37-42
18 A.M. Hodge, M.F. Lewis: "Detailed Investigations into SAW Convolvers", Proc. 1982 IEEE Ultrasonics Sympos., p. 175-180
19 K.L. Davis, J.F.Weller: "Surface Acoustic Wave Planar Prism Waveguide Coupler", J. Appl. Phys. Lett., 37, 1980, p. 985-987
20 T.J. Marynowski: Focusing Transducer for SAW Beamwidth Compression on YZ-Lithium Niobate", Proc. 1982 IEEE Ultrasonics Sympos., p. 160-165
21 J.B. Green, G.S. Kino: "SAW Convolvers Using Focused Interdigital Transducers", IEEE Trans on Sonics and Ultrason., SU-30, p. 43-50
22 W.R. Smith: "Broadband Amplitude and Phase-Compensating Transducers for Monolithic Convolvers", Proc. 1980 IEEE Ultrasonics Sympos., p. 59-64
23 D.P. Morgan, D.H. Warne, P.N. Naish, D.R. Selviah: "Monolithic SAW Convolvers Using Chirp Transducers", Proc. 1981 IEEE Ultrasonics Sympos., p. 186-191
24 H. Engan: "Focused, Dispersive Transducers for SAW Convolvers", Proc. 1984 IEEE Ultrasonics Sympos., p. 138-142
25 H. Engan: "Laser Probing of SAW Convolver Waveguide Loss", IEEE Trans. on Sonics and Ultrason., SU 30, p.321 323
26 E.L. Adler: "Spatial Uniformity and Broadband Matching in Multiply Tapped SAW Convolvers", IEEE Trans. on Sonics and Ultrason., SU-32, 1985, p.685-689
27 H.P. Grassl, H.W. Wörz: "Commercial Microwave CAD Matches SAW Convolver", Microwave Systems News, Sept. 1987, p. 123-128
28 A.R. Baghai-Wadji, S. Selberherr, F. Seifert: "On the Calculation of Charge, Electrostatic Potential, and Capacitance in General Finite SAW Structures", Proc. 1986 IEEE Ultrasonics Sympos., p. 44-48
29 L.A. Coldren, R.V.Schmidt: "Thin Film Acoustic Surface Waveguides on Anisotropic Media", IEEE Trans. on Sonics and Ultrason., SU-22, p.115-122
30 M. Anhorn, H. Engan, A. Rønnekleiv: "New SAW Velocity Measurements on Y-cut $LiNbO_3$", Proc. 1987 IEEE Ultrasonics Sympos.,
31 I. Schropp: Diploma thesis, Technische Universität München, 1988
32 L. Reindl, H.W. Wörz, H.P. Grassl: "CAD of Precise SAW Convolvers and Long Delay Lines", Proc. 1987 IEEE Ultrasonics Sympos.,

Multiple Scale Techniques Applied to Nonlinear Elastic and Piezoelectric Surface Waves

G.E. Tupholme

School of Mathematical Sciences, University of Bradford, Bradford, BD71DP, U.K.

1. INTRODUCTION

Theoretical investigations of the nonlinear propagation of a surface acoustic wave (SAW) have achieved renewed impetus recently by developments within the field of signal processing. There is currently interest in using advantageously a knowledge of the nonlinear effects of such waves to aid in the design of acoustic devices such as filters, transducers and convolvers.

Much of the earlier work was unsatisfactory because the field equations and boundary conditions were not included to the same order of approximation. An improved theory of nonlinear Rayleigh and Love waves was provided by KALYANASUNDARAM [1,2,3] using the method of multiple scales, as pioneered in fluid mechanics for studying surface waves on fluid layers and jets, but LARDNER [4] realized that these investigations also contain assumptions which render them unnecessarily restrictive. These limitations were overcome and the analyses extended in a series of papers by LARDNER [4,5,6,7], DAVID [8] and LARDNER and TUPHOLME [9] which give solutions which are uniformly valid in both spatial directions as well as time and model the nonlinear modulation of the wave amplitude.

The application of this technique to general anisotropic materials is outlined in Section 2, before in Section 3, its adaptation to cubic materials is discussed. However, piezoelectric solids are often used in the substrates of modern SAW devices and the way in which TUPHOLME and HARVEY [10] have modified the analysis to such materials is therefore indicated in Section 4.

2. ANISOTROPIC ELASTIC SOLIDS

2.1 Formulation and Basic Equations

We consider a half-space of homogeneous anisotropic elastic material which in its natural reference configuration has a uniform density ρ and occupies the region $x_2 < 0$, relative to a reference system of rectangular Cartesian coordinates x_1, x_2, x_3. A slowly modulated wave is supposed to be propagating subsonically and independently of x_3 in the positive x_1-direction over the free surface $x_2 = 0$.

The governing equations of motion are taken to be

$$\rho \ddot{u}_i = \sigma_{ij,j}, \tag{1}$$

where u_i and σ_{ij} are the components of the displacement vector and Piola-Kirchhoff stress tensor, respectively, referred to the x_1, x_2, x_3 coordinates. Throughout, dots are used to denote time derivatives, commas denote partial derivatives with respect to x_j and the subscripts take the values 1, 2, 3 with the summation convention applied to those which are repeated. The stress tensor components are assumed to satisfy nonlinear constitutive equations which can be written in the form

$$\sigma_{ij} = c_{ijk\ell}u_{k,\ell} + d_{ijk\ell mn}u_{k,\ell}u_{m,n} \tag{2}$$

up to quadratic order terms in the displacement gradients, with

$$d_{ijk\ell mn} = \frac{1}{2}\left[c^{(3)}_{ijk\ell mn} + c_{ijn\ell}\delta_{km} + c_{njk\ell}\delta_{im} + c_{\ell jmn}\delta_{ik}\right], \tag{3}$$

where δ_{ij} is the Kronecker delta, and $c_{ijk\ell}$ and $c^{(3)}_{ijk\ell mn}$ are the second-order and third-order elastic moduli of the medium. These tensor components satisfy the symmetry properties

$$c_{ijk\ell} = c_{k\ell ij} = c_{jik\ell}, \tag{4}$$

$$c^{(3)}_{ijk\ell mn} = c^{(3)}_{k\ell ijmn} = c^{(3)}_{ijmnk\ell} = c^{(3)}_{jik\ell mn} \tag{5}$$

and consequently the quantities $d_{ijk\ell mn}$ are symmetrical among the three pairs (ij), $(k\ell)$, (mn) but not within them, as indicated symbolically by

$$d_{ijk\ell mn} \equiv d_{(ij)(k\ell)(mn)}. \tag{6}$$

It is assumed that the boundary $x_2 = 0$ is free of traction, so we must fulfil the stress boundary condition

$$\sigma_{i2} = 0, \quad \text{on } x_2 = 0. \tag{7}$$

Substituting (2) into (1) and (7), using (6), we can rewrite the basic governing equations as

$$\rho\ddot{u}_i = c_{ijk\ell}u_{k,\ell j} + 2d_{ijk\ell mn}u_{k,\ell}u_{m,nj}, \tag{8}$$

with

$$c_{i2k\ell}u_{k,\ell} + d_{i2k\ell mn}u_{k,\ell}u_{m,n} = 0, \quad \text{on } x_2 = 0. \tag{9}$$

2.2. Solution

We first introduce a small positive scaling parameter ε to monitor the orders of magnitude of the various terms and make the replacements $u_i \to \varepsilon u_i$. Equations (8) and (9) then become

$$\rho\ddot{u}_i = c_{ijk\ell}u_{k,\ell j} + 2\varepsilon d_{ijk\ell mn}u_{k,\ell}u_{m,nj} \tag{10}$$

$$c_{i2k\ell}u_{k,\ell} + \varepsilon d_{i2k\ell mn}u_{k,\ell}u_{m,n} = 0, \quad \text{on } x_2 = 0. \tag{11}$$

To consider a deformation consisting of a plane surface wave of arbitrary profile travelling in the positive x_1-direction we look for a solution of these of the form

$$u_j(x, y, t) = \int_{-\infty}^{\infty} \tilde{u}_j(k, y, \xi, \eta, \tau)e^{ik(x-ct)}dk, \tag{12}$$

where we use (x, y) synonymously with (x_1, x_2) and the integration over k allows the inclusion of an arbitrary spectrum of wave numbers. The Rayleigh wave speed is denoted by c. We intend to generate a perturbation solution and so, for an expansion uniform at the lowest order, slow time and distance scales $\tau = \epsilon t$, $\xi = \epsilon x$, $\eta = \epsilon y$ are introduced which permit slow modulation of the wave in time and space.

We seek a solution as a perturbation series in ϵ,

$$\hat{u}_j = A_j(k, y, \xi, \eta, \tau) + \epsilon B_j(k, y, \xi, \eta, \tau) + \dots \quad . \tag{13}$$

Substituting (12) and (13) into (10) and (11) and comparing coefficients of the different powers of ϵ leads to a hierarchy of equations for A_j, B_j etc.

(a) First-order solution. The lowest order equations for the first-order amplitudes A_j yield the system

$$c_{i_2 j_2} A_{j,yy} + ik(c_{i_1 j_2} + c_{i_2 j_1}) A_{j,y} + k^2(\rho c^2 \delta_{ij} - c_{i_1 j_1}) A_j = 0, \tag{14}$$

$$c_{i_2 j_2} A_{j,y} + ikc_{i_2 j_1} A_j = 0, \quad \text{on } y = 0 \, , \tag{15}$$

which are the standard equations of linear surface wave theory except for the additional parameters ξ, η and τ. The general solution of (14) is

$$A_j = \begin{cases} \sum_{\ell=1}^{3} \alpha_\ell(k, \xi, \eta, \tau) a_j^{(\ell)} e^{is^{(\ell)}ky}, & k > 0, \\ \sum_{\ell=1}^{3} \alpha_\ell(k, \xi, \eta, \tau) \overline{a_j^{(\ell)}} e^{\overline{is^{(\ell)}}ky}, & k < 0, \end{cases} \tag{16}$$

where $s^{(\ell)}$ ($\ell = 1, 2, 3$) are the three roots with negative imaginary parts of the equation $\det[L_{ij}(s)] = 0$ and $L_{ij}(s^{(\ell)}) a_j^{(\ell)} = 0$ with

$$L_{ij}(s) = c_{i_2 j_2} s^2 + (c_{i_2 j_1} + c_{i_1 j_2}) s + c_{i_1 j_1} - \rho c^2 \delta_{ij}. \tag{17}$$

Using the boundary condition (15) we can write

$$\alpha_\ell(k, \xi, 0, \tau) = \gamma(k, \xi, \tau) \beta_\ell, \tag{18}$$

where β_ℓ is a particular solution of $M_{i\ell}\beta_\ell = 0$ with

$$M_{i\ell} = (c_{i_2 j_1} + s^{(\ell)} c_{i_2 j_2}) a_j^{(\ell)} \quad . \tag{19}$$

The Rayleigh wave space, c, is determined by $\det[M_{i\ell}] = 0$ and the function γ is directly related to the Fourier transform of the surface displacements by the representation

$$u_j(x, 0, t) = \left[\sum_{\ell=1}^{3} \beta_\ell a_j^{(\ell)}\right] \int_0^\infty \gamma(k, \xi, \tau) e^{ik(x-ct)} dk + cc. \tag{20}$$

(b) Second-order solution. Substitution of the linear solution (16) into (10), (11), using (13), leads to a second-order system satisfied by B_j in the form

$$c_{i_2j_2}B_{j,yy} + ik(c_{i_1j_2} + c_{i_2j_1})B_{j,y} + k^2(\rho c^2\delta_{ij} - c_{i_1j_1})B_j =$$

$$-ik\sum_{\ell=1}^{3}\Big[\Big\{2c_{i_2j_2}s^{(\ell)}\alpha_{\ell,\eta} + (c_{i_1j_2} + c_{i_2j_1})(s^{(\ell)}\alpha_{\ell,\xi} + \alpha_{\ell,\eta})$$

$$+ 2c_{i_1j_1}\alpha_{\ell,\xi}, + 2\rho c\delta_{ij}\alpha_{\ell,\tau}\Big\}a_j^{(\ell)} - N_{i\ell\ell}^{(1)}I_{\alpha_\ell}(k)\Big]e^{is^{(\ell)}ky}$$

$$+ 2i\sum_L \int R_{iL}(k, k')e^{iT_L(k,k')ky}dk' \quad , \tag{21}$$

$$c_{i_2j_2}B_{j,y} + ikc_{i_2j_1}B_j + \sum_{\ell=1}^{3}(c_{i_2j_2}\alpha_{\ell,\eta} + c_{i_2j_1}\alpha_{\ell,\xi})a_j^{(\ell)}$$

$$-C_i^{(1)}I_\gamma(k) - 2C_i^{(2)}J_\gamma(k) = 0, \quad \text{on } \eta = 0. \tag{22}$$

In order to avoid secular terms of the type $ye^{is^{(\ell)}ky}$ in the solution the method of multiple scales, as described, for example, by NAYFEH [11], imposes a constraint upon the right-hand side of (21). Incorporating this we finally find, after a lot of manipulation, that the function γ satisfies the nonlinear integro-differential equation

$$\frac{\partial\gamma}{\partial\xi} + \frac{1}{c}\frac{\partial\gamma}{\partial\tau} + \int_0^\infty H(k, k')k'(k - k')\gamma(k')\gamma(k - k')dk' = 0, \tag{23}$$

where $H(k, k')$ is given by

$$H(k, k') = \begin{cases} \dfrac{P'}{Q} + \dfrac{2}{Qk}\sum\limits_{\ell\neq n}\beta_i'\beta_\ell\beta_n(k-k')\Big[c_{i_2j_2}\dfrac{s^{(\ell)}k' + s^{(n)}(k-k')}{k} + c_{i_2j_1}\Big] \\[2ex] \quad \cdot N_{m\ell n}^{(1)}L_{jm}^{-1}\Big[\dfrac{s^{(\ell)}k' + s^{(n)}(k-k')}{k}\Big], \qquad 0 < k' < k, \\[2ex] \dfrac{Q'}{Q} + \dfrac{2}{Qk}\sum\limits_{\ell,n}\beta_i'\beta_\ell\bar{\beta}_n\Big[c_{i_2j_2}\dfrac{s^{(\ell)}k' + \overline{s^{(n)}}(k-k')}{k} + c_{i_2j_1}\Big] \\[2ex] \quad \cdot \Big[k'N_{mn\ell}^{(2)} + (k-k')N_{m\ell n}^{(3)}\Big]L_{jm}^{-1}\Big[\dfrac{s^{(\ell)}k' + \overline{s^{(n)}}(k-k')}{k}\Big], \qquad k' > k. \end{cases} \tag{24}$$

For the sake of brevity we refer the interested reader to the analysis of LARDNER [7], where more details of the derivation and explicit definitions of all the quantities occurring in (21) to (24) are given. It should be noted that (24) corrects equations (55) and (56) in [7].

3. CUBIC ELASTIC SOLIDS

The general analysis of the previous section has been specialized to the particular case of an elastic material which is cubically symmetrical in its elastic response by LARDNER and TUPHOLME [9]. Suppose the coordinate axes are situated so that the free surface $x_2 = 0$ is a plane of symmetry and the x_1-axis is a cubic axis of the material. Then, denoting the second-order and third-order moduli by c_{pq} and c_{pqr} ($p, q, r = 1, \ldots, 6$) respectively, using the abbreviated index notation whereby (11) → 1, (22) → 2, (33) → 3, (23) or (32) → 4, (13) or (31) → 5, (12) or (21) → 6, we find there are nine non-zero

c_{pq} and twenty non-zero c_{pqr} of which only c_{11}, c_{12}, c_{44}, and c_{111}, c_{112}, c_{123}, c_{144}, c_{155}, c_{456} are independent. This produces forty-five non-zero $d_{ijk\ell mn}$ of which ten are distinct.

The first-order analysis of section 2.2 then yields the equation

$$p\{(\delta - 1)^2 + r(1 - r - p^2)\} = (1 - p^2)\{r(r - 1 + p^2)\}^{\frac{1}{2}}, \tag{25}$$

with $p^2 = 1 - \rho c^2/c_{44}$, $r = c_{11}/c_{44}$ and $\delta = (c_{12} + c_{44})/c_{44}$, for the determination of p, and hence the Rayleigh wave speed c, and $s^{(1)}$ and $s^{(2)}$ are appropriate roots of

$$rs^4 + (p^2 + rp^2 + r^2 - r - \delta^2)s^2 + p^2(r - 1 + p^2) = 0\ , \tag{26}$$

while $s^{(3)} = -ip$. Explicit representations are given in [9] for the various quantities which must then be successively found in evaluating the kernel H(k, k') in (23) for such cubic materials, and a computer program has been written to numerically calculate it for any given values of c_{pq} and c_{pqr}.

For the basic problem of harmonic generation by an initially sinusoidal wave, the relevant parameters are calculated in [9] for a series of thirty-five materials for which experimental data is available for the moduli, and ultimately a measure of the magnitude of the nonlinear effects is discussed.

4. PIEZOELECTRIC SOLIDS

The techniques of section 2 can be modified to study the propagation of SAW on piezoelectric materials. The governing equations which must be solved consist of (1) and the additional charge equation

$$D_{i,i} = 0, \tag{27}$$

satisfied by the components of the electric displacement vector $\underset{\sim}{D}$. If all free-space stress and electric field effects are neglected, not only is the surface $x_2 = 0$ free of traction, so that (7) must be fulfilled, but the normal component of $\underset{\sim}{D}$ must be continuous across the boundary which implies that

$$D_2 = 0, \text{ on } x_2 = 0. \tag{28}$$

In the quasi-static approximation, the nonlinear constitutive equations can be written, up to quadratic order terms, in the form

$$\sigma_{ij} = c_{ijk\ell}u_{k\ell}u_{k,\ell} + e_{kij}\varphi_{,k}$$
$$+ \tfrac{1}{2}\left[c^{(3)}_{ijk\ell mn} + \delta_{ik}c_{\ell jmn} + \delta_{im}c_{njk\ell} + \delta_{km}c_{ij\ell n}\right]u_{k,\ell}u_{m,n}$$
$$+ \left[e^{(3)}_{kij\ell m} + \delta_{i\ell}e_{kmj}\right]u_{\ell,m}\varphi_{,k} - \tfrac{1}{2}\ell_{kmij}\varphi_{,k}\varphi_{,m}, \tag{29}$$

$$D_i = e_{ik\ell}u_{k,\ell} - \epsilon_{ik}\varphi_{,k} + \tfrac{1}{2}\epsilon^{(3)}_{ijk}\varphi_{,j}\varphi_{,k} - \ell_{ijk\ell}\varphi_{,j}u_{k,\ell}$$
$$+ \tfrac{1}{2}\left[e^{(3)}_{ijk\ell m} + e_{ikm}\delta_{j\ell}\right]u_{j,k}u_{\ell,m}, \tag{30}$$

where φ is the electric potential and e_{ijk}, ϵ_{ij}, $e_{ijk\ell m}$, $\epsilon^{(3)}_{ijk}$ and $\ell_{ijk\ell}$ are the second-order piezoelectric, second-order dielectric, third-order piezoelectric, third-order dielectric and electrostrictive moduli of the medium, having symmetry properties

$$e_{ijk} = e_{ikj}, \quad \epsilon_{ij} = \epsilon_{ji}, \quad e^{(3)}_{ijk\ell m} = e^{(3)}_{i\ell mjk} = e^{(3)}_{ikj\ell m},$$

$$\epsilon^{(3)}_{ijk} = \epsilon^{(3)}_{ikj} = \epsilon^{(3)}_{jik}, \quad \ell_{ijk\ell} = \ell_{jik\ell} = \ell_{ij\ell k}. \tag{31}$$

At this stage it is convenient to introduce some quasi third-order piezoelectric moduli $f_{ijk\ell m}$, analogous to $d_{ijk\ell mn}$ in the elastic case given by

$$f_{ijk\ell m} = \frac{1}{2}\left[e^{(3)}_{ijk\ell m} + e_{ikm}\delta_{j\ell}\right]. \tag{32}$$

If we then define $u_4 = \varphi$ and extend our notation by allowing all the subscripts to range over the values 1, 2, 3 and 4, with the additional values of the $c_{ijk\ell}$ and $d_{ijk\ell mn}$ incorporating the piezoelectric, dielectric and electrostrictive constants according to the schemes

$$\left.\begin{aligned} \hat{c}_{ijk\ell} &\equiv \hat{c}_{(ij)(k\ell)} = c_{ijk\ell}, \\ \hat{c}_{4jk\ell} &\equiv \hat{c}_{(4j)(k\ell)} = e_{jk\ell}, \\ \hat{c}_{4j4\ell} &\equiv \hat{c}_{(4j)(4\ell)} = -\epsilon_{j\ell}, \end{aligned}\right\} (i, j, k, \ell = 1, 2, 3)$$

$$\hat{c}_{i4k\ell} \equiv \hat{c}_{(i4)(k\ell)} = 0 \quad (i, k, \ell = 1, 2, 3, 4) \tag{33}$$

and

$$\left.\begin{aligned} \hat{d}_{ijk\ell mn} &\equiv \hat{d}_{(ij)(k\ell)(mn)} = d_{ijk\ell mn}, \\ \hat{d}_{4jk\ell mn} &\equiv \hat{d}_{(4j)(k\ell)(mn)} = f_{jk\ell mn}, \\ \hat{d}_{ij4\ell 4n} &\equiv \hat{d}_{(ij)(4\ell)(4n)} = -\frac{1}{2}\ell_{\ell nij}, \\ \hat{d}_{4j4\ell 4n} &\equiv \hat{d}_{(4j)(4\ell)(4n)} = \frac{1}{2}\epsilon^{(3)}_{j\ell n} \end{aligned}\right\} (i, j, k, \ell, m, n = 1, 2, 3)$$

$$\hat{d}_{i4k\ell mn} \equiv \hat{d}_{(i4)(k\ell)(mn)} = 0 \quad (i, k, \ell, m, n = 1, 2, 3, 4), \tag{34}$$

again using parentheses to indicate the symmetries, the governing equations obtained by combining (1) and (27), (7) and (28) with (29) and (30) take the compact forms

$$\rho u_j \hat{\delta}_{ij} = \hat{c}_{ijk\ell} u_{k,\ell j} + 2\hat{d}_{ijk\ell mn} u_{k,\ell} u_{m,nj}, \tag{35}$$

$$\hat{c}_{i2k\ell} u_{k,\ell} + \hat{d}_{i2k\ell mn} u_{k,\ell} u_{m,n} = 0, \quad \text{on } x_2 = 0, \tag{36}$$

where the modified Kronecker delta, $\hat{\delta}_{ij}$, is defined by

$$\hat{\delta}_{ij} = \delta_{ij} - \delta_{i4}\delta_{j4}, \quad (i, j = 1, 2, 3, 4). \tag{37}$$

The affinity between (35), (36) and (8), (9) suggests that a solution might be constructed by the type of techniques used in the purely elastic case in section 2. This has been investigated analytically by TUPHOLME and HARVEY [10] and numerical calculations are reported by HARVEY [12].

ACKNOWLEDGEMENTS

The author is indebted to Professor G.A. Maugin, Dr D.F. Parker, Dr R.E. Craine, Dr E.A. David and, especially, Professor R.W. Lardner and Mr A.P. Harvey for instructive discussions and correspondence.

REFERENCES

1. N. Kalyanasundaram: Int. J. Eng. Sci. 19, 279 (1981)
2. N. Kalyanasundaram: Int. J. Eng. Sci. 19, 287 (1981)
3. N. Kalyanasundaram: Int. J. Eng. Sci. 19, 435 (1981)
4. R.W. Lardner: Int. J. Eng. Sci. 21, 1331 (1983)
5. R.W. Lardner: J. Appl. Phys. 55, 3251 (1984)
6. R.W. Lardner: Int. J. Eng. Sci. 23, 113 (1985)
7. R.W. Lardner: J. Elast. 16, 63 (1986)
8. E.A. David: Int. J. Eng. Sci. 23, 699 (1985)
9. R.W. Lardner, G.E. Tupholme: J. Elast. 16, 251 (1986)
10. G.E. Tupholme, A.P. Harvey: Submitted for publication
11. A.H. Nayfeh: Perturbation Methods (Wiley, New York 1973)
12. A.P. Harvey: In Recent Developments in Surface Acoustic Waves, ed. by D.F. Parker, G.A. Maugin, Springer Ser. Wave Phen. (Springer, Berlin, Heidelberg 1988)

Nonlinear Evolution of Piezoelectric SAWs

E.A. David and D.F. Parker

Department of Theoretical Mechanics, University of Nottingham, Nottingham, NG72RD, U.K.

1. Introduction

Theoretical treatment of waveform evolution for nonlinear elastic surface acoustic waves (SAWs) has developed through work by KALYANASUNDARAM et al [1,2,3], LARDNER [4,5,6], DAVID [7], LARDNER and TUPHOLME [8] and PLANAT [9]. All treatments, using variants on the method of multiple scales, yield an evolution equation equivalent to that first given by LARDNER [4] for the Fourier transform of the waveform. Extensions of the techniques to nonlinear piezoelectric SAWs are outlined in [10], [11] in this volume. This paper outlines an alternative derivation, generalizing the elastic treatment of PARKER [12]. The procedure given in PARKER and DAVID [13] is summarized, as are the results of numerical computations showing the influence of the electric boundary condition on the distortion and subsequent 'break-up' of initially sinusoidal waveforms.

On a 'Y-cut' lithium niobate ($LiNbO_3$) crystal, nonlinear effects on an initially sinusoidal waveform depend significantly on propagation direction when the surface is earthed. Unlike previous computations for nonlinear SAWs, the surface elevation profile does not remain anti-symmetric. This example is used to illustrate differing types of waveform distortion for waves travelling in various directions across a particular surface and to show that anisotropy manifests itself strongly through nonlinear effects.

2. Formulation and Linear Analysis

In a homogeneous, anisotropic, electro-elastic half-space, Lagrangian coordinates (X_1, X_2, X_3) are taken so that the material occupies $X_2 \leqslant 0$. The OX_1 axis is aligned along the propagation direction and disturbances independent of X_3 are considered, so that the Eulerian coordinates are given by

$$x_j = \delta_{jJ} X_J + \varepsilon u_j(X_1, X_2, t) , \tag{1}$$

where δ_{jJ} is a Kronecker symbol and the small parameter ε is a representative displacement gradient. The electrostatic approximation is assumed, with potential $\varepsilon\varphi(X_1, X_2, t)$. The momentum equation and the Gauss equation in the absence of free charge then become

$$T_{Lj,L} = \varepsilon\rho_0 \ddot{u}_j , \qquad D_{L,L} = 0 , \tag{2}$$

where T_{Lj} and D_L are components of the Piola-Kirchhoff stress and material electric field, ρ_0 is the reference density, dots denote time derivatives, commas denote partial derivatives and the summation convention for repeated indices is used.

To simplify the analysis, the notation $\varphi = u_4$, $D_L = T_{L4}$ of TAYLOR and CRAMPIN [14] is introduced (see also [10]) so that (2) becomes

$$T_{Lj,L} = \varepsilon\rho_0 \, \eta_{jm} \, \ddot{u}_m , \tag{3}$$

where $\eta_{jm} \equiv \delta_{jm} - \delta_{j4}\delta_{m4}$ and lower case subscripts are now taken to range over the values 1,...,4. In this notation the constitutive laws correct to $O(\varepsilon^2)$, taken from MAUGIN [15], become

$$T_{Lj} = \varepsilon c_{jLmM}\, u_{m,M} + \varepsilon^2 d_{jLmMnN} u_{m,M}\, u_{n,N}\ , \tag{4}$$

where c_{jLmM} and d_{jLmMnN} are symmetric with respect to interchanges of the pairs (jL), (mM) and (nN) of indices. In terms of the usual elastic coefficients c_{jLmM}, piezoelectric cofficients $e_{M.jL}$ and permittivity coefficients ϵ_{LM} (j=1,2,3) of linear theory, we have

$$c_{jL4M} = e_{m.jL} = c_{4MjL}, \qquad c_{4L4M} = \epsilon_{LM}\ , \tag{5}$$

while, for j,m,n=1,2,3, the second-order coefficients are

$$\begin{aligned} d_{jLmMnN} &= \tfrac{1}{2}\{c_{jLmMnN} + c_{jLMN}\,\delta_{mn} + c_{NLmM}\,\delta_{jn} + c_{MLnN}\,\delta_{jm}\} \\ d_{jLmM4N} &= \tfrac{1}{2}\{\delta_{jm}\, e_{N.LM} + e_{N.jLmM}\} = d_{jL4NmM} = \text{etc.} \\ d_{jL4M4N} &= -\tfrac{1}{2}\ell_{MN.jL} + \tfrac{1}{2}\epsilon_0(\delta_{jM}\delta_{NL} + \delta_{jN}\delta_{ML} - \delta_{jL}\delta_{MN}) = \text{etc.} \\ d_{4L4M4N} &= \tfrac{1}{2}\chi_{LMN}. \end{aligned} \tag{6}$$

Here c_{jLmMnN}, $e_{N.jLmM}$, $\ell_{MN.jL}$ and χ_{LMN} are the second-order elastic, the electro-elastic, the electrostriction and the second-order permittivity coefficients respectively.

The simplest set of boundary conditions (Case E)

$$T_{2j}\,\eta_{jm} = 0, \qquad u_4 = 0 \quad \text{on } X_2 = 0 \tag{7}$$

arises when the surface $X_2 = 0$ is earthed and free of traction. It will be considered first.

Before describing nonlinear effects, we summarize pertinent results from the linear theory of SAWs. According to linear theory, harmonic waves

$$u_j(X_1, X_2, t) = A_j(kX_2)e^{ik\theta} + \text{c.c.}\qquad , \qquad \theta = X_1 - ct\ , \qquad X_2 < 0\ ,$$

are nondispersive, so that the propagation speed c is independent of wavenumber k. Consequently, Fourier superposition shows that waves having *arbitrary* surface elevation profile may be written as functions of only X_2 and θ. Quadratic nonlinearities introduce perturbations and cause distortion over ranges measured on the scale of $X \equiv \varepsilon X_1$. This motivates the change to new independent variables (θ, X_2, X) so that partial derivatives are replaced according to

$$\frac{\partial}{\partial X_1} \to \frac{\partial}{\partial \theta} + \varepsilon\frac{\partial}{\partial X}\ , \qquad \frac{\partial}{\partial X_2} \to \frac{\partial}{\partial X_2}\ , \qquad \frac{\partial}{\partial t} \to -c\frac{\partial}{\partial \theta}\ .$$

Then, to $O(\varepsilon^2)$, equation (4) is altered only by the addition of a term $\varepsilon^2 c_{jLm1} u_{m,X}$ to the right-hand side, while (3) becomes

$$T_{Lj,L} = \varepsilon c^2 \rho_0 \eta_{jm} u_{m,11} = -\varepsilon T_{1j,X}\ , \tag{8}$$

where $_{,1}$ and $_{,X}$ now denote $\partial/\partial\theta$ and $\partial/\partial X$, respectively.

To leading order, these equations are satisfied by

$$u_j = \int_{-\infty}^{\infty} C(k, X)\, A_j(kX_2)e^{ik\theta}\, dk = U_j(\theta, X_2, X), \tag{9}$$

where the functions $A_j(kX_2)$, which decay as $X_2 \to -\infty$, describe the depth dependence in a harmonic wavetrain. They have the form

$$A_j(kX_2) = \begin{cases} \sum_{p=1}^{4} B^{(p)} a_j^{(p)} \exp(iks^{(p)}X_2) & k > 0, \\ \\ \sum_{p=1}^{4} B^{(p)*} a_j^{(p)*} \exp(iks^{(p)*}X_2) & k < 0, \end{cases} \tag{10}$$

(c.f. equations (16) of the preceding article [10]) where * denotes a complex conjugate. The scalars $s^{(M)}$ and vectors $a_m^{(M)}$ are the eigenvalues (having Im s < 0) and corresponding eigenvectors satisfying

$$\{c_{j1m1} - \rho_0 c^2 \delta_{jm} + s(c_{j1m2} + c_{j2m1}) + s^2 c_{j2m2}\} a_m = 0 ,$$

while the multipliers $B^{(p)}$ are determined by the linearized form of the boundary conditions (7) which give

$$\sum_{p=1}^{4} M_{jp} B^{(p)} = 0, \tag{11}$$

where

$$M_{jp} \equiv \{\eta_{jm}(c_{m2n1} + s^{(p)}c_{m2n2}) + \delta_{j4}\delta_{n4}\} a_n^{(p)}. \tag{12}$$

When solutions are normalized by the condition

$$\sum_{p=1}^{4} B^{(p)} a_2^{(p)} = 1 ,$$

giving $A_2(0) = 1 = A_2^*(0)$, it is seen that C(k,X) is the Fourier transform

$$C(k,X) = \frac{1}{2\pi} \int_{-\infty}^{\infty} U_2(\theta,0,X) e^{-ik\theta} \, d\theta \tag{13}$$

of the surface elevation $U_2(\theta,0,X)$ which is an arbitrary function of the phase variable θ. The conditions $A_j(-kX_2) = A_j^*(kX_2)$ and $C(-k,X) = C^*(k,X)$ which follow from (10) and (13) ensure that all displacements and the potential in (9) are real.

The dependence of C(k,X) on the range of propagation X is determined only by analysis of higher order terms, as in the next section.

3. The Evolution Equation

Since expressions (9) satisfy the governing equations and boundary conditions to leading order for all choices of C(k,X), the solution correct to $O(\varepsilon^2)$ is sought as

$$\varepsilon \underset{\sim}{u} = \varepsilon \underset{\sim}{U}(\theta,X_2,X) + \varepsilon^2 \underset{\sim}{w}(\theta,X_2,X) . \tag{14}$$

Then, when (4) and (14) are substituted into (3) and (7), comparison of $O(\varepsilon^2)$ terms gives

$$\rho_0 c^2 \eta_{jm} w_{m,11} - c_{jLmN} w_{m,LN} = d_{jLmMnN} (U_{m,M} U_{n,N})_{,L} + (c_{jLm1} + c_{j1mL}) U_{m,LR} \tag{15}$$

and

$$\begin{aligned} c_{j2mM} w_{m,M} &= - d_{j2mMnN} U_{m,M} U_{n,N} - c_{j2m1} U_{m,X} \qquad (m \neq 4) \\ w_4 &= 0 \qquad \text{on} \qquad X_2 = 0. \end{aligned} \tag{16}$$

Solutions are sought in $X_2 < 0$, subject to $|w| \to 0$ as $X_2 \to -\infty$.

As in [13], the Fourier transform $\bar{w}_m(k,X_2,X) = \mathcal{F}\{w_m\}$ is taken with respect to θ, so replacing (15) and (16) by the ordinary differential system

$$\mathcal{L}_{jm}\bar{w}_m = \bar{P}_j \text{ for } -\infty < X_2 < 0, \qquad \mathcal{B}_{jm}\bar{w}_m = \bar{Q}_j \text{ on } X_2 = 0. \tag{17}$$

Here, $\bar{P}_j$ and $-\bar{Q}_j$ are the Fourier transforms of the right-hand sides of (15) and (16), while the operators $\mathcal{L}_{jm}$ and $\mathcal{B}_{jm}$ involve differentiation only with respect to X_2.

The *solvability condition* for the problem (17) is found via a Green identity (integration by parts) to be

$$\left[A_j^*(k,X_2)\bar{Q}_j(k,X)\right]_{X_2=0} = \int_{-\infty}^{0} A_j^*(kX_2)\bar{P}_j(k,X_2,X)\, dX_2 \qquad (k > 0). \tag{18}$$

This is because $\mathcal{L}^*_{mj}$ and $\mathcal{B}^*_{mj}$ are the operators adjoint to $\mathcal{L}_{jm}$ and $\mathcal{B}_{jm}$, so that the only decaying solutions to the homogeneous problem adjoint to (17) are constant multiples of the complex conjugate of the function $A_j(kX_2)$ occurring in (9). Equation (18) is formally the same as in the purely elastic analysis [12]. The function $C(k,X)$ enters only as $\partial C/\partial X$ or as terms of convolution type, so that (18) may be rewritten as the *evolution equation*

$$JC_{,X}(k,X) = i\int_0^{\infty} K\left[\frac{\ell}{k}\right](k-\ell)\ell C(k-\ell,X)C(\ell,X)\, d\ell \qquad (k > 0). \tag{19}$$

This equation is of the form derived by LARDNER [4,6] (see also TUPHOLME [10]), but the expressions for the 'reduced kernel' $J^{-1}K(k^{-1}\ell)$ are simpler.

The (real) constant J and the (complex) kernel $K(k^{-1}\ell)$ are respectively of second and third degree in the function $\underset{\sim}{A}(kX_2)$ defining the depth dependence in a linear, harmonic wavetrain. Indeed, they are given in terms of $s^{(p)}$, $a_j^{(p)}$ and $B^{(p)}$ by equations (A.1) and (A.2) of [12], with the summation range extended up to 4, and using appropriate constitutive coefficients (5) and (6). Unlike the corresponding expressions in [6], [9] and [10], they are computed without the need for inversion of a matrix function of ℓ/k. (Actually (A.3) of [12] gives $K(k^{-1}\ell)$ explicitly as a sum of partial fractions in ℓ/k.)

4. Alternative Versions

By exploiting the property

$$K\left[\frac{\ell}{k}\right] = -2\frac{k}{\ell}K^*\left[\frac{k}{\ell}\right] \qquad \text{for } \ell > k > 0, \tag{20}$$

it is possible to rewrite (19) as

$$\begin{aligned} JC_{,X}(k,X) = {} & i\int_0^{k} K\left[\frac{\ell}{k}\right](k-\ell)\ell C(k-\ell,X)C(\ell,X)\, d\ell \\ & + 2i\int_k^{\infty} K^*\left[\frac{k}{\ell}\right]k(\ell-k)C^*(\ell-k,X)C(\ell,X)\, d\ell. \end{aligned} \tag{21}$$

This form involves C only for positive values of its first argument. It also shows how the first integral gives contributions to $\partial C/\partial X$ arising from pairs of waves for which the *sum* of the wavenumbers equals k, while the second integral gives contributions from pairs for which the *difference* of the wavenumber equals k.

In [13], analysis is given also for two other boundary conditions on $X_2 = 0$. In one case (Case F) the exterior region $X_2 > 0$ is treated as free-space, so that no 'material coordinates' are naturally defined there. However, by introducing 'pseudo-displacements' $\varepsilon\hat{v}_j$ (j=1,2,3) such that

$$x_j = \delta_{jJ}X_J + \varepsilon\hat{v}_j(\theta,X_2,X) \quad \text{in } X_2 > 0, \qquad \hat{v}_j(\theta,0,X) = u_j(\theta,0,X)$$

and writing Laplaces's equation for the exterior potential $\varepsilon\hat{\varphi}$ in terms of coordinates (θ,X_2,X), field equations and conditions matching (u_1,u_2,u_3,u_4) to $(\hat{v}_1,\hat{v}_2,\hat{v}_3,\hat{\varphi})$ at the material surface $X_2 = 0$ are derived, to both $O(\varepsilon)$ and $O(\varepsilon^2)$. The procedure outlined above may then be modified to describe Fourier transforms of fields in both $X_2 < 0$ and $X_2 > 0$.

The leading order solution (9) is supplemented by

$$\hat{\varphi} = \int_0^\infty C(k,X) \sum_{p=1}^{4} B^{(p)} a_4^{(p)} e^{-kX_2} e^{ik\theta} dk + c.c.$$

and the definitions (12) of M_{jp} are modified because (7) is replaced by a statement that T_{2j} is continuous. The fields $\hat{v}_j$ are arbitrary, provided they are bounded and decay as $X_2 \to \infty$. The solvability conditions (18) involves extra terms concerning $\hat{\varphi}$ and an integral over $0 < X_2 < \infty$. However, an equation of type (19) is still obtained, with straight-forward modifications to the definitions of J and $K(k^{-1}\ell)$.

A third case (Case O) is the 'open circuit' limit, in which (7) is replaced by $T_{2j} = 0$ on $X_2 = 0$. Likewise, this yields the evolution equation (19), and J and $K(k^{-1}\ell)$ redefined to accommodate the changed definition of M_{jp}.

For periodic waveforms of fundamental wavelength 2π, expressions (9) are replaced by

$$U_j(\theta,X_2,X) = \sum_{n=1}^{\infty} C_n(X) A_j(nX_2)e^{in\theta} + c.c. \; . \tag{22}$$

The evolution equation (19) is then replaced by coupled amplitude equations (CAE's)

$$\frac{dC_n}{dX} = i \sum_{m=1}^{n-1} \frac{1}{J} K\left[\frac{m}{n}\right] (n-m)m\, C_{n-m} C_m$$

$$- i \sum_{m=n+1}^{\infty} \frac{1}{J} K\left[\frac{m}{n}\right] (m-n)m\, C^*_{m-n} C_m \tag{23}$$

(or equivalent forms for the real and imaginary parts $D_n(X)$ and $E_n(X)$ of $2C_n(X)$, see [12].

5. Computation of Waveform Evolution

To exploit the preceding theory, numerical procedures have been written to convert given crystal constants into coefficients c_{iLmM}, etc. appropriate for axes $0X_1,X_2,X_3$ at any orientation to crystallographic 'X','Y' and 'Z' axes. The routines evaluate the Rayleigh speed c and the quantities $s^{(p)}$, $\underset{\sim}{a}^{(p)}$, $B^{(p)}$, when all elastic constants, permittivities and piezoelectric coefficients are non-dimensionalized with respect to c_{66}, ϵ_1 and $(c_{66}\epsilon_1)^{\frac{1}{2}}$, respectively (using Voigt's contracted notation for coefficients referred to 'X', 'Y' and 'Z' directions). The reduced kernel $J^{-1}\dot{K}(k^{-1}\ell)$ is the *only* term in (19) which depends on the boundary conditions, the anisotropy or the propagation direction, is then computed.

To date, attention has been concentrated on propagation on a 'Y-cut' surface of $LiNbO_3$, using the central values of the data given in MAUGIN [15]. In [13], computations using (22) and (23) truncated to 40 complex coefficients show that initially sinusoidal signals propagating in the 'X' and 'Z' directions distort in different manners and at different rates. The distortion features for Cases F and O are similar, but different from the earthed case (E).

Here, we consider only Case E, but investigate propagation at various angles ψ to the 'X' axis. Figure 1 displays the reduced kernel $J^{-1}K(m^{-1}\ell)$ and non-dimensionalized wavespeed $\bar{c} \equiv c(\rho_0/c_{66})^{\frac{1}{2}}$. It should be remembered that experimental data are known with little precision, so the trends exhibited are indicative of the possibilities only (see the comparison in [10]). Although Figure 1 concerned only Case E, results for Cases F and O with $\psi = 0^0$ ('X propagating' waves) and $\psi = 90^0$ ('Z-propagating' waves) are given in [13], where it is also checked that in Case F for $\psi = 90^0$ the value of c agrees with that of SINHA and TIERSTEN [16].

It is seen that $J^{-1}K$ is real only for $\psi = 0^0$, which is the case analysed in [8]. Some consequences of non-real K are exhibited by considering initially sinusoidal waveforms of unit amplitude. These are given by (22) and (23) with initial conditions $C_1(0) = -\frac{1}{2}i$, $C_n(0) = 0$ $(n\neq1)$. Elevation profiles $U_2(\theta,0,X)$ remain anti-symmetric only when K is real. It is shown in [13] that 'Z-propagating' waves lose symmetry and break up at ranges shorter than for 'X-propagating' waves. Figure 2 shows how the elevation profile develops steep asymmetric peaks (c.f. the symmetric peaks in [12]) over a range X $\simeq$ 0.008. The lack of symmetry is shown more clearly by the corresponding plot of the horizontal displacement $U_1(\theta,0,X)$, Figure 3(a), which is a rather biased version of the plot arising for propagation along the 'X' axis (Figure 3(b)).

Calculations for propagation at a number of angles ψ exhibit somewhat similar wave distortion. In all cases other than $\psi = 0^0$ the wave steepening is asymmetric. The rate

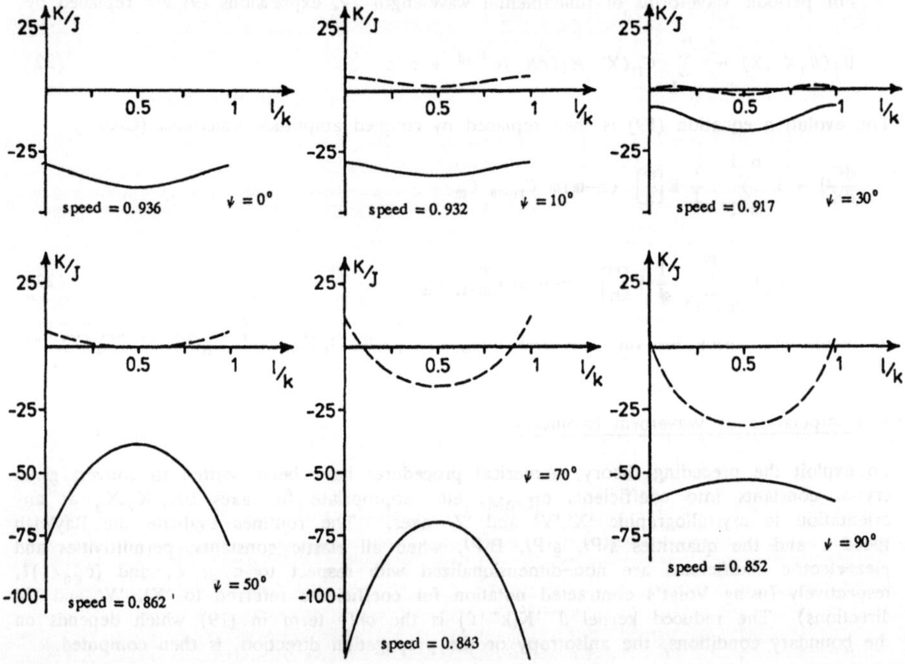

Figure 1 The reduced kernel $J^{-1}K(k^{-1}\ell)$ and dimensionless wavespeed $c(\rho_0/c_{66})^{\frac{1}{2}}$ for SAWs propagating at various angles ψ to the 'X' direction on 'Y-cut' $LiNbO_3$ having a traction-free earthed boundary. Real part ————, imaginary part — — — —. Results for $\ell > k > 0$ are obtained using (20)

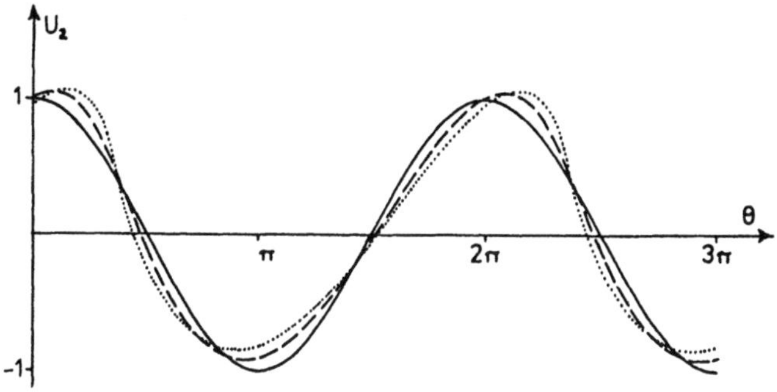

Figure 2 Distortion of an initially sinusoidal profile of surface elevation $U_2(\theta,0,X)$ at ranges X = 0 ————, X = 0.0045 — — — —, X = 0.008 ·················. Propagation along the 'Z' axis

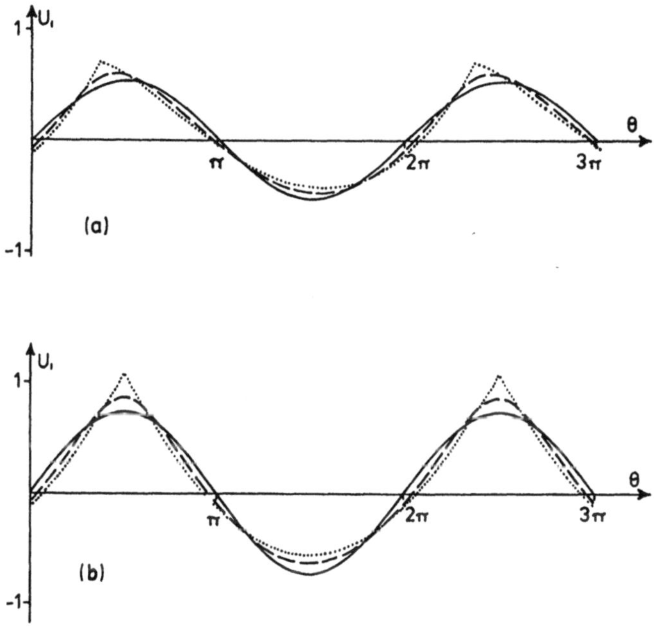

Figure 3 Horizontal displacement $U_1(\theta,0,X)$ at the free surface (a) for propagation along the 'Z' axis, at ranges X = 0 ————, X = 0.0045 — — — , X = 0.0085 ············ ; (b) for propagation along the 'X' axis at X = 0 ———— , X = 0.0075 — — —, X = 0.014 ············

of steepening depends fairly strongly on propagation distance and in all cases leads to formation of ripples on the waveform (which are probably an indication of error due to the truncation to 40 complex coefficients). This 'break-up' phenomenon is seen more clearly from the surface speed profiles. Figure 4 shows how the waveforms of vertical speed $-U_{2,\theta}$ develop, for propagation at angles ψ = 0°, 10°, 30°, 50°, 70° and 90°. From these it is possible to estimate the 'break-up range' as given in Table 1.

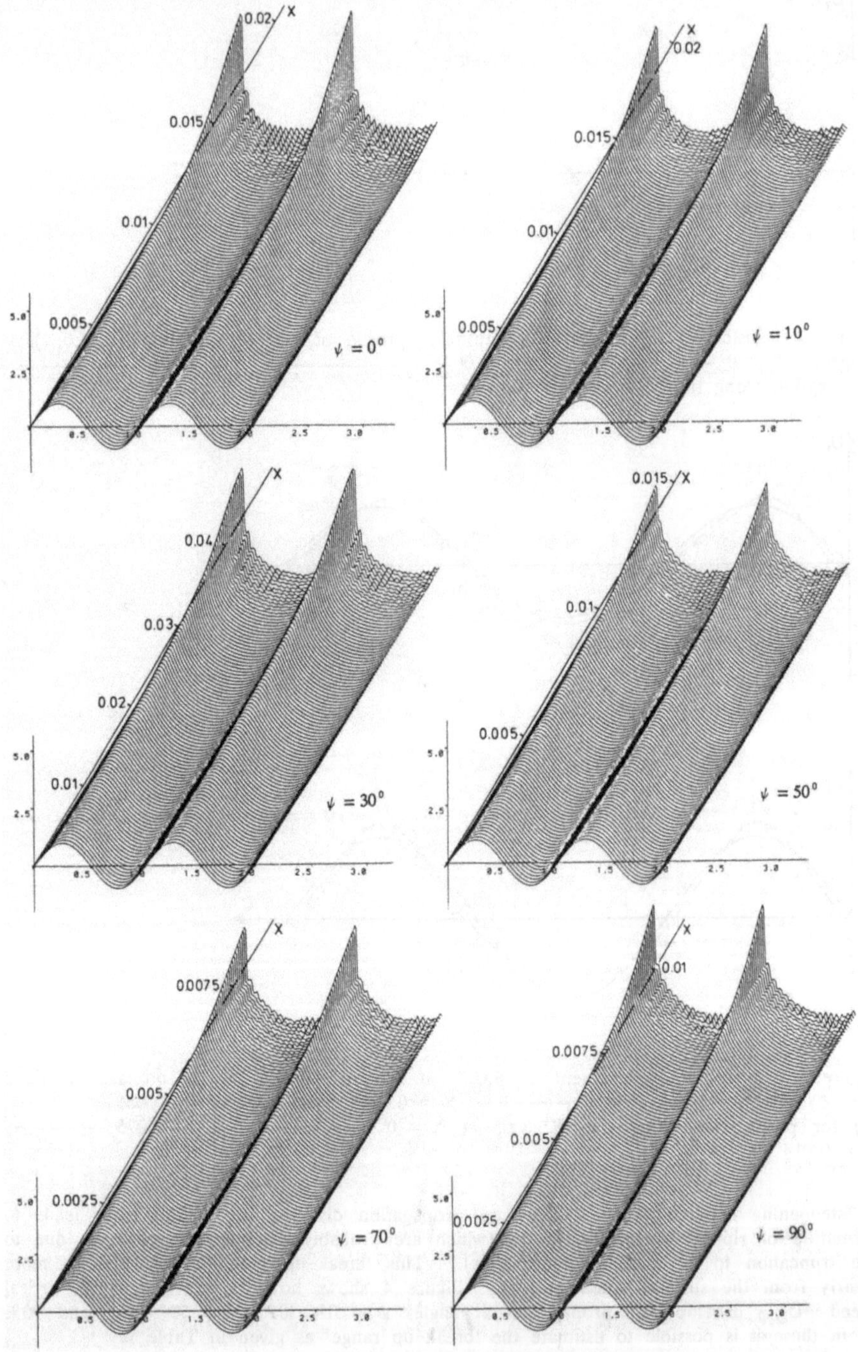

Figure 4 Evolution of the vertical speed distribution $-U_2(\theta,0,X)$ at the free surface at the propagation angles considered in Table 1.

Table 1

Dimensionless wavespeed and break-up range at angles ψ to the 'X-direction' on 'Y-cut' $LiNbO_3$ (Case E)

ψ	0^0	10^0	30^0	50^0	70^0	90^0
$c(\rho_0/c_{66})^{\frac{1}{2}}$	0.936	0.932	0.917	0.862	0.843	0.852
Range	0.0132	0.0144	0.0342	0.0106	0.0062	0.0079

Although it must be emphasized that detailed conclusions may depend strongly on the values of the material coefficients (and also may be different for Cases F and 0) the strong angular dependence of nonlinear effects (particularly around $\psi = 30^0 - 50^0$ should be noted.

Acknowledgement

This work was supported by Grant D 76899 from the Mathematics Committee of the Science and Engineering Research Council.

References

1. N. Kalyanasundaram: Int. J. Engng. Sci. 19, 279 (1981)
2. N. Kalyanasundaram, R. Ravindran, P. Prasad: J. Acoust. Soc. Am. 72, 488 (1982)
3. N. Kalyanasundaram: J. Sound Vib. 96, 411 (1984)
4. R. W. Lardner: Int. J. Engng. Sci. 21, 1331 (1983)
5. R. W. Lardner: Int. J. Engng. Sci. 23, 113 (1985)
6. R. W. Lardner: J. Elast, 16, 63 (1986)
7. E. A. David: Int. J. Engng. Sci. 23, 699 (1985)
8. R. W. Lardner and G. E. Tupholme: J. Elast. 16, 251 (1986)
9. M. Planat: J. Appl. Phys. 57, 4911 (1985)
10. G. E. Tupholme: In Recent Developments in Surface Acoustic Waves, ed. by D.F. Parker, G.A. Maugin, Springer Ser. Wave Phen. (Springer, Berlin, Heidelberg 1988)
11. A. P. Harvey, R. E. Craine, S. Syngellakis: In Recent Developments in Surface Acoustic Waves, ed. by D.F. Parker, G.A. Maugin, Springer Ser. Wave Phen. (Springer,Berlin, Heidelberg 1988)
12. D. F. Parker: Int. J. Engng. Sci. 26, 59 (1988)
13. D. F. Parker and E. A. David: In preparation
14. D. B. Taylor and S. G. Crampin: Proc. Roy. Soc. A364, 161 (1978)
15. G. A. Maugin: Nonlinear Electromechanical Effects and Applications (World Scientific, Singapore 1985)
16. B. K. Sinha and H. F. Tiersten: J. Appl Phys. 44, 4831 (1973)

Nonlinear Piezoelectric Surface Acoustic Waves Using the Multiple Scales Technique

A.P. Harvey[1], *R.E. Craine*[1], *and S. Syngellakis*[2]

[1]Faculty of Mathematical Studies, University of Southampton, Southampton, SO9 5NH, U.K.

[2]Department of Civil Engineering, University of Southampton, Southampton, SO9 5NH, U.K.

1. INTRODUCTION

The recent developments in surface acoustic wave (SAW) devices such as convolvers, filters and transducers have increased interest in theoretical studies of the nonlinear propagation of surface waves. Although many signal processing devices rely upon nonlinear coupling in their piezoelectric substrates, much of the early theoretical work [1,2] failed to give solutions valid far from the source and neglected modulation of the wave amplitude with propagation. Recent attempts to remedy these deficiencies, through the use of multiple scale techniques, have been discussed by TUPHOLME [3].

A method for reducing the field and constitutive equations appropriate to a piezoelectric solid to the same form as those for an anisotropic elastic solid has been presented in [3]. A solution to these equations using the multiple scales technique is outlined in section 2 and some numerical results are discussed in section 3.

In signal processing the principal use of piezoelectric SAW convolvers is the modulation of an input signal through interaction with a second signal. To model this process we require a nonlinear formulation for the harmonic coupling of two SAWs propagating in opposite directions. This model, in conjunction with an expression for the power input of both waves, is used in section 4 to give estimates for the nonlinear coupling constant, or figure of merit. The numerical method developed allows calculation of the figure of merit for the interaction of two oppositely propagating harmonic waves in any anisotropic piezoelectric material in any orientation and it is hoped that this information will prove useful in the design of efficient convolvers.

2. FORMULATION AND SOLUTION

The field equations, constitutive equations and boundary conditions for a deformed piezoelectric half space, neglecting free space stress and electric field effects, are set out in [3], equations (35) to (37), and are not repeated in this paper. We shall consider solutions to this system of equations obtained with a method similar to that used by Tupholme for the anisotropic elastic case.

A solution is sought in which the displacement vector, u_j, depends on the small scaling parameter ε through the perturbation series

$$u_j = \varepsilon u_j^{(0)} + \varepsilon^2 u_j^{(1)} + \dots \qquad (j=1,\dots 4), \tag{1}$$

where, as in [3], u_4 denotes the electric potential ϕ .

On substituting (1) into the system in [3] ((35) and (36)) leading order field equations and boundary conditions can be written

$$\hat{c}_{ijk\ell}\, u^{(0)}_{k,\ell j} - \rho\hat{\delta}_{ij}\, \ddot{u}^{(0)}_j = 0 \tag{2}$$

$$\hat{c}_{i2k\ell}\, u^{(0)}_{k,\ell} = 0 \quad \text{on} \quad x_2 = 0 \ , \tag{3}$$

where repeated subscripts are assumed to satisfy the summation convention, taking values from 1 to 4 , a superposed dot denotes a time derivative and comma denotes a partial derivative with respect to x_j .

Introduce Cartesian coordinates (x,y,z) and consider a plane wave propagating in the positive x direction in the half space $y \leq 0$, where $y = 0$ is a free surface and there is no dependence on z . The quantities $u^{(0)}_j$, $u^{(1)}_j$... are allowed to depend upon x, y and t through the slow variables $\xi(=\varepsilon x)$, $\eta(=\varepsilon y)$ and $\tau(=\varepsilon t)$. The field equations (2) yield a solution in the form

$$u^{(0)}_j = \sum_{n=1}^{\infty} \sum_{\ell=1}^{4} \alpha^{n(\ell)}\, a^{(\ell)}_j \exp[ins^{(\ell)}y]\exp[in(x-ct)] + \text{complex conjugate}\ , \tag{4}$$

where ℓ denotes the four modes obtained as solutions to the characteristic equation $\det[L_{ij}(s)] = 0$ that decay as $y \to -\infty$. Here, $L_{ij}(s)$ is defined by

$$L_{ij}(s) = \hat{c}_{i2j2}\, s^2 + (\hat{c}_{i2j1} + \hat{c}_{i1j2})s + \hat{c}_{i1j1} - \rho c^2\hat{\delta}_{ij} \ . \tag{5}$$

All spatial variables are non-dimensionalised with respect to the fundamental wave number k , and the summation (4) includes all harmonics.

In the standard way the Rayleigh wave speed, c , is obtained from the determinantal condition for a non-trivial solution to conditions (3). For a general piezoelectric material no analytical solution for c is possible and for these materials c must be determined through an iterative solution of the characteristic equation and determinantal equation.

The linear boundary conditions also provide the particular solutions $\beta^{(\ell)}$ which are related to the surface amplitudes γ^n via the equation

$$\alpha^{n(\ell)} = \beta^{(\ell)}\gamma^n \qquad \text{on} \quad y = \eta = 0 \ . \tag{6}$$

The complete first order solution $u^{(0)}_j$ is now known except for the dependence of γ^n on the slow variables ξ, η and τ . This dependence is determined from the second order system of equations

$$\hat{c}_{ijk\ell}\, u^{(1)}_{k,\ell j} - \rho\hat{\delta}_{ij}\, \ddot{u}^{(1)}_j = \sum_{n=0}^{\infty} \sum_{\ell=1}^{4} \left\{ \sum_{m=1}^{\infty} \sum_{r=1}^{4} \left[G_i(m,n)\alpha^{m(\ell)}\alpha^{n-m(r)} \right] \right.$$

$$+ F_i^1 \frac{\partial\alpha^{n(\ell)}}{\partial\xi} + F_i^2 \frac{\partial\alpha^{n(\ell)}}{\partial\eta} + F_i^3 \frac{\partial\alpha^{n(\ell)}}{\partial\tau}\Bigg\}\exp[in(x-ct)]$$

$$+ \text{complex conjugate,} \qquad (7)$$

$$\hat{c}_{i2k\ell}\, u^{(1)}_{k,\ell} = \sum_{n=0}^{\infty}\sum_{\ell=1}^{4}\Bigg\{\sum_{m=1}^{\infty}\sum_{r=1}^{4}\Big[g_i(m,n)\alpha^{m(\ell)}\alpha^{n-m(r)}\Big]$$

$$+ f_i^1 \frac{\partial\alpha^{n(\ell)}}{\partial\xi} + f_i^2 \frac{\partial\alpha^{n(\ell)}}{\partial\eta}\Bigg\}\exp[in(x-ct)]$$

$$+ \text{complex conjugate,} \qquad (8)$$

where F,G,f and g, depending on the second and third order constants, are exponential in y and $\alpha^{-k(\ell)} = \bar{\alpha}^{k(\ell)}$, a superposed bar denoting a complex conjugate.

Investigation of a solution satisfying both (7) and (8) and elimination of secular terms of the form $y \exp[ins^{(\ell)}y]$ leads to a system of coupled amplitude equations in the surface amplitude γ^n. These equations may be written, for a monochromatic source, in the form

$$\frac{d\gamma^n}{d\xi} + \sum_{m=1}^{\infty} \hat{H}(m,n)\, \gamma^m \gamma^{n-m} = 0 , \qquad n = 1,2,\ldots . \qquad (9)$$

The coefficients $\hat{H}(m,n)$ depend in a complicated way on the second and third order material constants and, for convenience, are not stated explicitly. Further details of the solution derived in this section will appear in HARVEY [4].

3. NUMERICAL RESULTS

Numerical routines have been written to assign and rotate the crystal constants for various symmetry classes of piezoelectric material, to iteratively solve for the corresponding Rayleigh wave speed and to solve the resulting coupled amplitude equations using finite difference methods. In solving for these amplitudes the infinite series must be replaced by some finite number of harmonics.

These programs require numerical values for the complete set of second and third order elastic, piezoelectric, dielectric and electrostrictive constants. It appears that all the constants are known only for a small number of piezoelectric materials, but the list does include Lithium Niobate ($LiNbO_3$) which is used extensively in practical devices. However, the quoted values of the constants for $LiNbO_3$ differ significantly [5,6] and these variations produce major differences in the solutions.

Before calculating results for piezoelectric materials the numerical routines were tested using data corresponding to the cubic elastic materials investigated in [7]. Their numerical values for the growth of harmonics were reproduced exactly.

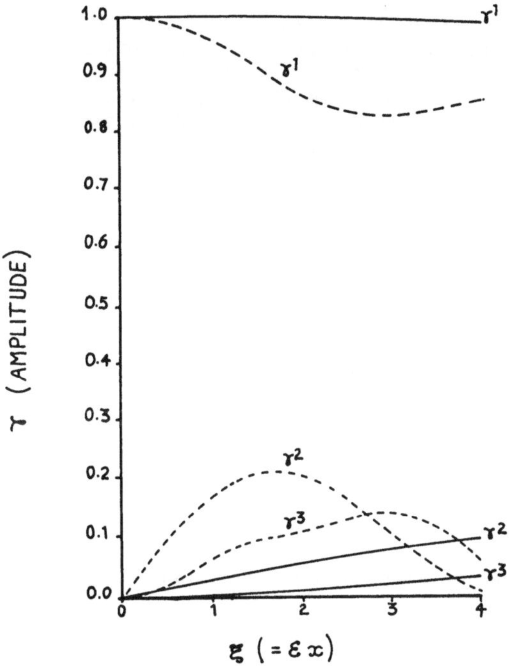

Figure 1 Growth and decay with distance of first three harmonics on Y-cut Lithium Niobate for a SAW propagating along the positive X-axis (calculated from a system of ten harmonics): ———— data from [5]; ------- data from [6].

Results were then obtained for $LiNbO_3$ using the data stated in [5]. The linear piezoelectric Rayleigh wavespeeds found in [8] were reproduced. Figure 1 shows the variation with ξ of the modulus of the amplitudes of the first three harmonics for a particular crystal orientation. Results have been calculated for systems including various numbers of harmonics. We note that Fig. 1 is qualitatively similar to that shown in [7] for the cubic material MgO, but that the variation of amplitudes with ξ is slower for $LiNbO_3$ for the particular cut chosen. However, preliminary results indicate that rotation away from the axis of symmetry in $LiNbO_3$ can significantly affect the growth rate of harmonics.

As mentioned earlier, discrepancies exist in the data quoted in the literature for $LiNbO_3$. Running the programs with the numerical values for the constants given in [6] produced significantly different results for the growth and decay of harmonics, as shown in Fig. 1.

4. INTERACTION OF TWO SAWs

A multiple scales approach to the interaction of the fundamental modes of two oppositely propagating SAWs on isotropic elastic substrates has been given by Kalyanasundaram [9]. We have extended this work by considering the interaction of any two harmonics of a pair of SAWs propagating in an anisotropic piezoelectric material.

Particular attention has been given to the interaction of the fundamental modes of two identical waves propagating in opposite directions on a piezoelectric half space. Here, as in the general situation, we have no interaction effects at first order due to phase matching but, unlike the general case, the characteristic equation obtained when determining the complementary function for the second order system has a pair of zero roots. This leads to the appearance of bulk waves which we must assume to be absorbed at the base of the convolver.

It is found that the surface amplitude of the interaction wave at a particular frequency can be calculated from knowledge of the surface amplitudes of the appropriate harmonics in the two opposite propagating constituent waves.

The power vector, P_j, of a propagating acoustic wave is defined [10] as

$$P_j = - \dot{u}_i \sigma_{ij} + \phi \dot{D}_j \, , \tag{10}$$

where σ_{ij} and D_j denote the Piola-Kirchhoff stress tensor and electric displacement vector respectively.

From (9) we deduce that the average power density of a plane surface wave of the type considered in section 2 is

$$P_j^{av} = \frac{1}{2} \operatorname{Re}\left[\iint\left(-\bar{\dot{u}}_i \sigma_{ij} + \phi \bar{\dot{D}}_j\right) dy \, dz\right] . \tag{11}$$

It is easily seen that the linear terms in the series expansions for u_j, σ_{ij} and D_j lead to a value for P_j^{av} which is second order in ε.

The nonlinear coupling constant, M, or figure of merit, for a particular crystal and cut is defined as the ratio of the root mean square open circuit surface potential at twice the input frequency to the power density input to the two waves of a degenerate convolver (identical waves) [11], i.e.

$$M = \frac{\phi^{RMS}(0)}{\left(\dfrac{P(1)P(2)}{W^2}\right)^{\frac{1}{2}}} \, , \tag{12}$$

where W is the width of the beam.

The surface potential is obtained by considering the interaction of all harmonics that yield sum or difference frequencies equal to twice the input frequency. It seems reasonable to assume that the greatest contribution will arise from the standing wave produced by the interaction of the two fundamental modes. With this postulate and equations (11) and (12) we can then estimate the figure of merit.

Leading order results give expressions analogous to those in [12] and numerical results using the data in [5] exhibit the same qualitative behaviour. Work is in progress to assess the effects on the calculated values of M arising from the interaction of other harmonics.

Even when more accurate values for M are obtained it seems doubtful that they will agree with experimentally measured values since we have not attempted in our analysis to model either the waveguides used in practice or the excitation of the crystal by a transducer.

Acknowledgements

The authors wish to thank Dr. D.F. Parker, Dr. E.A. David and especially Dr. G.E. Tupholme for useful discussions. APH wishes to thank the S.E.R.C. for funding and support throughout the duration of study.

References

1. D.F. Nelson: Electric Optic and Acoustic Interactions in Dielectrics (Wiley, New York, 1979)
2. H.F. Tiersten, J.C. Baumhauer: J. Appl. Phys. 45(10), 4272 (1974)
3. G.E. Tupholme: In Recent Developments in Surface Acoustic Waves, ed. by D.F.Parker, G.A. Maugin, Springer Ser. Wave Phen. (Springer, Berlin, Heidelberg 1988)
4. A.P. Harvey: Ph.D. Thesis, University of Southampton
5. G.A. Maugin: Nonlinear Electromechanical Effects and Applications (World Scientific, Singapore, 1985)
6. Y. Cho, K. Yamanouchi: J. Appl. Phys. 61(3), 875 (1987)
7. R.W. Lardner, G.E. Tupholme: J. Elast. 16, 251 (1986)
8. J.J. Campbell, W.R. Jones: IEEE Trans. Sonics. Ultrasonics, SU-15, 209 (1968)
9. N. Kalyanasundaram: J. Acoust. Soc. Am. 73(6), 1956 (1983)
10. B.A. Auld: Acoustic Fields and Waves in Solids, Vol. I (Wiley, New York, 1973)
11. T.C. Lim, E.A. Kraut, R.B. Thompson: Appl. Phys. Lett. 20, 127 (1972)
12. A.K. Ganguly, K.L. Davis: J. Appl. Phys. 51(2), 920 (1980)

Linear and Nonlinear Surface Waves on Magnetostrictive Crystals in a Bias Magnetic Field

Abo-El-Nour Abd-Alla and G.A. Maugin*

Université Pierre et Marie Curie, Laboratoire de Modélisation en Mécanique, Tour 66, 4 place Jussieu, F-75252 Paris Cedex 05, France

The problem of the propagation, from a source, of linear and nonlinear surface magnetoacoustic waves on a magnetostrictive centrosymmetric substrate placed in a bias magnetic field is investigated. The nonlinear nature of magnetostrictive couplings means that the problem examined is quite different from the corresponding one in piezoelectric electroelastic solids. In particular, the linearized coupled theory necessarily involves a bias magnetic field while, in the nonlinear theory, the results obtained, either through a direct Poincaré expansion (for harmonic SAW generation and anischronism) or by using a multiple-scale technique (for studying the spatial evolution of coupled complex amplitudes of harmonics), are markedly different from those of the corresponding purely (nonlinear) elastic and piezoelectric cases.

1. INTRODUCTION

There are only a very few works dealing with nonlinear SAWs in coupled fields (e.g., [1]-[2]), and to the knowledge of the authors, none in the case of the magnetoelasticity of magnetostrictive materials. The present work has the purpose of filling this gap by considering the propagation of linear and nonlinear surface waves of a mixed nature on a magnetoelastic anisotropic substrate. This is done on the basis of rotationally invariant nonlinear equations built in a previous work [3]. The essential bulk-wave characteristics have also been established as a prerequisite [4]. The interest for this magnetoacoustic case stems from several facts. First, magnetostrictive materials can be used in a number of signal-processing devices (so-called EMATs [5]). Second, magnetostriction in itself is a nonlinear coupling effect. This has several consequences, one being that the linearized version of coupled magnetoelasticity implies the existence of a bias magnetic field in magnetostrictive materials. Another consequence is that in solving a nonlinear magnetoelastic wave problem by successive approximations (e.g., in Ref. 4 for bulk waves), in the hierarchy of problems there always is a discrepancy of one order between mechanical and magnetic entities, the magnetic field at order n being generated by the magnetomechanical problem at order $n-1$. It must also be noticed that there does not exist for magnetic processes the equivalent of grounding at a surface so that, of necessity, a matching of internal and external solutions is involved for the magnetoacoustic surface problem. All these imply that the nonlinear SAW problem on magnetostrictive materials presents original features which are examined here below.

* On leave from Sohag University (Egypt)

2. EQUATIONS OF NONLINEAR MAGNETOELASTICITY

In a regular region D_0 of E^3 in the reference configuration K_R the equation of motion and Maxwell's equations in insulators read [3]

$$\rho_R \frac{\partial^2 u_i}{\partial t^2} = T^t_{Ki,K} \quad , \quad B_{K,K} = 0 \quad , \quad H_K = -\emptyset_{,K} \quad , \tag{2.1}$$

while at the regular boundary ∂D_0 of unit outward normal N_K we have the following jump conditions :

$$N_K(T^t_{Ki}) = 0 \quad , \quad N_K(B_K) = 0 \quad , \quad (\emptyset) = 0 \quad , \tag{2.2}$$

where (..) denotes the jump of the enclosed . The nonlinear stress and field constitutive equations have been obtained on a thermodynamical basis as (thermal effects are ignored, the material is centrosymmetric [3])

$$\begin{aligned} T^t_{Ki} &= \delta_{Ri}(C_{KRMN}\, u_{M,N} + \gamma'_{KRMNPQ}\, u_{M,N}\, u_{P,Q} \\ &\quad + \delta_{KRMNPQAB}\, u_{M,N}\, u_{P,Q}\, u_{A,B} + B'''_{KRMN}\, H_M\, H_N \\ &\quad + B''''_{KRMNPQ}\, H_M\, H_N\, u_{P,Q}) + \text{h.o.t}, \end{aligned} \tag{2.3}$$

$$\begin{aligned} B_K &= \mu_{KL}\, H_L - \overset{o}{B}_{KLMN}\, H_L\, u_{M,N} + \frac{1}{6}\chi_{KLMN}\, H_L\, H_M\, H_N \\ &\quad - \overset{o}{\lambda}_{KRMNPQ}\, H_R\, u_{M,N}\, u_{P,Q} + \text{h.o.t}, \end{aligned} \tag{2.4}$$

where upper and lower case Latin indices refer to material and present Cartesian coordinates, respectively. The symbol ρ_R, $u_i = x_i - X_K\delta_{Ki}$, x_i, B_K, H_K, $\emptyset$ and T^t_{Ki} denote, respectively, the matter density at K_R, the elastic displacement, the nonlinear motion, the "Lagrangian" magnetic induction and field, the magnetostatic scalar potential and the total Piola-Kirchhoff nonsymmetric stress tensor. δ_{Ki} is a shifter. C_{KLMN} and μ_{KL} are tensors of elasticity coefficients of the second order and of magnetic permeabilities, χ_{KLMN} a tensor of magnetic susceptibilities of the fourth order, while γ_{KRMNPQ} and $\delta_{KRMNPQAB}$ are <u>effective</u> tensors of elasticity coefficients of the third and fourth orders, B'''_{KRMN} and $\overset{o}{B}_{KLMN}$ <u>effective</u> tensors of magnetostriction coefficients of the first order and B''''_{KRMNPQ} and $\overset{o}{\lambda}_{KRMNPQ}$ effective tensors of magnetostriction of the second order [3]. The last two of these are neglected in the following analysis.

3. LINEAR SURFACE MAGNETOACOUSTIC WAVES

3.1. Linearized equations in presence of a bias magnetic field

For small elastic displacements $\tilde{u}$ and magnetic fields $\tilde{h}$ (with associated potential $\tilde{\emptyset}$) superimposed on a state of zero displacement and spatially uniform magnetic field H^o, one deduces from $(2.1)_{1-2}$ and (2.2) the following linearized nondimensional equations which resemble those of linear piezoelectricity (but for the fact that the coefficients $\bar{F}'_{KLM}$ and $\bar{F}^*_{KLM}$ are not equal) :

. in D_o :

$$\frac{\partial^2 \tilde{u}_L}{\partial t^2} = \bar{C}_{KLMN}\, \tilde{u}_{M,NK} - \bar{F}'_{KLN}\, \tilde{\phi}_{N,K} \quad ,$$
$$\bar{\mu}_{NK}\, \tilde{\phi}_{,NK} + \bar{F}^*_{KMN}\, \tilde{u}_{M,NK} = 0 \quad ; \tag{3.1}$$

. at ∂D_o :

$$N_K(\bar{C}_{KLMN}\, \tilde{u}_{M,N} - \bar{F}'_{KLN}\, \tilde{\phi}_{,N}) = 0 ,$$
$$N_K(\bar{\mu}_{KL}\, \tilde{\phi}_{,L} + \bar{F}^*_{KMN}\, \tilde{u}_{M,N}) = 0 \quad , \tag{3.2}$$
$$(\tilde{\phi}) = 0 \quad ;$$

wherein

$$F'_{KLN} = 2B'''_{KLMN}\, H^o_M \quad , \quad F^*_{KMN} = B^o_{KLMN}\, H^o_L \quad , \quad H^o_L = -\phi^o_{,L} \tag{3.3}$$

and the scaling based on characteristic length, time and magnetic field, L, T and H^o, is such that

$$\bar{C}_{KLMN} = C_{KLMN}\, T^2/\rho_R\, L^2 \quad , \quad \bar{F}'_{KLN} = F'_{KLN}\, H^o\, T^2/\rho_R\, L^2 \quad ,$$
$$\bar{\mu}_{NK} = \mu_{NK}\, H^o \quad , \quad \bar{F}^*_{KMN} = F^*_{KMN} \quad . \tag{3.4}$$

3.2. Surface wave modes in the linearized theory

Let $D_o = \{X_2 > 0\}$ (Figure 1). The half-space $\{X_2 < 0\}$ is a vacuum of unit magnetic permeability where Laplace's equation holds:

$$\delta_{KL}\, \tilde{\phi}_{,LK} = 0 \quad . \tag{3.5}$$

The general surface wave solution of (3.1) and (3.5) with jump conditions (3.2) at $X_2 = 0$ is sought as a linear combination of four waves that travel at the same velocity V_o and such that $(\alpha = 1,\ldots,4)$

$$\tilde{u}_K = \sum_\alpha A_\alpha\, \hat{u}^\alpha_K \exp[i(\omega t - k_o\, \lambda^\alpha_J\, X_J)] ,\ X_2 > 0 \quad ,$$
$$\tilde{\phi} = \sum_\alpha A_\alpha\, \hat{\phi}^\alpha \exp[i(\omega t - k_o\, \lambda^\alpha_J\, X_J)] ,\ X_2 > 0 \quad , \tag{3.6}$$

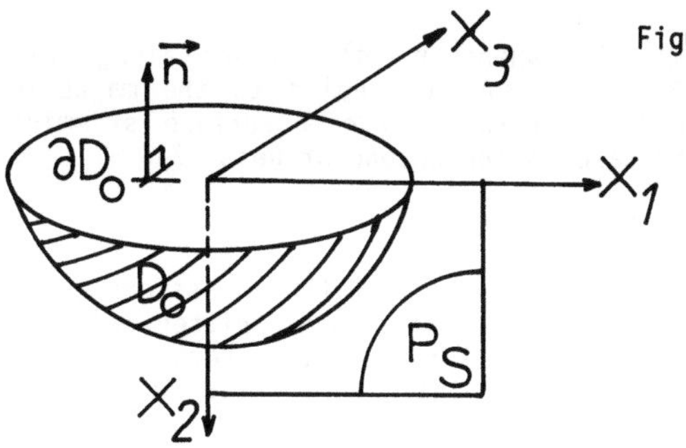

Figure 1 - Surface wave problem

$$\tilde{\phi} = \tilde{\phi}^{ex} = \sum_{\alpha} A_{\alpha}\, \tilde{\phi}^{\alpha}_{ex} \exp[i(\omega t - k_0\, \overline{\lambda}^{\alpha}_J X_J)] \, , \; X_2 < 0 \quad , \tag{3.7}$$

with $\overline{\lambda}^{\alpha}_1 = \lambda^{\alpha}_1 = \underline{1}$ and $\lambda^{\alpha}_3 = \overline{\lambda}^{\alpha}_3 = 0 \;\forall\, \alpha$ and $\lambda^{\alpha}_2 = i\chi^{\alpha}$, $\overline{\lambda}^{\alpha}_2 = i\overline{\chi}^{\alpha}$, where the χ's and $\overline{\chi}$'s are purely real and such that

$$\chi^{\alpha} < 0 \quad , \quad \overline{\chi}^{\alpha} > 0 \, . \tag{3.8}$$

Substituting from (3.6) and (3.7) into (3.2) we obtain a linear homogeneous system of four equations written in shortened notation as

$$\sum_{\alpha=1}^{4} A_{\alpha}\, \mathcal{D}^{\alpha}_{2\beta} = 0 , \tag{3.9}$$

where we have set

$$\mathcal{D}^{\alpha}_{2\beta} = (D^{\alpha}_{2L} \;\vdots\; D^{\alpha}_{2}) \quad , \quad L = 1,2,3 \tag{3.10}$$

and

$$\begin{aligned} D^{\alpha}_{2L} &= \overline{C}_{2LMN}\, \lambda^{\alpha}_N\, \hat{u}^{\alpha}_M + \overline{F}'_{2LN}(\overline{\lambda}^{\alpha}_N - \lambda^{\alpha}_N)\, \hat{\phi}^{\alpha} \quad , \\ D^{\alpha}_{2} &= \overline{F}^{*}_{2MN}\, \lambda^{\alpha}_N\, \hat{u}^{\alpha}_M - \overline{\mu}_{2M}(\overline{\lambda}^{\alpha}_M - \lambda^{\alpha}_M)\, \hat{\phi}^{\alpha} \quad . \end{aligned} \tag{3.11}$$

Equation (3.10) defines a 4 X 4 matrix with lines α and columns β. The compatibility condition for solving (3.9) for the A_{α}'s provides the following "dispersion relation" that determines V_0 :

$$\det |\mathcal{D}^{\alpha}_{2\beta}| = 0 \quad . \tag{3.12}$$

So much for the general case which compares directly to the case of <u>generalized</u> Rayleigh surface waves in piezoelectricity [6]. In the case of a centrosymmetric <u>cubic</u> material and when $\mathbf{H}^0$ is set along the X_3-axis, hence orthogonally to the sagittal plane, the above-set problem splits in two surface wave problems one being a pure Rayleigh wave problem and the other a problem of the Bleustein-Gulyaev type (SH elastic component coupled to the magnetic potential), a result similar to the one obtained in paramagnets by MAUGIN and HAKMI [7]. Whenever $\mathbf{H}^o$ is set along X_1, the problem splits into a pure Rayleigh problem coupled to the magnetic potential and a pure SH wave which degenerates into a surface skimming bulk wave (compare the piezoelectric case in the second of Refs. 7).

4. NONLINEAR SURFACE MAGNETOACOUSTIC WAVES

We now proceed to a more difficult problem when the full expressions (2.3) and (2.4) are kept. The resulting nonlinear equations are written entirely in nondimensional form and two infinitesimal small parameters ε_1 and ε_2 show up in the process, the first being related to elastic nonlinearities and the second to magnetostriction effects. We have $\varepsilon_1 = 0(\varepsilon_2) = 0(10^{-4} - 10^{-5})$.

4.1. Straightforward Poincaré expansion of the solution

In terms of ε_1 and ε_2 we consider the expansions (involving a bias magnetic field)

$$u_M = \varepsilon_1 u_M^{(o)} + \varepsilon_1^2 u_M^{(1)} + \varepsilon_1^3 u_M^{(2)} + \ldots , \quad \phi_{,R} = -H_R^{(o)} + \varepsilon_2 \phi_{,R}^{(1)} + \varepsilon_2^2 \phi_{,R}^{(2)} \ldots \quad (4.1)$$

Substituting from this into eqns. (2.1) and (2.2) produces the following hierarchy of surface wave problems in nondimensional form :

. at order one in ε_1 and ε_2 :

$$\begin{aligned} &\ddot{u}_R^{(o)} - \bar{C}_{KRMN}\, u_{M,NK}^{(o)} = 0 \quad , \quad X_2 > 0 \\ &\bar{C}_{2RMN}\, u_{M,N}^{(o)} = 0 \quad \text{at} \quad X_2 = 0 . \end{aligned} \quad (4.2)$$

. at order two in ε_1 and ε_2 :

$$\begin{aligned} &\ddot{u}_R^{(1)} - \bar{C}_{KRMN} u_{M,NK}^{(1)} = \bar{\gamma}'_{KRMNPQ} u_{M,NK}^{(o)} u_{P,Q}^{(o)} - \bar{B}'_{KRMN} H_N^{(o)} \phi_{,MK}^{(1)} \quad , \quad X_2 > 0 , \\ &\bar{\bar{\mu}}_{KR}\, \phi_{,RK}^{(1)} + \bar{B}_{KRMN} H_R^{(o)} u_{M,NK}^{(o)} = 0 \quad , \quad X_2 > 0 \end{aligned} \quad , \quad (4.3)$$

and

$$\begin{aligned} &N_K(\bar{C}_{KRMN} u_{M,N}^{(1)} + \bar{\gamma}'_{KRMNPQ} u_{M,N}^{(o)} u_{P,Q}^{(o)} - \bar{\bar{B}}'_{KRMN} H_N^{(o)} \phi_{,M}^{(1)}) = 0 \quad , \quad X_2 = 0 \quad , \\ &N_K(\bar{\bar{\mu}}_{KR}\, \phi_{,R}^{(1)} + \bar{\bar{B}}_{KRMN} H_R^{(o)} u_{M,N}^{(o)}) = 0 \quad , \quad X_2 = 0 \end{aligned} \quad , \quad (4.4)$$

$$\delta_{RK}\, \phi_{,KR}^{(1)} = 0 \quad , \quad X_2 < 0 . \quad (4.5)$$

At order one, eqns. (4.2) provide the Stoneley solution [8] for surface waves on anisotropic linear elastic substrates, i.e., a linear combination of three modes of the type $(\lambda_2^\alpha = \lambda_\alpha + i\chi^\alpha)$

$$u_R^{(o)} = \sum_{\alpha=1}^{3} A_\alpha \hat{u}_R^\alpha \exp[i(\omega t - k_o X_1) - ik_o \lambda_2^\alpha X_2]$$

$$= U_1 \sum_{\alpha=1}^{3} C_R^\alpha \exp(\chi^\alpha X_2) \exp[i(\omega t - k_o X_1 - ik_o \lambda_\alpha X_2)] \tag{4.6}$$

with (satisfying the field equations)

$$\mathfrak{C}_R^\alpha \stackrel{\text{def}}{=} G_{RM}^\alpha C_M^\alpha = 0 \quad \text{or} \quad (\overline{C}_{KRMN} \lambda_N^\alpha \lambda_K^\alpha - V_o^2 \delta_{RM})\hat{u}_M^\alpha = 0 \tag{4.7}$$

and (boundary conditions)

$$\mathfrak{D}_R \stackrel{\text{def}}{=} \sum_{\alpha=1}^{3} D_{2R}^\alpha = 0 \tag{4.8}$$

with

$$D_{2R}^\alpha \equiv (\overline{C}_{2RM1} + i\, \overline{C}_{2RM2}\, \chi^\alpha) C_M^\alpha$$

$$C_M^\alpha = A_\alpha \hat{u}_M^\alpha \quad , \quad A_1 = U_1 \quad . \tag{4.9}$$

At order two, on account of (4.6) one can solve for $(4.3)_2$ and $(4.4)_2$ and then rewrite $(4.3)_1$ and $(4.4)_1$ as follows :

$$\ddot{u}_R^{(1)} - \overline{C}_{KRMN}\, u_{M,NK}^{(1)} = \frac{1}{2} k_o^3 U_1^2 \sum_{\alpha,\beta} \{D_R^{(\alpha,\beta)} \exp(-i\, \psi^{\alpha\beta})$$

$$+ \overline{D}_R^{(\alpha,\beta)} \exp(-i\, \theta^{\alpha\beta})\} - k_o^2 \sum_\alpha J_R^\alpha \exp(-i\, \psi^\alpha) \tag{4.10}$$

and

$$\overline{C}_{2RMN}\, u_{M,N}^{(1)} = -\frac{1}{2} k_o^2 U_1^2 \sum_{\alpha,\beta} \{H_{2R}^{(\alpha,\beta)} \exp(-i\, \psi^{\alpha\beta}) + \overline{H}_{2R}^{(\alpha,\beta)} \exp(-i\, \theta^{\alpha\beta})\}$$

$$- k_o \sum_\alpha [I_{2R}^{(\alpha)} \exp(-i\, \psi^\alpha) + \overline{I}_{2R}^{(\alpha)} \exp(-\, i\, \overline{\psi}^\alpha)], \tag{4.11}$$

where we have set (* = complex conjugate)

$$\psi^\alpha = k_o X_1 + k_o \lambda_2^\alpha X_2 - \omega t \quad , \quad \psi^\alpha = k_o X_1 + k_o \overline{\lambda}_2^\alpha X_2 - \omega t \, ,$$

$$\psi^{\alpha\beta} = \psi^\alpha + \psi^\beta \quad , \quad \theta^{\alpha\beta} = \psi^\alpha - \psi^\beta \, , \tag{4.12}$$

$$F_{KRM} = \overline{\overline{B}}'_{KRMN} H_N^o \quad , \quad J_R^\alpha = F_{KRM}\, \Phi^\alpha \lambda_M^\alpha \lambda_K^\alpha \, ,$$

$$I_{KM}^{(\alpha)} = i\, F_{KMN}\, \Phi^{\alpha}\, \lambda_N^{\alpha} \quad , \quad \bar{I}_{KM}^{(\alpha)} = i\, F_{KMN}\, \bar{\Phi}^{\alpha}\, \bar{\lambda}_N^{\alpha} \quad , \tag{4.13}$$

and

$$H_{KR}^{(\alpha,\beta)} = \bar{\gamma}_{KRMNPQ} C_M^{\alpha} \lambda_N^{\alpha} C_P^{\beta} \lambda_Q^{\beta} \quad , \quad \bar{H}_{KR}^{(\alpha,\beta)} = \bar{\gamma}_{KRMNPQ} C_M^{\alpha *} \lambda_N^{\alpha *} C_P^{\beta} \lambda_Q^{\beta} \quad ,$$

$$D_R^{(\alpha,\beta)} = i\{2H_{2R}^{(\alpha,\beta)} + H_{2R}^{(\alpha,\beta)}(\lambda_2^{\alpha}+\lambda_2)\} \;,\; \bar{D}_{2R}^{(\alpha,\beta)} = (-i)\bar{H}_{2R}^{(\alpha,\beta)}(\lambda_2^{\beta}-\lambda_2^{\alpha *}). \tag{4.14}$$

The system (4.10)-(4.11) for $u_R^{(1)}$ is solved first by looking for a general solution of the propagation equation ; this solution must check the boundary condition (4.11) and possibly the condition of excitation at the source. The complete solution of (4.10) accounting for the right hand side is found in the form

$$\begin{aligned} u_M^{(1)} &= U_1^2 \sum_{\alpha} [k_o^2\, X_1\, C_M^{\alpha}\, y^{(1)} + k_o\, a_M^{(\alpha)}\, z^{(\alpha)}] \exp(-2i\, \Psi^{\alpha}) \\ &+ \sum_{\alpha} [k_o\, X_1\, C_M^{\alpha}\, g^{(1)} + a_M^{(\alpha)}\, h^{(\alpha)}] \exp(-i\, \Psi^{\alpha}) \\ &+ U_1^2 \sum_{\alpha,\beta} k_o\, a_M^{(\alpha,\beta)\prime} \exp(-i\, \Psi^{\alpha\beta}) \\ &+ U_1^2 \sum_{\forall\alpha,\beta} k_o\, b_M^{(\alpha,\beta)} \exp(-i\, \Theta^{\alpha\beta}) \quad , \end{aligned} \tag{4.15}$$

where all coefficients can be determined (see further work). The above given solution satisfies exactly the boundary condition and asymptotically the equations of motion on the condition that the coefficients in the general solution of the inhomogeneous equations $(4.3)_1$ vary slowly with X_1. This solution (4.15) contains terms at a frequency of 2ω which correspond to the phenomenon of harmonic production and terms at frequency ω which combine with those already present in $u_M^{(0)}$ to yield anisochronism. Indeed, the rate of production of second harmonic is given by

$$U_{2\omega}/U_{\omega} \simeq k_o^2\, X_1\, U_1\, y^{(1)}, \tag{4.16}$$

while by writing the global solution at frequency ω at order two in ε as

$$u_M^{(\omega)} = U(X_1) \sum_{\alpha=1}^{3} C_M^{\alpha} \exp[i\omega(t - \frac{X_1}{V(X_1)} - \frac{\lambda_2^{\alpha} X_2}{V_o})] \quad , \tag{4.17}$$

we find that

$$U(X_1) = U_1(1 + k_o\, U_1\, X_1\, e_1) \quad , \quad \frac{\Delta V}{V_o} = U_1\, f_1 \tag{4.18}$$

if $g^{(1)} = e_1 + if_1$. The result (4.16) is of the same type as in pure elasticity [9] while the result $(4.18)_2$ shows that the relative change in velocity of the fundamental component, which is due solely to magnetostriction, does not depend on the travelled distance, an X_1-dependence of $\Delta V/V_0$ appearing in the pure elastic case only at the next order of approximation [9].

4.2. Uniformly valid far-field solution

The amplitudes obtained in the foregoing paragraph grow unphysically with travelled distance. These solutions, therefore, are not valid far from the source where there is need to obtain a uniformly valid solution, all the more since surface waves can propagate over many wavelengths in distance. To obtain such a solution we apply a multiple-scale technique with two auxiliary (slow) spatial variables ξ and η introduced by [10] :

$$\xi = \varepsilon X_1 \quad , \quad \eta = \varepsilon X_2 \quad , \quad \varepsilon = O(\varepsilon_1) = O(\varepsilon_2) \tag{4.19}$$

and consider

$$u_M(X_1,X_2,t) = \tilde{u}_M(X_1,X_2,\xi,\eta,t) \ , \ H_M(X_1,X_2,t) = \tilde{H}_M(X_1,X_2,\xi,\eta,t) \tag{4.20}$$

with the expansions

$$\tilde{u}_M = \varepsilon_1 \tilde{u}_M^{(o)} + \varepsilon_1^2 \tilde{u}_M^{(1)} + \ldots \ , \ \tilde{\phi}_{,M} = - H_M^{(o)} + \varepsilon_2 \tilde{\phi}_{,M}^{(1)} + \ldots \tag{4.21}$$

Substituting from (4.21) into $(2.1)_{1-2}$ and $(2.2)_{1-2}$, we obtain the following hierarchy of surface-wave problems in nondimensional form :

. at order one in ε :

$$\begin{aligned} &\ddot{\tilde{u}}_R^{(o)} - \bar{C}_{KRMN} \tilde{u}_{M,NK}^{(o)} = 0 \quad , \quad X_2 > 0 \\ &\bar{C}_{2RMN} \tilde{u}_{M,N}^{(o)} = 0 \qquad\qquad , \quad X_2 = 0 . \end{aligned} \tag{4.22}$$

. at order two in ε :

$$\begin{aligned} \ddot{\tilde{u}}_R^{(1)} - \bar{C}_{KRMN} \tilde{u}_{M,NK}^{(1)} = {} & \bar{\gamma}'_{KRMNPQ} \tilde{u}_{M,NK}^{(o)} \tilde{u}_{P,Q}^{(o)} + \bar{C}_{1RMN} \tilde{u}_{M,N\xi}^{(o)} \\ & + \bar{C}_{KRM2} \tilde{u}_{M,K\eta}^{(o)} + \bar{C}_{KRM1} \tilde{u}_{K,M\xi}^{(o)} \\ & + \bar{C}_{2RMN} \tilde{u}_{M,N\eta}^{(o)} - \bar{\bar{B}}'_{KRMN} H_N^{(o)} \tilde{\phi}_{,MK}^{(1)} \ , \end{aligned} \tag{4.23$_1$}$$

$$\bar{\bar{\mu}}_{KR} \tilde{\phi}_{,RK}^{(1)} + \bar{\bar{B}}^{o}_{KRMN} H_R^{o} \tilde{u}_{M,NK}^{(o)} = 0 \quad , \tag{4.23$_2$}$$

for $X_2 > 0$, $\delta_{KR} \tilde{\phi}_{R,K}^{(1)} = 0$ for $X_2 = 0$ and, at $X_2 = 0$

$$\overline{C}_{2RMN}\,\tilde{u}^{(1)}_{M,N} = -\overline{\gamma}_{2RMNPQ}\,\tilde{u}^{(o)}_{M,N}\,\tilde{u}^{(o)}_{P,Q} - \overline{C}_{2RM1}\,\tilde{u}^{(o)}_{M,\xi}$$

$$- \overline{C}_{2RM2}\,\tilde{u}^{(o)}_{M,\eta} - F'_{2RN}(\tilde{\phi}^{(1)}_{ex,N} - \tilde{\phi}^{(1)}_{,N}) \qquad (4.24)_1$$

$$F^{o}_{2MN}\,\tilde{u}^{(o)}_{M,N} + F^{o}_{2M1}\,\tilde{u}^{(o)}_{M,\xi} + F^{o}_{2M2}\,\tilde{u}^{(o)}_{M,\eta} - \overline{\overline{\mu}}_{2R}(\tilde{\phi}_{ex,R} - \tilde{\phi}^{(1)}_{,R}) = 0 \quad , \qquad (4.24)_2$$

where we set $F'_{KMN} = \overline{\overline{B}}_{KRMN}H^{(o)}_{R}$, $F_{KMN} = \overline{\overline{B}}^{o}_{KRMN}H^{(o)}_{R}$.

The solution of the homogeneous system (4.22) can be chosen in the form of a superposition of multiple harmonics of slowly varying (in ξ and η) complex amplitude, i.e., (c.c. = complex conjugate)

$$\tilde{u}^{(o)}_{M} = \frac{1}{2}\,U_1 \sum_{\alpha=1}^{3} \{C^{\alpha}_{M}\,A^{\alpha}_{n}\exp(in\,\psi^{\alpha}) + c.c.\}$$

$$\psi^{\alpha} = \omega t - k_o X_1 - k_o \lambda^{\alpha}_2 X_2 \;,\quad A^{\alpha}_{n} = \mathcal{A}^{\alpha}_{n}(\zeta) \;,\quad \zeta = \xi + \mu^{\alpha}_{n}\,\eta \;, \qquad (4.25)$$

which satisfy identically (4.22) since these read

$$\mathfrak{C}^{\alpha}_{R}[n^2 A^{\alpha}_{n}\exp(in\,\psi^{\alpha})] = 0 \quad , \quad \forall\, \alpha, n, \quad , X_2 > 0$$

$$\mathfrak{D}_{R}[n\,A^{\beta}_{n}\exp(in\,\psi^{\beta})] = 0 \quad , \quad \forall\, \beta, n \;;\; \forall\, X_1, t;\; X_2 = 0 \qquad (4.26)$$

upon use of the definitions in (4.7) and (4.8). The freedom left to the $\tilde{u}^{(o)}$ solution is in the functions $\mathcal{A}^{\alpha}_{n}$, where μ^{α}_{n} is the depth parameter for the α-th component and the n-th harmonic. The $\mathcal{A}^{\alpha}_{n}$'s are to be determined at the next order by a secularity condition. Indeed, $\tilde{\phi}^{(1)}$ being generated by $\tilde{u}^{(o)}$, we can take $\tilde{\phi}^{(1)}$ and $\tilde{\phi}^{(1)}_{ex}$ of the same form as $\tilde{u}^{(o)}$ but with coefficients $\Phi^{\alpha} A^{\alpha}_{n}$ and $\overline{\Phi}^{\alpha} \overline{A}^{\alpha}_{n}$, which must be equal at $X_2 = 0$ according to $(2.3)_3$. Now we can theoretically proceed to the $\tilde{u}^{(1)}$ solution of $(4.23)_1$ and $(4.24)_2$. The complete solution is a sum of a general solution of the homogeneous equation and a particular solution, the whole satisfying the condition $(4.24)_2$. In the process one must be careful to distinguish between terms in phase with the solution of the homogeneous equation and those which are out of phase. This follows the same method as in pure elasticity [11]-[12]. The whole solution must check the boundary condition. The compatibility condition for solving the resulting linear system, after a somewhat lengthy algebra to be given elsewhere, is shown to read

$$A^{\alpha}_{n,\zeta} = \sum_{p,q} \Gamma^{\alpha}_{npq}\,A^{\alpha}_{p}\,A^{\alpha}_{q} + \sum_{\beta\neq\alpha}\sum_{r,s} \Gamma^{\alpha\beta}_{nrs}\,A^{\alpha}_{r}\,A^{\beta}_{s}$$

$$- \sum_{\beta}\sum_{c,d} \overline{\Gamma}^{\alpha,\beta}_{ncd}\,A^{\alpha}_{c}\,A^{\beta *}_{d} - \Gamma^{\alpha}_{n}\,A^{\alpha}_{n} \qquad (4.27)$$

where $p + q = n$, $r + s = n$, $c - d = n$, and Γ^{α}_{npq}, $\Gamma^{\alpha,\beta}_{nrs}$ and $\overline{\Gamma}^{\alpha,\beta}_{ncd}$ depend on the linear and nonlinear elasticity coefficients while Γ^{α}_{n} is essentially determined by magnetostriction. The system of nonlinear ordinary diffe-

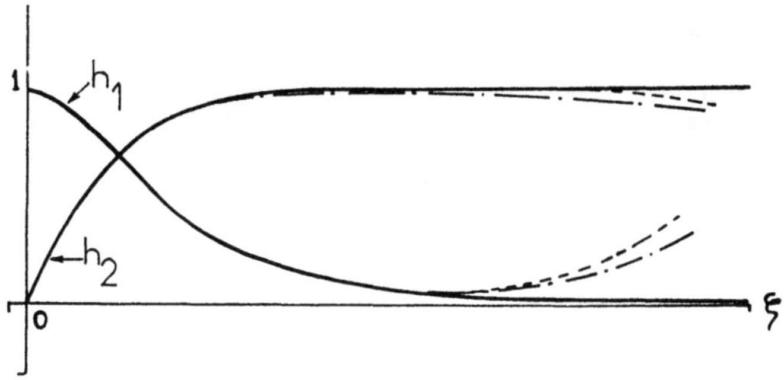

Figure 2 - Evolution of amplitude of the first two harmonics with distance (— purely elastic case ; -·-·- positive magnetostriction ; ---- negative magnetostriction)

rential equations (4.27) is a system of complex coupled-amplitude equations as often encountered in nonlinear optics [13] and second-harmonic SAW generation [14]-[15]. For cubic symmetry and $\mathbf{H}^{(o)}$ set along X_3 (i.e., orthogonally to the sagittal plane), the elastic component of the problem reduces to an SH wave. Moreover, considering the second-harmonic generator, we have $\alpha = \beta = 3$ only (that we omit) and $n = 1,2$. We are left with the following coupled (complex) amplitude equations

$$\frac{dA_1}{d\zeta} = - \Gamma_{121} A_1^* A_2 - \Gamma_1 A_1 \quad , \quad \frac{dA_2}{d\zeta} = \bar{\Gamma}_{211} A_1^2 - \Gamma_2 A_2 \quad , \tag{4.28}$$

where Γ_{121} and $\bar{\Gamma}_{211}$ are due to nonlinear elasticity and Γ_1 and Γ_2 depend on the magnetostriction coefficient (which may be either positive or negative) and on the magnitude of the bias magnetic field. Contrary to the case where Γ_1 and Γ_2 are zero [9], [12] for which an exact analytic solution can be found, the solution of (4.28) for sensible values of the coefficients has been obtained numerically (Figure 2) with the source conditions $h_1(0) = 1$ and $h_2(0) = 0$ for the real amplitudes of the first two harmonics. The purely elastic solution corresponds to solutions in h_1 and h_2 of the sech and tanh types. In Figure 2, the influence of (exaggerately strong) magnetostrictive couplings (of either sign) is felt only for rather large travelled distances. Notice that $|\Gamma_{121}|$ and $|\bar{\Gamma}_{211}|$ are $O(10^{-1})$ while $|\Gamma_1|$ and $|\Gamma_2|$ are $O((10^{-3} - 10^{-4})$ for usual magnetoelastic materials and bias fields [4].

REFERENCES

1. G. Goussot and E. Bridoux : J. Physique, C6, 276-280 (1972).
2. N. Kalyanasundaram : J. Sound Vibr., 96, 411-420 (1984).
3. A.N. Abd-Alla and G.A. Maugin : J. Acoust. Soc. Amer., 82, 1746-1752 (1987).
4. A.N. Abd-Alla and G.A. Maugin : S.I.A.M. Jl. Appl. Math., (submitted in 1987).
5. V.M. Ristič : Principles of Acoustic Devices (J. Wiley-Interscience, New-York, 1983).
6. E. Dieulesaint and D. Royer : Elastic Waves in Solids (J. Wiley, U.K., 1980) ; and in : Techniques de l'Ingénieur, Part E3210, pp. 1-26 (Editions "Techniques de l'Ingénieur", Paris, 1983).

7. G.A. Maugin and A. Hakmi : J. Acoust. Soc. Amer., 76, 826-840, (1984).
8. R. Stoneley : Proc. Roy. Soc. London, A.232, 447-452, (1955).
9. M. Planat, Doctoral Thesis, LPMO-CNRS, Univ. of Besançon, France, (1984).
10. N. Kalyanasundaram : Int. J. Engng. Sc., 19, 279-289, (1981).
11. M. Planat : J. Appl. Phys., 57, 4911-4915, (1985).
12. G.A. Maugin : Nonlinear Electromechanical Effects and Applications : A Series of Lectures (World Scientific, Singapore, 1985).
13. J.A. Amstrong, N. Bloembergen, J. Ducuing and P.J. Pershan : Phys. Rev., 127, 1918 (1962).
14. E.G. Lean and C.C. Tseng : J. Appl. Phys., 41, 3912-3917, (1970).
15. N. Kalyanasundaram, R. Ravindran and P. Prasad : J. Acoust. Soc. Amer., 72, 488, (1982).

Nonlinear Mode Coupling Between Rayleigh and Love Waves on an Isotropic Layered Half-Space

N. Kalyanasundaram

Department of Electronics and Communication Engineering,
Regional Engineering College, Tiruchirapalli 620015, India

1 Introduction

In this paper we use a multiscale perturbation method to study the nonlinear mode coupling between (initially) monochromatic Rayleigh and Love waves on a half-space of homogeneous isotropic elastic solid with a thin superficial layer of another such solid. This coupling between a Rayleigh-wave mode and a Love-wave mode is possible only at those frequencies, called critical frequencies [1], for which the phase speeds of the two modes in question coincide for the given (favourable) combination of material constants and layer thickness. For this problem the solvability conditions at the third stage of the perturbation analysis take the form of a pair of coupled nonlinear ordinary differential equations governing the slow variations in the complex amplitudes of the two modes with propagation distance. Exact solution of these coupled amplitude equations (CAEs), in terms of Jacobian elliptic functions, shows that there is, in general, a continuous exchange of energy back and forth between the two interacting modes.

2 Formulation

Making use of the notation set out in Table 1 the nondimensional governing differential equations and the boundary conditions of the problem may be written down, correct to the third order in displacement gradients, as

$$L_j(u,v) = N_{j1_x}(u_x,\ldots,w_y) + N_{j2_y}(u_x,\ldots,w_y) \quad , \quad j = 1,2,$$
$$B_j(u,v)_{y=h} = -N_{j2}(u_x,\ldots,w_y)_{y=h} \,/\, c_{s1}^2 \quad , \quad j = 1,2, \tag{1a}$$

$$[\![\sigma(y)B_j(u,v)]\!] = -[\![\sigma(y)N_{j2}(u_x,\ldots,w_y)/c_s^2]\!] \quad , \quad j = 1,2,$$
$$[\![u]\!] = [\![v]\!] = 0 \quad , \tag{1b}$$

$$L_3(w) = N_{31_x}(u_x,\ldots,w_y) + N_{32_y}(u_x,\ldots,w_y) \quad ,$$
$$B_3(w)_{y=h} = -N_{32}(u_x,\ldots,w_y)_{y=h} \,/\, c_{s1}^2 \quad , \tag{2a}$$

$$[\![\sigma(y)B_3(w)]\!] = -[\![\sigma(y)N_{32}(u_x,\ldots,w_y) \,/\, c_s^2]\!] \quad ,$$
$$[\![w]\!] = 0 \quad , \tag{2b}$$

where the linear operators L_j, B_j ($j = 1,2,3$) are defined by

$$L_1(u,v) \triangleq u_{tt} - c_d^2 u_{xx} - (c_d^2 - c_s^2)v_{xy} - c_s^2 u_{yy} \quad ,$$
$$L_2(u,v) \triangleq v_{tt} - c_s^2 v_{xx} - (c_d^2 - c_s^2)u_{xy} - c_d^2 v_{yy} \quad , \tag{3a}$$
$$L_3(w) \triangleq w_{tt} - c_s^2(w_{xx} + w_{yy}) \quad ,$$

$$B_1(u,v) \triangleq u_y + v_x \quad ,$$

$$B_2(u,v) \triangleq (c_d^2/c_s^2 - 2)u_x + (c_d^2/c_s^2)v_y \quad , \tag{3b}$$

$$B_3(w) \quad \triangleq w_y \quad ,$$

and $[\![\quad]\!]$ denotes the jump in the enclosed quantity across the interface from the layer to the substrate. In (1)-(3) and in the sequel, literal subscripts denote partial differentiation with respect to appropriate independent variables. In addition to the boundary conditions (1b) and (2b) the displacement components u,v and w are also required to decay sufficiently rapidly as $y \to -\infty$.

Table 1: Notation for the various constants and variables

Dimensional quantities	Nondimensional quantities
a) Constants	
ρ_{0k} (k=1,2) :densities of the layer and the substrate materials in the unstrained state	$c_{dk}^2 \triangleq (\lambda_k + 2\mu_k)/\rho_{0k}v_0^2$, k = 1,2 $c_{sk}^2 \triangleq \mu_k/\rho_{0k}v_0^2$, k = 1,2
$\rho_0 \triangleq \rho_{01}$ in the layer, ρ_{02} in the substrate	$c_s \triangleq c_{s1}$, in the layer, c_{s2} , in the substrate
λ_k,μ_k (k=1,2) :Lamé constants of the layer and the substrate materials	$c_d \triangleq c_{d1}$, in the layer, c_{d2} , in the substrate
ω_0, v_0: common angular frequency and common phase speed of the two interacting modes	
$k_0 \triangleq \omega_0/v_0$	
h' :layer thickness	$h \triangleq k_0h'$
b) Independent variables	
a_1 :coordinate along the direction of propagation	$x \triangleq k_0a_1$
a_2 :coordinate along the direction of the outward normal to the substrate boundary	$y \triangleq k_0a_2$
t' :time	$t \triangleq \omega_0t'$
c) Dependent variables	
u_j (j=1,2,3) :dilatational, shear vertical and shear horizontal components of the displacement vector	$u = k_0u_1$ $v = k_0u_2$ $w = k_0u_3$
$L_{jk}^{(N)}$ (j=1,2,3 k=1,2) :the part of the Lagrangian stress tensor containing terms quadratic and cubic in the displacement gradients	$N_{jk}(u_x,\ldots,w_y) = L_{jk}^{(N)}/\rho_0v_0^2$
	$\sigma(y) = 1$, for $0<y\leq h$, $\sigma(y) = (\mu_2/\mu_1)^{\frac{1}{2}}$, for $-\infty<y\leq 0$

3 Multiscale Method

The displacement components u,v and w are assumed to admit power series expansions in a nondimensional perturbation parameter ϵ of the form

$$
\begin{aligned}
u(x,y,t) &\triangleq \tilde{u}(x,y,t;\xi,\eta) = \sum_{n=1}^{\infty} \epsilon^n u_n(x,y,t;\xi,\eta) \quad , \\
v(x,y,t) &\triangleq \tilde{v}(x,y,t;\xi,\eta) = \sum_{n=1}^{\infty} \epsilon^n v_n(x,y,t;\xi,\eta) \quad , \qquad (4) \\
w(x,y,t) &\triangleq \tilde{w}(x,y,t;\xi,\eta) = \sum_{n=1}^{\infty} \epsilon^n w_n(x,y,t;\xi,\eta) \quad ,
\end{aligned}
$$

where ξ and η are slow variables defined by $\xi = \epsilon^2 x$ and $\eta = \epsilon^2 y$. Substituting the expansion (4) into the differential equations (1a) and (2a) and the boundary conditions (1b) and (2b) expressed in terms of the independent variables x,y,t,ξ and η and equating separately the coefficients of ϵ^n for n = 1,2,3, on both sides we get a homogeneous boundary value problem for n = 1 and nonhomogeneous boundary value problems for n = 2,3 for the linear differential operators L_j, B_j, j = 1,2,3.

A convenient form for the solution of the first order problem consistent with monochromatic signalling conditions at x = y = 0 is

$$
u_1(x,y,t;\xi,\eta) = \begin{cases} A_R(\xi)[c(y) + p_2\delta_2 \cos p_2(y+\psi_2)]\exp i\varphi + c.c. \;, & \text{in the layer,} \\ [B_1(\xi,\eta)\exp q_1 y - r q_2 B_2(\xi,\eta)\exp q_2 y]\exp i\varphi + c.c. \;, & \text{in the substrate,} \end{cases}
$$

$$
v_1(x,y,t;\xi,\eta) = \begin{cases} -iA_R(\xi)[p_1 s(y) + \delta_2 \sin p_2(y+\psi_2)]\exp i\varphi + c.c. \;, & \text{in the layer,} \\ -i[q_1 B_1(\xi,\eta)\exp q_1 y - r B_2(\xi,\eta)\exp q_2 y]\exp i\varphi + c.c. \;, & \text{in the substrate,} \end{cases} \qquad (5)
$$

$$
w_1(x,y,t;\xi,\eta) = \begin{cases} A_L(\xi)\cos p_2(y-h)\sec p_2 h \exp i\varphi + c.c. \;, & \text{in the layer,} \\ B_3(\xi,\eta)\exp(q_2 y + i\varphi) + c.c. \;, & \text{in the substrate,} \end{cases}
$$

where

$$
p_1 = +(1 - c_{d1}^{-2})^{\frac{1}{2}} \;, \quad p_2 = +(c_{s1}^{-2} - 1)^{\frac{1}{2}} \;, \quad q_1 = +(1 - c_{d2}^{-2})^{\frac{1}{2}} \;,
$$

$$
q_2 = +(1 - c_{s2}^{-2})^{\frac{1}{2}} \;, \quad \varphi = x - t \;, \quad c(y) = \sigma_1 \exp p_1 y + \nu_1 \exp -p_1 y \;,
$$

$$
s(y) = \sigma_1 \exp p_1 y - \nu_1 \exp -p_1 y \;,
$$

and σ_1, ν_1, δ_2, ψ_2 and r are constants depending only on the material parameters and layer thickness. In (5) the complex amplitudes $A_R(\xi)$ and $A_L(\xi)$ of the Rayleigh- and Love-wave modes are to be taken to be functions only of the slow variable ξ. However, the arbitrary constants B_1, B_2 and B_3 for the solution in the semi-infinite substrate are to depend on both the slow variables ξ and η in order to permit elimination of terms proportional to y exp $q_1 y$ and y exp $q_2 y$, which render the perturbation solution, correct to the third order, nonuniform with respect to the y-coordinate [2]. The functions $B_1(\xi,\eta)$, $B_2(\xi,\eta)$ and $B_3(\xi,\eta)$ are subject to the boundary conditions

$$
B_1(\xi,0) = B_2(\xi,0) = A_R(\xi) \;, \qquad B_3(\xi,0) = A_L(\xi) \;. \qquad (6)
$$

Substitution of the first-order solution (5) into the right hand sides of the second-order differential equations reveals that the nonhomogeneous terms for these equations are of the form

$$P_0(y) + P_2(y)\exp 2i\varphi + \text{c.c.} \quad ,$$

where we have suppressed the functional dependence of $P_0(y)$ and $P_2(y)$ on ξ and η. (We shall, henceforth, resort to the above short notation to denote functions of y, ξ and η). We may, therefore, assume solution of the second order problem in the form

$$\begin{aligned} u_2(x,y,t;\xi,\eta) &= if_2(y)\exp 2i\varphi + \text{c.c.} \;, \\ v_2(x,y,t;\xi,\eta) &= v_{20}(y) + g_2(y)\exp 2i\varphi + \text{c.c.} \;, \qquad (7) \\ w_2(x,y,t;\xi,\eta) &= iw_{20}(y) + ih_2(y)\exp 2i\varphi + \text{c.c.} \;. \end{aligned}$$

The functions $f_2(y)$, $g_2(y)$, $h_2(y)$, $v_{20}(y)$ and $w_{20}(y)$ can be uniquely determined by solving the ordinary differential equations satisfied by them along with the associated boundary conditions and the decay condition at infinite depth. This is possible because, the modes being dispersive, the phase-matching condition will be satisfied only by the fundamental-frequency terms which reappear for the first time only at the third stage of a straightforward perturbation expansion. This fact, incidently, determines the choice of scaling of the slow variables.

Substituting the first and second order solution into the right hand sides of the third order differential equation we find the nonhomogeneous terms are of the following two types:

$$P_1(y)\exp i\varphi + \text{c.c.} \qquad \text{and} \qquad P_3(y)\exp 3i\varphi + \text{c.c.} \quad .$$

Since terms of the latter type do not satisfy the phase matching condition the unique solution corresponding to these third harmonic terms can be readily written down following the procedure adopted for obtaining the second order solution.

Assuming a solution corresponding to the fundamental-frequency terms in the form

$$\begin{aligned} u_3^{(1)}(x,y,t;\xi,\eta) &= f_3(y)\exp i\varphi + \text{c.c.} \;, \\ v_3^{(1)}(x,y,t;\xi,\eta) &= -ig_3(y)\exp i\varphi + \text{c.c.} \;, \\ w_3^{(1)}(x,y,t;\xi,\eta) &= h_3(y)\exp i\varphi + \text{c.c.} \;, \end{aligned}$$

substituting into the second order differential equations and denoting derivatives with respect to y by a prime we get the following ordinary differential equations for f_3, g_3 and h_3:

$$\begin{aligned} &(c_d{}^2 - 1)f_3 \\ &\;-(c_d{}^2 - c_s{}^2)g_3' \\ &\qquad - c_s^2 f_3'' = \end{aligned} \begin{cases} i[(c_{d1}^2 + 1 + p_1^2 c_{s1}^2)c(y) \\ \qquad + \delta_2 p_2 (c_{d1}^2 + c_{s1}^2)\cos p_2(y+\psi_2)]A_{R\xi} \\ \qquad + F_1(y; A_R|A_R|^2,\ A_R|A_L|^2,\ A_R^* A_L{}^2) \;, \\ \qquad\qquad \text{in the layer} \;, \\ C_1 \exp q_1 y - rC_2 \exp q_2 y \\ \qquad + F_2(y; B_1|B_1|^2,\ B_1{}^2 B_2^*,\ |B_1|^2 B_2,\ B_1^* B_2{}^2, \\ \qquad\qquad B_1|B_3|^2,\ B_2|B_3|^2,\ B_1^* B_3{}^2,\ B_2^* B_3{}^2) \;, \\ \qquad\qquad \text{in the substrate,} \end{cases}$$

$$
(c_s{}^2-1)g_3 - (c_d{}^2-c_s{}^2)f_3' - c_d{}^2 g_3'' = \begin{cases} i[p_1(c_{d_1}^2 + c_{s_1}^2)s(y) \\ \quad + \delta_2(c_{s_1}^2 + 1 - p_2^2 c_{d_1}^2)\sin p_2(y+\psi_2)]A_{R_\xi} \\ \quad + G_1(y; iA_R|A_R|^2,\ iA_R|A_L|^2,\ iA_RA_L{}^2)\ , \\ \qquad \text{in the layer}\ , \\ D_1 \exp q_1y - rD_2 \exp q_2y \\ \quad + G_2(y; iB_1|B_1|^2,\ iB_1{}^2B_2^*,\ i|B_1|^2B_2,\ iB_1^*B_2{}^2, \\ \quad iB_1|B_2|^2,\ iB_2|B_2|^2,\ iB_1|B_3|^2,\ iB_2|B_3|^2, \\ \quad iB_1^*B_3{}^2,\ iB_2^*B_3{}^2)\ , \qquad \text{in the substrate,} \end{cases} \tag{8}
$$

$$
(c_s{}^2-1)h_3 - c_s{}^2 h_3'' = \begin{cases} 2ic_{s_1}^2[\cos p_2(y-h)\ \sec p_2h]A_{L_\xi} \\ \quad + H_1(y; A_R{}^2A_L^*,\ |A_R|^2A_L,\ |A_L|^2A_L)\ , \\ \qquad \text{in the layer,} \\ D_3 \exp q_2y + H_2(y; B_1{}^2B_3^*,\ B_1B_2B_3^*,\ B_2{}^2B_3^*, \\ \quad |B_1|^2B_3, |B_2|^2B_3,\ |B_1B_2|B_3,\ |B_3|^2B_3)\ , \\ \qquad \text{in the substrate}\ , \end{cases}
$$

where

$$
\begin{aligned}
C_1 &= i(c_{d_2}^2 + 1 + q_1^2 c_{s_2}^2)B_{1_\xi} + q_1(c_{d_2}^2 + c_{s_2}^2)B_{1_\eta} \quad , \\
C_2 &= iq_2(c_{d_2}^2 + c_{s_2}^2)B_{2_\xi} + (c_{d_2}^2 + c_{s_2}^2 - 2)B_{2_\eta} \quad , \\
D_1 &= iq_1(c_{d_2}^2 + c_{s_2}^2)B_{1_\xi} + (c_{d_2}^2 + c_{s_2}^2 - 2)B_{1_\eta} \quad , \\
D_2 &= i(c_{s_2}^2 + 1 + q_2^2 c_{d_2}^2)B_{2_\xi} + q_2(c_{d_2}^2 + c_{s_2}^2)B_{2_\eta} \quad , \\
D_3 &= 2c_{s_2}^2\ (iB_{3_\xi} + q_2B_{3_\eta}) \quad ,
\end{aligned}
$$

and the functions F_k, G_k and H_k (k=1,2,3) depend linearly on the set of variables after the semicolon. Also F_2, G_2 and H_2 do not involve any terms proportional to $\exp q_1y$ and $\exp q_2y$ which are solutions of the linear homogeneous problem for the substrate at the frequency ω_0. In order that terms like $y \exp q_1y$ and $y \exp q_2y$ be absent from the third order solution we need the compatibility conditions:

$$
C_1 = q_1D_1 \quad , \quad q_2C_2 = D_2 \quad , \quad D_3 = 0 \quad , \tag{9}
$$

which may be rewritten as

$$
iB_{1_\xi} + q_1B_{1_\eta} = 0 \quad , \quad iB_{2_\xi} + q_2B_{2_\eta} = 0 \quad , \quad iB_{3_\xi} + q_2B_{3_\eta} = 0 \quad . \tag{10}
$$

The solution of (10) subject to the boundary condition (6) is

$$
\begin{aligned}
&B_1(\xi,\eta) = A_R(\xi - i\eta/q_1) \quad , \quad B_2(\xi,\eta) = A_R(\xi - i\eta/q_2) \quad , \\
&B_3(\xi,\eta) = A_L(\xi - i\eta/q_2) \quad .
\end{aligned} \tag{11}
$$

Once the compatibility conditions (9) are satisfied a uniformly valid solution of the ordinary differential equations (8) decaying exponentially with depth may be expressed as the sum of a complementary function and a particular integral. Substituting this solution into the associated boundary conditions we get the following two systems of linear algebraic equations for determining the arbitrary constants appearing in the complementary function:

$$W_R(\omega_0,k_0)X_R = Y_R \quad , \qquad W_L(\omega_0,k_0)X_L = Y_L \quad , \tag{12}$$

where $W_R(\omega_0,k_0)$ and $W_L(\omega_0,k_0)$ are 6x6 and 3x3 matrices, X_R and X_L are column vectors of arbitrary constants and Y_R and Y_L are column vectors of nonhomogeneous terms, which are functions only of the slow variable ξ. However the coefficient matrices W_R and W_L are singular since det $W_R = 0$ and det $W_L = 0$ are precisely the dispersion relations satisfied by the interacting Rayleigh- and Love-wave modes at the critical frequency ω_0. Hence a solution for X_R and X_L will exist only if the compatibility conditions

$$N_R Y_R = 0 \quad , \qquad N_L Y_L = 0 \quad , \tag{13}$$

are satisfied, where N_R and N_L are any left null vectors of the singular matrices W_R and W_L respectively. These compatibility conditions take the form of a pair of coupled amplitude equations for A_R and A_L:

$$\begin{aligned} iA_{R\xi} &= \alpha_R A_R|A_R|^2 + \beta_R A_R|A_L|^2 + \gamma_R A_R^* A_L^2 \quad , \\ iA_{L\xi} &= \alpha_L A_L|A_L|^2 + \beta_L A_L|A_L|^2 + \gamma_L A_L^* A_R^2 \quad , \end{aligned} \tag{14}$$

where the coefficients $\alpha_R,\ldots,\gamma_L$ depend only on the material constants of the media and the layer thickness.

4 Solution of the Coupled Amplitude Equations

Representing $A_R(\xi)$ and $A_L(\xi)$ in polar form i.e.,

$$A_R(\xi) = \rho_R(\xi) \exp i\theta_R(\xi) \quad , \qquad A_L(\xi) = \rho_L(\xi) \exp i\theta_L(\xi) \quad ,$$

substituting into (14) and separating into real and imaginary parts we obtain the following coupled system of four equations for the (real) amplitudes ρ_R and ρ_L and phases θ_R and θ_L:

$$\begin{aligned} \rho_{R\xi} &= -\gamma_R \rho_R \rho_L^2 \sin\theta_{RL} \quad , \\ \rho_{L\xi} &= \gamma_L \rho_L \rho_R^2 \sin\theta_{RL} \quad , \end{aligned} \tag{15a}$$

$$\begin{aligned} \theta_{R\xi} &= -\alpha_R \rho_R^2 - \beta_R \rho_L^2 - \gamma_R \rho_L^2 \cos\theta_{RL} \quad , \\ \theta_{L\xi} &= -\alpha_L \rho_R^2 - \beta_L \rho_L^2 - \gamma_L \rho_R^2 \cos\theta_{RL} \quad , \end{aligned} \tag{15b}$$

where we have put

$$\theta_{RL} = 2(\theta_R - \theta_L) \quad .$$

Two first integrals of (15) are

$$I(\xi) = I(0) \tag{16a}$$

$$J(\xi) = J(0) \tag{16b}$$

where

$$I(\xi) = \rho_R^2(\xi)/\gamma_R + \rho_L^2(\xi)/\gamma_L \quad ,$$

$$J(\xi) = \rho_R^2(\xi)\rho_L^2(\xi)\cos\theta_{RL}(\xi) + (\alpha_R-\alpha_L)\rho_R^4(\xi)/2\gamma_R + (\beta_L-\beta_R)\rho_L^4(\xi)/2\gamma_L \quad .$$

Eliminating ρ_L, ρ_R and θ_{RL} from (15) with the help of (16) we obtain the following nonlinear ordinary differential equation for

$$R(\xi) \triangleq \rho_R^2(\xi) - \rho_R^2(0) :$$

$$\begin{aligned}(dR/d\xi)^2 = & \{\gamma_L(2+a+b)R^2 + 2[\gamma_L(1+a)\rho_R^2(0)-\gamma_R(1+b)\rho_L^2(0)]R \\ & -2\gamma_R(1+\delta)\rho_R^2(0)\rho_L^2(0)\} \times \{\gamma_L(2-a-b)R^2 \\ & +2[\gamma_L(1-a)\rho_R^2(0)-\gamma_R(1-b)\rho_L^2(0)]R - 2\gamma_R(1-\delta)\rho_R^2(0)\rho_L^2(0)\} \\ = & \; U(R) \qquad \text{(say)} \end{aligned} \tag{17}$$

in which

$$a = (\alpha_R-\alpha_L)/\gamma_L \quad , \quad b = (\beta_L-\beta_R)/\gamma_R \quad \text{and} \quad \delta = \cos\theta_{RL}(0) \quad .$$

Solution of (17) is possible only in the region $\{R : U(R)>0\}$. If $U(R) = 0$ has no real roots for R then no bounded solutions exist. If $U(R) = 0$ has 4 real roots for a suitable choice of $\rho_R(0)$, $\rho_L(0)$ and $\theta_R(0)-\theta_L(0)$ then (17) will have a bounded periodic solution of the form

$$\rho_R^2(\xi) = \alpha_0 + \alpha_1[1 + \beta_1 \mathrm{sn}^2(\gamma\xi;K)]^{-1}$$

where α_0, α_1, β_1 and γ are real constants and K is the modulus of the elliptic sine. In the case when $U(R) = 0$ has only two real roots the solution for ρ_R takes the form

$$\rho_R^2(\xi) = \bar{\alpha}_0 + \bar{\alpha}_1[1 + \bar{\beta}_1 \mathrm{cn}(\bar{\gamma}\xi;\bar{K})]^{-1}$$

where $\bar{\alpha}_0$, $\bar{\alpha}_1$, $\bar{\beta}_1$ and $\bar{\gamma}$ are again real constants and $\bar{K}$ is the modulus of the elliptic cosine. With $\rho_R(\xi)$ available the amplitude $\rho_L(\xi)$ of the Love-wave mode can be immediately obtained from (16a). It is also a periodic function with the same period as that of $\rho_R(\xi)$. Thus the modes exchange energy continuously back and forth as they propagate. Once the solution for ρ_R and ρ_L is known, the solution for the phase angles θ_R and θ_L can be obtained by integrating (15b) after substituting for $\cos\theta_{RL}$ in terms of ρ_R and ρ_L from (16b).

Going back to (17) we observe that, if $\cos\theta_{RL} = \pm 1$, zero will be a root of the equation $U(R) = 0$ and the solution for ρ_R and ρ_L degenerates to

$$\rho_R(\xi) \equiv \rho_R(0) \quad , \quad \rho_L(\xi) \equiv \rho_L(0)$$

and consequently the two modes undergo only linear phase shift to the leading order. Also, if the starting amplitudes of either of the modes is zero, the corresponding mode will not be excited and the amplitude of the other mode remains constant at its initial value with its phase varying linearly with propagation distance [3].

The preceding presentation, of necessity, has been very brief and sketchy; it is hoped that this drawback will be remedied with the publication of a detailed version in the near future.

References

1. N. Sugimoto, M. Hirao: J. Acoust. Soc. Am. **62**, 23 (1977)
2. R.W. Lardner: Int. J. Engng. Sci. **21**, 133 (1983)
3. N. Kalyanasundaram: Int. J. Engng. Sci. **19**, 287 (1981)

Finite Amplitude Love Waves on an Incompressible Elastic Layered Half Space

M. Teymur

Department of Mathematics and Computer Science, The University, Dundee, DD1 4HN, Scotland, U.K.

By employing a singular perturbation expansion the nonlinear modulation of Love waves on a single layered isotropic incompressible hyperelastic half space is investigated. A nonlinear Schrödinger equation is derived to govern this motion asymptotically. It is shown that the surface solitary wave may exist depending on the nonlinear constitution of the layered half space.

1. INTRODUCTION

The study of the effects of nonlinearity on the propagation characteristics of surface elastic waves has recently attracted considerable attention in view of their applications in seismology, non-destructive testing of materials and electronic signal processing devices [1,2,3]. Nonlinear propagation of Rayleigh and Love waves on an isotropic half space have been investigated by several authors. In the following some related works will be reviewed. For other studies on these topics the reader may consult the references given below.

Nonlinear Rayleigh waves are examined in [4,5,6,7,8] by utilizing multiple scales procedures. Approximate coupled amplitude equations governing the propagation of a quasi-monochromatic Rayleigh wave are presented by KALYANASUNDARAM [4]. He also analysed the nonlinear mode coupling between two co-directional quasi-harmonic waves in [5], whereas in [6] he and his colleagues derived an infinite set of coupled amplitude equations governing the propagation of a quasi-harmonic wave. The non-uniformities arising in these analysis are eliminated by LARDNER [7] and DAVID [8].

A completely different approach has been applied for analytical treatments of nonlinear Rayleigh waves of permanent form, by PARKER [9], and PARKER and TALBOT [10,11]. Also PARKER [12] introduced a new form of nonlinear dispersion relation for Rayleigh waves.

The propagation characteristic of finite amplitude quasi-monochromatic Love waves are studied by KALYANASUNDARAM [13], employing the method of multiple scales. A semi-linear partial differential equation governing the slow variation in complex amplitude has been derived. This equation is reduced to a pair of semi-linear hyperbolic equations for the real amplitude and phase of the quasi-monochromatic waves. The solution of these equations for a signalling problem shows that the amplitude of the wave remains constant along its characteristics and the wave is subject only to a phase shift . In this analysis the effect of dispersion does not appear. In the work done by BATAILLE and LUND [14], a model equation is derived via an intuitive approach guided by physical arguments to take into consideration the dispersive nature of Love waves as well as the nonlinearity. It is then shown that this equation has an exact solitary wave and an approximate modulated solitary wave solution which provide mechanisms for efficient localized energy propagation along the surface of a nonlinear layered medium.

Later TEYMUR [15] considered the propagation of small but finite amplitude weakly nonlinear waves in a layered compressible half space utilizing a multiple scales perturbation expansion. It is found that the nonlinear self-modulation of a quasi-monochromatic Love wave is governed by the nonlinear Schrödinger (NLS) equation which takes into consideration the effect of dispersion and results from the balance between weak nonlinearity and dispersion. Thus, balancing the nonlinearity and dispersion, not only the phase shift but also the long time slow modulation of Love waves is taken into account. Then, since the behaviour of solutions of the NLS equation depends on the sign of the product of its coefficients, the variation of these coefficients with the nondimensional wave number has been performed numerically giving appropriate values to the material constants of the layer and the half space, so as to account for softening and hardening behaviour of them. For fixed linear material properties it has been observed that the type of solitary wave depends on the nonlinear constitution of the layered medium. In this work a parallel investigation is presented for a layered isotropic incompressible hyperelastic half space.

2. GENERALISED SHEAR MOTION IN NEO-HOOKEAN MATERIALS

First we consider a generalised shear motion, (see for example ERINGEN and SUHUBI [16]), described by equations

$$x_k = X_k\,\delta_{kK} + u_3(X_\Delta,t)\,\delta_{k3}\,, \qquad \Delta = 1,2\,, \tag{2.1}$$

where x_k and X_K are respectively the spatial and material co-ordinates of a point referred to the same fixed rectangular Cartesian system of axes; u_3 is the displacement of a particle in the X_3 direction, t is the time and δ_{kK} is the usual Kronecker symbol. The summation convention on repeated indices is implied in (2.1) and in the sequel.

Let t_{kl} be the Cauchy stress tensor field accompanying the deformation field (2.1); in the absence of body forces, the equations of motion in the reference state are

$$t_{13,1} + t_{23,2} = \rho_o\,\ddot{u}_3,\; t_{11,1} + t_{12,2} = t_{12,1} + t_{22,2} = 0\,. \tag{2.2 a,b,c}$$

From here on subscripts preceded by a comma indicate partial differentiation with respect to the material co-ordinates, a dot above represents partial differentiation with respect to t.

Let us now assume that the motion takes place in a homogeneous isotropic incompressible hyperelastic material whose strain energy function depends on only the first principal invariants of the associated deformation tensor. That is $\Sigma = \Sigma\,(I)$ where, say, $I = \mathrm{tr}c^{-1}$ and $c^{-1}_{kl} = x_{k,K}\,x_{l,K}$, the Finger deformation tensor. This class of materials is called generalised neo-Hookean materials. For the deformation field (2.1) the components of the Cauchy stress tensor are given as follows:

$$t_{11} = t_{12} = t_{22} = 0,\; t_{33} = \Phi\,K^2,\; t_{\alpha 3} = \Phi\,u_{3,\alpha},\quad \alpha = 1,2\,, \tag{2.3}$$

where

$$\Phi = d\Sigma/dI,\ I = 3 + K^2,\ K^2 = u_{3,1}^2 + u_{3,2}^2 \tag{2.4}$$

and the unknown pressure function p is chosen in such a way that $t_{11} = 0$. Thus, it is seen that (2.2 b,c) are satisfied identically and (2.2a) becomes

$$(\Phi u_{3,1})_{,1} + (\Phi u_{3,2})_{,2} = \rho_o \ddot{u}_3 \,. \tag{2.5}$$

Note that if $d\Phi/dK < 0$ the material is softening whereas if $d\Phi/dK>0$ it is hardening. Also, the hyperbolicity condition for the quasi-linear equation (2.5) requires that $\Phi\ d\tau/dK>0$, where $\tau = \Phi K$. On the other hand, because of the Baker-Ericksen inequality, $\Phi>0$ [16]. Therefore the hyperbolicity condition is satisfied if and only if $d\tau/dK>0$, or $d(\Phi K)/dK>0$.

3. NONLINEAR MODULATION OF WAVES

In this section the propagation of small but finite amplitude Love waves in an incompressible elastic half-space $X_2 \leq 0$ covered by a different incompressible elastic layer of uniform thickness h is considered. The positive X_1 - axis is taken along the direction of wave propagation. It is assumed that the boundary $X_2 = h$ is free of tractions and the stresses and displacements are continuous at the interface $X_2 = 0$. Hence,

$$T_{2k} = 0 \quad \text{on } X_2 = h \tag{3.1}$$

$$[T_{2k}] = 0 \text{ on } X_2 = 0 \tag{3.2}$$

$$[u_3] = 0 \quad \text{on } X_2 = 0 \tag{3.3}$$

where [.] signifies the jump of the enclosed quantity across $X_2 = 0$.

Let $X = X_1$, $Y = X_2$ and $Z = X_3$ and, u and v represent the displacements in the layer and half-space, respectively. Assume that the constituent materials are neo-Hookean, then the governing equations and the boundary conditions of the problem can be written as (~ denoting quantities in Y<0)

$$(\Phi u_{,X})_{,X} + (\Phi u_{,Y})_{,Y} = \rho_o \ddot{u} \tag{3.4}$$

$$(\tilde{\Phi} v_{,X})_{,X} + (\tilde{\Phi} v_{,Y})_{,Y} = \tilde{\rho}_o \ddot{v} \tag{3.5}$$

$$\Phi u_{,Y} = 0 \text{ on } Y = h \tag{3.6}$$

$$\Phi u_{,Y} = \tilde{\Phi} v_{,Y} \text{ on } Y = 0 \tag{3.7}$$

$$u = v \text{ on } Y = 0 \tag{3.8}$$

$$v \to 0 \text{ as } Y \to -\infty \,. \tag{3.9}$$

Since we consider the propagation of small but finite amplitude waves let us express u and v as asymptotic power series in a small parameter ε, representing the strength of nonlinearity, as follows:

$$u \sim \sum_{n=1}^{\infty} \varepsilon^n u_n (x_o, x_1, x_2, \ldots, t_o, t_1, t_2, \ldots, y),$$

$$v \sim \sum_{n=1}^{\infty} \varepsilon^n v_n (x_o, x_1, x_2, \ldots, t_o, t_1, t_2, \ldots, y), \tag{3.10}$$

where

$$x_i = \varepsilon^i X,\ t_i = \varepsilon^i t,\ y = Y,\ i = 0,1,2,\ ..... \tag{3.11}$$

Accordingly, the derivatives are considered as (see for example JEFFREY and KAWAHARA [17])

$$\frac{\partial}{\partial t} = \sum_{i=1}^{N} \varepsilon^i \frac{\partial}{\partial t_i},\ \frac{\partial}{\partial X} = \sum_{i=1}^{N} \varepsilon^i \frac{\partial}{\partial x_i},\ \frac{\partial}{\partial Y} = \frac{\partial}{\partial y} \quad . \tag{3.12}$$

On the other hand, since this work is centered around the weakly nonlinear waves it is aimed to obtain just the uniformly valid first order solution, therefore in the analysis we will take N = 2.

If the expansions (3.10) are used in the governing equations and in the boundary conditions, after transformation (3.12) equating co-efficients of like powers of ε, we obtain a hierarchy of equations and boundary conditions from which it is possible to obtain u_n and v_n successively. Since we are interested in the modulation of quasi-monochromatic waves, for the solutions u_n and v_n of these perturbation problems we let

$$u_n = \{ A_n^{(1)} \exp (ilkpy) + B_n^{(1)} \exp (-ilkpy) \} \exp (il\phi) + c.c. \tag{3.13}$$

$$v_n = C_n^{(1)} \exp (lk\nu y + il\phi) + c.c. \tag{3.14}$$

where $A_n^{(1)}$, $B_n^{(1)}$ and $C_n^{(1)}$ are functions of x_1, x_2, t_1, t_2 and

$$\phi = kx_o - wt_o,\ p = (c^2/c_1^2 - 1)^{1/2},\quad \nu = (1-c^2/c_2^2)^{1/2},$$

$$c_1 = \Phi(3)/\rho_o,\ c_2 = \tilde{\Phi}(3)/\tilde{\rho}_o,\quad c_1 < c < c_2 \quad . \tag{3.15}$$

Also k is the wave number, ω is the angular frequency, $c = \omega/k$ is the phase velocity and c.c. denotes the complex conjugate of the preceding terms. Here v_n is chosen by taking the condition (3.9) into consideration.

The first order equations and boundary conditions govern the propagation of classical linear Love waves [15]. If (3.13) and (3.14), with n = 1, are substituted in these equations and boundary conditions we get

$$\underset{\sim}{W}_1 \underset{\sim}{U}_1^{(1)} = \underset{\sim}{0}, \tag{3.16}$$

where

$$\underset{\sim}{W}_1 = \begin{pmatrix} ilkp \exp(ilkph) & -ilkp \exp(-ilkph) & 0 \\ ilkp & -ilkp & -\sigma^2 lk\nu \\ 1 & 1 & -1 \end{pmatrix},\ \underset{\sim}{U}_1^{(1)} = \begin{pmatrix} A_1^{(1)} \\ B_1^{(1)} \\ C_1^{(1)} \end{pmatrix} \tag{3.17}$$

Note that $\det \underset{\sim}{W}_1 = \underset{\sim}{0}$ gives the classical Love wave dispersion relation $\tan(khp) = \sigma^2 \nu/p$, $\sigma^2 = \tilde{\Phi}(3)/\Phi(3)$, and $\det \underset{\sim}{W}_o$ vanishes. From now on we consider the self-modulation of a group of waves centred around k and the frequency ω. To exclude the self-resonance it is assumed that for $l \geq 2$ det $\det \underset{\sim}{W}_l \neq 0$ Then the solutions of the homogeneous algebraic equations (3.16) may be written as

$$\underset{\sim}{U}_1^{(1)} = \bar{A}_1 \underset{\sim}{R},\ \underset{\sim}{U}_1^{(o)} = \bar{B}_1 \underset{\sim}{P} + \bar{C}_1 \underset{\sim}{S},\ \underset{\sim}{U}_1^{(l)} \equiv \underset{\sim}{0} \text{ for } l \geq 2\ , \tag{3.18}$$

where $\bar{A}_1$ is a complex function of the slow variables representing the first order slowly varying amplitude of the modulated waves while $\bar{B}_1$ and $\bar{C}_1$ are real functions of the slow variables and are first order slowly varying long wave components. The column vectors $\underset{\sim}{R}$, $\underset{\sim}{P}$ and $\underset{\sim}{S}$ satisfy

$$\underset{\sim}{W}_1 \underset{\sim}{R} = \underset{\sim}{0}, \; \underset{\sim}{W}_0 \underset{\sim}{P} = \underset{\sim}{0}, \; \underset{\sim}{W}_0 \underset{\sim}{S} = \underset{\sim}{0} \; . \tag{3.19}$$

$\bar{A}_1$, $\bar{B}_1$ and $\bar{C}_1$ are to be determined in higher order perturbation problems. But here, since we only consider the self-modulation of waves excluding all the other nonlinear interaction processes, the non-existence of the long wave components in the initial disturbances will be assumed. Therefore the solution $\underset{\sim}{U}_1^{(0)}$ will be neglected in the first order. Hence u_1 and v_1 are written explicitly as

$$u_1 = \bar{A}_1 \, (R_1 \exp(ikpy) + R_2 \exp(-ikpy)) \exp(i\phi) + \text{c.c.},$$

$$v_1 = \bar{A}_1 \, R_3 \exp(k\nu y + i\phi) + \text{c.c.} \tag{3.20}$$

Note that the first order solution and the classical Love wave solution (see for example ERINGEN and SUHUBI [18]) are of the same form. The only difference is that in the classical solution A_1 is a constant but here it is a complex function representing the nonlinear modulation of waves.

If the first order solutions are employed in the second order equations and boundary conditions and if the solutions in this order are also taken in the forms (3.13,14) one gets

$$\underset{\sim}{W}_1 \, \underset{\sim}{U}_2^{(1)} = \underset{\sim}{b}_2^{(1)}, \tag{3.21}$$

where

$$\underset{\sim}{b}_2^{(1)} = -i \left(\frac{\partial \bar{A}_1}{\partial t_1} \frac{\partial W_1}{\partial w} - \frac{\partial \bar{A}_1}{\partial x_1} \frac{\partial W_1}{\partial k}\right) \underset{\sim}{R}, \text{ and } \underset{\sim}{b}_2^{(l)} \equiv 0 \text{ for } l \neq 1 \; . \tag{3.22}$$

Since $\det \underset{\sim}{W}_1 = 0$ and $\underset{\sim}{b}^{(1)} \neq \underset{\sim}{0}$, in order that (2.14) is algebraically solvable for $\underset{\sim}{U}_2^{(1)}$, the compability condition $\underset{\sim}{L} . \underset{\sim}{b}_2^{(1)} = 0$ must be satisfied, where $\underset{\sim}{L}$ is a vector defined by $\underset{\sim}{L}\underset{\sim}{W}_1 = \underset{\sim}{0}$. This condition gives

$$\frac{\partial \bar{A}_1}{\partial t_1} + V_g \frac{\partial \bar{A}_1}{\partial x_1} = 0, \tag{3.23}$$

where V_g is the group velocity. (3.23) implies that the complex amplitude $\bar{A}_1$ remains constant in a frame of reference moving with the group velocity V_g. That is $\bar{A}_1$ depends on x_1 and t_1 only through the combination $x_1 - V_g \, t_1$. Then the solution of (3.21) for $l = 1$ is found to be

$$\underset{\sim}{U}_2^{(1)} = \bar{A}_2 \, \underset{\sim}{R} - i \frac{\partial \bar{A}_1}{\partial x_1} \left(\frac{\partial \underset{\sim}{R}}{\partial k} + V_g \frac{\partial \underset{\sim}{R}}{\partial w}\right), \tag{3.24}$$

where $\bar{A}_2$ is a complex function of the slow variables representing the second order slowly varying amplitude modulation and it may be determined in higher-order perturbation problems. For $l = 0$, since $\det \underset{\sim}{W}_0 = 0$ and $\underset{\sim}{b}_2^{(0)} \equiv 0$, the solution of (3.21) can be written, as in the first order, in the form

$$\underset{\sim}{U}_2^{(0)} = \bar{B}_2 \, \underset{\sim}{P} + \bar{C}_2 \, \underset{\sim}{S}, \tag{3.25}$$

where $\bar{B}_2$ and $\bar{C}_2$ are real functions of the slow scales, representing the second order long wave components, to be determined in higher order perturbation problems like $\bar{A}_2$. But neither $\bar{A}_2$ nor the pair $\bar{B}_2$ and $\bar{C}_2$ need to be calculated explicitly. Since this work is centred around the weakly nonlinear waves, therefore it is aimed to obtain just the uniformly valid first order solution. Moreover it covers also only the self-modulation of waves excluding all the other processes like harmonic resonances, long wave-short wave interactions, etc. Therefore to obtain a first order solution, the determination of $\bar{A}_1$ will be sufficient and it will be done in the third order perturbation problem. A close examination of the third order problem also shows that during the performance of operations in this order there is no need to know the explicit forms of $\bar{B}_2$ and $\bar{C}_2$. This is due to the fact that they do not depend on x_o and t_o so the effect of operators on them yield zero at this order. For $l \geq 2$, since it is assumed that det $\underset{\sim}{W}_l \neq 0$ for $l>1$, the solutions of (3.21) are $\underset{\sim}{U}_2^{(l)} = 0$. Thus, the solution of the second-order perturbation problem is completed. If we use the first and second order solutions in the third order problem and if we choose the solutions in the forms (3.13,14), after a lengthy algebra we get

$$\underset{\sim}{W}_l \, \underset{\sim}{U}_3^{(l)} = \underset{\sim}{b}_3^{(l)}, \tag{3.26}$$

where $\underset{\sim}{b}_3^{(1)} \neq \underset{\sim}{0}$, $\underset{\sim}{b}_3^{(3)} \neq \underset{\sim}{0}$ and $\underset{\sim}{b}_3^{(l)} \equiv \underset{\sim}{0}$ for all $l \neq 1,3$. The solution of (3.26) for $l = 0$ can be found as in the previous orders. For $l = 1$, in order that (3.26) is algebraically solvable for $\underset{\sim}{U}_3^{(1)}$, the compatibility condition

$$\underset{\sim}{L} \cdot \underset{\sim}{b}_3^{(1)} = 0 \tag{3.27}$$

must be satisfied. If we assume that $\bar{A}_2$ depends on x_1 and t_1 through the combination $x_1 - V_g t_1$ as $\bar{A}_1$, and if we define the following non-dimensional variables and constants

$$\tau = wt_2, \quad \xi = k\varepsilon^{-1}(x_2 - V_g\, t_2), \quad A = k\bar{A}_1, \quad \Gamma = k^2\bar{\Gamma}/w, \quad \Delta = \bar{\Delta}/wk \tag{3.28}$$

from the compatibility condition (3.27) we obtain the following nonlinear Schrödinger equation for A:

$$i\frac{\partial A}{\partial \tau} + \Gamma\frac{\partial^2 A}{\partial \xi^2} + \Delta\, |A|^2 A = 0, \tag{3.29}$$

where

$$\bar{\Gamma} = (1/2)\, d^2w/dk^2, \quad \bar{\Delta} = -\underset{\sim}{L}\cdot\underset{\sim}{F}/(\underset{\sim}{L}\,(\partial \underset{\sim}{W}_1/\partial w)\underset{\sim}{R}) \tag{3.30}$$

and the vector $\underset{\sim}{F}$ is defined as in [15], but for this work the nonlinear constants q_1^2 and q_2^2 are

$$q_1^2 = d_1^2/c_1^2, \quad d_1^2 = (1/\rho_o)\,(d\Phi(3)/dI); \quad q_2^2 = d_2^2/c_2^2, \quad d_2^2 = (1/\tilde{\rho}_o)\,(d\tilde{\Phi}(3)/dI). \tag{3.31}$$

Thus, once a solution for A is derived from (3.29) for a given initial value $A_o(\xi,0)$ for A, (note that on the surface Y = h, at t = o one can write $u(X,h,o) = \varepsilon A_o(\varepsilon kX)\exp(ikX)/k\cosh kh + c.c.$), then the first order solutions for u_1 and v_1 can easily be constructed by (3.20).

Note that for $l = 3$ since det $\underset{\sim}{W}_3 \neq 0$ by assumption, then (3.26) can be solved uniquely for $\underset{\sim}{U}_3^{(3)}$ as

$$\underset{\sim}{U}_3^{(3)} = \underset{\sim}{W}_3^{-1}\, \underset{\sim}{b}_3^{(3)} \tag{3.32}$$

and for $l \neq 0,1,3$, since det $\underset{\sim}{W}_l \neq 0$ and $\underset{\sim}{b}_3^{(1)} \equiv \underset{\sim}{0}$, then $\underset{\sim}{U}_3^{(1)} \equiv \underset{\sim}{0}$. When the phase velocity of the fundamental wave coincides with that of its third harmonic at a critical wave number k_c, since in that case the pairs (k_c,w_c) and $(3k_c,3w_c)$ simultaneously satisfy the linear dispersion relation, i.e., both det $\underset{\sim}{W}_1 = 0$ and det $\underset{\sim}{W}_3 = 0$ for (k_c,w_c) then $\underset{\sim}{U}_3^{(3)}$ tends to infinity as $k \to k_c$ according to (3.32). The term $\varepsilon^3 \underset{\sim}{U}_3^{(3)}$ becomes the same order as $\underset{\sim}{U}_1^{(1)}$ when $k-k_c = O(\varepsilon^2)$ and becomes dominant when $k-k_c < O(\varepsilon^2)$, contrary to the implicit assumption used in deriving the perturbation expansion. Hence this expansion ceases to be uniformly valid. This phenomenon is called the third harmonic resonance. In general, harmonic resonance may exist between the fundamental mode and its any mth harmonic component, m=3,5,7,9,... , if (k_c,w_c) and (mk_c,mw_c) simultaneously satisfy the linear dispersion relation. When this is the case a uniform expansion to the problem may be obtained by including the interacting harmonic in the first order solution (3.18). Since also det $\underset{\sim}{W}_m = 0$ in this case, the first order solution will be of the form

$$\underset{\sim}{U}_1^{(1)} = \bar{A}_1^{(1)} \underset{\sim}{R}_1, \quad \underset{\sim}{U}_1^{(m)} = \bar{A}_1^{(m)} \underset{\sim}{R}_m, \quad \underset{\sim}{U}_1^{(n)} \equiv \underset{\sim}{0} \text{ for } n \neq 1,m . \tag{3.33}$$

If the perturbation analysis is advanced on this line, for third harmonic resonance for example, the following pair of equations are obtained for $\bar{A}^{(1)} = A$ and $\bar{A}^{(3)} = B$:

$$i\left(\frac{\partial A}{\partial t_2} + V_g^{(1)} \frac{\partial A}{\partial x_2}\right) + \Gamma^{(1)} \frac{\partial^2 A}{\partial x_1^2} = \Delta_1^{(1)} |A|^2 A + \Delta_2^{(1)} A|B|^2 + \Delta_3^{(1)} B (A^*)^2 \exp(i\delta) \tag{3.34}$$

$$i\left(\frac{\partial B}{\partial t_2} + V_g^{(3)} \frac{\partial B}{\partial x_2}\right) + \Gamma^{(3)} \frac{\partial^2 B}{\partial x_1^2} = \Delta_1^{(3)} |A|^2 B + \Delta_2^{(3)} B|B|^2 + \Delta_3^{(3)} A^3 \exp(i\delta),$$

where δ is a parameter which measures the nearness of the resonance, when $k_3 = 3k_1$ and $w_3 = 3w_1$, $\delta = 0$; and $\Gamma^{(\nu)}$ and $\Delta_\alpha^{(\nu)}$, $\nu = 1,3$, $\alpha = 1,2,3$, are certain constants which depend on the frequency of motion and the material constitution of half space. For fifth harmonic resonance a similar set of equations are valid but the terms multiplied by $\exp(i\delta)$ disappear. This is also the case for the equations describing the interaction of any two wave trains centred at (k_n,w_n) and (k_m,w_m). Detailed investigation on these topics will appear elsewhere.

4. CONCLUSIONS

The criterion whether $\Gamma\Delta > 0$ or $\Gamma\Delta < 0$ is important in determining how given initial data will evolve for long times for the asymptotic wave field governed by the NLS equation. It is well known that an initial disturbance vanishing as $|\xi| \to \infty$; tends to become a series of envelope solitary waves if $\Gamma\Delta > 0$, while it evolves into decaying oscillations if $\Gamma\Delta < 0$. On the other hand for the initial disturbances that tend to the uniform solution

$A = (ka_o/2) \exp(i\, k^2 a_o^2 \tau/4)$ as $|\xi| \to \infty$, the envelope dark soliton and the Ma soliton solutions exist [19,20,21].

To see the variation in $\Gamma\Delta$ with nonlinearity for certain layer and half space models, we perform the numerical evaluation of this term with respect to kh, for the lowest branch of the linear dispersion relation, by fixing the linear material properties. For the parameters q_1^2 and q_2^2 connected with the nonlinear material properties of the layer and the half space, respectively; we choose the pairs ($q_1^2 = 0.5$, $q_2^2 = 2.0$) ($q_1^2 = 2.0$, $q_2^2 = 0.5$) for hardening layer and hardening substrate (H-H) models and (q_1^2 =

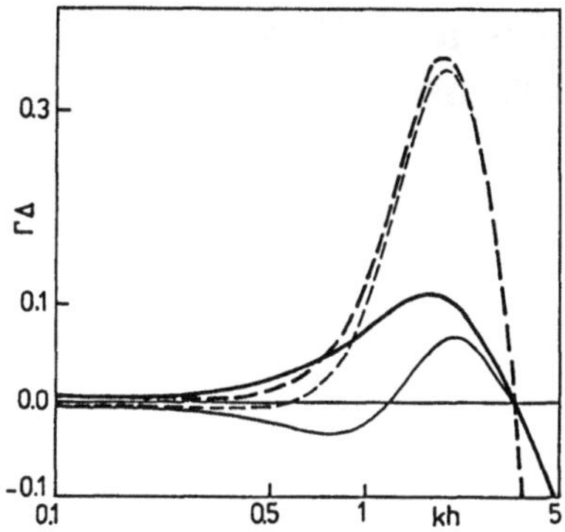

Fig.1 Variation of $\Gamma\Delta$ v.s. kh for the (H-H) models with ($q_1^2 = 0.5$, $q_2^2 = 2.0$) (——) and ($q_1^2 = 2.0$, $q_2^2 = 0.5$) (– – –), and for the (H-S) models with ($q_1^2 = 0.5$, $q_2^2 = -2.0$) (——) and ($q_1^2 = 2.0$, $q_2^2 = -0.5$) (– – –); $\sigma = 2.159$ and $c_2/c_1 = 1.297$

0.5, $q_2^2 = 2.0$), ($q_1^2 = 2.0$, $q_2^2 = 0.5$) for hardening layer and softening substrate (H-S) models. The set of curves is presented in Figure 1. From these results we may conclude that the type and behaviour of solutions for a given initial data at a specific wave number depend on the nonlinear constitution of the layered half space.

References

1. G.S. Kino, H. Mathews : In I.E.E.E. Spectrum, Vol.8, p.22 (1971)
2. J. DeKlerk : In Phys. Today, Vol.11, p.32 (1972)
3. G.A. Maugin : In Advances in Applied Mechanics, Vol.23, p.373 (Academic Press, New York 1983)
4. N. Kalyanasundaram : In Int. J. Engng. Sci., Vol.19, p.279 (1981)
5. N. Kalyanasundaram : In Int. J. Engng. Sci., Vol.19, p.435 (1981)
6. N. Kalyanasundaram, R. Ravindran, P. Prasad : In J. Acoust. Soc. Amer., Vol.72, p.488 (1982)
7. R.W. Lardner : In Int. J. Engng. Sci. Vol.21, p.1331 (1981)
8. E.A. David : In Int. J. Engng. Sci. Vol.23, p.699 (1985)
9. D.F. Parker : In Technical Report 78-26, I.A.M.S., Univ. of British Columbia (1978)
10. D.F. Parker, F.M. Talbot : In Nonlinear Deformation Waves, ed. by U. Nigul, J. Engelbrecht (Springer, Berlin, 1983)
11. D.F. Parker, F.M. Talbot : In Jou. Elasticity, Vol.15, p.389 (1985)
12. D.F. Parker : In Physica D, Vol.16, p.385 (1985)
13. N. Kalyanasundaram : In Int. J. Engng. Sci., Vol.19, p.287 (1981)
14. K. Bataille, F. Lund : Physica D, Vol.6, p.95 (1982)
15. M. Teymur : In Int. J. Engng. Sci (submitted)
16. A.C. Eringen, E.S. Suhubi : Elastodynamics, Vol.I, (Academic Press, New York 1979)
17. A. Jeffrey, T. Kawahara : Asymptotic Methods in Nonlinear Wave Theory (Pitman, Boston 1982)
18. A.C. Eringen, E.S. Suhubi : Elastodynamics, Vol.II, (Academic Press, New York 1975)
19. V.E. Zakharov, A.B. Shabat : In Sov. Phys. JETP, Vol.34, p.62 (1972)
20. V.E. Zakharov, A.B. Shabat : In Sov. Phys. JETP, Vol.37, p.823 (1973)
21. D.H. Peregrine, In J. Austral. Math. Soc. Ser.B, Vol.25, p.16 (1983)

Nonlinear Surface Acoustic Waves and Their Associated Surface Acoustic Solitons

A.A. Maradudin

Department of Physics, University of California,
Irvine, CA 92717, USA

We obtain the nonlinear dispersion relation, and the associated displacement field, for nonlinear surface acoustic waves of shear horizontal polarization that propagate without change of form in a system that consists of a film of thickness d of a linear, cubic elastic medium bonded to a semi-infinite, nonlinear, cubic elastic medium. From the nonlinear dispersion relation we obtain the nonlinear Schrödinger equation governing the propagation of a surface acoustic soliton in this system by methods developed originally by Whitham, Yuen and Lake, and Karpman and Krushkal´. It is found that stable solutions of this equation can exist in the system under consideration.

All elastic media are nonlinear to a greater or lesser extent. Like linear media, nonlinear media can support surface acoustic waves that propagate along their surfaces without change of form [1-6]. In this paper we consider a particular example of such nonlinear surface acoustic waves and a consequence of their existence that has been little discussed in the literature until now.

The energy in a surface acoustic wave is localized in the vicinity of the surface supporting it, and in general is much higher than that of the bulk elastic waves in this region. The effects of the nonlinearity of the medium on the propagation of these waves are correspondingly enhanced. If the surface acoustic wave in the linear approximation is also dispersive, one may expect that a balance between the spreading tendency of the dispersion and the contracting tendency of the nonlinearity can result in the formation of a stable solitary surface acoustic wave that travels along the surface without change of form and has some localized shape. The study of such a surface acoustic soliton [7] in a particular physical system is our principal goal in this paper.

We consider what appears to be the simplest type of surface acoustic soliton possible, viz. one based on a displacement field of shear horizontal (SH) polarization which, therefore, has only a single, nonzero component. Surface acoustic solitons of this polarization have been studied earlier, by BATAILLE and LUND [8]. However, these authors introduced the needed dispersion of the linear surface acoustic wave into the problem in an unnecessarily ad hoc fashion. (Surface acoustic solitons based on Rayleigh waves have been studied in a series of papers by SAKUMA and his colleagues [9-12] who proceeded from a quantum mechanical formulation of the problem. These authors also introduced dispersion into the problem in an ad hoc fashion. We will not consider nonlinear surface acoustic waves of this polarization here.) Our approach to obtaining surface acoustic solitons requires obtaining first the dispersion relation for the nonlinear surface acoustic waves supported by the physical system we will be studying, and obtaining the nonlinear evolution equation governing the propagation of a surface acoustic soliton from this dispersion relation. It appears to be more straightforward than the approaches taken in earlier studies of this problem, and free from some of the objections that can be raised against the work cited above.

The physical system we consider consists of a film of a linear elastic medium that occupies the region $-d < x_3 < 0$ that is bonded to a nonlinear elastic medium that fills the half-space $x_3 > 0$. The linear medium is assumed to be cubic, with the cube axes along the coordinate axes. It is characterized by a mass density $\rho^{(2)}$ and elastic moduli $\{c_{ij}^{(2)}\}$ in the contracted Voigt notation. The nonlinear medium is also assumed to be cubic with its cube axes along the coordinate axes. Its mass density is $\rho^{(1)}$, and its linear and nonlinear elastic moduli are $\{c_{ij}^{(1)}\}$, $\{c_{ijk}^{(1)}\}$, $\{c_{ijk\ell}^{(1)}\}$,... in the Voigt notation. The reasons for our choice of this system will emerge in what follows.

The elastic displacement field in this system is assumed to be SH-polarized, i.e.

$$\vec{u}(\vec{x};t) = (0,\ u_2(x_1,x_3;t),0)\ , \tag{1}$$

and describes a straight-crested $(\partial/\partial x_2 = 0)$ wave propagating in the x_1-direction. In Lagrangian coordinates the equation of motion of the nonlinear medium is $(u_2^{(1)}(x_1,x_3;t) \to u(x_1,x_3;t))$ [13]

$$\frac{1}{c_t^2}\frac{\partial^2 u}{\partial t^2} = \frac{\partial^2 u}{\partial x_1^2} + \frac{\partial^2 u}{\partial x_3^2} + \left[\alpha\left(\frac{\partial u}{\partial x_1}\right)^2 + \beta\left(\frac{\partial u}{\partial x_3}\right)^2\right]\frac{\partial^2 u}{\partial x_1^2}$$

$$+ 4\beta\,\frac{\partial u}{\partial x_1}\frac{\partial u}{\partial x_3}\frac{\partial^2 u}{\partial x_1 \partial x_3} + \left[\beta\left(\frac{\partial u}{\partial x_1}\right)^2 + \alpha\left(\frac{\partial u}{\partial x_3}\right)^2\right]\frac{\partial^2 u}{\partial x_3^2} \qquad x_3 > 0, \tag{2}$$

where

$$c_t^2 = c_{44}^{(1)}/\rho_o^{(1)} \tag{3a}$$

$$\alpha = \left[\tfrac{1}{2}\,c_{4444}^{(1)} + 3c_{344}^{(1)} + \tfrac{3}{2}\,c_{11}^{(1)}\right]/c_{44}^{(1)} \tag{3b}$$

$$\beta = \left[\tfrac{1}{2}\,c_{4466}^{(1)} + c_{144}^{(1)} + 2c_{456}^{(1)} + c_{44}^{(1)} + \tfrac{1}{2}\,c_{12}^{(1)}\right]/c_{44}^{(1)}\ , \tag{3c}$$

and $\rho_o^{(1)}$ is the mass density of the nonlinear medium in its undeformed state. We note that quadratic nonlinear terms are absent from Eq. (2). The equation of motion of the linear film is $(u_2^{(2)}(x_1,x_3;t) \to v(x_1,x_3;t))$

$$\rho^{(2)}\,\frac{\partial^2 v}{\partial t^2} = c_{44}^{(2)}\left(\frac{\partial^2 v}{\partial x_1^2} + \frac{\partial^2 v}{\partial x_3^2}\right) \qquad -d < x_3 < 0. \tag{4}$$

The interface and boundary conditions satisfied by $u(x_1,x_3;t)$

and $v(x_1 x_3;t)$ are

$$u\Big|_{x_3=0} = v\Big|_{x_3=0}; \quad T_{23}^{(1)}\Big|_{x_3=0} = c_{44}^{(2)} \frac{\partial v}{\partial x_3}\Big|_{x_3=0} \tag{5a}$$

$$c_{44}^{(2)} \frac{\partial v}{\partial x_3}\Big|_{x_3=-d} = 0, \tag{5b}$$

where $T_{23}^{(1)}$ is the Piola-Kirchhoff stress tensor in the nonlinear medium [13]

$$T_{23}^{(1)} = c_{44}^{(1)}[1 + \beta(\frac{\partial u}{\partial x_1})^2 + \frac{1}{3}\alpha(\frac{\partial u}{\partial x_3})^2] \frac{\partial u}{\partial x_3} \,. \tag{6}$$

We now proceed to simplify the problem by eliminating the displacement field $v(x_1,x_3;t)$ in the linear film from consideration, at the expense of a more complicated boundary condition at $x_3 = 0$ satisfied by $u(x_1,x_3;t)$. We begin by expanding the boundary condition (5b) about $x_3 = 0$,

$$c_{44}^{(2)} \frac{\partial v}{\partial x_3}\Big|_{x_3=0} - d\, c_{44}^{(2)} \frac{\partial^2 v}{\partial x_3^2}\Big|_{x_3=0} + O(d^2) = 0. \tag{7}$$

When Eq. (7) is combined with the second of Eqs. (5a), we obtain the boundary condition satisfied by $u(x_1,x_3;t)$ in the form

$$T_{23}^{(1)}\Big|_{x_3=0} = d\, c_{44}^{(2)} \frac{\partial^2 v}{\partial x_3^2}\Big|_{x_3=0} + O(d^2) \tag{8a}$$

$$= d[\rho^{(2)} \frac{\partial^2 v}{\partial t^2} - c_{44}^{(2)} \frac{\partial^2 v}{\partial x_1^2}]_{x_3=0} + O(d^2). \tag{8b}$$

In obtaining Eq. (8b) from Eq. (8a) we have used the equation of motion (4). Now, since $u(x_1,0;t) = v(x_1,0;t)$ for all x_1 and t, according to the first of Eqs. (5a), the nth derivative of $u(x_1,0;t)$ with respect to x_1 must equal the nth derivative of $v(x_1,0;t)$ with respect to x_1 for all x_1. It follows that the boundary condition satisfied by $u(x_1,x_3;t)$ at $x_3 = 0$ can finally be written as

$$c_{44}^{(1)}[1 + \beta(\frac{\partial u}{\partial x_1})^2 + \frac{1}{3}\alpha(\frac{\partial u}{\partial x_3})^2] \frac{\partial u}{\partial x_3}\Big|_{x_3=0} = d[\rho^{(2)} \frac{\partial^2 u}{\partial t^2}$$

$$- c_{44}^{(2)} \frac{\partial^2 u}{\partial x_1^2}]_{x_3=0} + O(d^2) \,, \tag{9}$$

when we use Eq. (6).

To obtain the nonlinear dispersion relation for the SH-polarized nonlinear surface acoustic waves of interest to us we must now solve Eq. (2) and use the solution in the boundary condition (9). We seek a solution of Eq. (2) in the form of a wave that propagates in the x_1-direction without changing its form, viz.

$$u(x_1,x_3;t) = f(\theta;x_3), \tag{10}$$

where $\theta = kx_1-\omega t$. The equation satisfied by $f(\theta;x_3)$ is

$$-\left(k^2 - \frac{\omega^2}{c_t^2}\right)\frac{\partial^2 f}{\partial\theta^2} - \frac{\partial^2 f}{\partial x_3^2} = \alpha k^4\left(\frac{\partial f}{\partial\theta}\right)^2\frac{\partial^2 f}{\partial\theta^2} + \beta k^2\Big[\left(\frac{\partial f}{\partial x_3}\right)^2\frac{\partial^2 f}{\partial\theta^2} +$$

$$+ 4\frac{\partial f}{\partial\theta}\frac{\partial f}{\partial x_3}\frac{\partial^2 f}{\partial\theta\partial x_3} + \left(\frac{\partial f}{\partial\theta}\right)^2\frac{\partial^2 f}{\partial x_3^2}\Big] + \alpha\left(\frac{\partial f}{\partial x_3}\right)^2\frac{\partial^2 f}{\partial x_3^2}, \tag{11}$$

while the boundary condition it satisfies is

$$c_{44}^{(1)}\left[1 + \beta k^2\left(\frac{\partial f}{\partial\theta}\right) + \frac{1}{3}\alpha\left(\frac{\partial f}{\partial x_3}\right)^2\right]\frac{\partial f}{\partial x_3}\Big|_{x_3=0}$$

$$= d\left[\rho^{(2)}\omega^2 - c_{44}^{(2)}k^2\right]\frac{\partial^2 f}{\partial\theta^2}\Big|_{x_3=0} + O(d^2). \tag{12}$$

To solve these equations we set

$$f(\theta;x_3) = \varepsilon f_1(\theta;x_3) + \varepsilon^3 f_3(\theta;x_3) + \ldots \tag{13a}$$

$$\omega^2 = \omega_0^2 + \varepsilon^2\omega_2^2 + \ldots, \tag{13b}$$

where ε is an ordering parameter that can be set equal to unity at the end of the calculation. When Eqs. (13) are substituted into Eq. (11) and terms of the same order in ε on both sides of the resulting equation are equated, we obtain the following system of equations:

$$-\gamma_o^2\frac{\partial^2 f_1}{\partial\theta^2} - \frac{\partial^2 f_1}{\partial x_3^2} = 0 \tag{14a}$$

$$-\gamma_o^2\frac{\partial^2 f_3}{\partial\theta^2} - \frac{\partial^2 f_3}{\partial x_3^2} = -\frac{\omega_2^2}{c_t^2}\frac{\partial^2 f_1}{\partial\theta^2} + \alpha k^4\left(\frac{\partial f_1}{\partial\theta}\right)^2\frac{\partial^2 f_1}{\partial\theta^2}$$

$$+ \beta k^2\Big[\left(\frac{\partial f_1}{\partial x_3}\right)^2\frac{\partial^2 f_1}{\partial\theta^2} + 4\frac{\partial f_1}{\partial\theta}\frac{\partial f_1}{\partial x_3}\frac{\partial^2 f_1}{\partial\theta\partial x_3}$$

$$+ \left(\frac{\partial f_1}{\partial\theta}\right)^2\frac{\partial^2 f_1}{\partial x_3^2}\Big] + \alpha\left(\frac{\partial f_1}{\partial x_3}\right)^2\frac{\partial^2 f_1}{\partial x_3^2} \tag{14b}$$

. . . .

where

$$\gamma_o^2 = k^2 - (\omega_o^2/c_t^2). \tag{15}$$

The solutions of Eqs. (14) that vanish as $x_3 \to \infty$, as must be the case for surface acoustic waves, are

$$f_1(\theta;x_3) = a\cos\theta\, e^{-\gamma_o x_3} \tag{16a}$$

$$f_3(\theta;x_3) = \cos\theta\left[\frac{a}{2\gamma_o}\frac{\omega_2^2}{c_t^2}x_3 e^{-\gamma_o x_3} + \frac{a^3}{32\gamma_o^2}(\alpha k^4 - 2\beta k^2\gamma_o^2 - 3\alpha\gamma_o^4)e^{-3\gamma_o x_3}\right]$$

$$+ \cos 3\theta\left[C e^{-3\gamma_o x_3} + \frac{a^3}{24\gamma_o}(\alpha k^4 - 6\beta k^2\gamma_o^2 - \alpha\gamma_o^4)\, x_3 e^{-3\gamma_o x_3}\right] \tag{16b}$$

. . . .

where a is a constant amplitude, and C is an as yet arbitrary constant. We see from Eqs. (15) and (16) that if $f(\theta;x_3)$ is to vanish as $x_3 \to 0$ the condition

$$k^2 > \omega_o^2/c_t^2 \tag{17}$$

must be satisfied. We also note that although the secular terms in Eq. (16b) might be removed by the use of a multiple scales approach [14] (in x_3) to the solution of Eq. (11), their presence does not violate the boundary condition at infinity [15].

When the expansions (13) are substituted into the boundary condition (12) and terms of the same order in ε on both sides of the resulting equation are equated, we obtain the following system of equations at $x_3 = 0$:

$$c_{44}^{(1)}\frac{\partial f_1}{\partial x_3} - d[\rho^{(2)}\omega_o^2 - c_{44}^{(2)}k^2]\frac{\partial^2 f_1}{\partial\theta^2} = 0 \tag{18a}$$

$$c_{44}^{(1)}\frac{\partial f_3}{\partial x_3} - d[\rho^{(2)}\omega_o^2 - c_{44}^{(2)}k^2]\frac{\partial^2 f_3}{\partial\theta^2}$$

$$= - c_{44}^{(1)}\left[\beta k^2\left(\frac{\partial f_1}{\partial\theta}\right)^2 + \frac{1}{3}\alpha\left(\frac{\partial f_1}{\partial x_3}\right)^2\right]\frac{\partial f_1}{\partial x_3} + d\rho^{(2)}\omega_2^2\frac{\partial^2 f_1}{\partial\theta^2} \tag{18b}$$

. . . .

The substitution of Eq. (16a) into Eq. (18a) yields the value of $\omega_o^2(k)$, viz.

$$\omega_o^2(k) = c_t^2 k^2 [1 - (\frac{\rho^{(2)}}{\rho^{(1)}} - \frac{c_{44}^{(2)}}{c_{44}^{(1)}})^2 (kd)^2], \tag{19}$$

to lowest nonzero order in d. In then follows from Eqs. (15) and (19) that

$$\gamma_o = d(\frac{\rho^{(2)}}{\rho^{(1)}} - \frac{c_{44}^{(2)}}{c_{44}^{(1)}})k^2 + O(d^3). \tag{20}$$

Thus, in order that γ_o be real and positive so that Eqs. (10), (13a), and (16) describe the displacement field of a wave localized to the surface $x_3 = 0$, we must have that $(\rho^{(2)}/\rho^{(1)}) > (c_{44}^{(2)}/c_{44}^{(1)})$. Equation (19) is just the dispersion relation for the lowest frequency Love wave, to the order in k indicated. The substitution of Eq. (16b) into Eq. (18b) yields both the coefficient C and $\omega_2^2(k)$, viz.

$$C = -\frac{a^3}{144\gamma_o^2}\alpha(k^4 - \gamma_o^4) \tag{21a}$$

$$\omega_2^2(k) = c_t^2 \frac{a^2}{8[1-2(\rho^{(2)}/\rho^{(1)})\gamma_o]}(\alpha k^4 + 2\beta k^2\gamma_o^2 + \alpha\gamma_o^4) . \tag{21b}$$

When we combine Eqs. (19) and (20b) we obtain for the nonlinear dispersion relation the result that

$$\omega(k) = c_t k[1 - \frac{1}{2}(\frac{\rho^{(2)}}{\rho^{(1)}} - \frac{c_{44}^{(2)}}{c_{44}^{(1)}})^2 (kd)^2]$$

$$+ a^2 \frac{c_t(\alpha k^4 + 2\beta k^2\gamma_o^2 + \alpha\gamma_o^4)}{16k[1-2(\rho^{(2)}/\rho^{(1)})\gamma_o][1 - \frac{1}{2}(\frac{\rho^{(2)}}{\rho^{(1)}} - \frac{c_{44}^{(2)}}{c_{44}^{(1)}})^2 (kd)^2]} , \tag{22}$$

to the lowest nonzero order in a. If we now use Eq. (20) and work only to the lowest order in d, we can simplify Eq. (22) to

$$\omega(k) = c_t k[1 - \frac{1}{2}(\frac{\rho^{(2)}}{\rho^{(1)}} - \frac{c_{44}^{(2)}}{c_{44}^{(1)}})^2 (kd)^2] + a^2 c_t k \frac{\alpha k^2}{16} . \tag{23}$$

Thus, our artifice of applying a thin film of a linear elastic material to a nonlinear substrate has created a surface acoustic wave in the linear approximation, Eq. (16a), which does not exist in the absence of the film. It is, moreover, a dispersive surface acoustic wave, as we see from Eq. (19). With these results, and the nonlinear dispersion relation given by Eq. (23), in hand we now turn to the extraction of a surface acoustic soliton from them.

In obtaining the preceding results we have assumed that the amplitude a appearing initially in Eq. (16a) is a constant. Thus the nonlinear surface acoustic wave we have just obtained may be termed the CW surface acoustic wave solution of the equation of motion and the associated boundary conditions. We now relax this assumption and allow $a = a(x,t)$ to be a slowly varying function of the coordinate x ($\equiv x_1$) and the time t, and look for solutions for a(x,t) that describe surface acoustic solitons. These are pulse envelope waves that have the nature of solitary waves propagating along the surface of our nonlinear system.

The fact that we have assumed that a(x,t) depends only on the coordinate in the direction of propagation of the wave along the surface and is independent of the transverse coordinate $z(\equiv x_3)$ normal to the surface means that the transverse variation of the elastic displacement field enters the final pulse evolution equation to be obtained here (Eq. (35) below) only through the nonlinear surface dispersion relation. This is an adiabatic assumption that at any point (x,t) the transverse variation of the elastic displacement field is given by the CW nonlinear surface acoustic wave solution. This means, physically, that as the wave propagates along the surface the system adjusts to the nonlinear field distribution at a rate faster than the time scale over which the wave changes [16].

The equations determining a(x,t) are [17-19]:

the energy equation

$$\frac{\partial a^2}{\partial t} + \frac{\partial}{\partial x}\left(\frac{\partial \omega}{\partial k} a^2\right) = 0; \tag{24}$$

the dispersion relation

$$\omega(k,a) = \omega_o(k) - \frac{1}{2}\frac{\partial^2\omega_o(k)}{\partial k^2}\frac{a_{xx}}{a} + \omega_2(k)a^2; \tag{25}$$

the consistency equation

$$k_t + \omega_x = 0. \tag{26}$$

In the absence of the second term on the right hand side of Eq. (25) the dispersion relation given by the latter has the form of the CW nonlinear dispersion relation to $O(a^2)$ given by Eq. (23). The term $-\frac{1}{2}(\partial^2\omega_o(k)/\partial k^2)a_{xx}/a$ has been shown by YUEN and LAKE [18] to be present when the amplitude a is a slowly varying function of x and t.

We now introduce a phase function $\theta(x,t)$ and require that

$$\theta_t = -\omega, \quad \theta_x = k, \tag{27}$$

so that Eq. (24) is automatically satisfied. We next set

$$\omega = \omega_o + \bar{\omega}, \quad k = k_o + \bar{k}, \tag{28}$$

where ω_o and k_o are constants. We can therefore write

$$\theta = \theta_o + \bar{\theta}, \tag{29}$$

where

$$\theta_o = k_o x - \omega_o t, \quad \bar{\theta}_t = -\bar{\omega}, \quad \bar{\theta}_x = \bar{k}. \tag{30}$$

We next expand the dispersion relation (25) about $k = k_o$ in the following fashion:

$$\omega(k_o+\bar{k}) = \omega_o(k_o) + \bar{k}\left(\frac{\partial\omega_o}{\partial k}\right)_o + \frac{1}{2}\,\bar{k}^2\left(\frac{\partial^2\omega_o}{\partial k^2}\right)_o - \frac{1}{2}\left(\frac{\partial^2\omega_o}{\partial k^2}\right)_o \frac{a_{xx}}{a} + \omega_2(k_o)a^2, \tag{31}$$

where the subscript o means that the corresponding derivative is evaluated at $k = k_o$. It follows from this result, the first of Eqs. (27), and the second and third of Eqs. (30), that

$$\bar{\theta}_t = -\left(\frac{\partial\omega_o}{\partial k}\right)_o \bar{\theta}_x - \frac{1}{2}\left(\frac{\partial^2\omega_o}{\partial k^2}\right)_o \left(\bar{\theta}_x^2 - \frac{a_{xx}}{a}\right) - \omega^2(k_o)a^2. \tag{32}$$

In a similar way, we obtain from Eqs. (24) and (31) the result that

$$a_t = -\left(\frac{\partial\omega_o}{\partial k}\right)_o a_x - \frac{1}{2}\left(\frac{\partial^2\omega_o}{\partial k^2}\right)_o\left(\bar{\theta}_{xx}a+2\bar{\theta}_x a_x\right), \tag{33}$$

where we have used the result $(\partial^2\omega/\partial x\partial k) = \bar{\theta}_{xx}(\partial^2\omega_o/\partial k^2)_o$ that follows from Eq. (31) and the last of Eqs. (30).

Equations (32) and (33) can be combined into a single equation for the complex amplitude $A(x,t)$ defined by

$$A(x,t) = a(x,t)\exp i\bar{\theta}(x,t), \tag{34}$$

namely the nonlinear Schrödinger equation

$$i\left(A_t + \left(\frac{\partial\omega_o}{\partial k}\right)_o A_x\right) + \frac{1}{2}\left(\frac{\partial^2\omega_o}{\partial k^2}\right)_o A_{xx} - \omega_2(k_o)|A|^2A = 0. \tag{35}$$

The condition that Eq. (35) have stable soliton solutions is that $(\partial^2\omega_o/\partial k^2)_o\,\omega_2(k_o) < 0$ [20]. A (single soliton) solution of Eq. (35) is

$$A(x,t) = \left(-\frac{(\partial^2\omega/\partial k^2)_o}{\omega_2(k_o)}\right)^{1/2} \frac{e^{i\delta t}}{|v_g\Delta|}\,\mathrm{sech}\left(\frac{t-(x/v_g)}{\Delta}\right), \tag{36}$$

where

$$\delta = \frac{(\partial^2\omega_o/\partial k^2)_o}{2v_g^2\Delta^2}, \quad v_g = \left(\frac{\partial\omega_o}{\partial k}\right)_o, \tag{37}$$

and Δ is the half-width of the pulse represented by Eq. (36). It follows from Eq. (23) that

$$v_g = c_t\left[1 - \frac{3}{2}\left(\frac{\rho^{(2)}}{\rho^{(1)}} - \frac{c_{44}^{(2)}}{c_{44}^{(1)}}\right)^2 (kd)^2\right] \tag{38a}$$

$$\left(\frac{\partial^2\omega_o}{\partial k^2}\right)_o = - 3c_t\left(\frac{\rho^{(2)}}{\rho^{(1)}} - \frac{c_{44}^{(2)}}{c_{44}^{(1)}}\right)^2 kd^2 < 0. \tag{38b}$$

Thus, since we have identified a(x,t) with the amplitude of the displacement field of the nonlinear shear horizontal surface acoustic wave at the surface $x_3 = 0$, whose CW value was denoted by a in the derivation of Eqs. (16) and (23), we see that a stable surface acoustic soliton is possible if $\alpha > 0$.

The work reported here can be improved and extended by obtaining a solution of Eq. (11) free of secular terms in x_3 on the one hand, and calculating the evolution distance of the surface acoustic solitons obtained here, i.e. the distance traveled by a pulse of arbitrary form established at t = 0 before it evolves into the form given by Eq. (36), on the other [21]. It is hoped that both of these topics will be addressed in the near future.

Acknowledgements

The work described here was supported in part by Army Research Office Grant No. DAAG29-85R-0025. I am grateful to the Graduate Council of the University of California, Irvine, for travel funds that made my participation in this Symposium possible.

References

1. N. Kalyanasundaram: Int. J. Engng. Sci. 19, 279 (1981).
2. N. Kalyanasundaram: Int. J. Engng. Sci. 19, 287 (1981).
3. N. Kalyanasundaram: J. Acoust. Soc. Am. 72, 1518 (1982).
4. D. F. Parker and F. M. Talbot: In Nonlinear Deformation Waves, eds. U. Nigul and J. Engelbrecht (Springer-Verlag, New York, 1983), p. 397.
5. D. F. Parker: Physica 16D, 385 (1985).
6. D. F. Parker and F. M. Talbot: J. Elasticity 15, 389 (1985).
7. Strictly speaking, a soliton is a solitary wave that on colliding with another solitary wave passes through the latter without change of form and with only a small change in the phase of each. However, since the term "surface acoustic soliton" has entered the literature, and since it is less clumsy than the term "surface acoustic solitary wave," we will use the former here.
8. K. Bataille and F. Lund: Physica 16D, 95 (1985).
9. T. Sakuma and Y. Kawanami: Phys. Rev. B29, 869 (1984).
10. T. Sakuma and Y. Kawanami: Phys. Rev. B29, 880 (1984).
11. T. Sakuma and T. Miyazaki: Phys. Rev. B33, 1036 (1986).
12. T. Sakuma and O. Saito: Phys. Rev. B35, 1294 (1987).
13. See, for example, D. R. Bland: Nonlinear Dynamic Elasticity (Blaisdell, Waltham MA, 1969), p. 2ff.
14. See, for example, J. D. Cole: Perturbation Methods in Applied Mathematics (Blaisdell, Waltham MA, 1968).
15. Some authors, however, object to the presence of such terms. See, for example, R. W. Lardner: J. Appl. Phys. 55, 3251 (1984).
16. A. D. Boardman, G. S. Cooper, A. A. Maradudin, and T. P. Shen: Phys. Rev. B34, 8273 (1986).

17. G. B. Whitham: Linear and Nonlinear Waves (John Wiley and Sons, New York, 1974), p. 363.
18. H. C. Yuen and B. M. Lake: Phys. Fluids 18, 956 (1975).
19. V. I. Karpman and E. M. Krushkal′: Zh. Eksp. Teor. Fiz. 55, 530 (1968) [Soviet Physics - JETP 28, 277 (1969)].
20. R. K. Dodd, J. C. Eilbeck, J. D. Gibbon, and H. C. Morris: Solitons and Nonlinear Wave Equations (Academic Press, New York, 1982), p. 505 ff.
21. V. E. Zakharov and A. B. Shabat: Zh. Eksp. Teor. Fiz. 61, 118 (1971) [Soviet Physics - JETP 34, 62 (1972)].

Experiments on Nonlinear Effects in Ground-Surface Vibration

P.P. Dimitriu

Egnatia 311, Thessaloniki 54249, Greece

1 Summary

Results are presented of an experimental study of the nonlinear evolution and interaction of sinusoidal ground-surface vibration. The method is the same as used in acoustics: two vertical vibrators — sources of vertical harmonic excitation — were installed on the ground and driven separately and simultaneously, the resulting surface motion being picked up by an array of 8 vertical seismometers placed at various distances from the vibrators. The seismometer signals, representing the vertical component of the ground (particle) velocity, were recorded, then digitized and Fourier analysed. The novelty of the experiments in comparison with the similar work done in the USSR lies in the use of very small vibrators. This permitted generation of vibration with Mach numbers both larger and smaller than 10^{-5} – the strain level regarded in geomechanics as the boundary between linear and nonlinear soil behaviour.

In addition to harmonic generation and nonlinear interaction — effects observed earlier by other workers — the obtained data indicate the phenomena of self-modulation and Fermi-Pasta-Ulam recurrence in ground-surface vibration of sufficiently small amplitude.

2 Introduction

Although acoustics and seismics have much in common, the remarkable progress made by acoustics in the study of nonlinear elastic properties and their effects in the 60's and 70's did not affect geophysics. Thus, even though as early as 1965 GROOTENHUIS and AWOJOBI [1] provided evidence from sinusoidal testing that the ground exhibits considerable nonlinear behaviour, and two years later NIKOLAEV [2] proposed a simple nonlinear model to explain the seismic effect of earthquakes, it was not until 1979 that the first systematic investigation of nonlinear effects in seismic waves was started in the USSR. The main reason for this delay is the belief, deeply rooted in geophysicists and geomechanicians, that for common dynamic strains ($<10^{-5}$) nonlinear effects are negligible (e.g., see [3]), enforced by the lack of conclusive evidence to the contrary. The lack of experimental evidence is the result of the small number of sinusoidal tests made (because of their high costs) and the difficulty of interpreting their results owing to the strong nonlinearity of the ground-vibrator system and the complexity of the — often unknown — ground structure.

The original aim of the field tests, which I made in partial fulfilment of the requirements for a doctorate at Tokyo University, was to confirm the Soviet workers' results on harmonic generation and nonlinear interaction in vibrator-induced ground-surface vibration. The tests — probably the first of their kind to be performed outside the USSR — were carried out for a range of excitation levels and frequencies. Whereas at medium and high excitation levels the results obtained generally agree with those of Soviet workers (e.g., see SOLOV'EV [4]), at low levels the ground vibration behaviour suggests modulational instability and the Fermi-Pasta-Ulam recurrence phenomenon (see [5,6]).

3 Ground Vibration: Generation and Propagation

Workers in various fields attach great importance to the study of the dynamic behaviour of soils and the related problem of ground-vibration generation and propagation. Thus, one of the main tasks of environmental engineers is to minimize vibration due to man-made sources such as traffic and machinery. Moreover, in investigating the earth's interior geophysicists widely use seismic receivers, on the ground surface; the efficiency of this technique depends on the structure and properties of the underlying ground. Lastly, both seismologists and engineers are well aware of the fact that ground response is largely responsible for the seismic effect of earthquakes.

Indicative of the complexity of the problem of modelling wave propagation through the ground is that, despite the large amount of theoretical and experimental work done, issues as essential to the understanding of soil behaviour as the frequency dependence of absorption and the importance of nonlinearity remain controversial. Thus, in environmental engineering and geophysics, there seems to be no consensus whether absorption is frequency independent or is linearly or logarithmically dependent on frequency [7]. At the same time, in geomechanics soil response during earthquakes is analysed (and modelled) by the linear method if (shear) strains do not exceed 10^{-5}, by the quasi-linear method in the range 10^{-5}–10^{-3} and by the nonlinear method if strains exceed 10^{-3} [3].

At present there is little doubt that the simple linear models currently used generally fail to explain the ground-vibration behaviour observed in the field, the discrepancy between theory and experiment being particularly conspicuous in the data from sinusoidal tests [7,8]. Also, although a more exact modelling of the various vibration sources, the resulting wavefield and the influence on it of ground inhomogeneities may reduce the observed discrepancies, evidence is growing that these are due to ground nonlinearity. What is needed then, to obtain a better understanding of soil behaviour, is to combine thorough testing with seasoned theoretical judgement. It is essential in this respect that a standard testing procedure be used to allow comparison between sets of empirical data.

For a number of reasons, field test methods involving vibrators seem to be the ones best suited for the above purpose. Thus, vibrators are the only available sources that combine controllability and reproducibility of action with the possibility of choosing the direction (vertical or horizontal), amplitude and temporal dependence (harmonic or sweep) of the force to be applied to the ground, enabling one to control the amplitude, duration and spectral content of the seismic signal produced. But most importantly, sinusoidal testing is ideal for studying nonlinear effects.

In vibrator-based techniques, the vibrator(s) is installed on the earth's surface, and receivers (seismometers) are placed on the surface or beneath it at various distances from the source(s) to permit examination of the spatial evolution of a steady-state wavefield or phase identification in a finite signal. The kind of seismometers to be used (vertical, horizontal or three-component) and their optimum positioning are determined by the experiment's goal (e.g., study of a nonlinear effect), the kind of excitation to be used (vertical or horizontal), the ground characteristics and the kind of wave motion expected to dominate. Thus, according to theory (the solution of the problem of an elastic half-space excited sinusoidally by a point load applied to the surface, see MILLER and PURSEY [9]), surface motion is likely to be dominated by surface waves (Rayleigh waves for vertical excitation), whereas inside the earth shear waves are more intensive than longitudinal waves.

Although field measurements generally confirm the above rule, BERESNEV et al. [10] provide evidence that, at large distances, surface motion may be dominated by body waves scattered from inhomogeneities well below the surface. There are two possibilities for this to happen. The first is related to the fact that material damping increases with depth, making it possible for the smaller absorption of body waves, which penetrate deeper the larger the source-receiver distance, to compensate, at large enough surface distances, for the difference in initial energy and spreading attenuation. The second possibility is linked with the (likely) presence of surface disturbances (roads, embankments, trenches, etc.) and subsurface inhomogeneities, which can cause the above effect by reflecting and scattering surface as well as body waves.

4 Experiments

A brief description of the experimental site, instrumentation and method used follows. For details the reader is referred to [5,6].

The experiments were performed at Chiba Experimental Station of Tokyo University on a 60×30 m open area surrounded by one– and two–storied buildings. Borehole measurements of longitudinal and shear wave speeds (a and b), made by OHBO [11] for a nearby site, indicate a soil composed of a 5m–thick layer of loam (a = 350 m/s, b = 130 m/s) on top of a clayey layer (a = 500 m/s, b = 230 m/s). Based on these values, Ohbo computed the group and phase dispersion curves for the first two Rayleigh–wave modes.

Two electrodynamic vibrators, with a peak force of 4441 N the larger and 835 N the smaller, were used for producing vertical steady–state sinusoidal excitation with a force amplitude given by $F = W(2\pi f)^2$, where W (kg m) is the moment and f (Hz) is the frequency of excitation. The larger vibrator had a variable force arm, permitting the force amplitude to be varied while the frequency was held fixed. 8 identical vertical seismometers, with a flat frequency response curve in the range 3–60 Hz, were used for measuring the ground–surface vibration, as velocity.

In the first set of tests, the vibrators, installed on the ground 11 m apart (Fig.1), were driven separately and simultaneously at various frequencies in the range 7–14 Hz the larger vibrator and 10–23 Hz the smaller one. In another set of tests, the larger vibrator was relocated some 6 m to the left and driven alone, the use of four moments (W = 0.5, 0.35, 0.24 and 0.19 kg m) allowing generation of ground vibration with different amplitudes at the same frequency, varied in the range 7–14.7 Hz. The amplitude range covered was approximately 0.015–0.6 cm/s at 1 m from the vibrator.

The ground vibration was monitored on the surface along four lines (Fig.1). During each test, the signals from all 8 seismometers, placed along one line at a time, were recorded on a tape recorder, then digitized and Fourier analysed by a FFT routine.

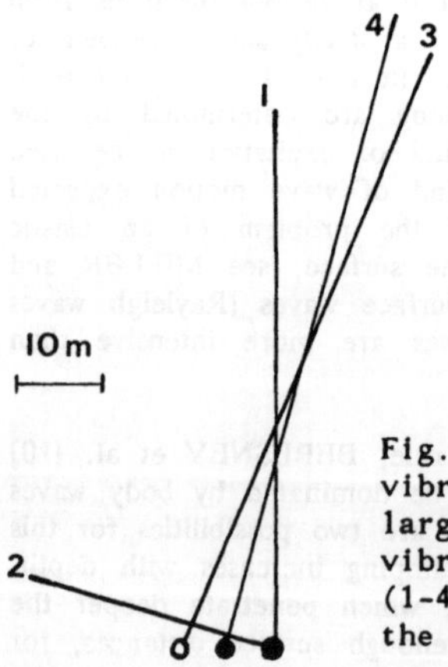

Fig.1. The experimental arrangement: vibrator locations (filled dots - larger vibrator, circle - smaller vibrator) and measurement lines (1-4) along which, in turn, the 8 seismometers were placed

5 Results

5.1 Nonlinear Evolution

The purpose of the evolution tests was to investigate, independently of each other, the effects of the excitation level and frequency upon the evolution of sinusoidal ground-surface vibration. Dependence on the amplitude would mean nonlinear ground behaviour, which in a dispersive medium such as the test ground should also be manifested in the characteristic spatial 'beating' of the second harmonic (e.g., see [12]), provided, of course, the surface motion is indeed dominated by the Rayleigh wave, as expected.

The results obtained for comparatively large excitation levels, corresponding to Mach numbers typically greater than 10^{-5}, generally agree with the expected behaviour. Thus harmonic distortion, clearly visible in nearly all spectra, from points both next to and away from the vibrator, was stronger for large excitation. (Harmonics up to the sixth appear in the spectra from the tests at $f \geqslant 13$ Hz and $W = 0.5$ kg m). Also, ground vibration of not too large amplitude ($M \sim 10^{-5}$—2×10^{-5}) exhibited a spatial evolution of the relative amplitude of the second harmonic that, allowing for the effects of material and geometric attenuation, is typical of nonlinear wave propagation in dispersive media. Figure 2 shows an example of such evolution. The Q curve (the quality factor Q is the reciprocal of the frequency-independent damping coefficient), computed for the fundamental amplitude with correction for cylindrical spreading (surface waves),

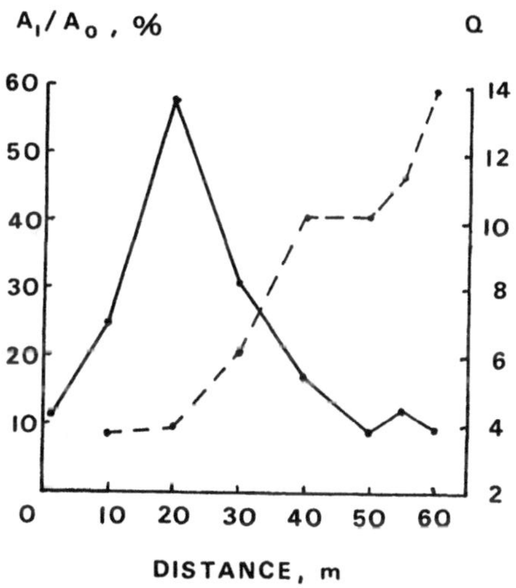

Fig.2. Evolution of the amplitude ratio second/first harmonic, A_1/A_0, (continuous line) and the quality factor Q ($Q = \pi f/\alpha c$, $\alpha(f)$ - frequency-dependent damping coefficient, c - wave speed) computed for the first harmonic with correction for cylindrical spreading; f = 11 Hz, line 1. Had the ground response been linear, Q would have been independent of distance; note also the good correlation between the two quantities

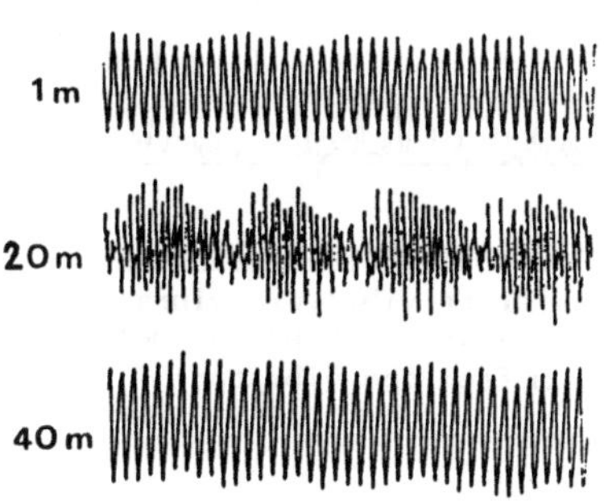

Fig.3. Recurrence via self-modulation and demodulation; f = 11 Hz, line 1

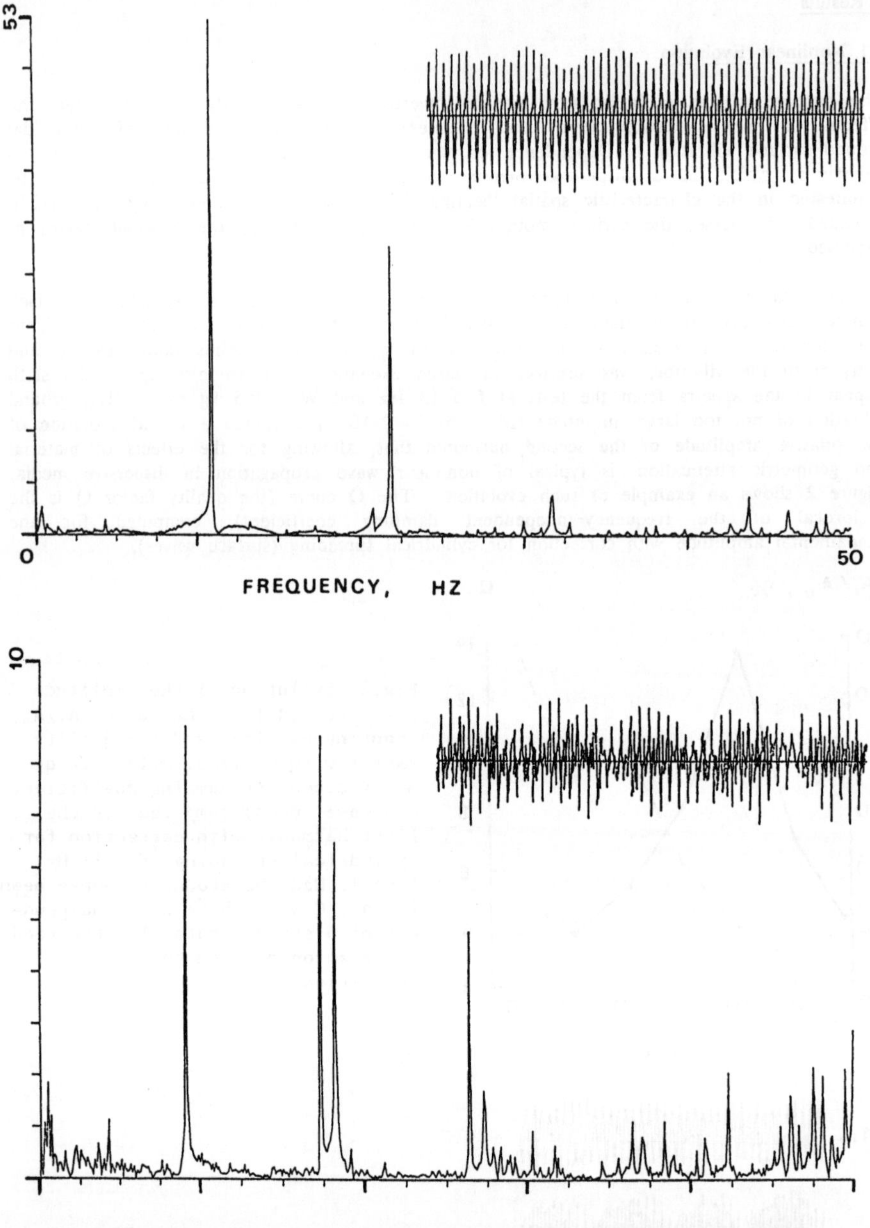

Fig.4. 'Stable' (top) and 'unstable' (bottom) ground-surface vibration, recorded at the same point 20 m away from the larger vibrator; vibration of the first kind was observed for $M>10^{-5}$, of the second kind for $M<10^{-5}$ (M – amplitude of Mach number, evaluated at 1 m from the vibrator)

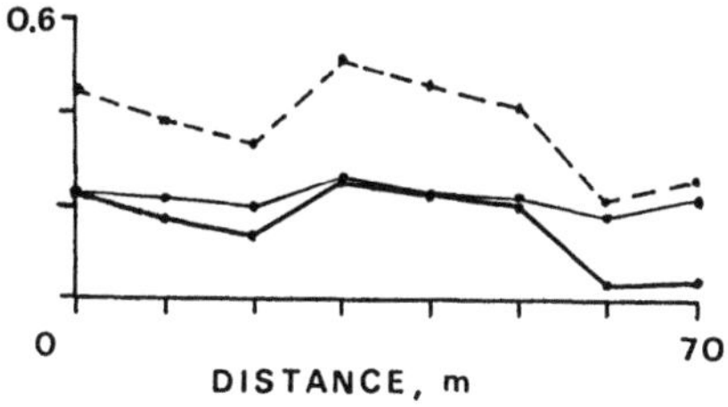

Fig.5. Dependence of the evolution of the fundamental amplitude, A, at 10 Hz on force amplitude [kN]: log A(2) - log A(0.8) (top line), log A(1.4) - log A(0.8) (middle line) and log A(2) - log A(1.4) (bottom line); the curves would have been parallel to the x-axis had the ground behaviour been linear

demonstrates the dependence of damping on distance and correlates well with the second/first harmonic amplitude curve.

Decreasing the excitation below the level corresponding to $M = 10^{-5}$, I expected to observe a gradual passage from nonlinear to linear ground-vibration behaviour. Indeed, as expected, the harmonic distortion reduced considerably, but only next to the vibrator. What happened farther from it is shown in Figs.3 and 4 (bottom). (The top plot in Fig.4 is the 'normal' spectrum, from the same point — at 20m from the vibrator — as the 'abnormal' spectrum.) This kind of vibration evolution was named by me 'unstable' or 'recurrent' [5,6] because of the striking similarity between it and the evolution of nonlinear deep-water wavetrains, studied experimentally (in a water tank) by Lake et al. [13].

Thus whenever sinusoidal ground vibration was induced with a level below that approximately corresponding to $M = 10^{-5}$, it evolved in a recurrent manner (Fig.3): with increasing distance the vibration seemed to be in the process of losing coherence and uniformity, often becoming strongly modulated (usually at a distance between 20 m and 30 m), after which it gradually recovered its original form (usually at a distance between 40 and 60 m). As the Fourier spectra show, the 'splitting' of the higher harmonics (second and/or third) and the dramatic growth of the amplitudes of the dominant harmonic-sideband pair relative to the fundamental amplitude are responsible for the strong modulation observed (Fig.4). As the observed evolution is amplitude dependent (Fig.5), it is nonlinear.

Behaviour of this kind is long known and well studied theoretically and numerically in nonlinear physics (the Fermi-Pasta-Ulam recurrence phenomenon), where the paradox of the spontaneous modulation and demodulation of a uniform coherent wavetrain is explained through solitons, arising as a result of the interplay between the effects of modulational instability, nonlinearity and dispersion (e.g., see [13]).

5.2 Nonlinear Interaction

The aim of the interaction tests, with both vibrators driven simultaneously at different frequencies, was only to observe generation of combination components. More detailed study of the interaction process, such as testing the directionality criterion which nonlinear interactions must satisfy for non-collinear mixing measurements, as in acoustics (e.g., see [14]), is difficult in view of our current inability to generate collimated beams.

The test in which the larger vibrator was driven at 9 Hz and the smaller at 22 Hz was chosen for presentation not because it is the most representative one, but because alone it illustrates the whole variety of combinational interactions observed: f_1+f_2, $2f_1+f_2$ and $2f_2-2f_1$ (see Fig.6). Remarkably, none of the tests indicated the f_2-f_1 interaction.

6 Conclusion

The experimental evidence presented here strongly supports the contention that the dynamic behaviour of the ground is nonlinear. Whereas this might have been expected

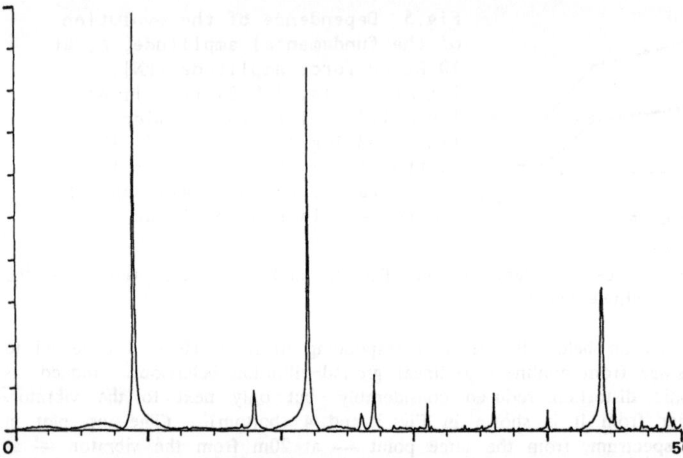

Fig.6. Nonlinear interaction: the larger vibrator driven at 9 Hz, the smaller at 22 Hz; measurement at 2 m along line 2. Note the combinational components at 31, 40 and 26 Hz

for large enough excitations, the result that ground vibration with Mach numbers below 10^{-5} also exhibits nonlinear behaviour (though of a qualitatively different kind than observed for $M > 10^{-5}$) is believed to be new. Thus in addition to the effects of harmonic generation and nonlinear interaction, observed for $M > 10^{-5}$, ground-vibration evolution suggestive, despite some important pecularities, of the Fermi-Pasta-Ulam recurrence phenomenon was observed for $M < 10^{-5}$.

Acknowledgements

I am grateful to T. Katayama, my thesis supervisor, for his support, and to the members of his laboratory for their help during the field work. I thank my brother, Yuri, for drawing the figures.

References

1. P. Grootenhuis, A.O. Awojobi: Symposium – Vibration in Civil Engineering, British Nat. Sec., Int. Assoc. Earthquake Eng. (Imperial College, London 1965).
2. A.V. Nikolaev: Izv. Earth Physics **2** (Moskva 1967)
3. K. Ishihara: In Numerical Methods in Geomechanics, ed. by R. Dungar, G.N. Pande, J.A. Studer (A.A.Balkema, Rotterdam 1982)
4. V.S. Solov'ev: In Problemy Nelinejnoj Sejsmiki, ed. by A.V. Nikolaev, I.N. Galkin (Nauka, Moskva 1987)
5. P.P. Dimitriu: Doctoral Thesis, Dept. Civil Eng., Tokyo University (Tokyo 1986)
6. P.P. Dimitriu: Phys. Earth Plan. Inter. **50**, 74 (1988)
7. T.G. Gutowski, C.L. Dym: J. Sound Vibr. **49**, 179 (1976)
8. R.G. White, M.E.J. Mannering: J. Soc. Environ. Eng. **14**, 3 (1975)
9. G.F. Miller, H. Pursey: Proc. Roy. Soc. **A223**, 521 (1955)
10. I.A. Beresnev et al.: Izv. Earth Physics **10** (Moskva 1986)
11. N. Ohbo: Doctoral Thesis. Dept. Civil Eng., Tokyo University (Tokyo 1985)
12. O.V. Rudenko, S.I. Soluyan: Theoretical Foundations of Nonlinear Acoustics, (Consultants Bureau, New York, London 1977)
13. B.M. Lake et al.: J. Fluid Mech. **83**, 49 (1977)
14. P.A. Johnson et al.: J. Geoph. Res. **92B**, 3597 (1987)

Part II

Surface Waves on Anisotropic Substrates

Aspects of the Behaviour of Surface Waves in Transversely Isotropic Elastic Media

P. Chadwick

School of Mathematics, University of East Anglia, Norwich, NR4 7TJ, U.K.

1. Introduction

A major outcome of the theory of elastic surface waves developed by Barnett and Lothe and their coworkers is a perception of the crucial role of plane body waves in the basic understanding of surface-wave behaviour. As pointed out by CHADWICK and SMITH [1], this connection entails a systematic procedure for determining the nature of surface-wave propagation in elastic media with specified symmetry, beginning with an elastodynamic classification of the materials, proceeding to a characterization of the exceptional transonic states, and culminating in a study of the secular equation for the configurations in which the possibility of transmitting a subsonic surface wave is not guaranteed by the fundamental existence theorem. The detailed implementation of the programme is described in [1] for elastic media with cubic symmetry. A parallel treatment of transverse isotropy (equivalent, in respect of mechanical response to small deformations, to hexagonal symmetry) has subsequently been completed and the purpose of this paper is to outline the main results in advance of a full presentation to be published elsewhere. Despite the increased number of elastic moduli involved, a more complete realization of the programme has proved possible for transversely isotropic than for cubic materials, due essentially to the rotational symmetry of the slowness surface.

We use the same terminology and, as far as possible, the same notation, as in [1] and, to avoid repetition, Part I of [1] is taken as read. It should be noted, however, that the recent proof of BARNETT and LOTHE [2, sect.3] that an exceptional transonic state is necessarily of Type 1 allows the existence theorem quoted in [1, sect.3.3] to be given the following sharpened form.

(i) A unique subsonic surface wave exists when the transonic state is normal and also when the transonic state is exceptional and $F(\hat{v}) < 0$.

(ii) No subsonic surface wave exists when the transonic state is exceptional and $F(\hat{v}) \geq 0$.

The definitions of the *limiting speed* $\hat{v}$ and the function F are given in [1, sects 3.1,3.2; see also 3]. $F(v) = 0$ is the *secular equation*, determining the speed of propagation v_s of a surface wave, and the wave is called *subsonic* when $v_s < \hat{v}$.

The later sections of the paper are arranged in much the same sequence as in Parts II and III of [1], except that considerations relating to subsequent transonic states and body-wave polarization fields are omitted. Accounts of these topics are given in [4] and [5, ch.5].

2. An Elastodynamic Classification of Transversely Isotropic Elastic Materials

We deal throughout with a homogeneous body made of transversely isotropic elastic material. The linear elastic behaviour of the material is characterized by five numerically independent moduli c_{11}, c_{33}, c_{12}, c_{13}, c_{44} and a unit vector $\underline{e}$ directed along the axis of rotational symmetry. The constitutive equations are invariant under the replacement of $\underline{e}$ by $-\underline{e}$. Denoting by ρ the density of the material in its natural state, we use, in place of c_{11} etc., the constants

$$\left.\begin{aligned} a_1 &= c_{11}/\rho, \quad a_2 = \tfrac{1}{2}(c_{11} - c_{12})/\rho, \quad a_3 = c_{33}/\rho, \\ a_4 &= c_{44}/\rho, \quad a_5 = (c_{13} + c_{44})/\rho, \end{aligned}\right\} \tag{1}$$

each having the physical dimensions of $(\text{speed})^2$. Necessary and sufficient conditions for the strain energy of the material to be positive definite are

$$a_1 > a_2 > 0, \; a_4 > 0, \; (a_1 - a_2)a_3 > (a_4 - a_5)^2 \tag{2}$$

(cf. [6, p.33]).

The inequalities (2) imply that three plane body waves can propagate in the direction defined by $\underline{n}$, an arbitrary unit vector. The *acoustical tensor* $\underset{\sim}{Q}(\underline{n})$ governing the properties of these waves is given by

$$\begin{aligned} \rho^{-1}\underset{\sim}{Q}(\underline{n}) = {} & (a_2\sin^2\theta + a_4\cos^2\theta)\underset{\sim}{I} + (a_1 - a_2)\underline{n}\otimes\underline{n} \\ & - (a_1 - a_2 - a_5)\cos\theta(\underline{e}\otimes\underline{n} + \underline{n}\otimes\underline{e}) \\ & - \{(a_2 - a_4)\sin^2\theta - (a_1 - a_2 + a_3 - a_4 - 2a_5)\cos^2\theta\}\underline{e}\otimes\underline{e}, \end{aligned} \tag{3}$$

where $\underset{\sim}{I}$ is the identity tensor and

$$\theta = \cos^{-1}(\underline{e}\cdot\underline{n}) \tag{4}$$

the inclination of $\underline{n}$ to the axis. Because of the invariance mentioned above, θ can be confined to the interval $[0,\tfrac{1}{2}\pi]$.

Setting $\theta = 0$, then $\theta = \tfrac{1}{2}\pi$ in (3) gives

$$\rho^{-1}\underset{\sim}{Q}(\underline{e}) = a_4\underset{\sim}{I} + (a_3 - a_4)\underline{e}\otimes\underline{e}, \tag{5}$$

$$\rho^{-1}\underset{\sim}{Q}(\underline{n}) = a_1\underline{n}\otimes\underline{n} + a_2(\underline{e}\wedge\underline{n}) \otimes (\underline{e}\wedge\underline{n}) + a_4\underline{e}\otimes\underline{e}, \quad \underline{e}\cdot\underline{n} = 0. \tag{6}$$

The squared speeds of propagation of the plane body waves with *wave normal* $\underline{n}$ are the eigenvalues of $\rho^{-1}\underset{\sim}{Q}(\underline{n})$. The speeds of waves travelling along and at right angles to the axis are thus, from (5) and (6), $\sqrt{a_3}$, $\sqrt{a_4}$, $\sqrt{a_4}$ and $\sqrt{a_1}$, $\sqrt{a_2}$, $\sqrt{a_4}$ respectively. For simplicity we assume henceforth that a_1, a_2 and a_3 are all unequal to a_4 and that a_5 and

$$A = (a_1 - a_2)(a_3 - a_4) - a_5^2 \tag{7}$$

are non-zero. Since two of the body-wave speeds are equal when $\theta = 0$, the axis of material symmetry is an *acoustic axis* [6, p.94]. It can be deduced from (3) and (2) that other acoustic axes exist only if $a_2 - a_4$ and A have the same sign. The additional axes then form the cone $\mathscr{C}$ given by $\theta = \theta_{ab}$ where

$$\theta_{ab} = \cot^{-1}\{(a_1 - a_2)(a_2 - a_4)/A\}^{\frac{1}{2}}, \tag{8}$$

and the equal values are the smallest of the body-wave speeds.

In view of $(2)_1$ it follows from these observations that the ordering of the body-wave speeds is uniquely determined for all $\underline{n}$ by the signs of

$$a_1 - a_4,\ a_2 - a_4,\ a_3 - a_4,\ A, \tag{9}$$

(see Table 2 below). Moreover, the signs of $(a_1 - a_4,\ a_2 - a_4)$ can occur only in the combinations

$$1.\ (>0,\ >0),\ 2.\ (>0,\ <0),\ 3.\ (<0,\ <0), \tag{10}$$

and it is plain from the definition (7) that (10) are also the only possible signs of $(a_3 - a_4, A)$. We thus arrive at a subdivision of transversely isotropic elastic materials into 9 *main classes* C_{ij}, the values 1, 2 or 3 of the suffixes i and j specifying the signs of the first and second pairs of the quantities (9) according to the scheme (10). The classification is summarized in Table 1. None of the entries is incompatible with (2); neither does any of the classes automatically satisfy (2).

Class	$a_1 - a_4$	$a_2 - a_4$	$a_3 - a_4$	A
C_{11}	> 0	> 0	> 0	> 0
C_{12}	> 0	> 0	> 0	< 0
C_{13}	> 0	> 0	< 0	< 0
C_{21}	> 0	< 0	> 0	> 0
C_{22}	> 0	< 0	> 0	< 0
C_{23}	> 0	< 0	< 0	< 0
C_{31}	< 0	< 0	> 0	> 0
C_{32}	< 0	< 0	> 0	< 0
C_{33}	< 0	< 0	< 0	< 0

Table 1. Definition of the main classes of transversely isotropic elastic materials

Limiting cases, interfacial to the main classes, arise when one or more of the quantities (9) is allowed to be zero. There turn out to be 19 distinct *interjacent classes*: they are of rather more than academic interest, but are not considered further in this paper.

3. Properties of Plane Body Waves

On calculating the eigenvalues of $\rho^{-1}\underset{\sim}{Q}(\underline{n})$ from (3) we obtain the expressions

$$\left.\begin{aligned} c_a^2(\underline{n}) &= \tfrac{1}{2}\{(a_1 + a_4)\sin^2\theta + (a_3 + a_4)\cos^2\theta - d(\theta)\}, \\ c_b^2(\underline{n}) &= a_2\sin^2\theta + a_4\cos^2\theta, \\ c_3^2(\underline{n}) &= \tfrac{1}{2}\{(a_1 + a_4)\sin^2\theta + (a_3 + a_4)\cos^2\theta + d(\theta)\}, \end{aligned}\right\} \tag{11}$$

with

$$d(\theta) = [\{(a_1 - a_4)\sin^2\theta - (a_3 - a_4)\cos^2\theta\}^2 + 4a_5^2\sin^2\theta\cos^2\theta]^{\frac{1}{2}}, \tag{12}$$

for the squared body-wave speeds. The corresponding waves are referred to as W_a, W_b and W_3. For all $\underline{n}$, W_3 is the fastest, or joint fastest, wave. We denote by

$$c_1(\underline{n}) = \min\{c_a(\underline{n}),\ c_b(\underline{n})\} \text{ and } c_2(\underline{n}) = \max\{c_a(\underline{n}),\ c_b(\underline{n})\} \tag{13}$$

the speeds of the slowest and intermediate waves, W_1 and W_2. For each $\underline{n}$ either $W_1 = W_a$, $W_2 = W_b$ or $W_1 = W_b$, $W_2 = W_a$, depending on the class of the material concerned and also, when θ_{ab} exists, on the sign of $\theta - \theta_{ab}$. The correspondence between (W_1, W_2) and (W_a, W_b) is specified in Table 2.

Class	$c_1(\underline{n})$	$c_2(\underline{n})$
C_{11}	$c_b(\underline{n}),\ 0 \le \theta \le \theta_{ab}$ $c_a(\underline{n}),\ \theta_{ab} \le \theta \le \frac{1}{2}\pi$	$c_a(\underline{n}),\ 0 \le \theta \le \theta_{ab}$ $c_b(\underline{n}),\ \theta_{ab} \le \theta \le \frac{1}{2}\pi$
C_{12}, C_{13}	$c_a(\underline{n})$	$c_b(\underline{n})$
C_{22}, C_{23}, C_{32}, C_{33}	$c_a(\underline{n}),\ 0 \le \theta \le \theta_{ab}$ $c_b(\underline{n}),\ \theta_{ab} \le \theta \le \frac{1}{2}\pi$	$c_b(\underline{n}),\ 0 \le \theta \le \theta_{ab}$ $c_a(\underline{n}),\ \theta_{ab} \le \theta \le \frac{1}{2}\pi$
C_{21}, C_{31}	$c_b(\underline{n})$	$c_a(\underline{n})$

Table 2. Definition of the speeds of propagation of the slowest and intermediate plane body waves W_1 and W_2

The *polarizations* of W_a, W_b and W_3 are orthonormal eigenvectors of $\rho^{-1}\underset{\sim}{Q}(\underline{n})$ corresponding to the eigenvalues (11). For all $0 < \theta \le \frac{1}{2}\pi$, W_b is transverse with polarization $\operatorname{cosec}\theta\,\underline{e}\wedge\underline{n}$ and the polarizations of W_a and W_3 are coplanar with $\underline{e}$ and $\underline{n}$. When $\theta = 0$, it is seen from (5), (11) and (12) that W_3 is longitudinal and that W_a and W_b are transverse, forming a circularly polarized pair with speed $\sqrt{a_4}$, if $a_3 > a_4$, that is for materials in classes C_{11}, C_{12}, C_{21}, C_{22}, C_{31}, C_{32}. For materials in the other classes, W_a is longitudinal and W_3 forms with W_b a transverse circularly polarized pair. When $\theta = \frac{1}{2}\pi$, we conclude from (6), (11) and (12) that W_3 is longitudinal and W_a transverse with polarization $\underline{e}$ if $a_1 > a_4$ (classes C_{11}, C_{12}, C_{13}, C_{21}, C_{22}, C_{33}). For materials in the other classes the properties of W_a and W_3 are interchanged.

If $a_1 - a_4 - a_5$ and $a_3 - a_4 - a_5$ have the same sign there are additional directions of propagation, given by $\theta = \theta_d$ where

$$\theta_d = \cot^{-1}\{(a_1 - a_4 - a_5)/(a_3 - a_4 - a_5)\}^{\frac{1}{2}}, \tag{14}$$

for which W_3 is longitudinal and W_a, like W_b, is transverse.

4. The Slowness Surface

To the plane body waves W_a, W_b and W_3 there correspond closed surfaces $\mathcal{S}_a$, $\mathcal{S}_b$ and $\mathcal{S}_3$ in slowness space, given by

$$\underline{s} = c_a^{-1}(\underline{n})\underline{n},\quad \underline{s} = c_b^{-1}(\underline{n})\underline{n},\quad \underline{s} = c_3^{-1}(\underline{n})\underline{n}, \tag{15}$$

and each having rotational symmetry about the axis, reflexional symmetry about the *basal plane* $\theta = \frac{1}{2}\pi$ and centrosymmetry relative to the origin O of slowness vectors $\underline{s}$. The union $\mathcal{S}$ of $\mathcal{S}_a$, $\mathcal{S}_b$ and $\mathcal{S}_3$ is the *slowness surface* of the material transmitting the waves, $\mathcal{S}_3$ forming the *inner sheet*. In view of (15) the construction of the *outer* and *medial sheets* $\mathcal{S}_1$ and $\mathcal{S}_2$ from $\mathcal{S}_a$ and $\mathcal{S}_b$ can be inferred directly from Table 2. For a material in C_{11}, for example, $\mathcal{S}_b$ coincides with $\mathcal{S}_1$ and $\mathcal{S}_a$ with $\mathcal{S}_2$ from the axis to the cone $\mathcal{C}$ on which $\mathcal{S}_a$ and $\mathcal{S}_b$ intersect. Between $\mathcal{C}$ and the basal plane, $\mathcal{S}_a$ coincides with $\mathcal{S}_1$ and $\mathcal{S}_b$ with $\mathcal{S}_2$.

The only *singular points* of $\mathcal{S}$ are double points at which two sheets intersect: they are confined to the axis and $\mathcal{C}$. The two axial singularities are *tangential points*, the sheets which touch being $\mathcal{S}_1$ and $\mathcal{S}_2$ when the class is C_{11}, C_{12}, C_{21}, C_{22}, C_{31} or C_{32}, and $\mathcal{S}_2$ and $\mathcal{S}_3$ for the other classes. The singularities on $\mathcal{C}$ are *ridge points*, so called because they lie on two distinct tangent planes, one to each of $\mathcal{S}_a$ and $\mathcal{S}_b$. Ridge points occur on $\mathcal{S}$ only for materials in classes C_{11}, C_{22}, C_{23}, C_{32}, C_{33} (see Table 2) and occupy the two circles common to $\mathcal{S}_a$, $\mathcal{S}_b$ and $\mathcal{C}$.

In summary, the class to which a transversely isotropic elastic material belongs implies the structure of $\mathcal{S}$ and the location of its singular points.

It is apparent from $(11)_2$, $(15)_2$ and (4) that $\mathcal{S}_b$ is a spheroid with semi-axes $1/\sqrt{a_4}$ in the axial and $1/\sqrt{a_2}$ in basal directions. Since $\mathcal{S}_3$ meets $\mathcal{S}_2$ in at most two points, the inner sheet is also convex [7, sect.2]. The remaining part $\mathcal{S}_a$ of $\mathcal{S}$ is not normally convex. The points of concavity of a *zonal section* of $\mathcal{S}_a$ (that is the intersection of $\mathcal{S}_a$ with a *zonal plane* containing the axis) correspond to one or more subintervals of $0 \le \theta \le \frac{1}{2}\pi$. Subintervals $[0,\alpha)$, $(\beta,\frac{1}{2}\pi]$ and (γ,δ), with α, β, γ and δ acute and $\gamma < \delta$, represent in turn a pair of *axial concavities*, a *basal concavity* and a pair of *oblique concavities*.

The possible forms of $\mathcal{S}$ comprised in the preceding description admit transonic states of Types 1, 2 and 4. States of Type 6 can also appear when the constants $a_1, \ldots, a_5$ are suitably related, but Type 3 states are limited to three of the interjacent classes and the conditions for a state of Type 5 are incompatible with transverse isotropy. The incidence of transonic states of Types 1, 2, 4 and 6 is specified in Table 3.

Class	Type 1	Type 2	Type 4	Type 6
C_{11}	u	u	u	
C_{12}	u	c	c	
C_{13}	u		c	
C_{21}, C_{31}	u	u		
C_{22}, C_{32}	u	c	u	c
C_{23}, C_{33}	u		u	c

Table 3. Types of transonic states occurring for each of the main classes: u stands for unconditional and c for conditional

5. Exceptional Waves

An *exceptional wave* is a plane body wave which produces no traction on a family of parallel planes (the *immune planes*).

For all $0 < \theta \le \frac{1}{2}\pi$, the transverse wave W_b is exceptional with a unit normal $\underline{k}$ to the immune planes directed tangentially to the zonal section of $\mathcal{S}_b$ through the point representing W_b. The transverse waves which can propagate along the axis ($\theta = 0$) and in the basal plane ($\theta = \frac{1}{2}\pi$) are also exceptional, $\underline{k}$ in each case forming an orthonormal set with $\underline{n}$ and the polarization. When (14) defines a real value of θ_d, these properties extend to the additional transverse wave W_a which can then propagate if

$$a_1 - 2a_2 + a_4 - a_5 = 0. \tag{16}$$

There are up to three values of θ between 0 and $\frac{1}{2}\pi$ for which either W_a or W_3 is non-transverse and exceptional.

Let

$$t_1 = \frac{(2a_4 - a_5)b + a_5 r}{2(a_1 - a_4)\{a_4(a_5 - a_4)\}^{\frac{1}{2}}}, \quad t_2 = \frac{(2a_4 - a_5)b - a_5 r}{2(a_1 - a_4)\{a_4(a_5 - a_4)\}^{\frac{1}{2}}}, \tag{17}$$

with

$$b^2 = a_5^2 - (a_1 - a_4)(a_3 - a_4), \quad r^2 = b^2 - 4a_4(a_5 - a_4). \tag{18}$$

Then $\theta = \theta_{a1}$, $\theta = \theta_{a2}$ and $\theta = \theta_{31}$, $\theta = \theta_{32}$, where

$$\left.\begin{aligned} \theta_{a1} &= \tan^{-1}(-t_1), \quad \theta_{a2} = \tan^{-1}(-t_2), \\ \theta_{31} &= \tan^{-1} t_1, \quad \theta_{32} = \tan^{-1} t_2, \end{aligned}\right\} \tag{19}$$

define possible directions of propagation in which the non-transverse exceptional waves are W_a and W_3 respectively. At most two of (19) determine real angles in $(0,\frac{1}{2}\pi)$, so no more than two of the four possibilities can take effect. An exceptional wave propagating in a direction given by $\theta = \theta_{a1}$, θ_{a2}, θ_{31} or θ_{32} is polarized along the *ray*, that is the normal to $\mathcal{S}$ at the *representative point* of the wave (cf. [4, sects 3(a), 5(a)]). When the wave is W_a (respectively W_3), $\underline{k}$ is codirectional with the polarization of the wave W_3 (resp. W_a) travelling in the same direction.

The other value of θ for which a non-transverse exceptional wave may exist is

$$\beta_o = \tan^{-1}\left[\frac{(a_4 - a_5)\{(a_1 - 2a_2)(a_3 - a_4) + a_5(a_4 - a_5)\}}{(a_1 - 2a_2)\{(a_1 - a_4)(a_4 - a_5) + a_5(a_1 - 2a_2)\}}\right]^{\frac{1}{2}} \tag{20}$$

(see [4, sect.5(b), 8]). This wave is W_a or W_3 according as $a_1 - 2a_2$ and $a_5 - a_4$ have the same or opposite signs, the immune planes being in each case parallel to the zonal plane through the representative point. When the relation (16) holds, the wave is W_a and transverse. As can easily be confirmed from (14) and (20), β_o is then equal to θ_d.

6. Exceptional Limiting Waves

To qualify as an *exceptional limiting wave*, that is the limiting wave for an exceptional transonic state of Type 1, an exceptional wave E has to meet two further requirements: its representative point P must lie on the outer sheet $\mathcal{S}_1$ and the tangent L to $\mathcal{S}_1$ at P in the direction of the unit normal $\underline{k}$ to the immune planes must intersect $\mathcal{S}$ nowhere else. In relation to surface wave motion, $\underline{k}$ is directed along the outward normal to the traction-free boundary of the semi-infinite body carrying the wave and $\underline{j}$, the unit vector along the perpendicular from O, the *centre* of $\mathcal{S}$, to L, defines the direction of propagation. R = span($\underline{j}$, $\underline{k}$) is the *reference plane* of the surface wave (see Fig. 1 which shows a zonal section of $\mathcal{S}$ for a material belonging to class C_{22}).

Since $\mathcal{S}_1$ is formed from $\mathcal{S}_a$ and $\mathcal{S}_b$, an exceptional limiting wave is either W_a or W_b. For the transverse wave W_b, it has been noted in section 5 that $\underline{k}$ is directed tangentially to the spheroid $\mathcal{S}_b$ in the zonal plane through P. Provided $\mathcal{S}_b$ coincides with $\mathcal{S}_1$ at P and the tangent there does not meet $\mathcal{S}_a$, W_b is accordingly an exceptional limiting wave. In the surface wave context, R is a zonal plane and the unit vector $\underline{e}$ along the axis of symmetry is coplanar with the surface-wave and boundary normals $\underline{j}$, $\underline{k}$.

Of the three cases in which W_a is or may be a transverse exceptional wave, propagation along the axis ($\theta = 0$) can be ignored since the associated transonic state is of Type 2 and therefore normal [2, sect.3]. And, as pointed out at the end of section 5, the wave for which $\theta = \theta_d$ may be viewed as a degenerate form of the non-transverse exceptional wave with $\theta = \beta_o$. For the third wave, $\theta = \frac{1}{2}\pi$, the polarization is $\underline{e}$ and $\underline{k}$ is normal to the zonal plane through

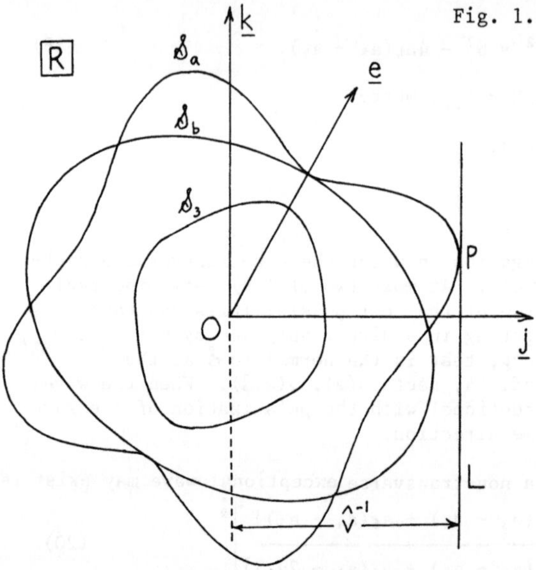

Fig. 1. The transonic state in R relative to $\underline{k}$

P. R is thus the basal plane in this case and $\underline{e}$, $\underline{j}$, $\underline{k}$ are orthonormal. On account of the axisymmetry of $\mathcal{S}$, the *slowness section* $\mathcal{S} \cap R$ consists of three concentric circles and the tangency condition stated in the opening paragraph is hence satisfied if $\mathcal{S}_a$ coincides with $\mathcal{S}_1$ on the basal plane.

Turning to the non-transverse exceptional waves, we first remark that the components of the polarization of W_a in the direction of $\underline{e}$ and in a direction radially outwards from the axis have opposite signs. It follows that when W_a is polarized along the ray, P lies in either an axial or a basal concavity of the zonal section of $\mathcal{S}_a$ containing $\underline{n}$. Since, for a non-transverse exceptional wave propagating in a direction given by $(19)_1$ or $(19)_2$, $\underline{k}$ acts along the tangent L to $\mathcal{S}_a$ in the zonal plane through P, we conclude that such a wave violates the tangency condition and is not, therefore, an exceptional limiting wave.

For the remaining non-transverse exceptional wave to be an exceptional limiting wave, $\mathcal{S}_a$ must again coincide with $\mathcal{S}_1$ at P. In this instance L is normal to the zonal plane through P, so the tangency condition is fulfilled if the curve in which $\mathcal{S}_a$ intersects this plane is convex at P. R is perpendicular to the zonal plane through P with $\underline{j}$ directed along OP, and the axis, being orthogonal to L, is parallel to the boundary of the transmitting body.

In each of the exceptional limiting wave solutions identified above, $\underline{e}$ is a member of one of the families of orientations given by

$$\left.\begin{aligned} \underline{e} &= \sin\alpha\underline{j} + \cos\alpha\underline{k}, && 0 \le \alpha \le \tfrac{1}{2}\pi \ (\alpha\ \textit{configurations}), \\ \underline{e} &= \cos\beta\underline{j} - \sin\beta\underline{j}\wedge\underline{k}, && 0 < \beta \le \tfrac{1}{2}\pi \ (\beta\ \textit{configurations}). \end{aligned}\right\} \quad (21)$$

In this framework, and apropos surface-wave propagation, we can place the exceptional transonic states in transversely isotropic elastic materials into three categories as follows.

Case 1 is a family of transonic states, attained in materials belonging to the classes for which $\mathcal{S}_a$ forms all or part of $\mathcal{S}_1$ as $\underline{e}$ ranges over a subset of the α configurations given by

$$\left.\begin{aligned} &\alpha_o < \alpha < \tfrac{1}{2}\pi && \text{for materials in } C_{11}, \\ &0 \le \alpha < \tfrac{1}{2}\pi && \text{for materials in } C_{21},\ C_{31}, \\ &0 \le \alpha < \alpha_o && \text{for materials in } C_{22},\ C_{23},\ C_{32},\ C_{33}. \end{aligned}\right\} \quad (22)$$

Here α_o is the inclination to the axis of a double tangent to a zonal section of $\mathcal{S}$ having one point of contact on $\mathcal{S}_a$ and the other on $\mathcal{S}_b$, and $(22)_3$ is subject to the proviso that the transonic state is of Type 1 when $\alpha = 0$. The exceptional limiting wave is W_b, transverse with polarization $\underline{j}\wedge\underline{k}$.
Case 2 is a discrete transonic state, realized in materials in classes C_{11}, C_{12}, C_{13} when $\underline{e}$ occupies the $\beta = \frac{1}{2}\pi$ configuration. The exceptional limiting wave is W_a, transverse and with polarization $\underline{e} = \underline{j}\wedge\underline{k}$.
Case 3 is also a discrete transonic state, arising in materials belonging to the classes C_{11}, C_{12}, C_{13}, C_{22}, C_{23}, C_{32}, C_{33} when $\underline{e}$ takes up the $\beta = \beta_o$ configuration. The exceptional limiting wave is W_a, normally non-transverse and polarized parallel to the boundary.

The class condition and, in Case 1, the limitations (22) on α are the only restrictions applying to Cases 1 and 2. In Case 3, however, the angle β_o must exist and $\mathcal{S}_a$ must be sufficiently exposed at P in the sense that L has no other point of intersection with $\mathcal{S}$. In sum these extra requirements are severe.

7. Survey of the Existence of Surface Waves

We are now able to conclude from the existence theorem stated in section 1 that a unique subsonic surface wave can propagate in a semi-infinite transversely isotropic elastic body except when the conditions of one of Cases 1 to 3 are satisfied and $F(\hat{v}) \ge 0$. The existence of a subsonic surface wave is thus the usual situation and effective computational procedures are then available for evaluating the speed of propagation and other properties of the motion (see, for example, [2, sect.5]). For a complete understanding of the behaviour of surface waves in transversely isotropic elastic media it remains to calculate and study $F(v)$ for the exceptional configurations contained in Cases 1 to 3.

In the remainder of the paper we sketch the investigation of $F(v)$ for Cases 1 and 3. A full theoretical treatment of Case 2 has been given by CHADWICK and SMITH [9, sects VI,E, VIII,A,C, IX,E] and LOTHE and AL'SHITS [10].

8. Surface-Wave Behaviour: α Configurations

When $\underline{e}$ is given by $(21)_1$ the integral involved in the representation of $F(v)$ [1, eq.(3.3)] can be expressed as the sum of contributions associated with the parts $\mathcal{S}_b \cap R$ and $(\mathcal{S}_a \cup \mathcal{S}_3) \cap R$ of the slowness section in R. The term connected with $\mathcal{S}_b \cap R$ is found to vanish, implying, *via* $(15)_2$ and $(11)_2$, that $F(v)$ does not depend on a_2. When $\underline{e}$ assumes an α configuration, the displacement produced by a surface wave is everywhere parallel to R which is a zonal plane. The polarization of a plane body wave W_b with direction of propagation in R is normal to R, and this decoupling of plane and antiplane motions may be held responsible for the elliptical part $\mathcal{S}_b \cap R$ of the slowness section having no influence on $F(v)$ in an α configuration. In general, $F(v)$ takes a value between 0 and 1 at $v = 0$ and has at most one zero in the subsonic interval

(0, $\hat{v}$) [3, sects 3,4]. For all α configurations, $F(v) \to -\infty$ as $v \to v_a$ where v_a is defined by the following extension of the construction of the transonic state illustrated in Fig. 1.

Imagine a line in R, parallel to $\underline{k}$ and initially far from O, moving in the direction $-\underline{j}$. Then $1/v_a$ is the perpendicular distance of O from the line when it first becomes tangent to $\mathcal{S}_a$, and $1/v_b$ is similarly taken to be the perpendicular distance of O from the line when it is tangent to $\mathcal{S}_b$. The limiting speed $\hat{v}$ of the transonic state is the smaller of v_a and v_b : $\hat{v} = v_b$ when the conditions of Case 1 are satisfied and $\hat{v} = v_a$ (as in Fig. 1) for the other α configurations.

We deduce that $F(v)$ has a zero v_s in $(0,\hat{v})$, and hence that a unique subsonic surface wave exists with speed of propagation v_s, when Case 1 does not apply and also when Case 1 holds and $F(v_b) < 0$. When Case 1 applies and $F(v_b) > 0$, $F(v)$ has a zero larger than $\hat{v}$ and there then exists a *supersonic surface wave* [1, sect.3.4]. These conclusions are clearly in accord with the existence theorem in section 1.

Since C_{11} is by far the most heavily populated class of transversely isotropic elastic solids, Case 1 is of common occurrence in actual materials. As an example we consider the piezoelectric crystal cadmium telluride (CdTe) for which measured values of c_{11}, c_{33}, c_{12}, c_{13}, c_{44} are, in turn, 62.2, 68.9, 35.9, 29.1, 11.6 GPa [11, p.40] and ρ = 6200 kgm^{-3} [12, p.B-80]. The surface-wave speed v_s and the limiting speed $\hat{v}$ have been computed at 10^0 intervals in α and the results, extending over the full range of α configurations, are represented in Fig.2. For this material, Case 1 holds for $22.65^0 < \alpha < 90^0$ (see $(22)_1$). Between $\alpha = 0$ and $\alpha = \alpha_o = 22.65^0$, $\hat{v} = v_a$ and the surface wave is subsonic as shown above. When $\alpha = \alpha_o$, v_a and v_b are equal and the transonic state is of Type 4. Thereafter, as α increases, $\hat{v} = v_b$ and the surface wave is seen from Fig. 2 to be subsonic at first, then supersonic, then subsonic again, the variations of v_s and $\hat{v}$ with α being continuous. When $\alpha = \frac{1}{2}\pi$, v_a and v_b once more coincide, the transonic state now being of Type 2.

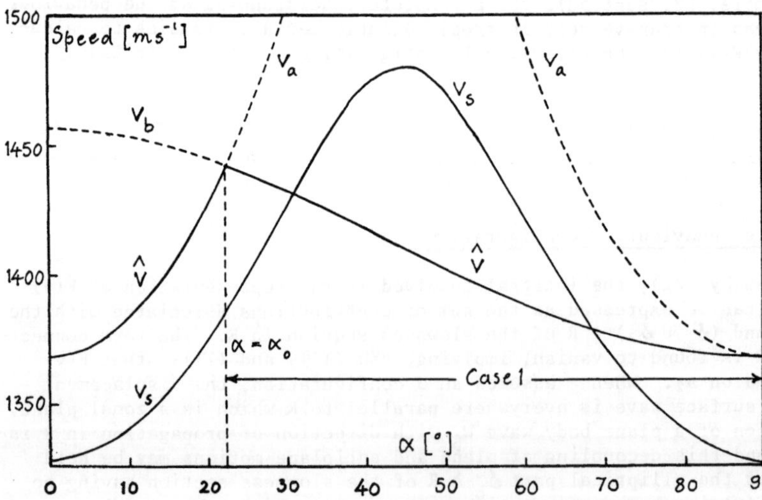

Fig. 2. Variation of v_s and $\hat{v}$ over the α configurations for Cd Te

The continuous transition between subsonic and supersonic surface-wave propagation typified by these results seems not to have been reported previously and, in particular, does not occur in media with cubic symmetry. It should be noted, however, that when an orientation of $\underline{e}$ belonging to Case 1 is approached through placements of the axis which are not α configurations, the progression from subsonic to supersonic propagation has the discontinuous character encountered in cubic materials [13, sect.V].

9. Surface-Wave Behaviour: β Configurations $(0 < \beta < \frac{1}{2}\pi)$

For the β, as for the α configurations, the integral representation of the tensor $\underline{S}_1(v)$ in terms of which $F(v)$ is defined can be evaluated in closed form. The contribution of the elliptical portion $\mathcal{S}_b \cap R$ of the slowness section is, however, no longer zero. R is now perpendicular to a zonal plane and the construction described in section 8 again yields speeds v_a and v_b relating to the parts $\mathcal{S}_a \cap R$ and $\mathcal{S}_b \cap R$ of $\mathcal{S} \cap R$. When the transonic state occurs on $\mathcal{S}_a \cap R$, $\hat{v} = v_a$ and it transpires that $F(v)$ has the limiting form

$$F(v) = B(\hat{v}^2 - v^2)^{-\frac{1}{2}} + C + D(\hat{v}^2 - v^2)^{\frac{1}{2}} + O(\hat{v}^2 - v^2) \tag{23}$$

as $v \to \hat{v}$.

When $\beta \neq \beta_o$, Case 3 does not apply and $B < 0$. From (23), $F(v) \to -\infty$ as $v \to \hat{v}$, implying, in agreement with the existence theorem, that a unique subsonic surface wave can propagate. When $\beta = \beta_o$ and $a_5 > 0$, B is zero. It is then seen from (23) that a unique subsonic surface wave continues to exist when $C < 0$, but not when $C \geq 0$. Due to the presence of the square root of $\hat{v}^2 - v^2$ in (23), $F(v)$ has no real continuation beyond $v = \hat{v}$, whence, in contrast to Case 1, no supersonic surface wave can propagate in Case 3: when $C \geq 0$ there is no surface wave at all.

One of the comparatively few known materials which meet all the conditions of Case 3 is the crystal titanium boride (TiB_2), belonging to class C_{22}. Measured values of c_{11}, c_{33}, c_{12}, c_{13}, c_{44} for TiB_2 are 690, 440, 410, 320, 250 GPa, respectively [11, p.45] and $\rho = 4500\ \mathrm{kgm}^{-3}$ [12, p.B-153]. The surface-wave and limiting speeds have been computed at 10^0 intervals in β and the results are plotted in Fig. 3. The outer sheet $\mathcal{S}_1$, coinciding with $\mathcal{S}_a$ for $0 < \beta < 62.35^0$, has axial concavities and the transonic state is consequently of Type 4 when $0 \leq \beta < \theta_a$. θ_a, equal to 30.56^0 for TiB_2, is the half-angle of the cone through O and the points of contact of the two tangent planes orthogonal to the axis which touch $\mathcal{S}_a$ in circles. For $\theta_a < \beta < \frac{1}{2}\pi$, the transonic state is of Type 1, but the passage from $\mathcal{S}_a \cap R$ to $\mathcal{S}_b \cap R$ at $\beta = \theta_{ab} = 62.35^0$ of the representative point of the limiting wave is marked by a kink in the $\hat{v}$ versus β curve. In fact,

$$\hat{v} = \begin{cases} [a_4 - \{(a_1^{\frac{1}{2}} + a_4^{\frac{1}{2}})^{-1}|a_5| - a_1^{\frac{1}{2}}(a_3 - a_4)(|a_5| + b)^{-1}\}^2]^{\frac{1}{2}}\cos\beta, & 0 \leq \beta \leq \theta_a, \\ c_a(\underline{j}), \quad \theta_a \leq \beta \leq \theta_{ab}, & \\ c_b(\underline{j}), \quad \theta_{ab} \leq \beta \leq \frac{1}{2}\pi & \end{cases} \tag{24}$$

(cf.[8]), where c_a and c_b are given by $(11)_{1,2}$ and (12) and b by $(18)_1$. The surface wave is subsonic when $\beta \neq \beta_o$, but v_s is indistinguishable from $\hat{v}$ in Fig. 3 over an interval of about 15° surrounding $\beta_o = 43.14^0$. The v_s and $\hat{v}$ curves touch at $\beta = \beta_o$ and the subsonic surface wave degenerates there into an exceptional wave. The variation of $F(v)$ with v at $\beta = \beta_o$ takes the form of a gradual decrease from $v = 0$ to $v = 0.996\hat{v}$. The curve then makes a rapid upturn terminating at $v = \hat{v}$ with positively infinite gradient.

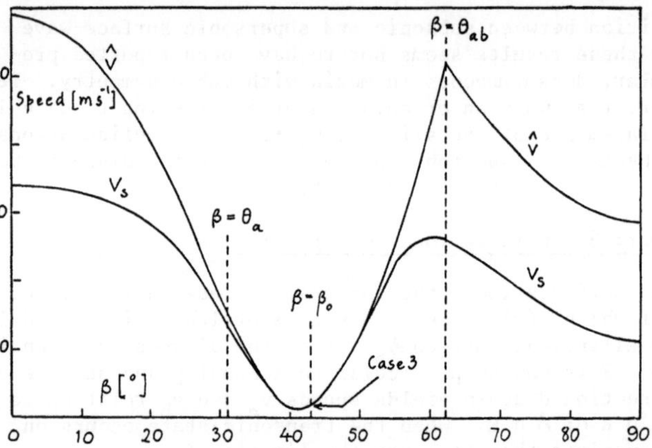

Fig. 3. Variation of v_s and $\hat{v}$ over the β configurations for TiB_2

The existence, under "fairly strict" conditions, of isolated directions of propagation in a zonal plane in which a subsonic surface wave collapses into an exceptional wave seems to have first been established by AL'SHITS and LOTHE [8]. A detailed analysis of the effect has not been given previously and this is the first case to have come to light of the non-existence of a subsonic surface wave not being accompanied by the appearance of a supersonic surface wave.

Acknowledgement

The computations on which Figs 2 and 3 are based were carried out by Dr. P. B. Heppell.

References

1. P. Chadwick, G. D. Smith: In Mechanics of Solids. The Rodney Hill 60th Anniversary Volume, ed. by H. G. Hopkins, M. J. Sewell, pp. 47 - 100 (Pergamon Press, Oxford, etc. 1982)
2. D. M. Barnett, J. Lothe: Proc. R. Soc. Lond. A 402, 135 - 152, 1985
3. P. Chadwick, T. C. T. Ting: Quart. Appl. Math. 45, 419 - 427, 1987
4. P. Chadwick: Proc. R. Soc. Lond. A 401, 203 - 223, 1985
5. P. B. Heppell: Wave Propagation in Hexagonal Elastic Media, Ph.D. thesis, University of East Anglia, Norwich 1986
6. F. I. Fedorov: The Theory of Elastic Waves in Crystals (Plenum Press, New York 1968)
7. G. F. D. Duff: Phil. Trans. R. Soc. Lond. A 252, 249 - 273, 1960
8. V. I. Al'Shits, J. Lothe: Soviet Phys. Crystallogr. 23, 509 - 515, 1978
9. P. Chadwick, G. D. Smith: In Advances in Applied Mechanics, ed. by C. -S. Yih, Vol. 17, pp. 303 - 376 (Academic Press, New York, etc. 1977)
10. J. Lothe, V. I. Al'Shits: Soviet Phys. Crystallogr. 22, 519 - 525, 1977
11. H. H. Landolt, R. Börnstein: Zahlenwerte und Funktionen aus Naturwissenschaften und Technik. Neue Serie, ed. by K. H. Hellwege, Vol. III/11 (Springer, Berlin 1979)
12. Handbook of Chemistry and Physics, 56th ed., ed. by R. C. Weast (CRC Press, Cleveland 1975)
13. G. W. Farnell: In Physical Acoustics, ed. by W. P. Mason and R. N. Thurston, Vol. V1, pp. 109 - 166 (Academic Press, New York, London 1970)

A New Method for Surface Impedance and Surface Wave Calculations

S.A. Gundersen and J. Lothe

Department of Physics, University of Oslo,
P.O. Box 1048, N-0316 Oslo, Norway

Abstract: In the integral formalism for surface waves in elastically anisotropic media as developed by Barnett and Lothe and by Chadwick and Smith, a certain 6×6 matrix $\overline{N}$ plays a central role. An iteration method for calculation of $\overline{N}$ is presented. The efficiency of the method is illustrated by a sample calculation.

1 Introduction

BARNETT and LOTHE [1] have developed an integral theory on the basis of STROH's [2] sextic theory for dislocations and surface waves in anisotropic elastic media. As shown by GUNDERSEN and LOTHE [3] the central matrix $\overline{N}$ occurring in the integral theory can conveniently be calculated by an iteration scheme in the static case. In the present paper we will consider application of the method to dynamic problems and in particular to subsonic surface waves.

2 General Theory

Consider partial waves of the form

$$\vec{u} = \vec{A}_\alpha e^{ik(\vec{m}_0\cdot\vec{x}+p_\alpha\vec{n}_0\cdot\vec{x}-vt)} \tag{1}$$

travelling with velocity

$$\vec{v} = v\vec{m}_0 \tag{2}$$

along a surface with inward unit normal $\vec{n}_0$. Surface waves satisfying the condition of a free surface will have to be built up from partial solutions for which $\mathrm{Im}p_\alpha > 0$.

The partial waves (1) must satisfy the condition of dynamic equilibrium,

$$C'_{ijkl}\frac{\partial^2 u_k}{\partial x_i \partial x_l} = 0 \tag{3}$$

where

$$C'_{ijkl} = C_{ijkl} - \rho v_i \delta_{jk} v_l \tag{4}$$

are the dynamic elastic coefficients. C_{ijkl} are the static elastic coefficients, and ρ is mass density.

The requirement that (1) satisfy (3) leads, according to Stroh, to the eigenvalue problem

$$N_0 \xi_\alpha = p_\alpha \xi_\alpha \tag{5}$$

where

$$N_0 = -\begin{pmatrix} (n_0 n_0)^{-1}(n_0 m_0) & (n_0 n_0)^{-1} \\ (m_0 n_0)(n_0 n_0)^{-1}(n_0 m_0) - (m_0 m_0) & (m_0 n_0)(n_0 n_0)^{-1} \end{pmatrix}. \tag{6}$$

Here (ab) is an abbreviation for a matrix associated with two vectors $\vec{a}$ and $\vec{b}$ by the prescription that

$$(ab)_{jk} = a_i C'_{ijkl} b_l. \tag{7}$$

ξ_α is a six-dimensional vector

$$\xi_\alpha = \begin{pmatrix} \vec{A}_\alpha \\ \vec{L}_\alpha \end{pmatrix} \tag{8}$$

where $\vec{L}_\alpha$ is related to $\vec{A}_\alpha$ as

$$\vec{L}_\alpha = -((n_0 m_0) + p_\alpha (n_0 n_0))\vec{A}_\alpha. \tag{9}$$

From this formulation Stroh derives the orthogonality relation

$$\xi_\alpha T \xi_\beta = 0 \quad \text{when } p_\alpha \neq p_\beta \tag{10}$$

where in block notation

$$T = \begin{pmatrix} 0 & 1 \\ 1 & 0 \end{pmatrix}. \tag{11}$$

In more detail, (10) is

$$\vec{A}_\alpha \cdot \vec{L}_\beta + \vec{A}_\beta \cdot \vec{L}_\alpha = 0 \quad \text{when } p_\alpha \neq p_\beta. \tag{12}$$

We will consider eigenvectors ξ_α to be normalized so that

$$\xi_\alpha T \xi_\alpha = 1 \quad (\text{no summation over } \alpha). \tag{13}$$

Consider now a set of orthogonal unit vectors $(\vec{m}, \vec{n})$ tilted relative to $(\vec{m}_0, \vec{n}_0)$ by φ (fig. 1). Form an operator $N(\varphi)$ in terms of $(\vec{m}, \vec{n})$ in the same way as N_0 was formed from $(\vec{m}_0, \vec{n}_0)$ by the prescription equations (6) and (7). $N(\varphi)$ will depend on $\vec{m}_0$ only implicitly through C'_{ijkl}, eqs. (4) and (2).

As shown by Barnett and Lothe, $N(\varphi)$ obeys the differential equation

$$\frac{\partial N}{\partial \varphi} = -(1 + N^2), \tag{14}$$

which can be formally integrated to

$$N(\varphi) = \frac{N_0 - \tan\varphi}{1 + N_0 \tan\varphi}. \tag{15}$$

That is, $N(\varphi)$ is a function of N_0 and therefore has the *same* eigenvectors ξ_α,

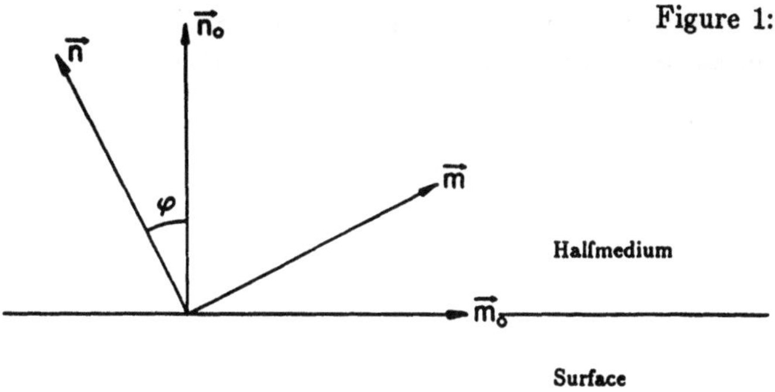

Figure 1: Surface geometry.

$$N(\varphi)\xi_\alpha = p_\alpha(\varphi)\xi_\alpha. \tag{16}$$

The eigenvalues depend on φ according to

$$p_\alpha(\varphi) = \tan(\psi_\alpha - \varphi) \tag{17}$$

where ψ_α is a complex angle. The form (17) follows from (15).

Since

$$\overline{p}_\alpha = \frac{1}{2\pi}\int_0^{2\pi} p_\alpha(\varphi)d\varphi = \pm i \tag{18}$$

according to whether $\mathrm{Im} p_\alpha \gtrless 0$, and since ξ_α is invariant to φ, it follows from (16) that

$$\overline{N}\xi_\alpha = \pm i\xi_\alpha \tag{19}$$

where

$$\overline{N} = \frac{1}{2\pi}\int_0^{2\pi} N(\varphi)d\varphi. \tag{20}$$

This is the integral expression for the important matrix $\overline{N}$, which, assuming that the ξ_α form a complete set, must satisfy the equation

$$\overline{N}^2 + 1 = 0 \tag{21}$$

according to (19).

The sextic expression for $\overline{N}$ is

$$\overline{N} = \pm i\xi_\alpha \otimes (\xi_\alpha T) \tag{22}$$

where $\otimes$ denotes dyadic multiplication. (22) is just the average of

$$N(\varphi) = p_\alpha(\varphi)\xi_\alpha \otimes (\xi_\alpha T). \tag{23}$$

The block structure of $\overline{N}$ is

$$\overline{N} = \begin{pmatrix} S & Q \\ B & S^T \end{pmatrix} \tag{24}$$

where Q and B are symmetric matrices. S^T is the transpose of S.

In terms of these blocks one may construct a surface impedance Z

$$Z = -(Q^{-1} + iQ^{-1}S), \tag{25}$$

which relates the surface traction amplitude $\vec{T}$ and the displacement amplitude $\vec{A}$ in the surface,

$$\vec{T} = -kZ\vec{A} \tag{26}$$

(INGEBRIGTSEN and TONNING [4], LOTHE and BARNETT [5], BARNETT and LOTHE [6]).

In terms of Z, the condition for a free surface wave of velocity v_S is

$$\|Z\| = 0 \quad \text{when } v = v_S, \tag{27}$$

i.e. nonzero displacement amplitude $\vec{A}$ for a free surface, $\vec{T} = 0$. $\| \ \|$ means determinant.

Alternatively, and equivalently (MALÉN and LOTHE [7], BARNETT and LOTHE [8], CHADWICK and SMITH [9]), the condition determining v_S can be stated as

$$\|B\| = 0 \quad \text{when } v = v_S. \tag{28}$$

Actually, as is well known, a zero of B will be a double zero so that

$$\|B^m_{(ij)}\| = 0 \quad \text{when } v = v_S \tag{29}$$

where $B^m_{(ij)}$ is the ij-th minor obtained by deleting the i-th column and the j-th row.

Either criterion requires calculation of $\overline{N}$ as function of v. If one knows $\overline{N}(v)$ the impedance $Z(v)$ is readily found from (25).

3 The Iteration Method

The central question is calculation of $\overline{N}(v)$, equation (20). Since from (15), $N(\varphi + \pi) = N(\varphi)$, a first approximation to the integral for $\overline{N}$ would be

$$N_1 = \frac{1}{2}(N_0 + N(\frac{\pi}{2})). \tag{30}$$

However, according to (15),

$$N(\frac{\pi}{2}) = -N_0^{-1}$$

so that (30) can be written as

$$N_1 = \frac{1}{2}(N_0 - N_0^{-1}).$$

Next form

$$\begin{aligned} N_2 &= \frac{1}{2}(N_1 - N_1^{-1}) \\ N_3 &= \frac{1}{2}(N_2 - N_2^{-1}) \\ &\text{etc.} \end{aligned} \tag{31}$$

If this iteration sequence gives a well defined limit $\overline{N}$, $\overline{N}$ must satisfy

$$\overline{N} = \frac{1}{2}(\overline{N} - \overline{N}^{-1}) \tag{32}$$

i.e.

$$\overline{N}^2 + 1 = 0$$

in agreement with (21).

It can be proved that when all p_α's have a nonzero imaginary part the above iteration scheme gives convergence to the sought $\overline{N}$ (GUNDERSEN and LOTHE [3]). A proof is easily constructed on the basis of (23) and (22).

At the limiting velocity a pair of complex conjugate roots coalesce into one doubly degenerate real root, and the method then breaks down. However, the imaginary parts of the subsonic roots corresponding to the real root at the limiting velocity $\hat{v}$ increase rapidly with decreasing velocity, and the iteration method should be practical up to the velocities very close to the limiting velocity and well beyond the Rayleigh velocity in typical cases.

4 A Numerical Test of the Iteration Method

We consider Rayleigh waves on the (001) plane in the [100] direction in diamond. This case has been considered before by BARNETT, LOTHE, NISHIOKA and ASARO [10], who calculated the Rayleigh velocity to be

$$v_R = 1.0970 \times 10^4 \text{ms}^{-1}. \tag{33}$$

The limiting velocity, not given by the above authors, is readily calculated to be

$$\hat{v} = 1.2802 \times 10^4 \text{ms}^{-1}. \tag{34}$$

Here we use the iteration method to calculate $\overline{N}$, from which we calculate $B^m_{(zx)}$, (x-axis along [100], y-axis along [001], z-axis along $[0\bar{1}0]$). The Rayleigh velocity is determined from criterion (29). The constants used are the same as used by Barnett et al. : $\rho = 3.512 \times 10^3 \text{kg m}^{-3}$, and in units of $10^{11} N\text{m}^{-2}$ we have $c_{11} = 10.76$, $c_{12} = 1.250$, and $c_{44} = 5.758$.

Figures 2 and 3 present results as a function of the number of iterations used. It is seen that already after 4 or 5 iterations an accurate result in agreement with the previously obtained result (33) is obtained.

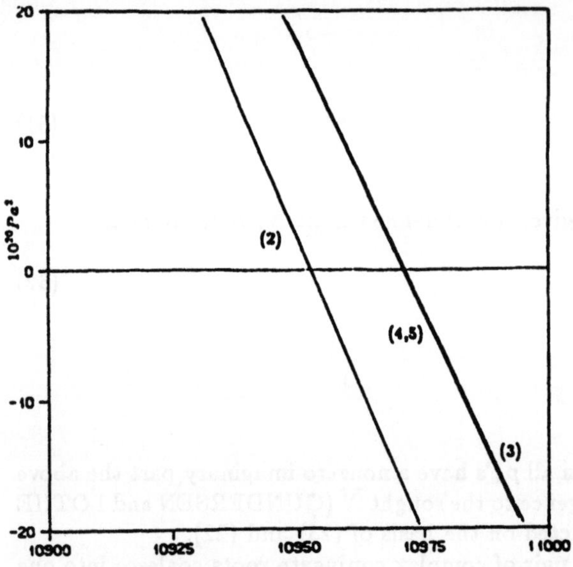

Figure 2: The subdeterminant in the vicinity of the Rayleigh velocity in the [100] direction as a function of velocity for 2,3,4, and 5 iterations

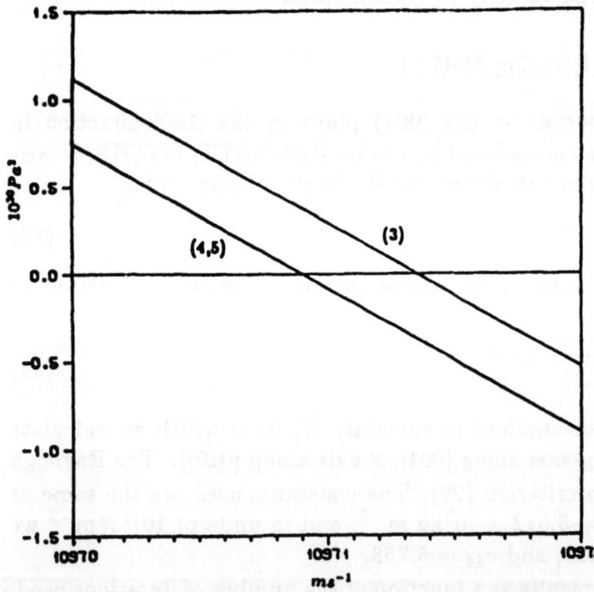

Figure 3: The subdeterminant in the vicinity of the Rayleigh velocity in the [100] direction as a function of velocity for 3,4, and 5 iterations

With 5 iterations results were accurate to 3 figures up to $v = 12420 ms^{-1}$. With 6 iterations results were accurate to 3 figures to within $100 ms^{-1}$ of the limiting velocity.

5 Concluding Remarks

From the standpoint of accuracy to say 4 places, the iteration method and the sextic method are about equally time consuming as regard computer time. In high accuracy work the iteration scheme is superior: A few more iterations greatly increases the accuracy.

Perhaps the greatest advantage of the iteration scheme is economy with respect to human time. Expressions and programming procedures are extremely simple. This aspect becomes even more apparent when one considers derivatives of $\overline{N}$, say $\frac{\partial \overline{N}}{\partial v}$ and $\frac{\partial \overline{N}}{\partial \theta}$, which are required in group velocity calculations. A change in velocity or direction gives a change in the input N_0

$$N_0 \rightarrow N_0 + \delta N_0 \tag{35}$$

where $\delta N_0 = \frac{\partial N_0}{\partial v}\delta v$ or $\delta N_0 = \frac{\partial N_0}{\partial \theta}\delta\theta$. The change δN_0 propagates with a change in δN_1,

$$\delta N_1 = \frac{1}{2}(\delta N_0 + N_0^{-1}\delta N_0 N_0^{-1}), \tag{36}$$

which again gives a change in δN_2,

$$\delta N_2 = \frac{1}{2}(N_1 + N_1^{-1}\delta N_1 N_1^{-1}) \tag{37}$$

etc.

By this iteration scheme, carried out in *parallel* with the iterations that give $\overline{N}$ itself, $\delta\overline{N}$ is also obtained.

Acknowledgement: Research support from STATOIL is acknowledged.

References

1. D. M. Barnett and J. Lothe, *Physica Norvegica*, **3**, 13 (1973)
2. A. N. Stroh, *J.Math.&Phys.* **41**, 77 (1962)
3. S. A. Gundersen and J. Lothe, *phys.stat.sol.(b)*, **143**, 73 (1987)
4. K. A. Ingebrigtsen and A. Tonning, *Phys.Rev.* **184**, 942 (1969)
5. J. Lothe and D. M. Barnett, *Journ.App.Phys.* **47**, 428 (1976)
6. D. M. Barnett and J. Lothe, *Proc.Roy.Soc.Lond.A*, **402**, 135 (1985)
7. K. Malén and J. Lothe, *phys.stat.sol.(b)* **43**, K139 (1971)
8. D. M. Barnett and J. Lothe, *J.Phys.F*, **4**, 671 (1974)
9. P. Chadwick and G. D. Smith, *Advances in App. Mech.* **17**, 303 (1977)
10. D. M. Barnett, J. Lothe, K. Nishioka and R. J. Asaro, *J.Phys.F*, **3**, 1083 (1973)

Part III

Surface Waves on Rough Surfaces

Surface Acoustic Waves on Rough Surfaces

A.A. Maradudin

Department of Physics, University of California,
Irvine, CA 92717, USA

A survey is given of recent work on acoustic excitations localized in the vicinity of a rough surface. The rough surfaces considered, which would all be planar in the absence of the roughness, are of three kinds: (a) a planar surface with an isolated protuberance on it; (b) a periodically corrugated surface; and (c) a randomly rough surface. The first kind of roughness gives rise to surface shape resonances, i.e. to modes that are spatially localized in the vicinity of the surface perturbation and that are characterized by discrete frequencies. Waves localized to the surface can propagate across surfaces of the second and third kinds. The effects of the roughness on their dispersion curves are described. Experimental results bearing on the theoretical results obtained are presented.

1. Introduction

In this paper we study acoustic excitations that are spatially localized in the vicinity of three types of rough surface on an elastic medium, each of which would be planar in the absence of the roughness. These surfaces are all defined by an equation for the position of the surface of the form $x_3 = \zeta(x_1,x_2)$. They differ in the nature of the surface profile function $\zeta(x_1,x_2)$. For the first type of rough surface $\zeta(x_1,x_2)$ is deterministic and is nonzero only for a finite range of values of x_1 and is independent of x_2 (an isolated ridge on an otherwise planar surface), or is nonzero only for a limited range of values of both x_1 and x_2 (an isolated protuberance on an otherwise planar surface.) For the second type of rough surface $\zeta(x_1,x_2)$ is again deterministic, but is a periodic function of x_1 and independent of x_2 (a classical grating), or is a periodic function of both x_1 and x_2 (a bigrating). For the third type of rough surface $\zeta(x_1,x_2)$ is a stochastic function of x_1 and x_2, or of x_1 alone, and describes a randomly rough surface.

The first type of roughness, by destroying the infinitesimal translational invariance of the planar surface on which it is found, gives rise to acoustic surface shape resonances. These are vibrational modes that are spatially localized in the vicinity of the surface perturbation and that are characterized by discrete frequencies that depend on the shape of that perturbation. Waves localized to the surface can propagate across surfaces of the second and third kinds. Because rough surfaces can support nonpropagating, acoustic surface shape resonances, as well as propagating surface acoustic waves, we will often use the more capacious term, surface acoustic excitations, in referring to the full range of vibrational modes associated with a rough surface.

There are several reasons why the study of acoustic excitations localized to rough surfaces is of interest. All real surfaces, even the most carefully prepared ones, are rough to some degree. It is therefore at least interesting, and often important, to know how the properties of surface acoustic excitations are modified by that roughness. A second reason for studying surface acoustic excitations on rough surfaces is that surface roughness gives rise to new types of surface acoustic waves that cannot exist in the absence of the roughness. We will see examples of both of these consequences of surface roughness on surface acoustic excitations in what follows.

2. Acoustic Surface Shape Resonances

A rough surface consisting of an otherwise planar surface on a semi-infinite solid on which an isolated protuberance, fabricated from the same or a different material as the substrate, is placed can give rise to surface shape resonances. These are excitations that are spatially localized in the vicinity of the protuberance. Because of the destruction of infinitesimal translational invariance in directions parallel to the nominal planar surface of the solid caused by the presence of the surface perturbation, the frequencies of surface shape resonances are discrete (there is no continuously varying quantum number - a wave number - on which these frequencies can depend), and depend on the shape of the protuberance. Because these frequencies fall in the frequency range allowed the normal modes of the solid in the absence of the protuberance, the surface shape resonances can decay into the latter modes, and acquire a finite lifetime thereby. Put another way, the frequencies of surface shape resonances are complex, where the imaginary part of the frequency is the reciprocal of twice the lifetime of the energy in the mode.

At the present time the only surface shape resonances to have been studied theoretically are electrostatic [1-7], electromagnetic [6,7], and magnetostatic [8] surface shape resonances. In this section we describe some first steps in extending such studies to the case of acoustic surface shape resonances. Interest in acoustic surface shape resonances stems from the fact that their excitation, e.g. by thermal fluctuations, or by surface acoustic waves impinging on the protuberance supporting these excitations, can lead to stress concentrations in the vicinity of the protuberance that in turn can affect physical processes occurring at a solid surface, such as chemical reactions [9,10]. In addition, some rough surfaces, whether periodically or randomly corrugated, can be regarded as an array of protuberances on an otherwise planar surface. Knowledge of the excitations associated with an isolated protuberance could be valuable in understanding the excitations associated with such a rough surface.

We consider what is perhaps the simplest system that can support acoustic surface shape resonances [11]. This is a rectangular ridge of height d and width 2ℓ, oriented parallel to the x_2-axis, that is fabricated from a material of mass density $\rho^{(1)}$ and elastic moduli $\{c_{ij}^{(1)}\}$. It is bonded to a semi-infinite substrate fabricated from a material of mass density $\rho^{(2)}$ and elastic moduli $\{c_{ij}^{(2)}\}$. Each of these materials is assumed to possess cubic symmetry, with the cube axes aligned parallel to the coordinate axes.

The displacement field in this system is assumed to have the form

$$\vec{u}(\vec{x};t) = \left(0,u_2(x_1,x_3|\omega),0\right)\exp(-i\omega t), \tag{2.1}$$

where ω is the frequency of the modes we seek. The amplitude $u_2(x_1,x_3|\omega)$ inside the ridge ($-\ell < x_1 < \ell$, $0 < x_3 < d$) can be expressed in the form

$$u_2^{(1)}(x_1,x_3|\omega) = \sum_{n=0}^{\infty} \varepsilon_n B_n \cos\frac{n\pi}{2\ell}(x_1-\ell)\cos\beta_n(\omega)(x_3-d), \qquad (2.2)$$

where

$$\beta_n(\omega) = \left(\frac{\omega^2}{c_1^2} - \left(\frac{n\pi}{2\ell}\right)^2\right)^{1/2}, \qquad \mathrm{Re}\beta_n(\omega) > 0,\ \mathrm{Im}\beta_n(\omega) > 0, \qquad (2.3)$$

with $c_1 = \left(c_{44}^{(1)}/\rho^{(1)}\right)^{1/2}$. The quantity ε_n equals $\frac{1}{2}$ when $n = 0$ and 1 when $n > 0$. The displacement field defined by Eq. (2.2) satisfies stress-free boundary conditions on the top and on the vertical faces of the ridge. In the substrate ($-\infty < x_1 < \infty$, $x_3 < 0$) $u_2(x_1,x_3|\omega)$ takes the form

$$u_2^{(2)}(x_1,x_3|\omega) = \int \frac{dk}{2\pi} A(k\omega)\, e^{ikx_1+\alpha(k\omega)x_3}, \qquad (2.4)$$

where

$$\alpha(k\omega) = \left(k^2 - \frac{\omega^2}{c_2^2}\right)^{1/2}, \qquad \mathrm{Re}\alpha(k\omega) > 0,\ \mathrm{Im}\alpha(k\omega) < 0, \qquad (2.5)$$

with $c_2 = \left(c_{44}^{(2)}/\rho^{(2)}\right)^{1/2}$. The expression (2.4) satisfies vanishing or outgoing wave boundary conditions at $x_3 = -\infty$.

The coefficients $\{B_n\}$ and $A(k\omega)$ are determined by the boundary conditions at the plane $x_3 = 0$. The vanishing of the stress along the portion of this plane defined by $|x_1| > \ell$ and its continuity along the portion $|x_1| < \ell$ yields the equation

$$A(k\omega) = \ell\,\frac{c_{44}^{(1)}}{c_{44}^{(2)}} \sum_{n=0}^{\infty} \varepsilon_n \beta_n(\omega)\sin(\beta_n(\omega)d)\,\frac{S_n(k)}{\alpha(k\omega)}\,B_n. \qquad (2.6)$$

The continuity of the displacement across this plane along the portion $|x_1| < \ell$ leads to

$$\cos(\beta_m(\omega)d)B_m = \int \frac{dk}{2\pi} S_m^*(k)A(k\omega). \qquad (2.7)$$

The function $S_n(k)$ appearing in Eqs. (2.6) and (2.7) is defined by

$$S_n(k) = \frac{1}{\ell}\int_{-\ell}^{\ell} dx_1\, e^{-ikx_1} \cos\frac{n\pi}{2\ell}(x_1-\ell). \qquad (2.8)$$

When we eliminate $A(k\omega)$ from Eqs. (2.6) and (2.7), we obtain the dispersion relation for the frequencies of surface shape resonances in the form

$$D(\omega) = 0, \tag{2.9}$$

where

$$D(\omega) = \det\left\{\cos(\beta_m(\omega)d)\delta_{mn} - \frac{\ell}{d}\frac{c_{44}^{(1)}}{c_{44}^{(2)}} M_{mn}(\omega) \times \varepsilon_n(\beta_n(\omega)d)\sin(\beta_n(\omega)d)\right\}, \tag{2.10a}$$

with

$$M_{mn}(\omega) = \int \frac{dk}{2\pi}\frac{S_m^*(k)S_n(k)}{\alpha(k\omega)} . \tag{2.10b}$$

In practice the infinite determinant defined by Eq. (2.10) is made into an $N \times N$ determinant by restricting m and n to the values 0,1,2,...,N-1. N is then increased until convergence of the frequencies is achieved.

It can be shown that $M_{mn}(\omega)$ vanishes unless m and n have the same parity. This has the consequence that the determinant $D(\omega)$ factors into the product of two determinants, $D(\omega) = D_+(\omega)D_-(\omega)$, where $D_+(\omega)$ is obtained by restricting m and n to even values in Eq. (2.10a), while $D_-(\omega)$ is obtained by restricting m and n to odd values in that equation. The zeros of $D_+(\omega)$ are the frequencies of the surface shape resonances whose displacement fields are even functions of x_1; similarly, the zeros of $D_-(\omega)$ are the frequencies of surface shape resonances whose displacement fields are odd functions of x_1.

From Eqs. (2.5) we see that $M_{mn}(\omega)$ is complex even for real values of ω. This is due to the radiation of energy by these modes into the interior of the substrate represented by the pure imaginary value of $\alpha(k\omega)$ in the radiative region $|k| < \omega/c_2$. Consequently, the solutions of Eq. (2.9) are complex. To obtain them while calculating $D(\omega)$ for real values of ω only, we can proceed as follows. If we introduce the dimensionless frequency $\Omega = \omega\ell/c_2$, and define $\hat{D}_\pm(\Omega) = D_\pm(\omega)$, then in the vicinity of the complex zero $\Omega_R^\pm - i\Omega_I^\pm$ of $\hat{D}_\pm(\Omega) = 0$ with $\Omega_R > 0$, $\Omega_I > 0$ (where the sign of the imaginary part of the solution is determined by considerations of energy conservation) we have the relation

$$\frac{1}{|\hat{D}_\pm(\Omega)|^2} \cong \frac{A}{(\Omega-\Omega_R^\pm)^2+(\Omega_I^\pm)^2} , \tag{2.11}$$

for Ω real, where A is an irrelevant amplitude. Consequently, a plot of $|\hat{D}_\pm(\Omega)|^{-2}$ as a function of real Ω will have a peak at $\Omega_R^\pm$ and a full width at half-maximum of $2\Omega_I^\pm$, for each acoustic surface shape resonance, which allows $\Omega_R^\pm$ and $\Omega_I^\pm$ to be obtained.

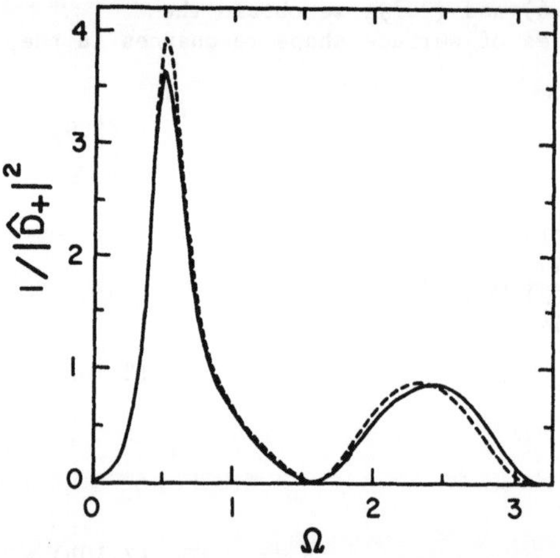

Fig. 1. A plot of $|\hat{D}_+(\Omega)|^{-2}$ vs Ω, where Ω is the dimensionless frequency $\Omega = \omega\ell/c_2$ for the case in which $\rho^{(1)} = \rho^{(2)}$ and $c_{ij}^{(1)} = c_{ij}^{(2)}$. The dashed curve is obtained from the approximation of $\hat{D}_+(\Omega)$ by a 1x1 determinant; the solid curve is obtained from the approximation of $\hat{D}_+(\Omega)$ by a 5 x 5 determinant. The peaks at $\Omega \cong 0.5$ and 2.4 correspond to the two lowest frequency acoustic surface shape resonances whose displacement fields are even functions of x_1 [Ref. 11].

A plot of $|\hat{D}_+(\Omega)|^{-2}$ vs Ω is shown in Fig. 1 for the case in which $\rho^{(1)} = \rho^{(2)}$, $c_{ij}^{(1)} = c_{ij}^{(2)}$, and $d = 2\ell$.[12] In obtaining this plot two approximations to $\hat{D}_+(\Omega)$ were used, viz. the one obtained by setting m = n = 0 in Eq. (2.10a) and the one obtained by setting m, n = 0,2,4,6,8 in that equation. The results are seen not to differ greatly. The peaks corresponding to the lowest frequency acoustic surface shape resonances of even symmetry (in x_1) are shown. Their frequencies are $\Omega_1 = 0.51-i0.15$ and $\Omega_2 = 2.36-i0.44$, respectively, so that they are rather strongly damped excitations. For typical values of c_2, e.g. 3×10^5 cm/sec, values of ℓ of several tens of micrometers yield frequencies of surface shape resonances in the range of 10^8 Hz.

3. Surface Acoustic Waves at Periodically Corrugated Surfaces and Interfaces

The propagation of surface acoustic waves across a grating has been studied by a number of authors. The reasons for such studies are several.

(1) A grating is a model for a randomly rough surface that can be studied exactly. The results obtained for a grating surface help to interpret those for a randomly rough surface.

(2) As a surface acoustic wave propagates across a grating surface a portion of the energy it carries can be converted by the surface corrugations into bulk waves, that radiate that energy into the interior of the medium on whose surface the grating is ruled. The results of theoretical studies of surface acoustic waves on gratings

are therefore useful in the design of transducers for the conversion of surface acoustic waves into bulk waves.

(3) A characteristic feature of the propagation of waves through periodic structures is the occurrence of gaps in the dispersion curves of these waves. These gaps define stop bands for wave propagation and the complementary pass bands. Knowledge of the positions and widths of these stop bands can be used in the design of filters for surface acoustic waves.

(4) Surface acoustic waves on gratings display the phenomenon of wave-slowing. That is, the phase (and group) velocity of a surface acoustic wave on a grating surface is slower than it is for the same wave on a planar surface.

(5) Grating surfaces can support surface acoustic waves that cannot exist in the absence of the surface corrugations.

We consider several aspects of the propagation of surface acoustic waves across a grating in this section. We assume an isotropic elastic medium that occupies the region $x_3 > \zeta(x_1)$, where the surface profile function $\zeta(x_1)$ is a single-valued function of x_1 that is periodic with period a: $\zeta(x_1+a) = \zeta(x_1)$. The unit vector normal to the surface at each point, and directed away from the solid, is given by

$$n = [1 + (\zeta'(x_1))^2]^{-1/2} (\zeta'(x_1),0,-1). \tag{3.1}$$

We will consider surface acoustic waves propagating in the x_1-direction. The sagittal plane in this case (viz. the plane defined by the direction of propagation and the normal to the surface), is then the x_1x_3-plane. It is convenient to study the propagation of surface acoustic waves of sagittal and shear horizontal polarization across such a surface separately, and we take up these cases in turn in the remainder of this section.

3.1. Sagittal Polarization [14-32]

The displacement field of a sagittally polarized surface acoustic wave propagating in the x_1-direction across the grating surface defined in the preceding subsection has the form

$$\vec{u}(\vec{x};t) = (u_1(x_1x_3|\omega),\ 0,\ u_3(x_1x_3|\omega))\exp(-i\omega t) \tag{3.2}$$

in the region $x_3 > \zeta(x_1)$. The amplitudes $u_{1,3}(x_1x_3|\omega)$ satisfy a pair of coupled equations of motion in this region:

$$-\omega^2 u_1 = \left(c_\ell^2 \frac{\partial^2}{\partial x_1^2} + c_t^2 \frac{\partial^2}{\partial x_3^2}\right)u_1 + \left(c_\ell^2 - c_t^2\right) \frac{\partial^2 u_3}{\partial x_1 \partial x_3} \tag{3.3a}$$

$$-\omega^2 u_3 = \left(c_\ell^2 - c_t^2\right) \frac{\partial^2 u_1}{\partial x_1 \partial x_3} + \left(c_t^2 \frac{\partial^2}{\partial x_1^2} + c_\ell^2 \frac{\partial^2}{\partial x_3^2}\right)u_3. \tag{3.3b}$$

The conditions that ensure that the surface $x_3 = \zeta(x_1)$ is stress-free take the forms

$$\{[-c_\ell^2\zeta'(x_1)\frac{\partial}{\partial x_1} + c_t^2\frac{\partial}{\partial x_3}]u_1$$

$$+ [-(c_\ell^2-2c_t^2)\zeta'(x_1)\frac{\partial}{\partial x_3} + c_t^2\frac{\partial}{\partial x_1}]u_3\}_{x_3=\zeta(x_1)} = 0 \tag{3.4a}$$

$$\{[-c_t^2\zeta'(x_1)\frac{\partial}{\partial x_3} + (c_\ell^2-2c_t^2)\frac{\partial}{\partial x_1}]u_1$$

$$+ [-c_t^2\zeta'(x_1)\frac{\partial}{\partial x_1} + c_\ell^2\frac{\partial}{\partial x_3}]u_3\}_{x_3=\zeta(x_1)} = 0. \tag{3.4b}$$

We also require that $u_{1,3}(x_1x_3|\omega)$ vanish as $x_3 \to \infty$. The solutions of Eqs. (3.3) that satisfy both the boundary conditions at infinity and the Bloch-Floquet condition $u_\alpha(x_1+a,x_3|\omega) = \exp(ika)u_\alpha(x_1x_3|\omega)$ $(\alpha=1,3)$, where k is the wave number of the wave, can be written in the form

$$u_1(x_1x_3|\omega) = \sum_{m=-\infty}^{\infty} e^{ik_mx_1}[A_m^{(\ell)}e^{-\alpha_\ell(k_m\omega)x_3}$$

$$+ A_m^{(t)}e^{-\alpha_t(k_m\omega)x_3}], \quad x_3 > \zeta(x_1)_{max} \tag{3.5a}$$

$$u_3(x_1x_3|\omega) = \sum_{m=-\infty}^{\infty} e^{ik_mx_1} i[\frac{\alpha_\ell(k_m\omega)}{k_m}A_m^{(\ell)}e^{-\alpha_\ell(k_m\omega)x_3} +$$

$$+ \frac{k_m}{\alpha_t(k_m\omega)}A_m^{(t)}e^{-\alpha_t(k_m\omega)x_3}], \quad x_3 > \zeta(x_1)_{max}, \tag{3.5b}$$

where we have defined $k_m = k + (2\pi m/a)$, while

$$\alpha_{\ell,t}(k_m\omega) = (k_m^2 - \frac{\omega^2}{c_{\ell,t}^2})^{1/2}, \qquad k_m^2 > \frac{\omega^2}{c_{\ell,t}^2} \tag{3.6a}$$

$$= -i(\frac{\omega^2}{c_{\ell,t}^2} - k_m^2)^{1/2}, \qquad k_m^2 < \frac{\omega^2}{c_{\ell,t}^2}. \tag{3.6b}$$

We now invoke the Rayleigh hypothesis [33] and use the solutions (3.5) in satisfying the boundary conditions (3.4) at the corrugated surface $x_3 = \zeta(x_1)$. In this way we obtain a pair of coupled, homogeneous matrix equations of infinite order for the coefficients $\{A_m^{(\ell,t)}\}$ (31):

$$\sum_{n=-\infty}^{\infty}\{\frac{I_{m-n}(\alpha_\ell(k_n\omega))}{k_n\alpha_\ell(k_n\omega)}[(k_m-k_n)\frac{\omega^2}{c_t^2} + 2k_m\alpha_\ell^2(k_n\omega)]A_n^{(\ell)}$$

$$+ \frac{I_{m-n}(\alpha_t(k_n\omega))}{\alpha_t(k_n\omega)} \left[2k_m k_n - \frac{\omega^2}{c_t^2}\right] A_n^{(t)}\} = 0, \tag{3.7a}$$

$$m = 0,\pm1,\pm2,\pm3...$$

$$\sum_{n=-\infty}^{\infty} \{\frac{I_{m-n}(\alpha_\ell(k_n\omega))}{k_n} \left[2k_m k_n - \frac{\omega^2}{c_t^2}\right] A_n^{(\ell)}$$

$$+ \frac{I_{m-n}(\alpha_t(k_n\omega))}{\alpha_t^2(k_n\omega)} \left[2k_m k_n^2 - (k_m + k_n) \frac{\omega^2}{c_t^2}\right] A_n^{(t)}\} = 0, \tag{3.7b}$$

$$m = 0,\pm1,\pm2,\pm3,...$$

where

$$I_m(\alpha) = \frac{1}{a} \int_0^a dx_1\, e^{-i\frac{2\pi m}{a} x_1} e^{-\alpha\zeta(x_1)} . \tag{3.8}$$

The dispersion relation for Rayleigh waves on a grating is obtained by equating to zero the determinant of the coefficients in Eqs. (3.7). In actual calculations the matrix of coefficients in Eqs. (3.7) is made to have a finite size by restricting m and n in these equations to run from -N to N. The value of N is increased until convergent results for the frequencies are obtained.

The results of the numerical calculations of the dispersion curves can be analyzed in terms of the flat-surface dispersion curves in both the extended- and reduced-zone schemes [31]. In the extended zone scheme of Fig. 2(a) the straight line $\omega = c_R k$ (shown with a solid and then a dotted segment) is the dispersion curve for a Rayleigh wave on a flat surface (c_R is the speed of such a wave), and the dashed lines are the lines $\omega = c_t|k+(2\pi/a)m|$, where either m = 0 (the transverse sound line) or m = ±1, ±2,... . Since any point (ω,k) in the region above the dashed lines, $\omega > c_t|k+(2\pi/a)m|$, will correspond to an imaginary $\alpha_t(k_m\omega)$, the displacement fields at such points will have radiative components. Thus, a Rayleigh wave initially on a flat surface with its (ω,k) on the dotted part of the dispersion curve in Fig. 2(a) will begin to radiate into the bulk at the moment the grating, however weak, is turned on. If a Rayleigh wave with (ω,k) on this radiative part of the dispersion curve is launched across the grating surface, it will decay with some characteristic lifetime or, equivalently, will acquire a finite mean free path. We will return to this point below in discussing the attenuation of Rayleigh waves on a grating. The Bloch-type surface waves that are true eigenmodes for the surface with a grating (i.e. possess an infinite lifetime) will thus be found only with (ω,k) beneath the dashed lines in Fig. 2(a). In other words, they originate out of the solid line part of the flat surface Rayleigh wave dispersion curve as the grating is turned on.

The flat surface dispersion curve with its solid and dotted line segments is folded back into the domain $-\pi/a < k < \pi/a$ (the first Brillouin zone for the grating) in the reduced zone scheme of Fig. 2(b). The nonradiative region is now beneath the single dashed line $\omega = c_t|k|$. At points like A, B, C,... on the flat surface dispersion curve of Fig. 2 we have two degenerate modes separated in wave vector by a multiple of $2\pi/a$, i.e. by a reciprocal lattice vector of the grating. If the Fourier coefficients of the grating profile function that correspond to these reciprocal lattice vectors are nonzero, then the degeneracy of the modes at the points A, B, C... will be lifted in the first order of perturbation theory

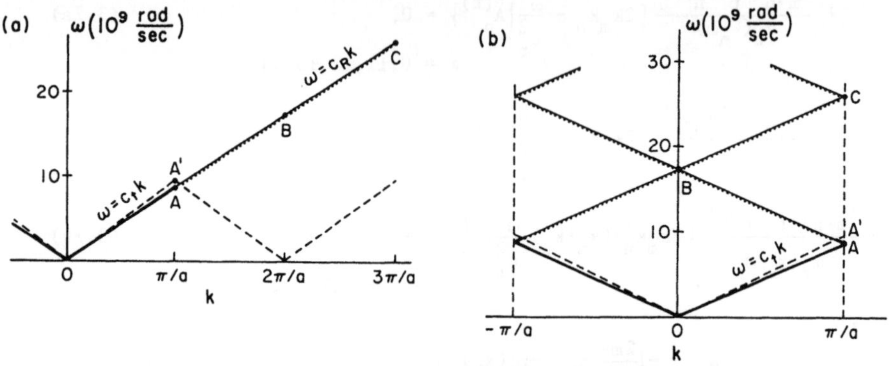

Fig. 2. The dispersion curve for Rayleigh waves on a flat surface, $\omega = c_R|k|$, in the nonradiative region (——) and in the radiative region (·······) as defined by the boundary lines $\omega = c_t|k+(2\pi m/a)|$ (– – –), where $m = 0, \pm 1, \pm 2, \ldots$, and where the period of the grating is $a = 10^{-6}$ m. $c_\ell = 5 \times 10^3$ m/sec and $c_t = 3 \times 10^3$ m/sec. (a) The extended zone scheme and (b) the reduced zone scheme [Ref. 30].

in the surface profile function, and gaps will occur in the dispersion curves at these points in both the extended and reduced zone schemes.

From Eqs. (3.7) it is easy to show that this dispersion curve is an even function of k, and is periodic in k with period $2\pi/a$. Thus we confine our attention to the region $0 \leqslant k \leqslant \pi/a$, i.e. to half the one-dimensional first Brillouin zone defined by the periodicity of the grating.

The preceding qualitative results are what is observed in a typical dispersion curve, such as the one plotted in Fig. 3 for a sinusoidal profile defined by $\zeta(x_1) = \zeta_o \cos(2\pi x_1/a)$, with $\zeta_o/a = 0.12$. Values of N of about 10 were used in obtaining this result. In the long wavelength limit the dispersion curve osculates the dispersion curve for Rayleigh waves on a flat surface. With increasing k the curve bends away from the latter curve, into the nonradiative region. Both its group and phase velocity decrease below c_R. This is the phenomenon of wave slowing that occurs for Rayleigh waves on a grating. Wave slowing occurs because in contrast with the dispersion curve for a Rayleigh wave on a flat surface, which has a constant slope for all k, the dispersion curve for a Rayleigh wave on a grating is required to have zero slope at the Brillouin zone boundary $k = \pi/a$ due to its periodicity and continuity. This forces it to bend over into the nonradiative region of the (ω,k)-plane, reducing its group velocity thereby. At the zone boundary a gap opens up in the dispersion curve, which defines a stop band for the propagation of Rayleigh waves on a grating. Finally, there is a high frequency branch to the dispersion curve above the gap, that terminates when it intersects the transverse sound line.

The method just described for the calculation of the dispersion curve of a sagittally polarized surface acoustic wave propagating across a grating can be extended and applied to a calculation of the attenuation of a Rayleigh wave on a grating [32]. It is now necessary to seek solutions of Eq. (3.7) in the radiative region of the (ω,k)-plane, i.e. for (ω,k) above the transverse sound line $\omega = c_t|k|$, where $\alpha_t(k_n\omega)$ is imaginary, and for

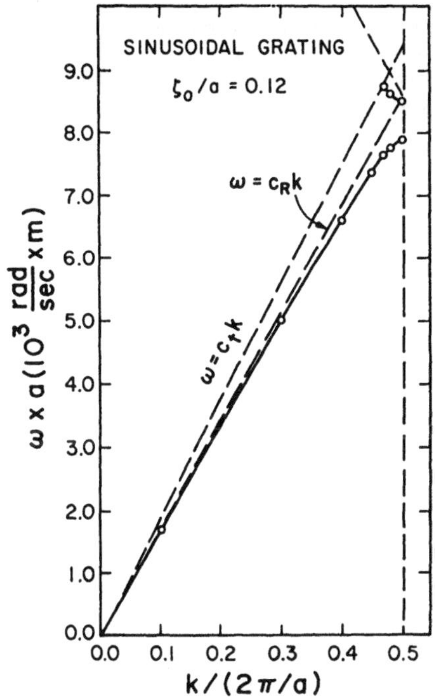

Fig. 3. Dispersion curve for Rayleigh waves propagating across a sinusoidal grating with corrugation strength $\zeta_0/a = 0.12$ on an isotropic elastic medium characterized by speeds of sound $c_\ell = 5 \times 10^3$ m/sec and $c_t = 3 \times 10^3$ m/sec [Ref. 30].

(ω,k) inside the gap in the Rayleigh wave dispersion curve at $k = \pi/a$ that is caused by the periodicity of the grating. In this region a Rayleigh wave can radiate energy into the bulk of the medium supporting it. By conservation of energy, the wave must be attenuated as it propagates across the grating. Consequently, if we assume that ω is real in these equations it is found that the corresponding wave number $k(\omega) = k_R(\omega) + ik_I(\omega)$ is complex, where $k_R(\omega)$ and $k_I(\omega)$ are real, positive functions of ω for a wave that travels in the $+x_1$-direction. The inverse attenuation length of the Rayleigh wave is then given by $\ell^{-1}(\omega) = 2k_I(\omega)$.

In solving Eqs. (3.7) for ω real and k complex one has to be careful about the manner in which the branch cut defining the square root of $\alpha_{\ell,t}(k_n\omega)$ is chosen. The correct choice turns out to be along the positive imaginary axis, and one seeks solutions on the first sheet of the corresponding Riemann surface [34]. With this choice of branch cut we obtain a wave that radiates energy into the interior, is attenuated as it propagates along the surface, but in addition has an amplitude that grows exponentially with increasing distance into the medium. The latter feature of the resulting wave is an unavoidable consequence of the exponential growth in the amplitude of the Rayleigh wave on the surface of the medium as $x_1 \to -\infty$, due to the fact that $k_I(\omega) > 0$ (35).

In Fig. 4 we show the experimental results of Rischbieter [36] for the attenuation of Rayleigh waves by a symmetric sawtooth grating ruled on an aluminum surface. In Fig. 5 we show a theoretical result for the acoustical attenuation ($20 \log_e \lambda_s k_I$) versus λ_s/a, where $\lambda_s/a = 2\pi c_t/\omega a$, for Rayleigh waves on a symmetric sawtooth grating with the same surface profile function as in the experiment of Rischbieter. This result was obtained by solving Eqs. (3.7) for $k_I(\omega)$ in the manner described above [32]. A comparison of Figs. 4 and 5 show that the theory described in

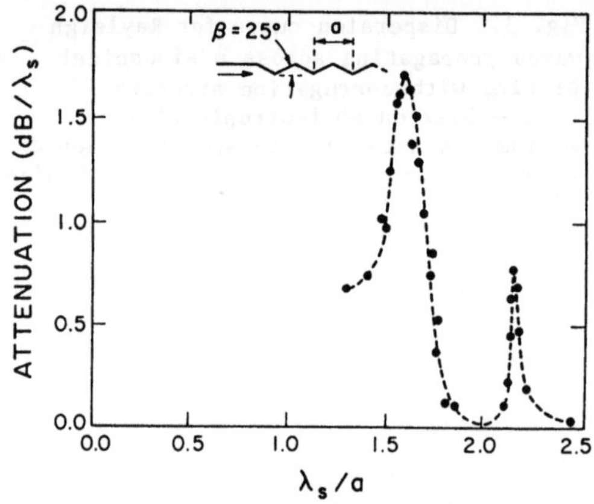

Fig. 4. Attenuation as a function of shear wavelength λ_s for Rayleigh waves on an aluminum sawtooth grating as measured by Rischbieter [Ref. 36].

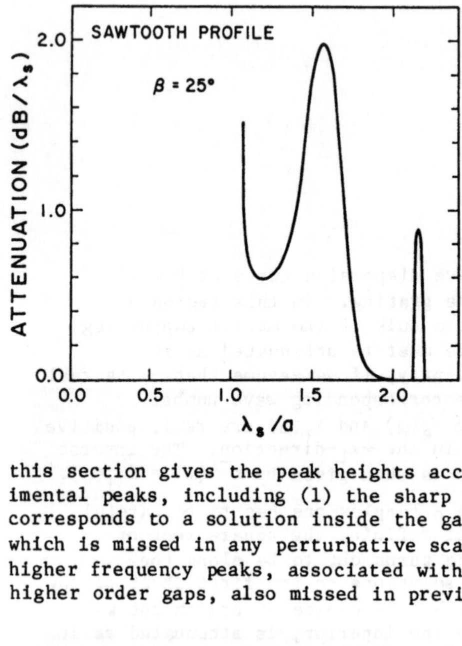

Fig. 5. Theoretical result for the attenuation as a function of shear wavelength λ_s of a Rayleigh wave on a symmetric sawtooth grating with $\beta = 25^o$, calculated for $c_t/c_\ell = 0.3$ [Ref. 31].

this section gives the peak heights accurately, and finds all the experimental peaks, including (1) the sharp peak at $\lambda_s/a = 2.15$, which corresponds to a solution inside the gap at the Brillouin zone boundary which is missed in any perturbative scattering theory [14,15]; and (2) the higher frequency peaks, associated with the edges of the successively higher order gaps, also missed in previous theories [14,15].

3.2. Shear Horizontal Polarization [37-39]

We now turn to a discussion of the propagation of a surface acoustic wave of shear horizontal polarization across a grating. This case is of interest because it is easy to show that a surface acoustic wave of this polarization cannot exist on a planar, stress-free surface of a semi-infinite, homogeneous isotropic elastic medium [40]: the displacement field that satisfies the equation of motion and the boundary condition at infinity cannot satisfy the condition that the surface be stress-free. A solution of the equation of motion that satisfies the stress-free boundary condition does exist, however. It is characterized by the dispersion relation $\omega = c_t k$, viz. the dispersion relation for bulk transverse waves with the same wave vector. However, this solution does not vanish as $x_3 \to +\infty$ (it is in fact independent of x_3), and describes a surface-skimming bulk transverse wave. We will see that the situation is quite different when the surface of the solid is periodically corrugated rather than planar.

The displacement field of a surface acoustic wave of shear horizontal polarization propagating in the x_1-direction across a grating surface has the form

$$\vec{u}(\vec{x};t) = (0,u_2(x_1,x_3|\omega),0)\exp(-i\omega t) \tag{3.9}$$

in the region $x_3 > \zeta(x_1)$. The time-independent equation of motion satisfied by $u_2(x_1,x_3|\omega)$ in this region is

$$-\omega^2 u_2 = c_t^2\left(\frac{\partial^2}{\partial x_1^2} + \frac{\partial^2}{\partial x_3^2}\right)u_2 . \tag{3.10}$$

The stress-free boundary condition at the surface $x_3 = \zeta(x_1)$ can be expressed in the form

$$\left(-\zeta'(x_1)\frac{\partial}{\partial x_1} + \frac{\partial}{\partial x_3}\right)u_2\Big|_{x_3=\zeta(x_1)} = 0. \tag{3.11}$$

In addition, we require that $u_2(x_1,x_3|\omega)$ vanish as $x_3 \to \infty$.

The solution of Eq. (3.10) that satisfies the boundary condition at infinity and the Bloch-Floquet condition can be written as

$$u_2(x_1,x_3|\omega) = \sum_{m=-\infty}^{\infty} A_m e^{ik_m x_1 - \alpha_t(k_m\omega)x_3}, \quad x_3 > \zeta(x_1)_{max}. \tag{3.12}$$

When this expansion is substituted into Eq. (3.11) the following infinite set of homogeneous linear equations for the coefficients $\{A_m\}$ is obtained [(39]:

$$\sum_{n=-\infty}^{\infty} \frac{I_{m-n}(\alpha_t(k_n\omega))}{\alpha_t(k_n\omega)}\left(k_m k_n - \frac{\omega^2}{c_t^2}\right)A_n = 0, \quad m = 0,\pm1,\pm2,\pm3,\ldots, \tag{3.13}$$

where $I_m(\alpha)$ has been defined in Eq. (3.8). The dispersion relation for shear horizontal surface acoustic waves on a grating is obtained by equating to zero the determinant of the coefficients in Eq. (3.13). Again, in practice the matrix of coefficients in Eq. (3.13) is made to have a finite size by restricting m and n to run from -N to N.

The dispersion curve for such waves is plotted in Fig. 6 for a sinusoidal surface profile defined by $\zeta(x_1) = \zeta_0\cos(2\pi x_1/a)$, with $\zeta_0/a = 0.6$. A value of N of the order of 20 was used to obtain this result. We see that in the long wavelength limit this curve is tangent to the dispersion curve for bulk transverse waves with the same wave vector, $\omega = c_t k$, viz. the dispersion curve for surface skimming bulk transverse waves. As k increases the dispersion curve bends away from the sound line and displays the phenomenon of wave slowing. There is only a single branch to the dispersion curve. Thus, the entire range of frequencies above the frequency at the zone boundary is a stop band for this wave.

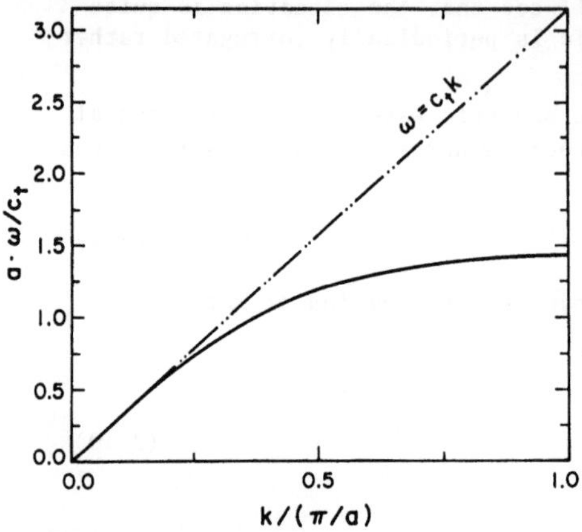

Fig. 6. Dispersion curve for a shear horizontal surface acoustic wave propagating across a sinusoidal grating with $\zeta_0/a = 0.6$ [Ref. 13].

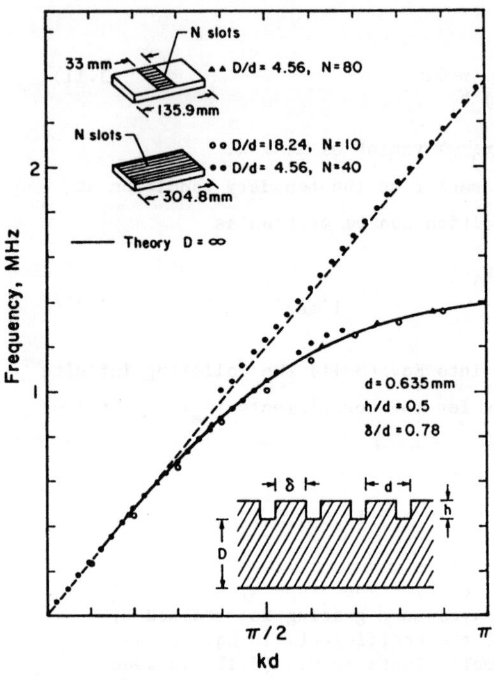

Fig. 7. Experimental dispersion curve for a shear horizontal surface acoustic wave propagating across a lamellar grating ruled on aluminum [Ref. 37].

In Fig. 7 we plot the experimental result of Auld et al. [37] for the dispersion curve of a shear horizontal surface acoustic wave propagating across a lamellar grating. The experimental curve displays all of the features of the theoretical curve displayed in Fig. 6.

We have here an example of surface roughness giving rise to surface acoustic waves that cannot exist in the absence of that roughness.

These shear horizontal surface acoustic waves are weakly bound, however, at least in the limit of small k. A perturbation calculation starting from Eq. (3.13) shows that in this limit

$$\alpha_t(k\omega) = \frac{\pi}{a}\left(\frac{\zeta_o}{a}\right)^2(ka)^2 , \tag{3.14}$$

for a sinusoidal grating. Since the decay length of the wave into the elastic medium is $\alpha_t^{-1}(k\omega)$, we see that for $\zeta_o/a = 0.05$ and $ka = 0.1\pi$, the decay length is $(4 \times 10^4/\pi^3)\ a \sim 1290a$. Thus it penetrates far into the medium. As the wave vector increases, however, the wave becomes more tightly bound to the surface.

Thus far in this subsection we have considered shear horizontal surface acoustic waves at the periodically corrugated interface between a solid and air. Shear horizontal surface acoustic waves at the periodically corrugated interface between two different isotropic elastic media have also been studied recently [41-43]. If the region $x_3 < \zeta(x_1)$ is now filled with an isotropic medium characterized by a mass density ρ' and speeds of longitudinal and transverse sound c_ℓ' and c_t', respectively, the displacement field in this medium that satisfies the boundary condition at $x_3 = -\infty$ can be written as

$$u_2(x_1,x_3|\omega) = \sum_{m=-\infty}^{\infty} B_m e^{ik_m x_1 + \beta_t(k_m\omega)x_3}, \qquad x_3 < \zeta(x_1)_{min} , \tag{3.15}$$

where

$$\beta_t(k_m\omega) = \left(k_m^2 - \frac{\omega^2}{c_t'^2}\right)^{1/2}, \qquad k_m^2 > \frac{\omega^2}{c_t'^2} \tag{3.16a}$$

$$= -i\left(\frac{\omega^2}{c_t'^2} - k_m^2\right)^{1/2}, \qquad k_m^2 < \frac{\omega^2}{c_t'^2} . \tag{3.16b}$$

When the solutions given by Eqs. (3.12) and (3.15) are used in the boundary conditions at the interface $x_3 = \zeta(x_1)$, which require that the displacement and the stress acting on the interface be continuous across it,

$$u_2\Big|_{x_3=\zeta(x_1)+} = u_2\Big|_{x_3=\zeta(x_1)-} \tag{3.17}$$

$$\rho c_t^2\left[-\zeta'(x_1)\frac{\partial}{\partial x_1} + \frac{\partial}{\partial x_3}\right]u_2\Big|_{x_3=\zeta(x_1)+}$$

$$= \rho' c_t'^2\left[-\zeta'(x_1)\frac{\partial}{\partial x_1} + \frac{\partial}{\partial x_3}\right]u_2\Big|_{x_3=\zeta(x_1)-} , \tag{3.18}$$

respectively, a pair of coupled, homogeneous, linear equations of infinite order for the coefficients $\{A_m(k\omega)\}$ and $\{B_m(k\omega)\}$ is obtained [43]:

$$\sum_{n=-\infty}^{\infty} \left[I_{m-n}(\alpha_t(k_n\omega))A_n - I_{m-n}(-\beta_t(k_n\omega))B_n\right] = 0, \tag{3.19a}$$

$$m=0,\pm1,\pm2,...$$

$$\sum_{n=-\infty}^{\infty} \left[\rho c_t^2 \frac{k_m k_n - (\omega^2/c_t^2)}{\alpha_t(k_n\omega)} I_{m-n}(\alpha_t(k_n\omega))A_n\right.$$

$$+ \rho' c_t'^{-2} \frac{k_m k_n - (\omega^2/c_t'^2)}{\beta_t(k_n\omega)} I_{m-n}(-\beta_t(k_n\omega))B_n] = 0, \qquad (3.19b)$$

$$m = 0, \pm 1, \pm 2, \ldots$$

The dispersion relation for shear horizontal surface acoustic waves localized to the interface $x_3 = \zeta(x_1)$ is obtained by equating to zero the determinant of the coefficients in this set of equations, after m and n have been restricted to the interval (-N, N).

In Fig. 8 is plotted the dispersion curve for a wave of this type for a sinusoidal surface profile function $\zeta(x_1) = \zeta_o \cos(2\pi x_1/a)$ with $\zeta_o/a = 2/\pi$. It is seen from this figure that the wave exists only for a limited range of values of k, in the nonradiative region of the (ω,k)-plane defined by the slower of the two speeds of transverse sound waves in the system. This result was obtained perturbatively by GULYAEV and PLESSKY [41,42].

Shear horizontal waves localized to the planar interface between two different isotropic elastic media cannot exist, because it is impossible to satisfy the boundary conditions at the interface by the solutions of the equations of motion that decay to zero with increasing distance into each medium from the interface. Thus, we have here a second example of

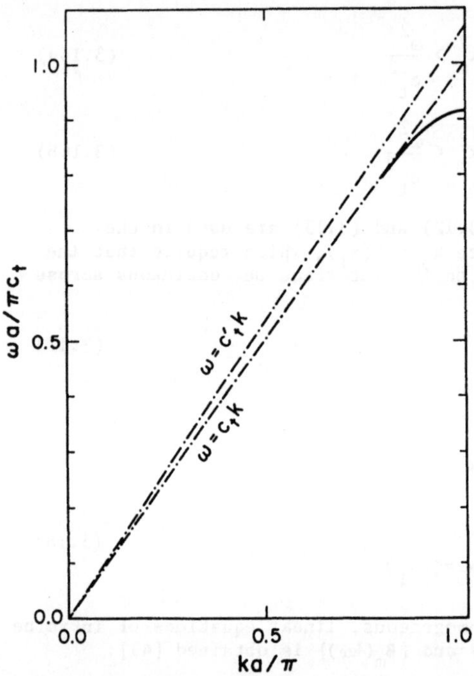

Fig. 8. The dispersion curve for surface acoustic waves of shear horizontal polarization propagating across the periodically corrugated interface between two different isotropic elastic media characterized by mass densities ρ and ρ′ and speeds of transverse sound waves c_t and c_t'. The interface profile function is $\zeta(x_1) = \zeta_o \cos(2\pi x_1/a)$, with $\zeta_o/a = 2/\pi$, while $\rho/\rho' = 0.5$, $\rho c_t^2/(\rho' c_t'^2) = 0.45$. [Ref. 43].

the way in which surface roughness can give rise to surface (or, more accurately, interface) acoustic waves that have no counterparts in the absence of the roughness.

Finally, we note that shear horizontal surface acoustic waves have been studied in a system that consists of a film of one elastic material bonded to a semi-infinite substrate of a different elastic material, when the interface between the film and air is periodically corrugated. For a description of the results obtained and references to the earlier literature on this problem the reader is referred to the paper by Dr. Wirgin at this Symposium.

4. Surface Acoustic Waves on Randomly Rough Surfaces [44-50]

An important class of rough surfaces is the class of randomly rough surfaces. Such surfaces are of interest because roughness is found in varying degrees on all real solid surfaces, even carefully prepared ones, and it is important to know how properties of surface acoustic waves obtained on the assumption of a perfectly planar surface are modified in the presence of surface roughness. For example, surface roughness attenuates a surface acoustic wave as it propagates across a randomly rough surface, by scattering energy out of the incident beam into other surface acoustic waves and into bulk acoustic waves. At the present time surface roughness appears to be the most important extrinsic mechanism responsible for the attenuation of Rayleigh waves [51]. In addition, just as a periodically corrugated surface does, a randomly rough surface gives rise to the phenomenon of wave slowing for surface acoustic waves propagating across it, and it can also support surface acoustic waves that cannot exist in the absence of the surface roughness. This is not surprising, because a randomly rough surface can be regarded as a superposition of an infinite number of gratings, each with its own period and amplitude, and with its own orientation with respect to the x_1- and x_2-axes in the nominal plane of the surface, viz. the plane $x_3 = 0$. Indeed, it is just in this way that we will characterize a randomly rough surface here (see Eq. (4.4) below).

In this section we sketch out a theory of the propagation of a surface acoustic wave across a randomly rough surface of a semi-infinite, isotropic elastic medium characterized by a mass density ρ and speeds of longitudinal and transverse waves c_ℓ and c_t, respectively [47].

We assume that the elastic medium occupies the region $x_3 > \zeta(\vec{x}_\parallel)$, where $\vec{x}_\parallel = \hat{x}_1 x_1 + \hat{x}_2 x_2$ while $\hat{x}_1$ and $\hat{x}_2$ are unit vectors along the x_1- and x_2-axes, respectively. Since the surface $x_3 = \zeta(\vec{x}_\parallel)$ is assumed to be randomly rough we do not know the surface profile function $\zeta(\vec{x}_\parallel)$ and are forced to characterize it by its statistical properties. Underlying this description is the notion that there is not a single rough surface but rather an ensemble of such surfaces. Physical observables are to be averaged over this ensemble, and it is assumed that this average coincides with the average over the surface in the limit of a very large surface area.

We do not need to know the complete probability distribution function of the surface profile function $\zeta(\vec{x}_\parallel)$. Knowledge of its first two moments suffices for our purposes here. Thus, we assume that $\zeta(\vec{x}_\parallel)$ is a single-valued function of $\vec{x}_\parallel$ and a stationary stochastic process characterized by the two properties

$$\langle\zeta(\vec{x}_\parallel)\rangle = 0 \qquad (4.1)$$

$$\langle\zeta(\vec{x}_\parallel)\zeta(\vec{x}'_\parallel)\rangle = \delta^2 W(|\vec{x}_\parallel - \vec{x}'_\parallel|), \tag{4.2}$$

where the angular brackets $\langle\cdots\rangle$ denote an average over the ensemble of realizations of the surface profile function. The result expressed by Eq. (4.1) simply states that the mean surface is the plane $x_3 = 0$. This is an assumption that can be made with no loss of generality. In Eq. (4.2) $\delta^2 = \langle\zeta^2(\vec{x}_\parallel)\rangle$ is the mean- square departure of the surface from flatness. The fact that the correlation function $W(|\vec{x}_\parallel - \vec{x}'_\parallel|)$ appearing on the right hand side of this equation depends on $\vec{x}_\parallel$ and $\vec{x}'_\parallel$ only through their difference (the stationarity assumption) has the consequence that averaging restores infinitesimal translational invariance in the plane $x_3 = 0$ to our system. The fact that it depends on the magnitude of the difference $\vec{x}_\parallel - \vec{x}'_\parallel$ has the consequence that if our system in the absence of the roughness is isotropic in the plane $x_3 = 0$ (as it is in the case considered here), then averaging will restore isotropy in this plane to our system.

The unit vector normal to the surface $x_3 = \zeta(\vec{x}_\parallel)$ at each point, and directed away from the solid, is given by

$$\hat{n} = [1 + (\nabla\zeta(\vec{x}_\parallel))^2]^{-1/2}\left(\frac{\partial\zeta(\vec{x}_\parallel)}{\partial x_1}, \frac{\partial\zeta(\vec{x}_\parallel)}{\partial x_2}, -1\right). \tag{4.3}$$

It is also convenient to introduce the Fourier integral representation of $\zeta(\vec{x}_\parallel)$,

$$\zeta(\vec{x}_\parallel) = \int \frac{d^2Q_\parallel}{(2\pi)^2}\hat{\zeta}(\vec{Q}_\parallel)e^{i\vec{Q}_\parallel\cdot\vec{x}_\parallel}, \tag{4.4}$$

where $\vec{Q}_\parallel = \hat{x}_1 Q_1 + \hat{x}_2 Q_2$. With the use of the Fourier inversion theorem and Eqs. (4.1)-(4.2) it is readily established that $\hat{\zeta}(\vec{Q}_\parallel)$ possesses the statistical properties

$$\langle\hat{\zeta}(\vec{Q}_\parallel)\rangle = 0 \tag{4.5}$$

$$\langle\hat{\zeta}(\vec{Q}_\parallel)\hat{\zeta}(\vec{Q}'_\parallel)\rangle = (2\pi)^2\delta(\vec{Q}_\parallel + \vec{Q}'_\parallel)\delta^2 g(|\vec{Q}_\parallel|), \tag{4.6}$$

where the <u>surface structure factor</u> $g(|\vec{Q}_\parallel|)$ is given by

$$g(|\vec{Q}_\parallel|) = \int d^2x_\parallel W(|\vec{x}_\parallel|)e^{-i\vec{Q}_\parallel\cdot\vec{x}_\parallel}. \tag{4.7}$$

Although the results to be obtained below will be presented for an arbitrary form of $g(|\vec{Q}_\parallel|)$, for numerical calculations it is necessary to choose a specific form for this function. We will choose an expression for $W(|\vec{x}_\parallel|)$ that is both physically reasonable and computationally convenient, viz.

$$W(|\vec{x}_\parallel|) = \exp(-x_\parallel^2/a^2). \tag{4.8}$$

The characteristic distance a over which $W(|\vec{x}_\parallel|)$ decreases to 1/e of its value at $|\vec{x}_\parallel| = 0$ is called the <u>transverse correlation length</u>. The surface structure factor corresponding to this choice of $W(|\vec{x}_\parallel|$ is

$$g(|\vec{Q}_\parallel|) = \pi a^2 \exp(-Q_\parallel^2 a^2/4). \tag{4.9}$$

Since surface roughness is a static perturbation of our system we can assume an elastic displacement field in the region $x_3 > \zeta(\vec{x}_\parallel)$ of the form $\vec{u}(\vec{x};t) = \vec{u}(\vec{x}|\omega)\exp(-i\omega t)$. The time-independent stress-tensor in this region can then be written in the form

$$T_{\alpha\beta}(\vec{x}|\omega) = \rho(c_\ell^2 - 2c_t^2)\delta_{\alpha\beta}\nabla\cdot\vec{u}(\vec{x}|\omega)$$

$$+ \rho c_t^2\left(\frac{\partial u_\alpha(\vec{x}|\omega)}{\partial x_\beta} + \frac{\partial u_\beta(\vec{x}|\omega)}{\partial x_\alpha}\right). \tag{4.10}$$

The solutions of the time-independent equations of motion

$$-\rho\omega^2 u_\alpha(\vec{x}|\omega) = \sum_\beta \frac{\partial T_{\alpha\beta}(\vec{x}|\omega)}{\partial x_\beta} = \rho(c_\ell^2 - c_t^2)\sum_\beta \frac{\partial^2 u_\beta(\vec{x}|\omega)}{\partial x_\alpha \partial x_\beta}$$

$$+ \rho c_t^2 \nabla^2 u_\alpha(\vec{x}|\omega), \tag{4.11}$$

that vanish at infinity or represent outgoing waves can be written as

$$\vec{u}(\vec{x}|\omega) = \int \frac{d^2 q_\parallel}{(2\pi)^2} e^{i\vec{q}_\parallel\cdot\vec{x}_\parallel}\Big[\Big(\hat{q}_\parallel + i\hat{x}_3 \frac{\alpha_\ell(q_\parallel\omega)}{q_\parallel}\Big)A_1(\vec{q}_\parallel\omega)$$

$$\times e^{-\alpha_\ell(q_\parallel\omega)x_3} + (\hat{q}_\parallel \times \hat{x}_\parallel)A_2(\vec{q}_\parallel\omega)e^{-\alpha_t(q_\parallel\omega)x_3}$$

$$+ \Big(\hat{q}_\parallel + i\hat{x}_3 \frac{q_\parallel}{\alpha_t(q_\parallel\omega)}\Big)A_3(\vec{q}_\parallel\omega)e^{-\alpha_t(q_\parallel\omega)x_3}\Big], \tag{4.12}$$

where

$$\alpha_{\ell,t}(q_\parallel\omega) = \Big(q_\parallel^2 - \frac{\omega^2}{c_{\ell,t}^2}\Big), \qquad q_\parallel^2 > \frac{\omega^2}{c_{\ell,t}^2} \tag{4.13a}$$

$$= -i\Big(\frac{\omega^2}{c_{\ell,t}^2} - q_\parallel^2\Big)^{1/2}, \qquad q_\parallel^2 < \frac{\omega^2}{c_{\ell,t}^2}. \tag{4.13b}$$

The coefficients $\{A_\alpha(\vec{q}_\parallel\omega)\}$ ($\alpha = 1,2,3$) are determined from the requirement that the surface $x_3 = \zeta(\vec{x}_\parallel)$ be stress-free,

$$\sum_\beta T_{\alpha\beta}(\vec{x}|\omega)\hat{n}_\beta\Big|_{x_3 = \zeta(\vec{x}_\parallel)} = 0, \quad \alpha = 1,2,3. \tag{4.14}$$

When Eqs. (4.3), (4.10), and (4.12) are substituted into Eqs. (4.14), and the resulting equations are expanded to first order in $\zeta(\vec{Q}_\parallel)$ (the small roughness approximation), the equations for the $\{A_\alpha(\vec{q}_\parallel\omega)\}$ take the form

$$\sum_\beta M^{(o)}_{\alpha\beta}(\vec{q}_\parallel|\omega)A(\vec{q}_\parallel\omega) + \sum_\beta \int \frac{d^2k_\parallel}{(2\pi)^2} \zeta(\vec{q}_\parallel-\vec{k}_\parallel)$$

$$\times X_{\alpha\beta}(\vec{q}_\parallel-\vec{k}_\parallel;\vec{k}_\parallel|\omega)A_\beta(\vec{k}_\parallel\omega) = 0. \tag{4.15}$$

The elements of the matrices $\overleftrightarrow{M}^{(o)}(\vec{q}_\parallel|\omega)$ and $\overleftrightarrow{X}(\vec{Q}_\parallel;\vec{q}_\parallel|\omega)$ are given in Ref. [47]. For reasons that will become clear later, we multiply the three equations (4.15) from the left by the matrix $\overleftrightarrow{S}(\hat{q}_\parallel)$ defined by

$$\overleftrightarrow{S}(\hat{q}_\parallel) = \begin{pmatrix} \hat{q}_1 & \hat{q}_2 & 0 \\ -\hat{q}_2 & \hat{q}_1 & 0 \\ 0 & 0 & 1 \end{pmatrix}. \tag{4.16}$$

The result is an equation for the $\{A_\alpha(\vec{q}_\parallel\omega)\}$ that can be written as

$$\sum_\beta \overline{M}^{(o)}_{\alpha\beta}(q_\parallel|\omega)A_\beta(\vec{q}_\parallel\omega) - \sum_\beta \int \frac{d^2k_\parallel}{(2\pi)^2} \zeta(\vec{q}_\parallel-\vec{k}_\parallel)$$

$$\times \overline{M}^{(1)}_{\alpha\beta}(\vec{q}_\parallel;\vec{k}_\parallel|\omega)A_\beta(\vec{k}_\parallel\omega) = 0. \tag{4.17}$$

The matrix elements $\overline{M}^{(o)}_{\alpha\beta}(q_\parallel|\omega)$ and $\overline{M}^{(1)}_{\alpha\beta}(\vec{q}_\parallel;\vec{k}_\parallel|\omega)$ are given in Ref. [40]. It is a consequence of multiplying Eqs. (4.15) from the left by the matrix $\overleftrightarrow{S}(\hat{q}_\parallel)$ that $\overline{M}^{(o)}_{\alpha\beta}(q_\parallel|\omega)$ depends on $\vec{q}_\parallel$ only through its magnitude, while $\overline{M}^{(1)}_{\alpha\beta}(\vec{q}_\parallel;\vec{k}_\parallel|\omega)$ depends on $\vec{q}_\parallel$ and $\vec{k}_\parallel$ only through their magnitudes and through the cosine $(\hat{q}_\parallel\cdot\hat{k}_\parallel)$ and sine $(\hat{q}_\parallel\times\hat{k}_\parallel)_3$ of the angle between them.

The coefficients $\{A_\alpha(\vec{q}_\parallel\omega)\}$ are random quantities because they satisfy a matrix integral equation, Eq. (4.17), with a random kernel. We will not seek the complete probability distribution function for these coefficients. As with the probability distribution function for $\zeta(\vec{x}_\parallel)$ we confine out attention to the low order moments of this distribution function, in particular to the first moment $\langle A_\alpha(\vec{q}_\parallel\omega)\rangle$. This describes the propagation of the mean surface acoustic wave across the randomly rough surface, which suffices for the determination of the dispersion relation of this wave.

To obtain the equation satisfied by $\langle A_\alpha(\vec{q}_\parallel\omega)\rangle$ from the equation (4.17) for $A_\alpha(\vec{q}_\parallel\omega)$ we introduce the smoothing operator P that averages everything it acts on over the ensemble of realizations of the surface profile: $PA(\vec{q}_\parallel\omega) \equiv \langle A_\alpha(\vec{q}_\parallel\omega)\rangle$. We also introduce the complementary operator Q = 1-P that yields the fluctuating part of everything it acts on. We apply P and Q to Eq. (4.17) in turn and obtain the pair of equations

$$\sum_\beta \overline{M}^{(o)}_{\alpha\beta}(q_\parallel|\omega)PA_\beta(\vec{q}_\parallel\omega) = \sum_\beta \int \frac{d^2k_\parallel}{(2\pi)^2} P\hat{\zeta}(\vec{q}_\parallel-\vec{k}_\parallel)$$

$$\times \overline{M}^{(1)}_{\alpha\beta}(\vec{q}_\parallel;\vec{k}_\parallel|\omega)(PA_\beta(\vec{k}_\parallel\omega)+QA_\beta(\vec{k}_\parallel\omega)), \tag{4.18a}$$

$$\sum_\beta \overline{M}^{(o)}_{\alpha\beta}(q_\| |\omega) QA_\beta(\vec{q}_\|\omega) = \sum_\beta \int \frac{d^2k_\|}{(2\pi)^2} Q\hat{\zeta}(\vec{q}_\|-\vec{k}_\|)$$

$$\times \overline{M}^{(1)}_{\alpha\beta}(\vec{q}_\|;\vec{k}_\| |\omega)(PA_\beta(\vec{k}_\|\omega)+QA_\beta(\vec{k}_\|\omega)). \tag{4.18b}$$

In writing these equations we have used the identity $A_\beta(\vec{k}_\|\omega) = PA_\beta(\vec{k}_\|\omega) + QA_\beta(\vec{k}_\|\omega)$. If we solve Eq. (4.18b) for $QA_\alpha(\vec{q}_\|\omega)$ to first order in $\hat{\zeta}(\vec{q}_\|-\vec{k}_\|)$ and use the fact that $P\hat{\zeta}(\vec{q}_\|-\vec{k}_\|) = 0$, Eq. (4.5), we obtain from Eq. (4.18a) the result that

$$\sum_\beta \overline{M}^{(o)}_{\alpha\beta}(q_\| |\omega)\langle A_\beta(\vec{q}_\|\omega)\rangle = \sum_{\beta\mu\nu} \int \frac{d^2k_\|}{(2\pi)^2} \int \frac{d^2p_\|}{(2\pi)^2}$$

$$\times \langle\hat{\zeta}(\vec{q}_\|-\vec{k}_\|)\hat{\zeta}(\vec{k}_\|-\vec{p}_\|)\rangle\, \overline{M}^{(1)}_{\alpha\mu}(\vec{q}_\|;\vec{k}_\| |\omega) M^{(o)-1}_{\mu\nu}(k_\|\omega)$$

$$\times \overline{M}^{(1)}_{\nu\beta}(\vec{k}_\|;\vec{p}_\| |\omega)\langle A_\beta(\vec{p}_\|\omega)\rangle. \tag{4.19}$$

With the use of Eq. (4.6) this set of equations reduces to a set of three homogeneous equations for the $\langle A_\alpha(\vec{q}_\|\omega)\rangle$:

$$\sum_\beta [\overline{M}^{(o)}_{\alpha\beta}(q_\| |\omega) - \Delta M_{\alpha\beta}(q_\| |\omega)]\langle A_\beta(\vec{q}_\|\omega)\rangle = 0, \tag{4.20}$$

where

$$\Delta M_{\alpha\beta}(q_\| |\omega) = \delta^2 \int \frac{d^2k_\|}{(2\pi)^2} g(|\vec{q}_\|-\vec{k}_\||) \sum_{\mu\nu} \overline{M}^{(1)}_{\alpha\mu}(\vec{q}_\|;\vec{k}_\| |\omega)$$

$$\times \overline{M}^{(o)-1}_{\mu\nu}(k_\| |\omega)\overline{M}^{(1)}_{\nu\beta}(\vec{k}_\|;\vec{q}_\| |\omega). \tag{4.21}$$

The fact that we obtain a set of algebraic equations for $\langle A_\alpha(\vec{q}_\|\omega)\rangle$ instead of a set of integral equations is due to the restoration of infinitesimal translational invariance to our system as a result of the averaging we have carried out.

The set of equations (4.20) is actually simpler than appears to be the case at first glance. When the angular integration on the right hand side of Eq. (4.21) is carried out the 12, 21, 23, and 32 elements of the matrix $\Delta M_{\alpha\beta}(q_\| |\omega)$ vanish. (The same elements are zero in the matrix $\overline{M}^{(o)}_{\alpha\beta}(q_\| |\omega)$.) That this occurs is due to the restoration of isotropy in the plane $x_3 = 0$ to our system by the averaging we have carried out. This is also the reason why the nonzero elements of the matrix $\Delta M_{\alpha\beta}(q_\| |\omega)$ depend on the wave vector $\vec{q}_\|$ only through its magnitude. It was to obtain these convenient results that the transition from Eq. (4.15) to Eq. (4.17) was carried out with the aid of the matrix $\overleftrightarrow{S}(q_\|)$ defined by Eq. (4.16).

Equations (4.20) thus break up into two sets of equations which have the forms

$$\begin{pmatrix} \overline{M}^{(o)}_{11}(q_\parallel|\omega)-\Delta M_{11}(q_\parallel|\omega) & \overline{M}^{(o)}_{13}(q_\parallel|\omega)-\Delta M_{13}(q_\parallel|\omega) \\ \overline{M}^{(o)}_{31}(q_\parallel|\omega)-\Delta M_{31}(q_\parallel\omega) & \overline{M}^{(o)}_{33}(q_\parallel|\omega)-\Delta M_{33}(q_\parallel|\omega) \end{pmatrix} \begin{pmatrix} \langle A_1(\vec{q}_\parallel\omega)\rangle \\ \langle A_3(\vec{q}_\parallel\omega)\rangle \end{pmatrix} = 0 \tag{4.22a}$$

and

$$[\overline{M}^{(o)}_{22}(q_\parallel|\omega) - \Delta M_{22}(q_\parallel|\omega)]\langle A_2(\vec{q}_\parallel\omega)\rangle = 0. \tag{4.22b}$$

The dispersion relation for surface acoustic waves polarized in the sagittal plane (Rayleigh waves) is then obtained by equating to zero the determinant of the coefficients in Eq. (4.22a); the dispersion relation for surface acoustic waves polarized perpendicular to the sagittal plane (shear horizontal surface acoustic wave) is obtained by equating to zero the coefficient of $\langle A_2(\vec{q}_\parallel\omega)\rangle$ in Eq. (4.22b).

The dispersion curves for surface acoustic waves of both polarizations are complex. In the case of sagittal polarization the frequency of a Rayleigh wave on a randomly rough surface can be written in the form

$$\omega_s(q_\parallel) = \omega_R(q_\parallel)[1 + \frac{\delta^2}{a^2}(\omega_1(aq_\parallel) - i\omega_2(aq_\parallel))], \tag{4.23}$$

where $\omega_R(q_\parallel) = c_R q_\parallel$ is the frequency of a Rayleigh wave on a planar surface, with c_R the speed of a Rayleigh wave on such a surface, while $\omega_{1,2}(aq_\parallel)$ are universal, dimensionless functions of $aq_\parallel$. Mathematically, the fact that the dispersion curve is complex is due to the fact that the matrices $\overline{M}^{(o)-1}_{\alpha\beta}(k_\parallel|\omega)$ and $\overline{M}^{(1)}_{\alpha\beta}(\vec{q}_\parallel;\vec{k}_\parallel|\omega)$ that enter the expression (4.21) for $\Delta M_{\alpha\beta}(q_\parallel|\omega)$ are functions of $\alpha_\ell(k_\parallel\omega)$ and $\alpha_t(k_\parallel\omega)$, as well as of the magnitudes of $\vec{q}_\parallel$ and $\vec{k}_\parallel$ and of the sine and cosine of the angle between them. Now, from Eqs. (4.13) we see that $\alpha_\ell(k_\parallel\omega)$ and $\alpha_t(k_\parallel\omega)$ are real for $k_\parallel > \omega/c_\ell$ and $k_\parallel > \omega/c_t$, respectively, while they are pure imaginary for $0 < k_\parallel < \omega/c_\ell$ and for $0 < k_\parallel < \omega/c_t$, respectively. Since all of these ranges of $k_\parallel$ contribute to the integral over $k_\parallel$ in Eq. (4.21), $\Delta M_{\alpha\beta}(q_\parallel|\omega)$ is complex for this reason. In addition, for $k_\parallel > \omega/c_t$, where $\alpha_\ell(k_\parallel\omega)$ and $\alpha_t(k_\parallel\omega)$ are both real, the matrix $\overline{M}^{(o)-1}_{\mu\nu}(k_\parallel|\omega)$ contributes a simple pole to the integrand on the right hand side of Eq. (4.21) at $k_\parallel = \omega/c_R$. The correct way of dealing with this pole is to give ω a positive, infinitesimal imaginary part, since from Eqs. (4.13) we see that this ensures the correct analytic continuation of $\alpha_\ell(k_\parallel\omega)$ and $\alpha_t(k_\parallel\omega)$ from the region where they are real to the region where they are pure imaginary. Thus this pole also makes a complex contribution to $\Delta M_{\alpha\beta}(q_\parallel|\omega)$. The solvability conditions for Eqs. (4.22a) and (4.22b) therefore yield complex values of ω even if $q_\parallel$ is real. The physical interpretation of these results is that the imaginary part of $\omega(q_\parallel)$ describes the damping of the Rayleigh wave due to its roughness-induced conversion into bulk waves or into other Rayleigh waves, depending on whether it arises from the integration in Eq. (4.21) over the interval $0 < k_\parallel < \omega/c_t$ or the interval $k_\parallel > \omega/c_t$, respectively.

The functions $\omega_{1,2}(aq_\parallel)$ are plotted in Figs. 9 and 10, respectively. From Fig. 9 we see that the roughness-induced shift in the frequency of the Rayleigh wave is always negative. This is the analogue for the pre-

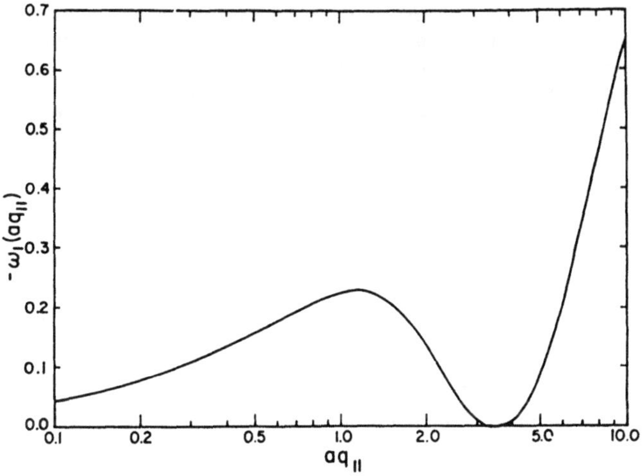

Fig. 9. Dimensionless shift in the frequency of a Rayleigh wave due to surface roughness, as a function of $aq_{\parallel}$. Note that the actual shift is given by $c_R q_{\parallel}(\delta^2/a^2)\omega_1(aq_{\parallel})$. It is assumed that $c_{\ell}^2 = 3c_t^2$. [Ref. 47].

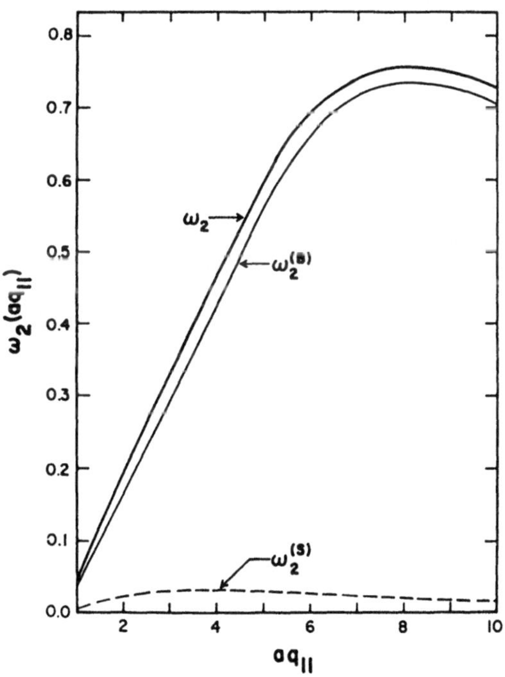

Fig. 10. The dimensionless imaginary part of the change in the frequency of a Rayleigh wave caused by surface roughness. The actual imaginary part is given by $c_R q_{\parallel}(\delta^2/a^2)\omega_2(aq_{\parallel})$. It is assumed that $c_{\ell}^2 = 3c_t^2$. [Ref. 47].

sent case of the phenomenon of wave slowing observed for Rayleigh waves on a grating. From Fig. 10 we see that the scattering of the Rayleigh wave into bulk waves is a much more effective attenuation mechanism than is scattering into other Rayleigh waves. The attenuation length of the Rayleigh wave, $\ell(q_\parallel)$, i.e. the distance over which the energy of the Rayleigh wave decays to 1/e of its initial value, is given in terms of $\omega_2(aq_\parallel)$ by

$$\ell^{-1}(q_\parallel) = 2\,\frac{\delta^2}{a^2}\,q_\parallel\omega_2(aq_\parallel). \tag{4.24}$$

It is found that the right hand side of this equation is proportional to $(aq_\parallel)^5 \propto \omega^5$ in the limit as $aq_\parallel \to 0$. The physical explanation for this dependence is that the frequency dependence of Rayleigh scattering in d dimensions is ω^{d+1}. The hills and valleys responsible for the scattering of a Rayleigh wave in the present case are three-dimensional, since they are defined by the equation $x_3 = \zeta(\vec{x}_\parallel)$. Thus the Rayleigh scattering law in the present case gives us an ω^4 frequency dependence in the low frequency, long wavelength limit. The remaining factor of ω arises because the penetration depth of the Rayleigh wave into the solid is of the order of its wavelength parallel to the surface [45].

If we assume the value $\delta/a = 0.3$, and the typical values $\omega = 10^8$ rad/sec and $c_R = 3 \times 10^5$ cm/sec, for which $q_\parallel = 333.33\ \mathrm{cm}^{-1}$ and the wavelength of the Rayleigh wave along the surface $\lambda_R = 2\pi/q_\parallel \cong 0.02$ cm, we find that $\ell = 2.6 \times 10^3$ cm for $aq_\parallel = 0.1$, $\ell = = 0.42$ cm for $aq_\parallel = 1$, and $\ell = 0.02$ cm for $aq_\parallel = 10$. (Note that although $\ell \gg \lambda_R$ in the first case, in the last case $\ell \cong \lambda_R$.) The relative downward shift in the frequency of the Rayleigh wave in these three cases is 0.4%, 2%, and 5.8%, respectively.

The attenuation of Rayleigh waves on a randomly rough surface has been measured for the first time very recently [52]. A plot of these results for a duraluminum surface is shown in Fig. 11. In view of the difficulty of the experiment, the agreement between theory and experiment has to be considered satisfactory.

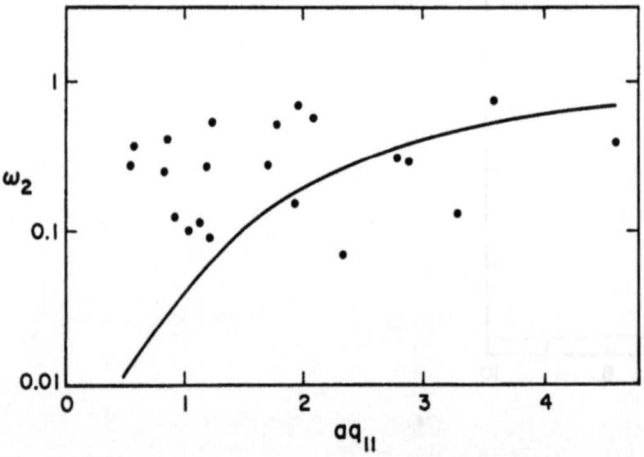

Fig. 11. An experimental result for the dependence of $\omega_2(aq_\parallel)$ on $aq_\parallel$ for a rough duraluminum surface. The points represent experimental data and the solid line represents the theoretical result of Ref. 47 [Ref. 52].

The dispersion curve for surface acoustic waves of shear horizontal polarization on a randomly rough surface is obtained from Eq. (4.22b) in the form

$$\omega_{SH}(q_\parallel) = c_t q_\parallel \left[1 - \frac{\Delta M_{22}^2(q_\parallel | c_t q_\parallel)}{2q_\parallel^2}\right]$$

$$= c_t q_\parallel \left[1 + \frac{\delta^4}{a^4} (\omega_1(aq_\parallel) - i\omega_2(aq_\parallel))\right]. \qquad (4.25)$$

In the absence of surface roughness the frequency of this wave is given by $\omega(q_\parallel) = c_t q_\parallel$, and the wave in fact is not a surface acoustic wave but instead is a surface skimming bulk transverse wave. In the presence of surface roughness the dimensionless frequency $\omega_1(aq_\parallel)$ is negative, so that a surface acoustic wave is created. In Fig. 12 is plotted $a\mathrm{Re}\,\omega_{SH}(q_\parallel)/c_t = aq_\parallel[1+(\delta^4/a^4)\,\omega_1(aq_\parallel)]$ as a function of $aq_\parallel$ for the choices $\delta/a = 0.05$, $c_t = 2.9 \times 10^5$ cm/sec. The attenuation length of this wave is given by

$$\ell^{-1}(q_\parallel) = 2\frac{\delta^4}{a^4} q_\parallel \omega_2(aq_\parallel). \qquad (4.26)$$

In the limit as $aq_\parallel \to 0$ the right hand side of this equation tends to zero as $(aq_\parallel)^6 \propto \omega^6$. This result can be understood as due to the ω^4 frequency dependence of Rayleigh scattering in the present case and to the fact that the penetration depth of this wave is proportional to the square of its wavelength along the surface. As a consequence of the last result the shear horizontal surface acoustic wave discussed here is only weakly bound to the randomly rough surface, i.e. it penetrates deeply into the medium.

We have here another example of surface roughness giving rise to a surface acoustic wave of a kind that cannot exist on a planar surface.

The propagation of surface acoustic waves of sagittal and shear horizontal polarization across a random grating has also been investigated theoretically [49,50)] and experimentally [52]. By a random grating we

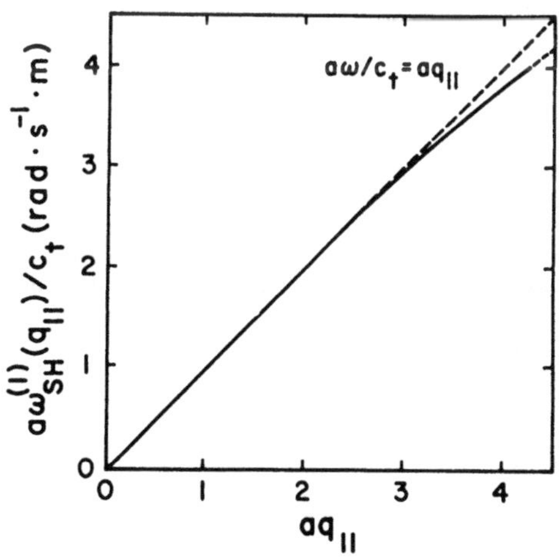

Fig. 12. The real part of the frequency of a shear horizontal surface acoustic wave propagating across a randomly rough aluminum surface ($c_t = 2.9 \times 10^3$ m/sec) with $\delta/a = 0.05$ [Ref. 48].

mean a surface defined by the equation $x_3 = \zeta(x_1)$, where the surface profile function $\zeta(x_1)$ is a function of x_1 alone. The sagittal plane is assumed to be the x_1x_3 plane, and the waves are assumed to propagate in the x_1-directions.

The frequencies of Rayleigh waves on such a surface are still given by Eq. (4.23); those of shear horizontal surface acoustic waves are still given by Eq. (4.25). The dependence of the dimensionless frequency shift $\omega_1(aq_\parallel)$ on the variable $aq_\parallel$ is qualitatively and quantitatively similar for both polarizations to the dependence of $\omega_1(aq_\parallel)$ on $aq_\parallel$ in the case of surface acoustic waves of the corresponding polarization on a randomly rough surface characterized by a surface profile function $\zeta(\vec{x}_\parallel)$ that is a function of both x_1 and x_2. There is, however, a significant qualitative difference between the dependence of the inverse attenuation length $\ell^{-1}(q_\parallel)$ on $aq_\parallel$ in the limit as $aq_\parallel \to 0$ for surface acoustic waves on a random grating and for surface acoustic waves on a two-dimensionally rough surface. For a random grating the right hand side of Eq. (4.24) (for a Rayleigh wave) is proportional to $(aq_\parallel)^4$ in this limit, while the right hand side of Eq. (4.26) (for a shear horizontal surface acoustic wave) is proportional to $(aq_\parallel)^5$. The difference from the results for a two-dimensionally rough surface are due to the fact that the frequency dependence of Rayleigh scattering from the hills and valleys of a random grating is ω^3, because these surface perturbations are two-dimensional, being defined by the equation $x_3 = \zeta(x_1)$. The penetration depths of Rayleigh waves and shear horizontal surface acoustic waves are still proportional to the wavelength and to the square of the wavelength, respectively, of these waves along the rough surface, which provide the additional factors of ω and ω^2 to the frequency dependence of the inverse attenuation lengths of these waves in the low frequency limit.

Experimental results that distinguish between the different frequency dependences of the inverse attenuation lengths of surface acoustic waves on these two types of randomly rough surfaces would certainly be welcome. The results of Ref. [52] are not sufficiently accurate for this purpose.

5. Some Directions for Future Research

To conclude this paper we offer some suggestions for directions in which future research into surface acoustic waves on rough surfaces could go. As in the body of this paper we consider deterministic, nonperiodic surfaces, periodically corrugated surfaces, and randomly rough surfaces in turn.

5.1. A Deterministic, Nonperiodic Surface

Because many, if not most, spatially localized perturbations of a planar surface are of bounded extent in both the x_1- and x_2-directions on the otherwise planar surface defined by the plane $x_3 = 0$, it would be of interest to extend the discussion of acoustic surface shape resonances presented in Section 2 to protuberances of such form. At the same time, it would be worthwhile to examine possible experimental signatures of such surface excitations. One technique that would seem to offer a way of detecting them is the scattering of surface acoustic waves from an isolated localized protuberance on an otherwise planar surface with the resonant excitation of the associated acoustic surface shape resonances. The scattering cross section as a function of the frequency of the incident surface acoustic wave should show a dip at the frequency of the acoustic surface shape resonances. Existing calculations of such scattering processes [8,53] have been carried out in first Born approximation, and hence cannot show such features.

5.2 A Periodically Corrugated Surface

In spite of the great deal of work that has been done on the propagation of surface acoustic waves across a classical grating, i.e. a surface defined by a surface profile function $\zeta(x_1)$ that is a periodic function of only one of the coordinates in the plane of the mean surface $x_3 = 0$, at least one gap remains to be filled. This is the lack of a non-perturbative calculation of the attenuation of shear horizontal surface waves propagating across such a surface, to complement the same calculations that have been carried out for sagittally polarized surface acoustic waves [32].

Such calculations would be carried out for a linear elastic medium. Interesting results can be expected, however, in the study of the propagation of sagittally and shear horizontally polarized surface acoustic waves across a classical grating when the elastic nonlinearity of the substrate is taken into account. The reason is that the dispersion curve of a surface acoustic wave on a nonlinear medium depends on the amplitude of the wave at the surface or, alternatively, on the power carried by the wave. This means that the position and width of a stop band in the dispersion curve of a surface acoustic wave on a nonlinear grating will become a function of these quantities. Consequently, one can think of the wave itself modulating the stop band as its intensity or power is varied, for example switching from transmission to reflection conditions when the frequency of the wave is close to the edge of the pass band. This is an effect that might be useful in device applications.

The overwhelming majority of theoretical studies of the propagation of surface acoustic waves across periodically corrugated surfaces have assumed a classical grating surface. For completeness, such calculations should be extended to bigratings, i.e. to surfaces defined by a surface profile function $\zeta(\vec{x}_\parallel)$ that is a periodic function of both x_1 and x_2. The only calculations of this kind of which we are aware [54] are perturbation-theoretic in nature, and are therefore limited to bigratings with a small corrugation strength. Nonperturbative calculations are required. Both the dispersion curves and attenuation of surface acoustic waves on such surfaces should be calculated. The results of such calculations could well be useful in the design of filters for surface acoustic waves and of transducers for the conversion of surface acoustic waves into bulk waves, and vice versa.

In the case of surface acoustic waves propagating across the periodically corrugated interface between two different elastic media, the case of waves of sagittal polarization propagating across a classical grating remains to be studied nonperturbatively. In view of the ability of such surfaces or interfaces to bind surface acoustic waves that cannot exist at planar surfaces or interfaces, one can expect that the conditions for the existence of a Stoneley wave, which are notoriously difficult to satisfy [55-57], will be relaxed in the presence of a corrugated interface between two different media.

5.3. A Randomly Rough Surface

Randomly rough surfaces that are prepared by distributing identical protuberances randomly and uniformly on the planar surface of a semi-infinite substrate that is formed from the same or a different medium as the protuberances offer some interesting possibilities in the theory of surface acoustic waves. One of them is the interaction of surface acoustic waves, e.g. Rayleigh waves, propagating across such a surface with the acoustic surface shape resonances associated with each protuberance. The dispersion curve of the former is essentially a linear function of its two-dimensional wave vector; that of the latter is essentially independent of this wave vector. These two dispersion curves cross, because the frequencies of acoustic surface shape resonances overlap those of the surface and bulk acoustic waves of the substrate. One therefore expects to see a hybridization of these two types of surface excitations with an attendant repulsion of levels, similar to what is found in the hybridization of the delocalized, dispersive s-bands with the localized, nondispersive d- or f-bands in the electronic structure of magnetic metals.

Although this paper is primarily concerned with thermally excited surface excitations, whose theoretical study constitutes a homogeneous problem (no external sources), there are two problems in the scattering of an acoustic wave from a randomly rough solid surface (an inhomogeneous problem) that warrant study, in our view. The first is the scattering of an acoustic wave, from the liquid side, onto a randomly rough, planar liquid-solid interface. The analagous calculation for an acoustic wave incident from within a solid onto the randomly rough interface of that solid with air, has been described at this Symposium by Professor McGurn. In these calculations it has been found that the angular dependence of the intensity of the diffuse component of the scattered waves has a peak in the retroreflection direction when the scattering occurs with no change of polarization, and in a somewhat different backward direction when the scattering takes place with a change in polarization. This result is interesting because under certain conditions this backscattering peak can be related to the surface roughness-induced localization of the surface acoustic waves that can be excited by the incident wave. The localization to which we refer is in directions parallel to the mean surface, and is characterized by the surface acoustic wave ceasing to be a propagating excitation, and having a displacement field that is sensibly nonzero within a region of the surface whose radius is called the localization length. The conditions for this to happen is that the inelastic mean free path of the surface acoustic wave must be much longer than the elastic mean free path, and that the wavelength of the surface acoustic wave be larger than the localization length. When these conditions are satisfied the angular width of the peak in the angular dependence of the diffuse component of the scattered radiation is a measure of the localization length. Similar effects are expected in the scattering of an acoustic wave from a randomly rough liquid-solid interface, with similar physical origins and interpretations. However, experimentally it is much easier to have the emitter and detector of the acoustic radiation in a fluid than in a solid, and it is for this reason that this investigation is proposed.

The second scattering calculation we think is worth carrying out also involves the scattering of an acoustic wave, incident from the liquid side, onto a randomly rough, planar liquid-solid interface. Here, what is to be calculated are the fluctuations in the intensity of the scattered acoustic waves, rather than in the amplitude. In effect, what is being proposed is a calculation of the speckle pattern produced by this type of scattering. Such calculations for other random systems [58] have yielded interesting results, and it is anticipated that equally interesting results will be obtained from the calculations proposed.

Finally, we point out that most studies of the effects of surface roughness on the propagation of surface acoustic waves have assumed that the surface studied is planar in the absence of the roughness. However, surface acoustic waves on curved surfaces, e.g. cylindrical and spherical surfaces, are of interest in several contexts [59]. The study of the effects of surface roughness on surface acoustic waves on curved surfaces would therefore also be of interest, to show how the properties of these waves on ideal curved surfaces are affected by roughness that is naturally or artificially present on these surfaces.

Acknowledgement

This work was supported in part by NSF Grant No. DMR 85-17634.

References

1. R. W. Rendell, D. J. Scalapino, and B. Mühlschlegel, Phys. Rev. Lett. 41, 1746 (1978).
2. R. W. Rendell and D. J. Scalapino, Phys. Rev. B24, 3276 (1981).
3. D. W. Berreman, Phys. Rev. 163, 855 (1967).
4. R. Ruppin, Solid State Commun. 39, 903 (1981).
5. P. C. Das and J. I. Gersten, Phys. Rev. B25, 6281 (1982).
6. A. G. Mal´shukov and Sh. A. Shekhmamet´ev, Fiz. Tver. Tela 25, 2623 (1983) [Sov. Phys. - Solid State 25, 1509 (1983)].
7. A. A. Maradudin and W. M. Visscher, Z. Phys. B60, 215 (1985).
8. A. A. Maradudin, in Surface Waves in Plasmas and Solids, ed. S. Vukovich (World Scientific, Singapore, 1986), p. 220.
9. M. Ciftan and E. Saibel, Solid State Commun. 27, 435 (1978).
10. M. Ciftan, V. Ruck, and E. Saibel, Solid State Commun. 27, 439 (1978).
11. A. A. Maradudin, P. Ryan, and A. R. McGurn (unpublished work).
12. The result presented in Fig. 1 corrects Fig. 3 of Ref. 13, which contained a preliminary account of the work presented here.
13. A. A. Maradudin, in Physics of Phonons ed. T. Paszkiewicz (Springer-Verlag, New York, 1987), p. 82.
14. L. M. Brekhovskikh, Akust. Zh. 5, 288 (1959) [Sov. Phys. - Acoustics 5, 288 (1960)].
15. P.V.H. Sabine, Electron. Lett. 6, 149 (1970).
16. A. Ronnekleiv and J. Souquet, IEEE Trans. Sonics and Ultrasonics SU-23, 188 (1976).
17. S. Datta and B. J. Hunsinger, IEEE Trans. Sonics and Ultrasonics SU-27, 333 (1980).
18. H. S. Tuan and J. P. Parekh, IEEE Trans. Sonics and Ultrasonics SU-24, 384 (1977).
19. H. A. Haus, IEEE Trans. Sonics and Ultrasonics SU-24, 259 (1977).
20. S. R. Seshadri, J. Acoust. Soc. Am. 65, 687 (1979).
21. A. D. Lapin, Akust. Zh. 25, 766 (1979) [Sov. Phys. - Acoustics 25, 432 (1979)].
22. A. D. Lapin, Akust. Zh. 26, 104 (1980) [Sov. Phys. - Acoustics 26, 55 (1980)].
23. V. I. Grigor´evskii and V. P. Plesskii, Pis´ma Zh. Tekh. Fiz. 5, 1398 (1979) [Sov. Tech. Phys. Lett. 5, 588 (1979)].
24. R. F. Humphryes and E. A. Ash, Electron. Lett. 5, 175 (1969).
25. I. D. Akhromeeva and V. V. Krylov, Akust. Zh. 23, 510 (1977) [Sov. Phys. - Acoustics 23, 292 (1977)].
26. Yu. V. Gulyaev, T. N. Kurach, and V. P. Plesskii, Pis´ma Zh. Tekh. Fiz. 5, 272 (1979) [Sov. Tech. Phys. Lett. 5, 111 (1979)].
27. Yu. V. Gulyaev, T. N. Kurach, and V. P. Plesskii, Pis´ma Zh. Eksp. Teor. Fiz. 29, 563 (1979) [Sov. Phys. - JETP Letters 29, 513 (1979).
28. Yu. V. Gulyaev, T. N. Kurach, and V. P. Plesskii, Akust. Zh. 26, 540 (1980) [Sov. Phys. - Acoustics 26, 296 (1980)].
29. Yu. V. Gulyaev and V. P. Plesskii, Fiz. Tver. Tela 21, 3479 (1979) (Sov. Phys. - Solid State 21, 2010 (1979)].

30. A. N. Avdeev and V. P. Plesskii, Akust. Zh. 28, 289 (1982) [Sov. Phys. - Acoustics 28, 173 (1982)].
31. N. E. Glass, R. Loudon, and A. A. Maradudin, Phys. Rev. B24, 6843 (1981).
32. N. E. Glass and A. A. Maradudin, J. Appl. Phys. 54, 796 (1983).
33. Lord Rayleigh, Phil. Mag. 14, 70 (1907); The Theory of Sound vol. II, 2nd ed. (Dover, New York, 1945), p. 89.
34. Equivalently, the branch cut can be chosen to coincide with the negative real axis, provided that one then seeks solutions on the lower sheet of the corresponding Riemann surface.
35. K. A. Ingebrigtsen and A. Tonning, Phys. Rev. 184, 942 (1968).
36. F. Rischbieter, Acustica 16, 75 (1965).
37. B. A. Auld, J. J. Gagnepain, and M. Tan, Electronics Lett. 12, 650 (1976).
38. Yu. V. Gulyaev and V. P. Plesskii, Zh. Tekh. Fiz. 48, 447 (1978) [Sov. Phys. - Technical Physics 23, 266 (1978)].
39. N. E. Glass and A. A. Maradudin, Electronics Lett. 17, 773 (1981).
40. See, for example, A. A. Maradudin, in Nonequilibrium Phonon Dynamics, ed. W. E. Bron (Plenum, New York, 1985), p. 395.
41. Yu. V. Gulyaev and V. P. Plessky, Electronics Lett. 15, 64 (1979).
42. Yu. V. Gulyaev and V. P. Plessky, Wave Electronics 4, 7 (1979).
43. B. Djafari-Rouhani and A. A. Maradudin (unpublished work).
44. E. I. Urazakov and L. A. Fal´kovskii, Zh. Eksp. Teor. Fiz. 63, 2297 (1972) [Sov. Phys. - JETP 36, 1214 (1973)].
45. A. A. Maradudin and D. L. Mills, Ann. Phys. (N.Y.) 100 262 (1976).
46. A. G. Eguiluz and A. A. Maradudin, Phys. Rev. B28, 711 (1983).
47. A. G. Eguiluz and A. A. Maradudin, Phys. Rev. B28, 728 (1983).
48. O. Hardouin Duparc and A. A. Maradudin, J. Electron. Spectr. and Rel. Phenom. 30, 145 (1983).
49. A. A. Bulgakov and S. I. Khankina, Solid State Commun. 44, 55 (1982).
50. Xuemei Huang and A. A. Maradudin, Phys. Rev. B (to appear).
51. R. F. Wallis, D. L. Mills, and A. A. Maradudin, Phys. Rev. B19, 3981 (1981).
52. M. de Billy, G. Quentin, and E. Baron, J. Appl. Phys. 61, 2140 (1987).
53. V. G. Polevoi, Akust. Zh. 29, 91 (1983) [Sov. Phys. - Acoustics 29, 52 (1983).
54. V. I. Pustovoit and L. A. Chernozatonskii, Proc. Int. Symp. on Surface Waves in Solids and Layered Structures (VINITI, Moscow Region, 1986), vol. I, p. 176.
55. K. Sezawa and K. Kanai, Bull. Earthquake Res. Inst. Tokyo Univ. 16, 504 (1938); 17, 1 (1939).
56. J. G. Scholte, Roy. Astron. Soc. London, Monthly Notices Geophys. Suppl. 5, 120 (1947).
57. G. W. Farnell and E. L. Adler, in Physical Acoustics, eds. W. P. Mason and R. N. Thurston, vol. IX, Chapter 2, (Academic Press, New York, 1972).
58. B. Shapiro, Phys. Rev. Lett. 57, 2168 (1986).
59. W. M. Ewing, W. S. Jardetzky, and F. Press, Elastic Waves in Layered Media (McGraw-Hill, New York, 1957).

Rayleigh Waves Along a Corrugated Surface

A. Boström

Division of Mechanics, Chalmers University of Technology,
S-41296 Göteborg, Sweden

1 Introduction

The Rayleigh wave on an elastic half-space and its various generalizations for layered media and nonplanar interfaces are of great importance for SAW devices and also for other applications in ultrasonics and earthquake engineering. In the present paper the implications of a periodically corrugated surface on an elastic half-space are investigated. Using the null field approach this problem has previously been studied by BOSTRÖM [1] although only propagating waves and mostly in the direction along the corrugations were explored numerically. Here a modal approach is used but as the results will also be presented elsewhere [2] the presentation is rather cursory with only a few numerical examples.

2 The Dispersion Relation

Consider an elastic half-space with a corrugated surface which for simplicity is assumed to be sinusoidally corrugated:

$$z = d \cos \pi y/b. \tag{1}$$

The elastic medium occupies the region $z < d \cos \pi y/b$ and has density ρ and Lamé constants λ and μ. Time harmonic conditions are assumed and the factor $\exp(-i\omega t)$ is suppressed. The equation of motion for the displacement field $\mathbf{u}$ in the elastic medium is thus

$$k_p^{-2} \nabla\nabla\cdot\mathbf{u} - k_s^{-2} \nabla\times\nabla\times \mathbf{u} + \mathbf{u} = 0, \tag{2}$$

where k_p and k_s are the longitudinal and transverse wavenumbers, respectively. On the boundary of the half-space the traction must vanish:

$$\mathbf{t} = \hat{\mathbf{n}}\lambda\, \nabla\cdot\mathbf{u} + 2\mu\, \partial\mathbf{u}/\partial n + \mu\, \hat{\mathbf{n}} \times (\nabla\times\mathbf{u}) = 0, \tag{3}$$

where $\hat{n}$ is the unit normal to the corrugated surface.

Due to the periodicity of the surface in the y direction the surface wave can be assumed to be a Floquet sum in the y coordinate and a simple plane wave in x:

$$u = \sum_{jn} \alpha_{jn} \phi_{jn} \quad , \tag{4}$$

where the mode index j is summed from 1 to 3 and n is summed over all integers, α_{jn} are to be determined, and the plane waves are

$$\phi_{1n} = \nabla\times[\hat{z} \exp(i\{qx + p_n y - h_{sn} z\})] \tag{5}$$

$$\phi_{2n} = i/h_{sn} \nabla\times\phi_{1n} \tag{6}$$

$$\phi_{3n} = \nabla[\exp(i\{qx + p_n y - h_{pn} z\})]. \tag{7}$$

These waves satisfy the equation of motion and are of SH, SV, and P type, respectively. q is the wavenumber in the x direction and $p_n = p + n\pi/b$ that in the y direction. The wavenumbers in the z direction are

$$h_{sn} = (k_s^2 - k_n^2)^{1/2} \tag{8}$$

$$h_{pn} = (k_p^2 - k_n^2)^{1/2}, \tag{9}$$

where $k_n^2 = q^2 + p_n^2$ is the horizontal wavenumber. To obtain a true surface wave Im $h_{sn} > 0$ and Im $h_{pn} > 0$. Also surface waves on other Riemann sheets are, however, of interest, as will be apparent by the numerical results in the next section. Instead of the wavenumbers q and p it is from a practical point of view better to employ the horizontal wavenumber k and the angle of propagation v:

$$q = k \cos v, \quad p = k \sin v. \tag{10}$$

It must be remembered that k and v are only the wavenumber and angle for the n = 0 term in the expansion (4). For small corrugations this term usually dominates and k and v will therefore be referred to as the wavenumber and the angle of propagation, but that this is a slight abuse of terminology should be remembered. It is only for propagation along the ridges ($v = 0^o$ and $\alpha_{j,n} = \alpha_{j,-n}$ for all j and n) and perpendicular to the ridges ($v = 90^o$) that there exists a well-defined angle of propagation.

To determine the expansion coefficients α_{jn} and the dispersion relation, the boundary condition (3) is applied to the expansion (4), and after projection on a plane wave and integration in y over one period, this results in a matrix equation

$$\sum_{jn} \alpha_{jn} Q_{jnj'n'} = 0 \ . \tag{11}$$

The matrix Q has rather simple elements that contain a Bessel function and the various wavenumbers. To conserve space the explicit expressions are not given (see [2] for these). For (11) to have a nontrivial solution

$$\det Q = 0 \tag{12}$$

and this is the desired dispersion relation. It should be remarked that this dispersion relation is essentially the same as that obtained by the null field approach [2], a result that may depend on the special form of the surface.

3 Numerical Examples

Before giving examples for the corrugated surface, it is appropriate to recall the characteristic features of the Rayleigh wave along a planar surface. The dispersion relation for the Rayleigh wave is

$$4k^2(k_s^2 - k^2)^{1/2}(k_p^2 - k^2)^{1/2} + (2k_s^2 - k^2)^2 = 0, \tag{13}$$

where k is the wavenumber. This equation has only one solution on the physical Riemann sheet, i.e. when both roots have positive imaginary parts. For $k_s/k_p = 2$ (corresponding to a Poisson ratio of $\nu = 1/3$) this root is $k = k_R = 1.072\ k_s$. Another solution is $k = k_R$ on the Riemann sheet with both roots having negative imaginary parts. Finally there are four solutions on the Riemann sheets with different signs for the imaginary parts of the roots which have the numerical values $k = (0.514 \pm i\ 0.074)\ k_s$.

Figures 1 and 2 show how the wavenumber evolves when the corrugation height d/b is increased from 0 to 0.5 for the fixed frequency $k_s b = 1$ and the angle of propagation $\nu = 0^o$. The figures show the complex wavenumber plane with d/b as a parameter along the curves. Figure 1 is for the solution which starts on the sheet with $\mathrm{Im} h_{p0} < 0$ and Fig. 2 for that which starts on the sheet with $\mathrm{Im} h_{s0} < 0$. The solutions in Figs. 1 and 2 start with the same numerical value but then evolve quite differently with increasing d/b. The wavenumber in Fig. 1 starts on the sheet with $\mathrm{Im} h_{p0} < 0$, then in turn crosses to the sheet with $\mathrm{Im} h_{p0} < 0$ and $\mathrm{Im} h_{s0} < 0$ for d/b = 0.23, to the physical sheet for d/b = 0.26, and to the sheet with $\mathrm{Im} h_{s0} < 0$ for d/b = 0.33. As the ordinary Rayleigh wave stays on the physical sheet for all d/b (see [1,2] for plots of how the wavenumber for the ordinary Rayleigh wave evolves as d/b increases), there are thus two physical solutions for 0.26 < d/b < 0.33, one that is undamped and the damped one in Fig. 1. The damping is moderate and it is noted that the wave speed of the solution in Fig. 1 is in the vicinity of the P wave speed and is thus about twice the ordinary Rayleigh wave speed. It should also be remarked that there is a second undamped solution on the physical Riemann sheet for d/b > 0.38 [1,2], but this solution does not evolve from the solutions in Figs. 1 and 2. The solution in Fig. 2 never enters the physical sheet, but from its start on the sheet with $\mathrm{Im} h_{s0} < 0$ it crosses to the sheet with $\mathrm{Im} h_{p0} < 0$ for d/b = 0.28 and to the sheet with $\mathrm{Im} h_{p0} < 0$ and $\mathrm{Im} h_{s0} < 0$ for d/b = 0.36. For d/b > 0.36 it has a vanishing imaginary part and a real part that stays close to $k = -k_s$.

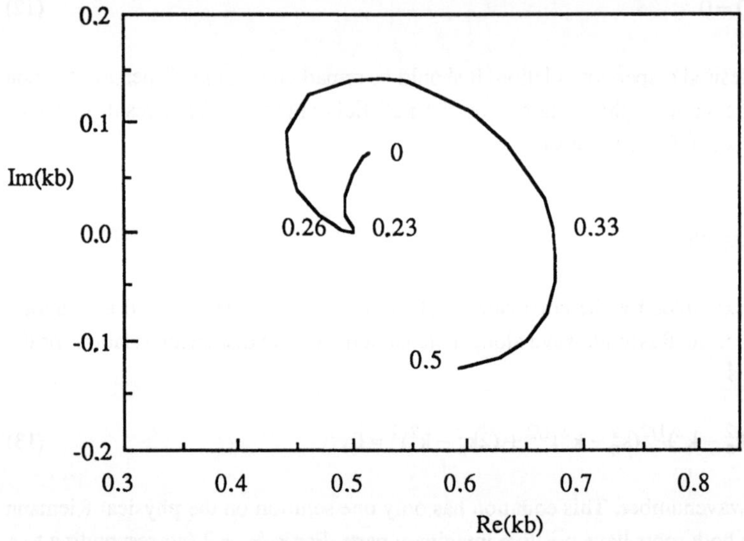

Fig. 1 The wavenumber of the mode which starts on the Riemann sheet with $\mathrm{Im} h_{p0} < 0$ for the frequency $k_s b = 1$ and angle of propagation $v = 0^o$ and with increasing values of d/b along the curve; the cut from the root h_{s0} runs along the real axis and the cut from the root h_{p0} starts at kb = 0.5 and runs to the left along the real axis

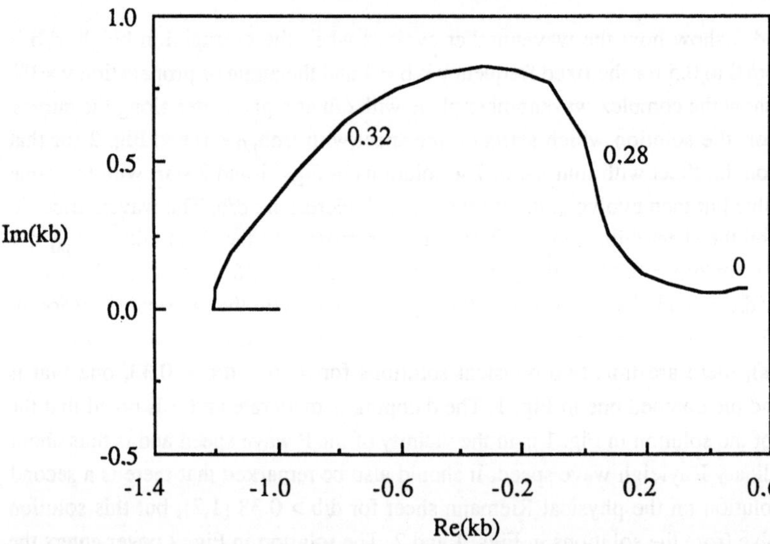

Fig. 2 The wavenumber of the mode which starts on the Riemann sheet with $\mathrm{Im} h_{s0} < 0$ for the frequency $k_s b = 1$ and angle of propagation $v = 0^o$ and with increasing values of d/b along the curve; the cuts from the root h_{s0} start at kb = ± 1, run to the origin, and up and down along the imaginary axis, and the cuts from the root h_{po} start at kb ± 0.5, run to the origin, and up and down along the imaginary axis

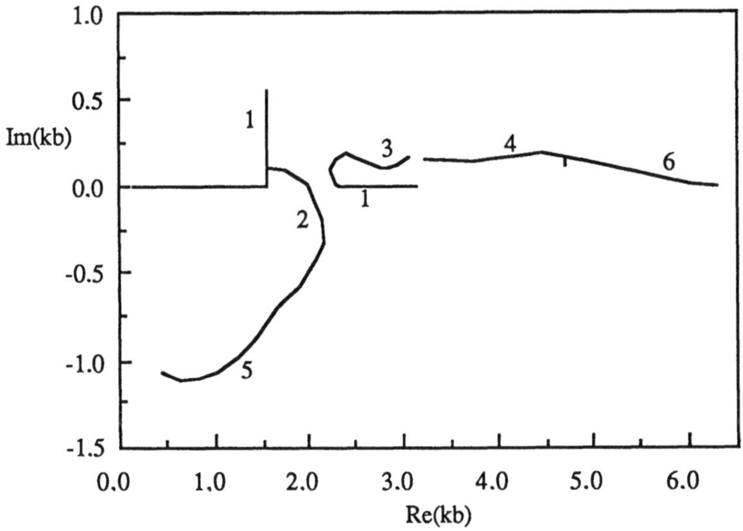

Fig. 3 The wavenumber of the physical mode for the angle of propagation $v = 90^0$ and corrugation height $d/b = 0.2$ and with increasing values of the frequency $k_s b$ along the curves; the cuts from the roots $h_{s,-n}$ and $h_{p,-n}$ start at $kb = n\pi \pm k_s b$ and $kb = n\pi \pm 0.5\, k_s b$, respectively, then run to $kb = n\pi$, and finally straight up and down

Figure 3 shows how the wavenumber evolves when the frequency $k_s b$ is increased from 0 to 6.23 for the corrugation height $d/b = 0.2$ and the angle of propagation $v = 90^0$. The figure shows the complex wavenumber plane kb with $k_s b$ as a parameter along the curves. The cuts from h_{p0}, h_{s0}, $h_{p,-1}$, $h_{s,-1}$, $h_{p,-2}$, and $h_{s,-2}$ are not shown (it must be remembered that the endpoints of the cuts move when $k_s b$ is varied). It is primarily the solutions on the physical Riemann sheet that are given. For low frequencies the wavenumber starts out real. For $k_s b = 1.18$ the wavenumber becomes $kb = 1.57$ ($= \pi/2$) and the first stopband then appears, inside which the real part remains constant and the imaginary part first increases and then decreases. Soon after leaving the stopband the solution enters the Riemann sheet with $\mathrm{Im} h_{s,-1} < 0$ for $k_s b = 1.72$ where it then stays for all frequencies considered. Another branch, which starts at $kb = \pi$ for low frequencies on the sheet with $\mathrm{Im} h_{s,-1} < 0$ and crosses to the physical sheet at $k_s b = 1.22$, instead takes over as the solution most closely resembling the ordinary Rayleigh wave. For low frequencies this solution, in fact, corresponds to one of the unphysical solutions to the Rayleigh equation for the flat surface [2]. For $1.22 < k_s b < 1.72$ there are thus two solutions on the physical sheet. At $k_s b = 2.90$ there is a discontinuity in the curve, and this is a resonance between the wave and its counterpart in the opposite direction that could appropriatly be called a cross-over resonance as the two solutions exchange identities. Further increasing the frequency there is another stopband around $k_s b = 4.6$ when $\mathrm{Re}(kb) = 4.71$ ($=3\pi/2$). When the solution reaches $kb = 2\pi$ at $k_s b = 6.23$ it crosses several cuts, but it has not been followed any further.

4 Concluding remarks

In the present paper a modal approach has been used to determine the dispersion relation for Rayleigh waves along the corrugated surface of an elastic half-space. The wavenumber of these waves show an intricate behaviour with stopbands, cross-over resonances, crossings between different Riemann sheets, and so on.

References

1. A. Boström: Appl. Sci. Res. 39, 129 (1982)
2. A. Boström: Comm. Division of Mechanics 1987:1 (Göteborg 1987)

Localization Effects in the Scattering of Acoustic Waves from a Random Stress-free Grating Surface

A.R. McGurn[1] *and A.A. Maradudin*[2]

[1]Department of Physics, Western Michigan University, Kalamazoo, MI 49008, USA

[2]Department of Physics, University of California, Irvine, CA 92717, USA

The angular dependence of the intensity of acoustic waves scattered diffusely from a random grating surface that is stress free is calculated. The waves are assumed to be incident on the surface from within the solid. The calculation proceeds by means of a diagrammatic expansion of the intensity of the scattered radiation in powers of the surface profile function, similar to the one used by Vollhardt and Wölfle in their study of electrons in a disordered medium. A striking feature of the resulting differential reflection coefficient is a narrow peak in the retroreflection direction, whose intensity is polarization-dependent. Under appropriate conditions this feature is related to the surface roughness-induced localization of Rayleigh waves along the surface. The width of the peak in the differential reflection coefficient under these conditions provides a determination of the localization length of the Rayleigh wave, i.e. the distance along the surface over which its displacement field is sensibly nonzero.

I. Introduction

We calculate the diffuse component of the elastic scattering of bulk acoustic waves incident on a grating that is ruled on a semi-infinite elastic medium whose mean surface is planar. The bulk elastic properties of the medium are taken to be isotropic, and the surface is considered to be stress free [1]. The acoustic scattering is found to arise from simple surface reflections as well as from the resonant excitation of Rayleigh waves.

This Rayleigh wave contribution to the diffuse scattering is of interest because the surface roughness causes the Rayleigh waves to become localized or confined to a subregion of the grating [2]. As the Rayleigh wave localization represents a drastic change in the Rayleigh wave functions from their extended planar surface forms, we expect that the localization will exhibit some distinctive signature in the diffuse scattering of bulk waves arising from resonant processes. Indeed, the Rayleigh wave localization is found to contribute a distinctive backscattering (anti-specular) peak in the angular dependence of the intensity of the diffuse scattering of bulk acoustic waves. This peak is found to have a full-width-at-half-maximum that is related to the linear size of the region perpendicular to the rulings of the grating over which the Rayleigh waves are confined. Our primary interest in the succeeding calculations will be to study the backscattering effects in the diffuse scattering which come from the localization of Rayleigh waves, and to obtain the size of the region of Rayleigh wave localization.

A very crude description of the events that give rise to localization-related backscattering proceeds as follows. An incident beam of radiation is sent from the bulk toward the rough surface. This beam can be taken to be in a state characterized by a wave vector whose component parallel to the surface is k. As the beam encounters the rough surface the loss of translational symmetry due to the surface roughness allows it to couple into Rayleigh wave modes of wavevector k_R propagating perpendicular to the grating grooves. These excited Rayleigh waves are essentially one-dimensional waves that move in the random potential created by the grooves of the grating and hence are localized by mechanisms originally explained by MOTT and TWOSE [3]. The Rayleigh waves are reflected from the edge of their region of localization. This reflection represents a rough time reversal of the Rayleigh wavetrain similar to that of a harmonic wave traveling on a rope that is fixed to a wall. The "time-reversed" Rayleigh wave states can then be rescattered by the surface roughness into outgoing radiative waves with wavevector -k opposite to that of the incident beam. These outgoing states then are responsible for the backscattering peaks.

Since in this process incoming states labeled by k are reflected into outgoing states labeled by -k, for scattering events that result in polarization changes the localization peak will appear at scattering angles that differ from the antispecular scattering angle, due to the polarization dependence of the speeds of sound. For scattering events in which the polarization remains unchanged, though, the backscattering peak is observed at the antispecular scattering angle.

Two theories for the backscattering effect are studied in this paper. One theory considers the weak localization limit in which Rayleigh wave attenuation due to internal friction present in the medium is large compared to attenuation by surface roughness. The second theory considers the opposite limit in which attenuation due to surface scattering dominates that from internal frictional losses. In the former case the width of the peak in the backscattering direction is related to both the localization length and the decay length of Rayleigh waves due to internal frictional losses. In the latter theory the peak width is inversely proportional to the localization length.

In Section II we shall formulate the acoustic scattering problem. Acoustic scattering in both the limits of large and small internal frictional attenuation is studied and a result for the localization length of surface Rayleigh waves is presented. Our conclusions are given in Section III.

II. Formulation

An isotropic elastic medium, characterized by mass density ρ and complex longitudinal and transverse speeds of sound c_ℓ and c_t, occupies the region $x_3 > \zeta(x_1)$, where the x_1- and x_3-axes are in and perpendicular to the mean surface, respectively. The surface profile function $\zeta(x_1)$ is assumed to be a Gaussianly distributed random variable [4]. The first two moments of its probability distribution function are

$$\langle\zeta(x_1)\rangle = 0 \qquad (1.1)$$

and

$$\langle\zeta(x_1)\zeta(x_1')\rangle = \sigma^2 W(|x_1-x_1'|) \qquad (1.2a)$$

$$\langle\zeta^2(x_1)\rangle = \sigma^2, \qquad (1.2b)$$

where $\langle\ldots\rangle$ indicates an average over the ensemble of realizations of $\zeta(x_1)$. In our calculations we will assume that $W(|x_1|) = \exp(-x_1^2/a^2)$.

The x_1x_3-plane is the plane of incidence as well as the plane of polarization of the acoustic waves. If we invoke the Rayleigh hypothesis [1] the scattering solutions are then of the standard form (a time dependence given by $\exp(-i\omega t)$ has been suppressed)

$$u_1(x_1,x_3|\omega) = e^{ikx_1}[kA_\ell(k)e^{-i\beta_\ell(k\omega)x_3} + \beta_t(k\omega)A_t(k)e^{-i\beta_t(k\omega)x_3}]$$

$$+ \int\frac{dq}{2\pi}\, e^{iqx_1}[qR_\ell(q)e^{i\beta_\ell(q\omega)x_3} + \beta_t(q\omega)R_t(q)e^{i\beta_t(q\omega)x_3}], \quad x_3 > \zeta(x_1)_{max} \tag{1.3}$$

$$u_3(x_1,x_3|\omega) = e^{ikx_1}[-\beta_\ell(k\omega)A_\ell(k)e^{-i\beta_\ell(k\omega)x_3} + kA_t(k)e^{-i\beta_t(k\omega)x_3}]$$

$$+ \int\frac{dq}{2\pi}\, e^{iqx_1}[\beta_\ell(q\omega)R_\ell(q)e^{i\beta_\ell(q\omega)x_3} - qR_t(q)e^{i\beta_t(q\omega)x_3}], \quad x_3 > \zeta(x_1)_{max} \tag{1.4}$$

with

$$\beta_{\ell,t}(q\omega) = [(\frac{\omega}{c_{\ell,t}})^2 - q^2]^{1/2}, \qquad q^2 < (\frac{\omega}{c_{\ell,t}})^2$$

$$= i[q^2 - (\frac{\omega}{c_{\ell,t}})^2]^{1/2}, \qquad q^2 > (\frac{\omega}{c_{\ell,t}})^2, \tag{1.5}$$

where u_1 and u_3 are the components along the x_1- and x_3-axes of the displacement field in the medium.

The time-averaged incident and reflected fluxes are

$$I_{in} = \int d^2x_\parallel |\langle P_3\rangle_{inc}|_{x_3 = \text{const}} = \frac{1}{2}\rho\omega^3 L_1L_2 \times$$

$$\times[\beta_\ell(k\omega)|A_\ell(k)|^2\Theta(\beta_\ell^2(k\omega))+\beta_t(k\omega)|A_t(k\omega)|^2\,\Theta(\beta_t^2(k\omega))] \tag{1.6}$$

and

$$I_{out} = \int d^2x_\parallel |\langle P_3\rangle_{scatt}|_{x_3 = \text{const}} = \frac{1}{2}\rho\omega^3 L_2 \times$$

$$\times \int\frac{dq}{2\pi}\,[\beta_\ell(q\omega)|R_\ell(q)|^2\Theta(\beta_\ell^2(q\omega))+\beta_t(q\omega)|R_t(q)|^2\,\Theta(\beta_t^2(q\omega))], \tag{1.7}$$

respectively, where P_3 is the 3-component of the time-averaged elastic Poynting vector, L_1 and L_2 are the lengths of the mean surface along the x_1- and x_2-axes, and $\Theta(x)$ is Heaviside's unit step function.

If the reflection amplitudes $R_\alpha(q)$ $(\alpha = \ell,t)$ are written in terms of the amplitudes of the incident waves $A_\gamma(k)$ $(\gamma = \ell,t)$ according to

$$R_\alpha(q) = \sum_\gamma R_{\alpha\gamma}(q|k)A_\gamma(k), \tag{1.8}$$

then from Eqs. (1.6) and (1.7) and the ratio I_{out}/I_{in} we can define the efficiency of the scattering of a wave (k) of polarization $\gamma(=\ell,t)$ into a wave (q) of polarization $\alpha(=\ell,t)$ to be

$$I_{\gamma\to\alpha}(q|k) = \frac{1}{L_1} \; |S_{\alpha\gamma}(q|k)|^2, \tag{1.9}$$

where

$$S_{\alpha\gamma}(q|k) = \left[\frac{\beta_\alpha(q\omega)}{\beta_\gamma(k\omega)}\right]^{1/2} R_{\alpha\gamma}(q|k)\Theta(\beta_\alpha^2(q\omega))\Theta(\beta_\gamma^2(k\omega)) \tag{1.10}$$

is the S-matrix.

The coefficients $\{R_{\alpha\beta}(q|k)\}$ are obtained when Eqs. (1.3) and (1.4) are used in the equations that express the condition that the surface $x_3 = \zeta(x_1)$ is stress free. They are found to satisfy the equation

$$\begin{pmatrix} I(-\beta_\ell(k\omega)|p-k)\,\dfrac{\omega^2(p-k) - 2c_t^2 p\beta_\ell^2(k\omega)}{\beta_\ell(k\omega)} & I(-\beta_t(k\omega)|p-k)(2c_t^2pk-\omega^2) \\ I(-\beta_\ell(k\omega)|p-k)(2c_t^2pk - \omega^2) & I(-\beta_t(k\omega)|p-k)\,\dfrac{c_t^2(\beta_t^2(k\omega)-k^2)p+k\omega^2}{\beta_t(k\omega)} \end{pmatrix}$$

$$= \int\frac{dq}{2\pi} \begin{pmatrix} I(\beta_\ell(q\omega)|p-q)\,\dfrac{\omega^2(p-q)-2c_t^2p\beta_\ell^2(q\omega)}{\beta_\ell(q\omega)} & I(\beta_t(q\omega)|p-q)(2c_t^2pq-\omega^2) \\ I(\beta_\ell(q\omega)|p-q)(\omega^2-2c_t^2pq) & I(\beta_t(q\omega)|p-q)\,\dfrac{c_t^2(q^2-\beta_t^2(q\omega))-q\omega^2}{\beta_t(q\omega)} \end{pmatrix}$$

$$\times \begin{pmatrix} R_{\ell\ell}(q|k) & R_{\ell t}(q|k) \\ R_{t\ell}(q|k) & R_{tt}(q|k) \end{pmatrix}, \tag{1.11}$$

where we use the notation

$$e^{i\alpha\zeta(x_1)} = \int\frac{dQ}{2\pi} e^{iQx_1} I(\alpha|Q) . \tag{1.12}$$

Following the methods developed in BROWN et al. [5], we can express the solutions to Eqs. (1.11) in the form

$$\bar{\bar{R}}(q|k) = 2\pi\delta(p-k)\bar{\bar{B}}(q)^{-1/2}\,\bar{\bar{R}}^{(o)}(q)\bar{\bar{B}}(q)^{1/2} + i\bar{\bar{F}}(q)\bar{\bar{T}}(q|k)\bar{\bar{F}}^T(k)\bar{\bar{B}}(k) \tag{1.13}$$

where

$$\overleftrightarrow{T}(p|k) = \overleftrightarrow{V}(p|k) + \int\frac{dq}{2\pi}\,\overleftrightarrow{V}(p|q)\overleftrightarrow{G}_o(q)\overleftrightarrow{T}(q|k) \tag{1.14}$$

and

$$\overleftrightarrow{R}(q|k) = \begin{pmatrix} R_{\ell\ell}(q|k) & R_{\ell t}(q|k) \\ R_{t\ell}(q|k) & R_{tt}(q|k) \end{pmatrix} \tag{1.15}$$

$$\overleftrightarrow{F}(q) = \begin{pmatrix} 1 & 0 \\ 0 & -i \end{pmatrix}\overleftrightarrow{G}_o(q) \tag{1.16}$$

$$\overleftrightarrow{G}_o(q) = \frac{4q}{D(q\omega)}\begin{pmatrix} 2iq\beta_t(q\omega) & q^2-\beta_t^2(q\omega) \\ q^2-\beta_t^2(q\omega) & 2iq\beta_\ell(q\omega) \end{pmatrix} \tag{1.17}$$

$$D(q\omega) = 4q^2\beta_\ell(q\omega)\beta_t(q\omega)+(q^2-\beta_t(q\omega))^2, \tag{1.18}$$

where

$$\overleftrightarrow{R}^{(o)}(q)$$

$$= \frac{1}{D(q\omega)}\begin{pmatrix} 4q^2\beta_t(q\omega)\beta_\ell(q\omega) - (q^2-\beta_t^2(q\omega))^2 & -4q\sqrt{\beta_t(q\omega)\beta_\ell(q\omega)}[q^2-\beta_t^2(q\omega)] \\ -4q\sqrt{\beta_\ell(q\omega)\beta_t(q\omega)}\,[q^2-\beta_t^2(q\omega)] & -4q^2\beta^2(q\omega)+(q^2-\beta_t^2(q\omega))^2 \end{pmatrix} \tag{1.19}$$

describes the reflection from a flat surface, and $B_{ij}(q) = \beta_i(q)\delta_{ij}$. To leading order in ζ we find

$$\overleftrightarrow{V}(p|k) = \hat{\zeta}(p-k)\begin{pmatrix} \frac{\omega^2}{8pk}\left\{\frac{\omega^2}{c_t^4} + 4pk\left(\frac{c_t^2-c_\ell^2}{c_t^2c_\ell^2}\right)\right\} & 0 \\ 0 & \frac{1}{8pk}\left(\frac{\omega}{c_t}\right)^4 \end{pmatrix}, \tag{1.20}$$

where $\hat{\zeta}(k)$ is the Fourier transform of $\zeta(x_1)$. This is the approximation to $\overleftrightarrow{V}(p|k)$ we shall employ in all calculations presented in this paper. The rough surface Rayleigh wave Green's function is then defined by

$$\overleftrightarrow{G}(p|k) = 2\pi\delta(p-k)\overleftrightarrow{G}_o(p) + \overleftrightarrow{G}_o(p)\overleftrightarrow{T}(p|k)\overleftrightarrow{G}_o(k), \tag{1.21}$$

and we can rewrite the scattering efficiency for diffuse scattering, averaged over the ensemble of realizations of $\zeta(x_1)$, in terms of this function as

$$\langle I_{\gamma\to\alpha}(q|k)\rangle_{diff} = \theta(\beta_\gamma^2(k\omega))\theta(\beta_\alpha^2(q\omega)) \frac{\beta_\alpha(q\omega)\beta_\gamma(k\omega)}{L_1} \langle|G_{\alpha\gamma}(q|k)|^2\rangle_{diff}, \tag{1.22}$$

where $\langle|G_{\alpha\gamma}(q|k)|^2\rangle_{diff}$ is that part of $\langle|G_{\alpha\beta}(q|k)|^2\rangle$ that contains no factor of $2\pi\delta(q-k)$.

The two particle Green's function in Eq. (1.22) is similar in form to the one encountered in the study of electronic conductivity in a disordered system [6,7], so that an expansion in powers of the surface profile pair correlation function, Eq. (1.2), similar to the expansions in powers of the impurity pair correlation function in the electron problem, can be made. In fact, the Feynman diagrams for this perturbation expansion are the same for both the electron and acoustic problems [6,7]. In both systems two subsets of diagrams are needed, in the limit of weak disorder, to determine the properties studied. These are the so-called ladder and maximally crossed diagrams [7] shown in Fig. 1, in which solid lines represent configurationally averaged single particle Green's functions and the dashed lines represent the two particle interaction potential which is proportional to the correlation function $\langle\zeta(x_1)\zeta(x_1')\rangle$. In the electron problem the ladder diagrams are responsible for the Drude conductivity [7], while in the acoustics problem they give rise to a general diffuse halo of scattered radiation at all scattering angles above the surface. The maximally crossed diagrams on the other hand are responsible for the localization of electrons in low-dimensional systems and for weak localization effects in three dimensional electron systems [7,8]. These maximally crossed diagrams are found in our problem to cause Rayleigh wave localization on the rough interface, which leads to backscattering (anti-specular) peaks in the acoustic scattering of bulk waves from the interface. The width of the backscattering peak is found to be inversely proportional to the Rayleigh wave decay length in lossy media and to the localization length in essentially lossless media.

In a lossy medium, i.e. one in which Rayleigh wave attenuation due to internal frictional losses is much larger than that due to scattering from the weak surface roughness, we find that the simple sum of the maximally crossed diagrams shown in Fig. 2 is sufficient to describe the backscattering effects. In this limit the two-particle Green's function in Eq. (1.22) is well represented by the sum of the ladder and the maximally crossed contributions of Fig. 1. Results in this approximation for the differential reflection coefficient $dR_{\gamma\to\alpha}/d\theta = (\omega/2\pi c_\alpha)\times \cos\theta\langle I_{\gamma\to\alpha}(q|k)\rangle_{diff}$ in $(\text{radian})^{-1}$ as a function of the outgoing wave angle θ measured from the normal to the mean surface are shown in Figs. 2 and 3 for $\omega a/(\text{Re } c_t) = 0.2$, $\omega\sigma/(\text{Re } c_t) = 0.01$, $(\text{Re } c_t)/(\text{Re } c_\ell) = 0.6$, and a medium in which the Rayleigh wave attenuation due to internal friction is six times greater than that due to scattering from surface

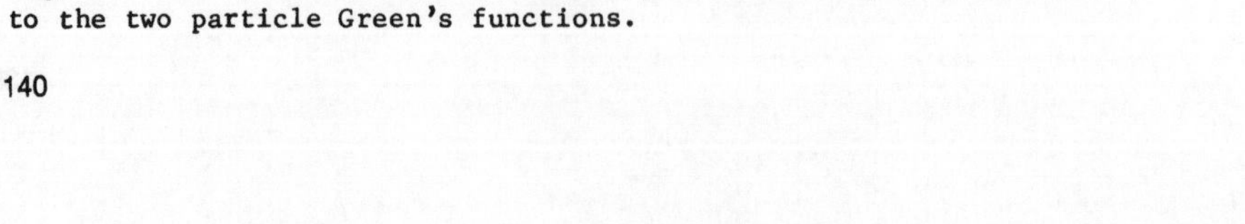

Fig. 1. The a) ladder and b) maximally crossed diagram contributions to the two particle Green's functions.

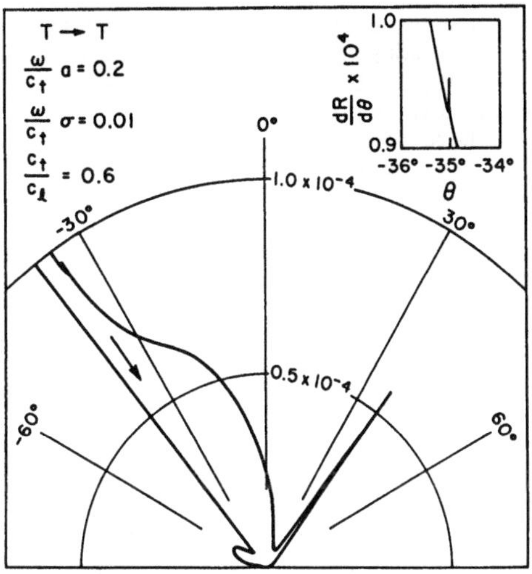

Fig. 2 Polar plot of $\frac{dR}{d\theta}$ versus θ for T → T scattering.

roughness. In these plots the angles of incidence of the incident beams are indicated by arrows, and results are presented for the differential reflection coefficient for the scattering of a transversely polarized wave with an angle of incidence of 35° into transversely polarized waves (Fig. 2) and for the scattering of a transversely polarized wave with an angle of incidence of 45° into longitudinally polarized waves (Fig. 3).

Insets in each figure present amplified views of the antispecular and non-antispecular backscattering peaks. Notice that the peaks are in the antispecular direction only when no polarization changes occur from incoming to outgoing states.

In a low loss or lossless medium in which surface scattering dominates attenuation of the Rayleigh waves due to internal friction, the sum of the maximally crossed diagrams in Fig. 1b becomes very large or even diverg-

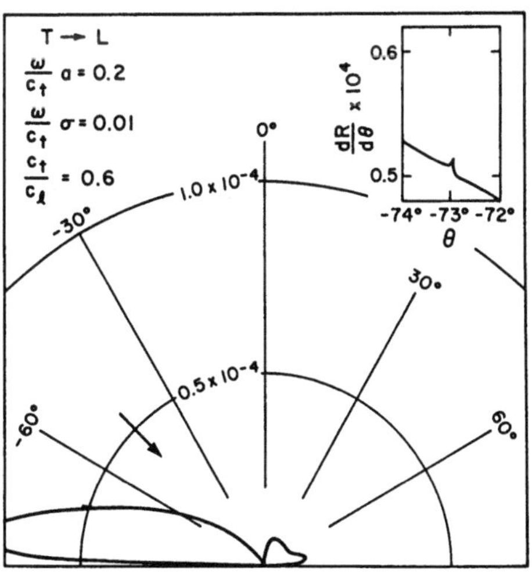

Fig. 3. Polar plot of $\frac{dR}{d\theta}$ versus θ for T → L scattering.

ent. In this case one must handle these contributions in the fully self-consistent manner developed by VOLLHARDT and WÖLFLE [7] for the electron system and later by CELLI et al. [10] in calculations of optical back-scattering effects. As the space here is limited, the fully self-consistent theory will be published elsewhere (see Ref. [9]). We only mention that the results for the differential reflection coefficient in the self-consistent theory are qualitatively similar to those for lossy media. Now, though, the backscattering peaks are higher and narrower. The Rayleigh wave localization length is found to be given by $\ell = 1/\Delta$ where Δ is the mean free path of the Rayleigh waves. For the parameters a and σ given in Fig. 2 we find $\ell \cong$ 200,000 wavelengths.

III. Conclusions

We have calculated various acoustic backscattering phenomena in the scattering of bulk waves from a random grating surface. These effects are related to the localization of Rayleigh waves by surface roughness.

Similar effects have been predicted by us in optics for the scattering of light from a rough metal surface on which surface polaritons are localized [10]. Recent experimental observations [11,12] of these optical phenomena encourage us to believe that the acoustic results described in this paper will also be confirmed experimentally.

Acknowledgements

The portion of the work of A.R.M. that was carried out at the University of California, Irvine, was supported by the Institute for Surface and Interface Science. The work of A.A.M. was supported in part by NSF Grant DMR 85-17634.

References

1. A. G. Eguiluz and A. A. Maradudin: Phys. Rev. B28, 711 (1983); Phys. Rev. B28, 728 (1983).
2. E. Abrahams, P. W. Anderson, D. C. Liciardello, and T. V. Ramakrishnan: Phys. Rev. Lett. 42, 673 (1979).
3. N. F. Mott and W. D. Twose: Adv. Phys. 10, 107 (1961).
4. A. A. Maradudin: in Surface Polaritons, eds. V. M. Agranovich and D. L. Mills (North-Holland, Amsterdam, 1982), p. 405.
5. G. C. Brown, V. Celli, M. Haller, and A. Marvin: Surf. Sci. 136, 381 (1984).
6. A. A. Abrikosov, L. P. Gor'kov and I. E. Dzyaloshinskii, Quantum Field Theoretical Methods in Statistical Physics ed. R. A. Silverman, (Pergamon Press, Oxford 1963), p. 327.
7. D. Vollhardt and P. Wölfle: Phys. Rev. B22, 4666 (1980); in Anderson Localization, eds. Y. Nagaoka and H. Fukuyama, (Springer-Verlag, New York, 1982), p. 26.
8. G. Bergmann: Phys. Rev. B28, 2914 (1983).
9. A. R. McGurn and A. A. Maradudin: (unpublished work).
10. A. R. McGurn, A. A. Maradudin, and V. Celli: Phys. Rev. B31, 4866 (1985); V. Celli, A. A. Maradudin, A. M. Marvin and A. R. McGurn: J. Opt. Soc. Am. A12, 2225 (1985); A. R. McGurn and A. A. Maradudin: J. Opt. Soc. Am. B6, 910 (1987).
11. A. A. Maradudin, A. R. McGurn, R. S. Dummer, Zu-Han Gu, A. Wirgin and W. Zierau: In Proceedings of the 3rd Bilateral US-USSR Symposium on Laser Optics of Condensed Matter, (Plenum, New York, 1987), (to appear).
12. E. R. Mendez and K. A. O'Donnell: Optics Commun. 61, 91 (1987).

A Comparison of Some Methods for Corrugated Waveguides

S.-E. Sandström

Institute of Theoretical Physics, Chalmers University of Technology, S-41296 Göteborg, Sweden

Abstract

Singly and doubly periodic parallel plate waveguides can be investigated by means of a number of methods. The techniques considered here are: the null-field method, the modal theory approach and the method of multiple scales. This is a brief outline of a forthcoming article [1] and will only address one aspect of the subject. Previous null-field results [2,3] are compared with the corresponding results obtained by means of the modal [4] and multiple scales [5] techniques. The first two methods are found to be essentially equivalent whereas the last one gives more of a qualitative indication as to the effects of a periodic corrugation. The most important of these effects is the resonance phenomenon that gives rise to stopbands. The first thing to compare is therefore the ability to accurately determine, for example, the stopband width. As seen in Fig. 1 the multiple scales technique does not agree with the other two methods. This discrepancy is not accounted for when the corrugation is weak since the accuracy of a perturbation technique should improve with decreasing corrugation.

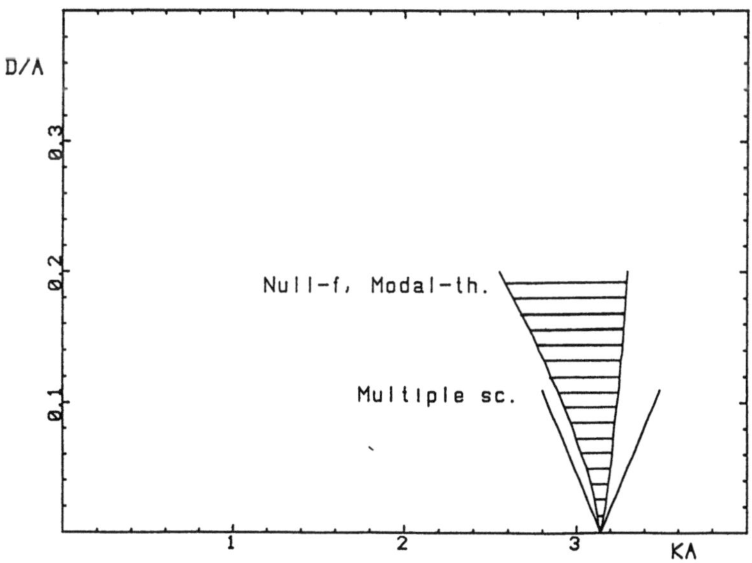

Figure 1. Stopband width versus corrugation height

References

1. S.-E. Sandström:"A comparison of some techniques for corrugated parallel plate waveguides," J. Acoust. Soc. Am., (to appear).
2. S.-E. Sandström: J. Acoust. Soc. Am. 79, 1293 (1986).
3. S.-E. Sandström: Wave Motion 9, 227 (1987).
4. K. Sakuda and A. Yariv: Opt. Commun. 8, 1 (1973).
5. O.R. Asfar and A.H. Nayfeh: SIAM Review 25, 455 (1983).

Love Waves in a Slab with Rough Boundaries

A. Wirgin

Laboratoire de Modélisation en Mécanique, Université Paris VI, 4 place Jussieu, F-75005 Paris Cedex 05, France

1. Introduction

The earth, like most natural material bodies, is a heterogeneous structure with a rough envelope. In dealing with the theoretical problems of earthquake damage and determining the earth's structure from seismograms, the usual procedure is to first model the earth by a series of concentric layers around a core, it being assumed that each boundary, including the outer envelope, is a perfect spherical surface. A more radical approximation, often used in connection with surface wave studies, is to take these boundaries to be flat, parallel planes (Fig. 1). In this context, the problem of earthquakes reduces to one of the propagation of elastic waves in a medium whose (piecewise constant) elastic constants are a function of the **vertical** coordinate only.

The two principal classes of surface waves for such a configuration are the **Rayleigh** waves and **Love** waves. The former involve particle motion in the vertical and longitudinal directions, whereas the latter are associated with particle motion in the horizontal direction. The properties of these waves have been analyzed in great detail by EWING, JARDETSKY and PRESS [1] and shown to account for many important features of experimentally-obtained seismograms. In Fig. 2 we compare the dispersion curve of group velocity, obtained from the theoretical dispersion equation for Love modes with the dispersion curve resulting from measurements, of arrival times as a function of wave periods, made from the seismogram recorded in Palisades, New York, of the earthquake which occurred below Fallon, Nevada on July 6, 1954. The theoretical curve was computed for a reasonable set of parameters, similar to those chosen in Ref. 1. It is shown, in the same reference, that varying these parameters within fairly large limits, and even introducing a velocity gradient in the mantle, does **not** alter the **fundamental discrepancy** between theory and experiment at the low-period ($T \lesssim 25$ sec) end of the dispersion diagram. In particular, the theory of Love waves predicts that **inverse dispersion** sets in for $T \lesssim 20$ sec, whereas the experiments reveal no such behavior.

A glance at a map of North America reveals that the stretch of land separating the earthquake epicenter from the seismograph location can hardly be considered to be flat as is assumed in the model for Love modes. In particular, there exist a number of mountains in Wyoming and Colorado with heights in excess of 4 km and many others with heights in the 2-4 km range. As the Love mode is associated with a guided wave in the crust and the latter is approximately 35 km thick, it is quite conceivable that a 10-15% crustal thickness irregularity could appreciably modify the dispersion of the Love modes. We have undertaken the present work in order to see if this possibility exists and is consistent with the experimental results, notably for small periods. Our analysis relies on the same basic

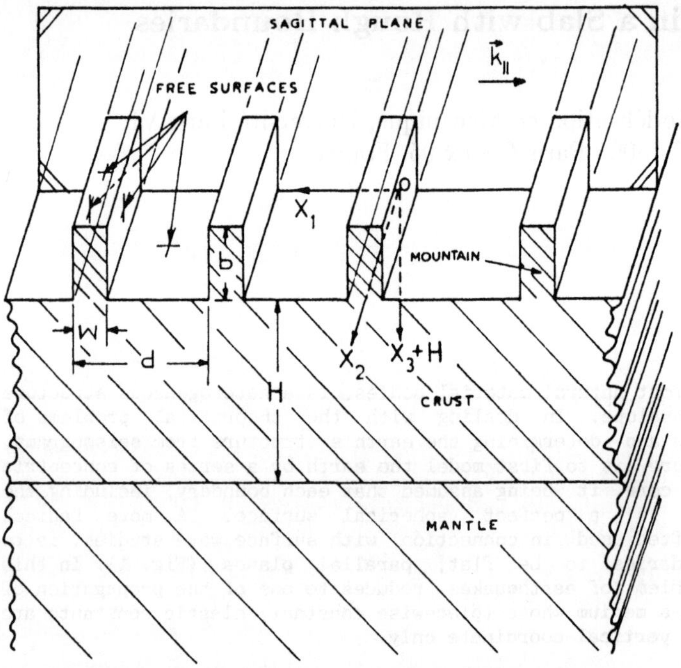

Fig. 1 The cylindrical geometry model of the earth used for the analysis in the present work. The mantle is of semi-infinite extent and in welded contact with the crust over a flat interface. The free upper boundary of the crust is flat and parallel to the first interface when no mountains are present such as in the model of pure Love waves. Otherwise, the mountains are periodically spaced cylinders of rectangular cross section. The SH wave displacement is perpendicular to the saggital plane, the latter being perpendicular to the cylinder axes.

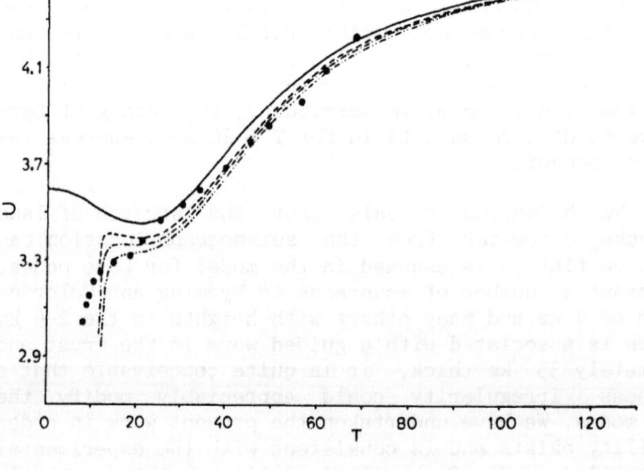

Fig.2

Caption see opposite page

model of the earth (Fig. 1) as the one used for Love waves except that we suppose the elastic constants to depend not only on the vertical coordinate x_3, but also on the **single** horizontal coordinate x_1 (a more realistic model would assume the elastic constants to depend on **both** horizontal coordinates).

A host of references to works which incorporate the earth surface roughness in the theoretical treatment of surface and bulk wave propagation at and near the earth's surface can be found in the book of AKI and RICHARDS [2]. All of these studies are concerned with the effect of the topography on boundary motion and stresses, which is not the topic herein. Works that are more in the spirit of our study are those [3,4] by MARADUDIN and co-workers, but these are principally concerned with Rayleigh waves and do not relate to the experimental aspects of earthquakes.

2. The SH Wave Two-dimensional Boundary Value Problem

We are concerned with harmonic wave displacement ($\exp(-i\omega t)$ implicit time factor) $\vec{u}(\vec{x})$ in homogeneous, isotropic, linear, elastic media $M^{(j)}$; $j=1,2,3,4$. We assume that **three** non-intersecting cylindrical boundaries, having axes that are parallel to the x_2 axis, divide $\mathbb{R}^3$ into **four** subdomains whose intersections with the $x_1 - x_3$ (sagittal) plane are designated by $\Omega^{(j)}$, $j=1,2,3,4$ ($\Omega^{(1)}$ and $\Omega^{(4)}$ are semi-infinite domains). Let Γ_j represent the common boundary of $\Omega^{(j)}$ and $\Omega^{(j+1)}$;$j=1,2,3$. Γ_1, Γ_2 and Γ_3 can be recognized as the curves of intersection of the cylinder boundaries with the x_1-x_3 plane. Let $\hat{n}_j$ denote the unit vector normal to Γ_j pointing towards $\Omega^{(j)}$. The fourth medium is air and its boundary with any other medium is (stress) free. $M^{(1)}$, $M^{(2)}$, $M^{(3)}$ designate the media occupying the mantle, crust, and mountainous extension of the crust respectively. Since we are dealing with an SH (shear horizontal or anti (sagittal)-plane displacement) problem, the displacements are tangent to the (cylindrical) boundaries and are governed by the Helmholtz equations (HE)

$$[k_T^{(j)}]^2 u_2^{(j)}(\vec{x}) + \partial_\alpha \partial_\alpha u_2^{(j)}(\vec{x}) = 0 \quad ; \forall \vec{x} \in \Omega^{(j)},\ j=1,2,3 \quad , \alpha=1,3 \quad ,$$

with

$$\vec{x} = x_\alpha \hat{x}_\alpha \quad ; \alpha = 1,3 \quad \text{(Einstein summation convention)} \quad ,$$

$$k_T^{(j)} = \omega / c_T^{(j)} \quad ,$$

$$c_T^{(j)} = (\mu^{(j)} / \rho^{(j)})^{1/2} \equiv \text{shear bulk wave velocity in } M^{(j)} \quad ,$$

Fig. 2 Experimental and theoretical group velocity dispersion (U, in km/sec, versus wave period T, in sec) for a continental path Fallon, Nevada ⟶ Palisades, New York, relative to the July 6, 1954 earthquake. The black points are the experimental results published in Fig. 4-52 (p. 212) of Ref. 1. The continuous curve is the m=0 pure Love wave dispersion curve for (b=0) H=35 km, $\mu^{(1)} / \mu^{(2)} = 1.8$, $c_T^{(1)} = 4.6$ km/sec, $c_T^{(2)} = 3.6$ km/sec. The dashed curve is for the m=0 quasi Love mode, with the same parameters as above and b = 3 km, d = 20 km, w = 3 km, $\mu^{(1)} / \mu^{(3)} = 1.8$, and $c_T^{(3)} = 3.6$ km/sec. The dash-dot curve is for the same configuration except that b =4 km. The dash-double-dot curve is for b = 5 km.

$\mu^{(j)}$, $\rho^{(j)}$ ≡ rigidity and mass density of $M^{(j)}$.

The non-vanishing stress tensor components are related to the displacement by

$$\tau^{(j)}_{\alpha 2} = \tau^{(j)}_{2\alpha} = \mu^{(j)}\partial_\alpha u^{(j)}_2 \quad .$$

The boundary conditions , expressing the continuity of particle displacement and traction (BC1-BC2) are

$$u_2^{(j)} - u_2^{(j)} = 0 \qquad ; \forall \vec{x} \in \Gamma_j \quad , j=1,2 \quad ,$$

$$\mu^{(j)}\hat{n}_j \cdot \nabla u_2^{(j)} - \mu^{(j+1)}\hat{n}_j \cdot \nabla u_2^{(j+1)} = 0 \qquad ; \forall \vec{x} \in \Gamma_j \quad , j=1,2 \quad .$$

On the free boundary we have the (traction-free) boundary condition (BC3)

$$\mu^{(3)}\hat{n}_3 \cdot \nabla u_2^{(3)} = 0 \qquad ; \forall \vec{x} \in \Gamma_3 \quad .$$

The condition of finite energy (FEC) implies that $|u_2^{(j)}|^2$ and $|\nabla u_2^{(j)}|^2$ are locally integrable in $\Omega^{(j)}$ for j=1,2,3. Uniqueness is assured by the requirement that the displacement field in the first (semi-infinite) domain vanish at points far removed from the first boundary (radiation condition RC).

3. Natural SH Modes of Vibration in a Crust-Mantle System with Free, Flat Upper Crustal Boundary and Flat, Crust-Mantle Interface

The geometry is that of Fig. 1 except that in the present case there are no mountains (b=0). The field representations, consistent with the HE, FEC and RC, are

$$u_2^{(j)}(\vec{x}) = T^{(j)}\exp[i(k_\parallel x_1 + k_\perp^{(j)-}x_3)] + R^{(j)}\exp[i(k_\parallel x_1 + k_\perp^{(j)+}x_3)] \quad ,$$

with

$$k_\perp^{(j)\pm} = \pm [(k_T^{(j)})^2 - k_\parallel^2] \equiv \pm \kappa^{(j)}; \; j=1,2 \; , \; T^{(1)} = 0 \quad ,$$

$$k_\parallel = \omega / c \quad ,$$

and $k_\parallel$, c the parallel component of the wavevector and the phase velocity of the normal mode wave. Since, in the present case, medium $M^{(3)}$ is absent, curves Γ_2 and Γ_3 coincide. Thus, we have type BC1 and BC2 boundary conditions on Γ_1 and a type BC3 boundary condition on Γ_2 which yield

$$\begin{bmatrix} 1 & -1 & -1 \\ \mu^{(1)}\kappa^{(1)} & \mu^{(2)}\kappa^{(2)} & -\mu^{(2)}\kappa^{(2)} \\ 0 & \exp(i\kappa^{(2)}H) & \exp(-i\kappa^{(2)}H) \end{bmatrix} \begin{pmatrix} R^{(1)} \\ T^{(2)} \\ R^{(2)} \end{pmatrix} = \begin{pmatrix} 0 \\ 0 \\ 0 \end{pmatrix} .$$

A non-trivial solution is obtained for det [ ] = 0 , so that

$$\mu^{(1)}\kappa^{(1)}\cos(\kappa^{(2)}H) - i\mu^{(2)}\kappa^{(2)}\sin(\kappa^{(2)}H) = 0 \quad .$$

A case of particular interest is

$$\kappa_r^{(1)} = 0 \quad , \quad \kappa_i^{(2)} = 0 \quad (\text{with } \kappa^{(j)} = \kappa_r^{(j)} + i\,\kappa_i^{(j)}) \quad .$$

Then

$$\kappa_i^{(1)} - (\mu^{(2)}/\mu^{(1)})\kappa_r^{(2)}\tan(\kappa_r^{(2)}H) = 0 \quad .$$

This (dispersion) equation, both of whose terms are real, has **real** solutions for tan( ) > 0, i.e.,

$$m\pi < \kappa_r^{(2)}H < (2m+1)\pi/2 \quad ; \ m = 0,1,2,\ldots ,$$

wherein m designates the the order of the guided modes in the crustal layer. These modes are none other than the so-called **Love modes**. Their guided (standing)-wave nature derives from the expression of the displacement field in the layer

$$u_2^{(2)}(\vec{x}) = 2R^{(2)}\exp[i(k_{\|}x_1 - \kappa^{(2)}H)]\cos[\kappa^{(2)}(x_3 + H)] \quad .$$

Associated with these modes we have an exponentially-damped plane wave in the mantle

$$u_2^{(1)}(\vec{x}) = R^{(1)}\exp[ik_{\|}x_1 - \kappa_i^{(1)}x_3] \quad ,$$

which, due to the fact that its amplitude is appreciable only in the vicinity of the surface separating the mantle from the crust, is termed a **surface wave**. The existence of the latter corresponds to $\kappa^{(1)}$ being imaginary, which fact is equivalent to $k_{\|} > k_T^{(1)}$ or $c < c_T^{(1)}$. This means that the surface wave is **slow** as compared to shear bulk waves in the mantle. The guided wave owes its existence to the fact that $\kappa^{(2)}$ is real, which means that $c > c_T^{(2)}$, i.e., the guided wave is **fast** as compared to shear bulk waves in the crust. Thus, the existence of Love modes is tied up with the conditions: $c_T^{(1)} > c_T^{(2)}$ and $c_T^{(1)} > c > c_T^{(2)}$.

We have computed the dispersion curve (c versus $k_{\|}$) numerically and obtained the group velocity U, defined by

$$U = d\omega/dk_{\|} = d(k_{\|}c)/dk_{\|} = c + k_{\|}dc/dk_{\|} \quad ,$$

by numerical differentiation of the dispersion curve. An example of such computations is given by the continuous curve in Fig. 2. The agreement with the experimental results is seen to be rather good except in the short period ($T = 2\pi/\omega$) range. This disagreement is the reason why the more elaborate analysis of the following section was undertaken.

4. Natural SH Modes of Vibration in a Crust-Mantle System with Free, Rough Upper Crustal Boundary and Flat Crust-Mantle Interface

The geometry is once again that of Fig. 1, with the mountains reinstated. The latter are visualized as cylinders of rectangular cross-section; this is, of course, an extremely simplified model of mountains. Furthermore, we have taken all of these mountains to be of the same size and shape, to be infinite in number and length (along the cylinder axes), and to be disposed in periodic manner, with period d, along the x_1 axis. The identity and periodic disposition of the "mountains" are not necessary requirements of our model, but they certainly simplify the analysis considerably. In particular, the displacement field can be shown to obey the Floquet relation (FR)

$$u_2^{(j)}(x_1+d,\ x_3) = u_2^{(j)}(x_1,\ x_3)\exp(ik_{\|}d) \quad ,$$

so that we need restrict our attention to the state of the field only in the interval $x_1 \in [-d/2,\ d/2]$.

The field representations which are consistent with the HE, FEC, RC and FR, are

$$u_2^{(j)}(\vec{x}) = \sum_{n=-\infty}^{+\infty} \{ T_n^{(j)}\exp[i(k_{n\|}x_1 + k_{n\perp}^{(j)-}x_3)]$$

$$+ R_n^{(j)}\exp[i(k_{n\|}x_1 + k_{n\perp}^{(j)+}x_3)]\} \quad ,$$

with

$$k_{n\|} = k_{\|} + 2n\pi/d \quad ,$$

$$k_{n\perp}^{(j)\pm} = \pm [(k_T^{(j)})^2 - k_{n\|}^{\,2}]^{1/2} \equiv \pm \kappa_n^{(j)} \equiv \zeta_n^{(j)} / \mu^{(j)}\ ;\ j=1,2 \quad ,\ T_n^{(1)}=0.$$

As concerns the field in the central mountain, we incorporate, at the outset, the free surface boundary condition BC3 on the protrusive portion of Γ_3 to obtain (via the HE and FEC)

$$u_2^{(3)}(\vec{x}) = \sum_{m=0}^{\infty} a_m \cos[\ \frac{m\pi}{w}(x_1 + \frac{w}{2})\ ]\{ \exp[ik_{m\perp}^{(3)+}(x_3+b+H)]$$

$$+ \exp[ik_{m\perp}^{(3)-}(x_3+b+H)]\ \} \quad ,$$

with

$$k_{m\perp}^{(3)\pm} = \pm [\ (k_T^{(3)})^2 - (\frac{m\pi}{w})^2]^{1/2} \equiv \pm \kappa_m^{(3)} \equiv \pm \zeta_m^{(3)} / \mu^{(3)} \quad .$$

We now apply the remaining boundary conditions to determine the wave coefficients in the above expressions. In the geometry of Fig. 1, Γ_1 is the straight line delineating the crust-mantle interface, Γ_2 the straight line delineating the upper boundary of a hypothetical parallel plane crust of thickness H, and Γ_3 a rectangular wave grating (mountain-valley structure) in contact with Γ_2. This means that Γ_2 and Γ_3 have the common segments : $\Gamma_4 = \{\ x_3=-H\ ;\ -w/2<x_1<w/2\ \}$, $\Gamma_5 = \{\ x_3=-H\ ;\ -d/2<x_1<-w/2\ \}$, $\Gamma_6 = \{\ x_3=-H\ ;\ w/2<x_1<d/2\ \}$. Thus, from BC2 and BC3 we have

$$\mu^{(2)}\partial_3 u_2^{(2)}(\vec{x}) = \begin{cases} \mu^{(3)}\ \partial_3\ u_2^{(3)}\ (\vec{x}) & ;\ \forall\ \ \vec{x} \in \Gamma_4 \\ 0 & ;\ \forall\ \ \vec{x} \in \Gamma_5\ ,\ \Gamma_6 \end{cases} \quad ,$$

from which it ensues, by projection onto $B = \{\ \exp(-ik_{j\|}x_1)\ ;\ j=0,\pm1,\pm2,\ldots\}$, that

$$\{\ -T_j^{(2)}\exp(i\kappa_j^{(2)}H) + R_j^{(2)}\exp(-i\kappa_j^{(2)}H)\ \}$$

$$= 2i\Upsilon \sum_{m=0}^{\infty} a_m \frac{\zeta_m^{(3)}}{\zeta_j^{(2)}} \sin(\kappa_m^{(2)} b)\ I_{jm}^- \quad ; j = 0, \pm1, \pm2, \ldots ,$$

where

$$\Upsilon = w/d \quad ,$$

$$I_{jm}^{\pm} = \int_{-w/2}^{w/2} \exp(\pm\, ik_{j\parallel} x_1)\ \cos[\frac{m\pi}{w} (x_1 + \frac{w}{2})]\ \frac{dx_1}{w} \quad .$$

From the application of BC1 to segment Γ_4 we find, after projection onto $\mathcal{B}= \{ \cos[(\ell\pi/w)(x_1+w/2)] \ ; \ \ell=0,1,2,\ldots\}$,

$$a_\ell = \frac{\epsilon_\ell}{2\cos(\kappa_\ell^{(3)}\ b)} \sum_{n=-\infty}^{\infty} [T_n^{(2)} \exp(i\kappa_n^{(2)} H) + R_n^{(2)} \exp(-i\kappa_n^{(2)} H)] I_{n\ell}^+$$

$$; \ \ell = 0,\ 1,\ 2, \ldots\ldots ,$$

where

$$\epsilon_0 = 1 \ , \ \epsilon_{\ell\neq 0} = 2 \quad \text{(Neumann factor)}.$$

The next step is to find expressions for { } and [ ] (in the above two linear systems of equations) in terms of the set $\{R_\ell^{(1)} ; \ \ell=0,1,2,\ldots\}$. To do this we apply BC1 and BC2 on Γ_1 , obtaining, after projection onto B,

$$T_j^{(2)} = \left(\frac{\zeta_j^{(2)} - \zeta_j^{(1)}}{2\zeta_j^{(2)}}\right) R_j^{(1)} \ , \ R_j^{(2)} = \left(\frac{\zeta_j^{(2)} + \zeta_j^{(1)}}{2\zeta_j^{(1)}}\right) R_j^{(1)} \quad ; \ j=0,\pm1,\pm2,\ldots \ .$$

A little bit of algebra then gives

$$R_j^{(1)} = \left\{\frac{2i\Upsilon}{[-i\zeta_j^{(2)} s_j^{(2)} + \zeta_j^{(1)} c_j^{(2)}]}\right\} \sum_{m=0}^{\infty} a_m \zeta_m^{(3)} s_m^{(3)} I_{jm}^- \quad ; \ j=0,\pm1,\pm2,\ldots ,$$

$$a_\ell = \left\{\frac{\epsilon_\ell}{2c_\ell^{(3)}}\right\} \sum_{n=-\infty}^{+\infty} R_n^{(1)} \left\{\frac{\zeta_n^{(2)} c_n^{(2)} - i\zeta_n^{(1)} s_n^{(2)}}{\zeta_n^{(2)}}\right\} I_{n\ell}^+ \quad ; \ \ell=0,1,2,\ldots ,$$

with

$$s_n^{(2)} = \sin(\kappa_n^{(2)} H) \quad , \quad c_n^{(2)} = \cos(\kappa_n^{(2)} H) \quad ,$$

$$s_n^{(3)} = \sin(\kappa_n^{(3)} b) \quad , \quad c_n^{(3)} = \cos(\kappa_n^{(3)} b) \quad .$$

By introducing the first of these systems into the second we obtain the following set of linear equations:

$$\sum_{m=0}^{\infty} G_{\ell m}\ a_m = 0 \quad ; \ \ell=0,1,2,\ldots. \quad ,$$

with

$$G_{\ell m} = \delta_{\ell m} - \eta_{\ell m} P_{\ell m} ,$$

$$\eta_{\ell m} = \gamma \epsilon_{\ell} \zeta_m^{(3)} s_m^{(3)} / c_{\ell}^{(3)} , \qquad \text{(summation convention does not apply)}$$

$$P_{\ell m} = \sum_{n=-\infty}^{\infty} F_n I^+_{n\ell} I^-_{nm}$$

$$F_n = \left\{ \frac{i\zeta_n^{(2)} c_n^{(2)} + \zeta_n^{(1)} s_n^{(2)}}{-i\zeta_n^{(2)} s_n^{(2)} + \zeta_n^{(1)} c_n^{(2)}} \right\} \frac{1}{\zeta_n^{(2)}} \equiv \frac{1}{f_n} ,$$

and $\delta_{\ell m}$ the Kronecker delta. This system, which can be written as the matrix equation $\mathbf{Ga} = \mathbf{0}$, has a non-trivial solution for $\det(\mathbf{G}) = 0$. This is the exact dipersion relation for the natural vibration modes of the configuration in Fig. 1.

The matrix equation can also be written as

$$G_{\ell\ell} a_{\ell} + \sum_{m=0,\ m\neq\ell}^{\infty} G_{\ell m} a_m = 0 \quad ; \ell = 0,1,2,...$$

In the first approximation we neglect the term containing the series so that the approximate dispersion relation we will deal with is $G_{\ell\ell} = 0$; $\ell=0,1,2,...$ There are actually as many equations to be solved as there are "mountain modes". Herein, we restrict our attention to the $\ell=0$ relation which, in fact, is the dispersion relation of the fundamental mode in the mountain (defined by $G_{00} a_0 = 0$). This amounts to searching for the roots of

$$g_{00} = \frac{G_{00}}{F_0} = \frac{1}{F_0} - \eta \frac{P_{00}}{F_0} = 0 \qquad (\eta \equiv \eta_{00}) .$$

This equation can be solved , as such, in numerical manner, but it is preferable to cast the infinite series P_{00} into a faster-converging form. We first note that

$$F_n \sim [\mu^{(2)} \frac{2\pi}{d} |n|]^{-1} \quad ; |n| \to \infty ,$$

so that

$$P_{00} = Q_{00} + F_0 I^+_{00} I^-_{00} + \sum_{n=-\infty,\ n\neq 0}^{\infty} (F_n - \frac{d}{2\pi\mu^{(2)} |n|}) I^+_{n0} I^-_{n0} ,$$

with

$$Q_{00} = \left(\frac{d}{2\pi\mu^{(2)}}\right) \int_{-w/2}^{w/2} \frac{dx'_1}{w} \int_{-w/2}^{w/2} \frac{dx}{w} S(x_1', x_1) ,$$

$$S(x_1',x_1) = \exp[ik_{\|}(x'_1-x_1)] \sum_{n=1}^{\infty} n^{-1}\{\exp[i\frac{2n\pi}{d}(x_1'-x_1)]$$

$$+ \exp[-i\frac{2n\pi}{d}(x_1'-x_1)]\} \quad .$$

Using a well-known identity we find

$$S(x_1',x_1) = -2 \exp[ik_{\|}(x_1'-x_1)] \ln [2|\sin\{\frac{\pi}{d}(x_1'-x_1)\}|] \;,$$

so that by assuming

$$\pi\gamma \ll 1 \quad ,$$

$$k_{\|}w \ll 1 \quad ,$$

we get

$$S(x_1',x_1) \cong -2 \ln [\frac{2\pi}{d}|x_1'-x_1|] \quad ; \; x_1', \; x_1 \in [-w/2, \; w/2].$$

The integrals can now be evaluated in closed form:

$$Q_{00} = \left(\frac{-d}{\pi\mu^{(2)}}\right)\{ \ln(2\pi\gamma) - \frac{3}{2} \} \; .$$

Furthermore,

$$I^{\pm}_{n0} = \frac{\sin(k_{n\|} \; w/2)}{k_{n\|} \; w/2} \equiv \operatorname{sinc}(k_{n\|}w/2) \;,$$

so that the dispersion relation finally becomes

$$g_{00}=f_0+ \eta\left(-1 - f_0Q_{00}-f_0 \sum_{n=-\infty,n\neq0}^{\infty}\left[\frac{1}{f_n} - \frac{d}{2\pi\mu^{(2)}\,|n|}\right][\operatorname{sinc}(k_{n\|}w/2)]^2\right) =0 \; .$$

When b=0, then η=0 so that $g_{00}= f_0$, which signifies that

$$-i\zeta_0^{(2)}s_0^{(2)}+ \zeta_0^{(1)}c_0^{(2)}= 0 \; .$$

This is none other than the dispersion relation (see Section 3) for the crust-mantle system with **flat** upper crustal boundary. Thus, the term η( ) represents the perturbation, due to the presence of the mountains, of the vibration modes of the flat, parallel plane system.

Now let us return to the field representations in $\Omega^{(1)}$ and $\Omega^{(2)}$ and compare them with those obtained for the flat,parallel plane system. We see that, in the absence of the mountains, all but the n=0 (order) plane waves are superfluous. Love modes then correspond to imaginary $\kappa_0^{(1)}$. The situation closest to this one, when the mountains are present, is

$$\kappa_n^{(1)}= \begin{cases} \text{imaginary} \; ; \; n = 0 \\ \text{imaginary} \; ; \; \forall n \neq 0 \end{cases} \qquad \text{(condition A)} \;,$$

(which signifies that the field in $\Omega^{(1)}$ is composed of surface waves only) and

$$\kappa_n^{(2)} = \begin{cases} \text{real} & ; n = 0 \\ \text{imaginary} & ; \forall n \neq 0 \end{cases} \qquad \text{(condition B)}$$

(which signifies that the field in the crust is composed of a (quasi) Love guided wave in addition to surface waves). We recall that $\kappa_0^{(3)} = k_T^{(3)}$ is real, so that the fundamental mode in the mountains is a propagative mode.

Conditions A and B imply

$$k_T^{(1)} < k_{\|} < k_T^{(2)} \quad \Leftrightarrow \quad c_T^{(2)} < c < c_T^{(1)} \quad \text{(condition C)} \ ,$$

(which was the previously-obtained condition of existence of (pure) Love modes) and

$$2\pi/d > k_{\|} \qquad \text{(condition D)} \ ,$$

$$| \, k_{\|} - 2\pi/d \, | > k_T^{(2)} > k_T^{(1)} \qquad \text{(condition E)} \ .$$

When conditions C, D, and E are fulfilled then g_{00} turns out to be **real** and $g_{00}=0$ has **real** roots. The latter lie, by virtue of condition C, in the same interval as those of the Love mode eigenvalues.

We computed the dispersion diagram arising from $g_{00}=0$ in numerical manner and also obtained the group velocity by numerical differentiation. The series (with the n=0 term excluded) in g_{00} is very rapidly convergent. The results of these computations are plotted in Fig. 2 for three mountain heights. We see that the effect of the mountains is to almost eliminate the inverse dispersion (this being complete for $b > 4$ km) for short ($T < 20$ sec) periods. The b=4 km and b=5 km curves are in qualitative agreement with the experimental curve for $T > 11$ sec. For smaller T the approximations made in the dispersion relation can no longer be relied on since $k_{\|} w$ is no longer small compared to one. Nevertheless, our results substantiate the hypothesis that the discrepancy between experimental and theoretical pure Love wave dispersion curves is due to large-scale geometrical irregularities (the Rocky mountains) on the earth's surface.

5. Conclusions

The model of periodically spaced rectangular cylinder mountains accounts, in satisfactory manner, for those features of the dispersion of SH waves in the earth's crust which cannot be explained by the excitation of pure Love modes. The constraints $\pi\gamma \ll 1$ and $k_{\|}w \ll 1$ can be easily dropped to widen the range of applicability and precision of the computations. Furthermore, it is not very difficult to reinstate at least some of the previously-neglected terms in the series $\Sigma \, G_{\ell m} a_m$.

A more interesting modification of the present analysis is to assume that the structure is **not** periodic. A stochastic distribution of mountain heights, base widths, separations and composition might furnish some interesting new insights.

References

1. W. M. Ewing, W.S. Jardetsky, F. Press: Elastic Waves in Layered Media, (McGraw-Hill, New York 1957)
2. K. Aki, P.G. Richards: Quantitative Seismology: Theory and Methods, (Freeman, San Francisco 1980) Vol.2, pp. 751-765
3. N.E. Glass, R. Loudon, A.A. Maradudin: Phys. Rev. B 24, 6843 (1981)
4. A.G. Eguiluz, A.A. Maradudin: Phys. Rev. B 28, 728 (1983)

Part IV

Surface Waves on Electromagnetic Bodies

Shear Horizontal Surface Acoustic Waves on Solids

G.A. Maugin

Université Pierre et Marie Curie, Laboratoire de Modélisation en Mécanique, Tour 66, 4 place Jussieu, F-75252 Paris Cedex 05, France

The existence of shear horizontal surface acoustic waves (which play a fundamental role in many areas of mechanical engineering, surface physics, acoustoelectronics, seismology, and material science) is viewed as the result of a perturbation in boundary conditions -the latter being understood in the large- in order to bind otherwise essentially skimming bulk SH waves to the limiting surface. Various means to produce this perturbation are considered in this spirit. They include spatial inhomogeneities, additional interfaces, curvature, gratings, roughness, spatial dispersion, and couplings with electric and magnetic fields for different types of materials.

1. INTRODUCTION

Surface acoustic waves are time-varying, spatially nonuniform perturbations that exhibit spatial variation in the field amplitude markedly confined to the vicinity of the bounding surface of a body and practically nil outside this relatively narrow zone. They represent thus a type of dynamical boundary-layer effect of which the wealth depends on both the constitution of the material and the shape and properties of the bounding surface itself. They correspond to one of the simplest dynamical boundary-value problems in which the bounding surface serves as a guide. The paragon of surface waves on solids is the Rayleigh surface acoustic wave which represents a nondispersive, mixed longitudinal-transverse, time-harmonic plane elastic mode of propagation that propagates at a speed less than that of volume shear elastic waves at the free surface of a semi-infinite isotropic homogeneous linear elastic space. Its amplitude, polarized in the sagittal plane (the plane spanned by the direction of the real wave vector and the unit normal to the limiting plane), decreases exponentially away from the surface in the elastic material, with a penetration depth of a few wavelength [1], [2]. In the above-enunciated conditions, there are no surface waves that correspond to an elastic displacement orthogonal to the sagittal plane, i.e., surface waves having the nature of SH (shear horizontal) waves in the jargon of seismology. The present contribution addresses the problem of the existence of such SH surface acoustic waves. Obviously, something must be added to allow for this existence and we follow Viktorov's point of view [3] that SH surface acoustic waves are SH bulk waves which are perturbed to meet specific boundary conditions at the limiting surface. This conversion of a "surface skimming" bulk wave into a true surface acoustic wave having SH polarization can be achieved by different means which are examined subsequently in this paper. Previous reviews along the same lines are due to MAUGIN [4] and MARADUDIN [5] - [7]. A natural emphasis is placed on the works of the author and co-workers and, for lack of space, only a selected bibliography is given. The interest for SH surface acoustic waves stems from several factors including their role in seismology studies, in modes of crack propagation, their use in acousto-electronics and nondestructive testing methods, their relative simplicity related to the one-dimensional nature of the associated elastic displacement, and their resemblance to surface electromagnetic waves, a subject of utmost importance in surface plasmon physics.

2. NOTATION, NOTION OF PERTURBATION

We consider the setting in Fig. 1. The propagation takes place along the $X_1 = x$ direction. The surface Rayleigh SAW of linear isotropic elasticity corresponds to an elastic displacement **u** parallel to the sagittal plane P_S, hence a linear combination of P (longitudinal = pressure) and SV (surface vertical) components. SH waves correspond to a travelling perturbation in the $u_3 = u$ component only [8]. Recently, ROYER and DIEULESAINT [9] have exhibited the cases for which pure Rayleigh waves exist when the material has a binary axis of symmetry orthogonal to P_S for monoclinic, orthorhombic, tetragonal, hexagonal and cubic crystals. For a general anisotropic linear elastic body and an arbitrary direction of propagation, the surface wave solution involves a <u>generalized</u> Rayleigh wave with a three-component elastic displacement and a decrease in amplitude with depth accompanied by oscillations. Concentrating on the case of SH wave we denote by $\Box$ the one-dimensional (d'Alembertian) wave operator and by a prime the partial derivative with respect to the depth coordinate so that

$$\Box = \frac{\partial^2}{\partial x^2} - \frac{1}{c^2}\frac{\partial^2}{\partial t^2} \quad , \quad f' = \frac{\partial f}{\partial y} \, . \tag{2.1}$$

We look for dynamical elastic solutions of the type $u_3 = u(x_1,x_2,t)$ and for homogeneous isotropic linear elasticity one obtains the following boundary-value problem ($y = 0$ is traction free):

$$\text{(P1)} \quad \boxed{\begin{aligned} \mathcal{D}u \equiv \Box_T\, u + u'' &= 0 \quad , \quad \text{for } y > 0 \; , \\ u' &= 0 \quad , \quad \text{at } \; y = 0 \; , \end{aligned}} \tag{2.2}$$

where $\Box_T$ is obtained by inserting $c_T = (\mu/\rho)^{\frac{1}{2}}$ instead of c. For <u>plane</u> waves propagating along $X_1 = x$, one takes $u = \tilde{u}(\zeta,y) = \hat{u}(y) \; F(\zeta)$, where $\zeta = x - ct$. Equation $(2.2)_1$ takes on the form

$$[1 - (\frac{c}{c_T})^2]\frac{\ddot{F}}{F} = - \frac{\hat{u}''}{\hat{u}} = \text{const.} \quad , \quad (\dot{F} = \frac{dF}{d\zeta}). \tag{2.3}$$

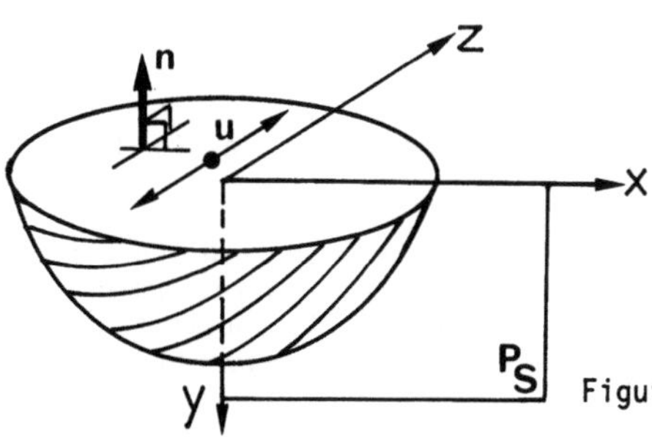

Figure 1 - Surface-wave problem

For a harmonic plane wave one considers the special case $F(\zeta) = \exp[i(kx - \omega t)] = \exp(-ik\zeta)$, $k = \omega/c$, and with $k_T := \omega/c_T$ the problem (2.2) reads

$$(P1)' \quad \begin{cases} \hat{u}'' - (k^2 - k_T^2)\hat{u} = 0 & \text{for} \quad y > 0 \ , \\ \hat{u}' = 0 & \text{at} \quad y = 0 \ . \end{cases} \tag{2.4}$$

The SAW problem consists in imposing the behaviour far from the limiting surface, thus $\hat{u} \to 0$ as $y \to +\infty$ and, more particularly, selecting a profile $\hat{u}(y) = u_0 \exp(-\chi y)$, real $\chi > 0$. The first of (2.4) imposes that $\chi^2 = k^2 - k_T^2 > 0$ or $c^2 < c_T^2$ while the second requires that either $u_0 = 0$ (no wave) or $\chi = 0$ (no decrease in profile). The second possibility rules out the existence of a pure SH surface wave but it also means that a "face" shear wave (also called "skimming bulk wave") satisfies identically the two equations (2.4) although it is not a surface wave per se. In some sense the boundary condition $(2.4)_2$ is too strong. If we want true SH SAWs to exist we must perturb this boundary condition to allow for $\hat{u}' \neq 0$ at $y = 0$. This can be achieved by various means, some being mechanical (e.g., inhomogeneities in the elastic properties), some other relying on coupling with another field (e.g., electric or magnetic) of a nonmechanical nature. Else, the geometry of the limiting surface (a plane in the example above) can be perturbed deterministically or randomly. For instance, the general boundary condition at $y = 0$ reads $\mathbf{t}.\mathbf{n} = \mathbf{0}$ in the absence of material on the $(y < 0)$-face or $\mathcal{B}_n := [\![\mathbf{t}]\!].\mathbf{n} = \mathbf{0}$ if $y = 0$ separates two material regions and there is no mass transfer across $y = 0$. This boundary condition can be symbolically perturbed by writing

$$(\mathcal{B}_n + \delta\mathcal{B}_n) = [\![\mathbf{t}]\!].\mathbf{n} + [\![\mathbf{t}]\!].\delta\mathbf{n} + [\![\delta\mathbf{t}]\!].\mathbf{n} = 0 \ . \tag{2.5}$$

Another way to perturb $\mathcal{B}_n = 0$ is to add a surface source $\delta\mathbf{F}$ at $y = 0$ all other things being kept unchanged ; then we obtain

$$(\mathcal{B}_n + \delta\mathcal{B}_n) = [\![\mathbf{t}]\!].\mathbf{n} + \delta\mathbf{F} = \mathbf{0} \ . \tag{2.6}$$

Practically all cases presented below fit into the pattern defined by (2.5) and (2.6). The perturbation in the boundary condition may be accompanied by one in the field equations for $y > 0$. In particular, the $\delta\mathbf{t}$ and $\delta\mathbf{F}$ terms in (2.5) and (2.6) may correspond to interactions with additional fields which are themselves governed by new field equations (e.g., Maxwell's equations, spin-precession equation, etc.).

3. "MECHANICAL" PERTURBATION

3.1. Love waves

The simplest way to perturb the surface condition $(2.2)_2$ at $y = 0$ was noted by A.E.H. Love who considered the case where an elastic layer L is glued on top of the substrate S (Fig. 2a).

This layer must have a transverse elastic speed c_T^L smaller than the corresponding speed c_T^S in the substrate for the SH surface wave, named Love wave, to exist (the combination gold layer on fused quartz substrate satisfies this constraint). The superimposed layer thus is referred to as a

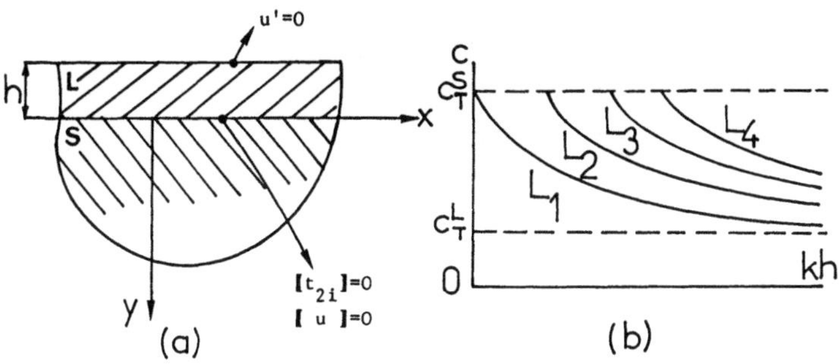

Figure 2 - Love wave ; **(a)** setting, **(b)** dispersion curves

slow layer (compared to the "faster" substrate). Three points must be emphasized. First, the slow layer serves as a guide and the notion of wave guide implies that of multimodes. Next, the thickness h of the layer introduces a characteristic length which yields dispersion, and this materializes in multiple reflexions in the course of propagation. Finally, the surface $y = 0$, now in fact an interface between two elastic solids, is no longer free since the continuity of displacement and traction must be checked at $y = 0$, i.e., $[\![u]\!] = 0$ and $[\![t_{i2}]\!] = 0$, respectively. In particular, the second of these, for SH waves ($u = u_3$), can be written as

$$\boxed{u'(S) - \beta\, u'(L) = u'(S) + \beta[-u'(L)] = 0 \quad \text{at} \quad y = 0 \quad ,} \tag{3.1}$$

where $\beta = \mu_L/\mu_S$, the ratio of shear moduli. The dispersion relation of Love waves is known to be ([10], p. 545)

$$D_{Love}(c,k) = \beta \tan\{[(\frac{c}{c_T})^2 - 1]^{\frac{1}{2}}kh\} - \{\frac{1 - \left(\frac{c}{c_T^S}\right)^2}{\left(\frac{c}{c_T^L}\right)^2 - 1}\}^{\frac{1}{2}} = 0\,,$$
$$c_T^L < c < c_T^S\,, \tag{3.2}$$

with the sequence of wave numbers

$$(k)_n = \pi(n-1)/4h[\left(\frac{c_T^S}{c_T^L}\right)^2 - 1]^{\frac{1}{2}} \quad , \quad n = 1,2,\ldots \tag{3.3}$$

The corresponding qualitative graph is given in Fig. 2b. All modes $n \geq 2$ exhibit a cutoff. It is of interest to note that as h goes to zero with fixed k, c goes to c_T^S along the first mode as

$$(kh)^2 \simeq \beta^{-2}[1 - \left(\frac{c}{c_T^S}\right)^2]/[1 - \left(\frac{c}{c_T^L}\right)^2]^2 \quad . \tag{3.4}$$

Obviously, the overall structure in Figure 2a is that of an inhomogeneous material (along the y direction) with piece-wise constant properties. This is a limit situation where the slowness $s_T = c_T^{-1}$ suffers a jump

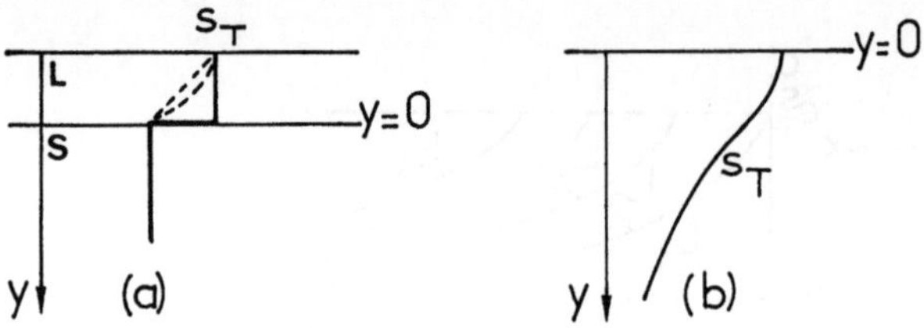

Figure 3 - Inhomogeneous substrate ; $S_T(y)$

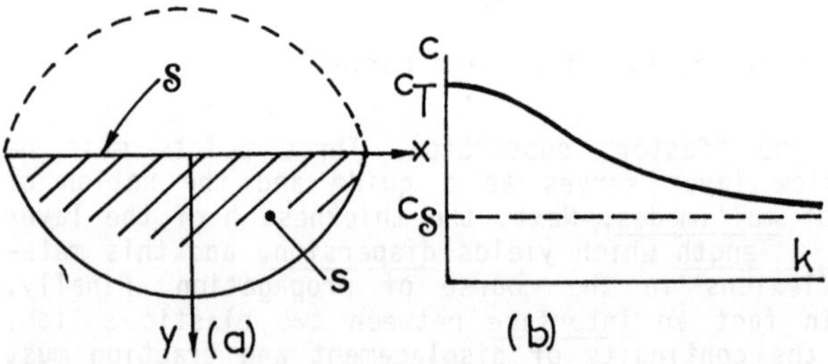

Figure 4 - {y=0}as an interface of zero thickness

discontinuity at $y = 0$, smoother variations (with the correct sense of variation, i.e., decreasing slowness with depth) on the intervals $\{-h < y < 0\}$ or $(-h,+\infty)$ corresponding probably more precisely to a physical reality such as the one that prevails in metallurgically or chemically treated surfaces and soil properties (Fig. 3). Another limit of Figure 2a corresponds to the case where h is infinitesimally small and the layer can be represented as an interface of zero thickness (Fig. 4) as understood now in the theory of thermodynamical interfaces (e.g. [11]). These two possibilities are briefly examined here below.

3.2. Inhomogeneous substrate

We refer to the works, among others, of VIKTOROV [12], KIELCZYNSKI [13] and BAKIRTAS and MAUGIN [14]. With a shear modulus varying with depth (and discarding variations in the density to simplify the presentation) the surface wave problem (2.2) is shown to be replaced by [notation of (2.1) and (2.2)]

$$\text{(P2)} \quad \boxed{\begin{array}{ll} \mathcal{D}u + \mu' u' = 0 & \text{for } y > 0 , \\ u' = 0 & \text{at } y = 0 . \end{array}} \qquad (3.5)$$

Notice that the perturbation is not in the boundary but in the field equation $(3.5)_1$ where a boundary layer effect of some sort is expected if the variation in $\mu^{-1}(y)$ is markedly confined to the vicinity of $y = 0$. However, one often uses the approximation of slow spatial variation of $\mu(y)$ to neglect the μ' contribution in $(3.5)_1$. Then there remains the problem

$$(P2)' \quad \boxed{\begin{aligned} \mathcal{D}(y)u &= 0 \quad \text{for} \quad y > 0 \ , \\ u' &= 0 \quad \text{at} \quad y = 0 \ , \end{aligned}} \tag{3.6}$$

with $\mathcal{D}(y)\ u = \Box_T(y)u + u''$, $c_T = c_T(y)$. For $\hat{u}$ we then have a simple Sturm-Liouville problem akin to those met in stationary solutions of the Schrödinger equation, as [compare (2.4)]

$$\begin{aligned} \hat{u}'' - [k^2 - k_T^2(y)]\hat{u} &= 0 \quad \text{for} \quad y > 0 \ , \\ \hat{u}' &= 0 \quad \text{at} \quad y = 0 \ , \\ \hat{u}(y \to +\infty) &= 0 \quad . \end{aligned} \tag{3.7}$$

It was noted by Viktorov that for simple variations of μ with depth the solution for $\hat{u}(y)$ is of the type

$$\hat{u} = u_0\ A(\alpha,\beta,\gamma,\xi)\ \exp(-\ \chi y) \quad , \tag{3.8}$$

where A is a four-parameter dependent hypergeometric function. As a matter of fact, systematic analyses have been conducted for related problems in light propagation (optics with spatially varying refraction index) when the exact solution of the surface wave problem based on (3.5) can be expressed in terms of hypergeometric, Whittaker, Bessel and exponential functions [15]-[17]. More recently, GUBERNATIS and MARADUDIN [18] have devised a method using expansions of the displacement in series of Laguerre polynomials, reducing the solution of the wave problem to an eigenvalue problem with excellent convergence properties in the accompanying numerical implementation for vertically inhomogeneous, anisotropic, linear elastic substrates. Considering a PZT-4 ceramic substrate, KIELCZYNSKI [13] has shown that the first SH mode obtained for an inhomogeneity in $s_T = c_T^{-1}$ of a Gaussian profile is practically undistinguisable from the first mode of Love waves. The same author and PAJEWSKI [19] have envisaged the application of SH surface waves in inhomogeneous solids as a nondestructive means for determining the depth of inhomogeneities. Data pertaining to seismic (Love) waves used to study crustal and upper mantle structures are to be found in a review [20].

3.3. Zero-thickness interface

The surface $y = 0$ may be considered as a flat thermodynamical interface S between the material region $y > 0$ and the exterior region $y < 0$. In the same way as in $y > 0$ one has the usual balance of linear momentum in the familiar Cartesian tensor notation

$$\rho \frac{dv_i}{dt} = t_{ij,j} + f_i \quad ., \quad y > 0 \quad , \tag{3.9}$$

one can show that for material particles that belong to $S = \{y = 0\}$ one has a general balance of linear momentum of the following type (generated simultaneously with (3.9) by using the principle of virtual power [11]) :

$$\hat{\rho}\,\frac{d\hat{v}_i}{dt} + [\![m(v_i - \hat{v}_i)]\!] = [\![t_{ij}]\!] n_j + (\hat{\nabla}_j + 2\Omega\, n_j)\hat{t}_{ij} + \hat{f}_i \tag{3.10}$$

at S, where a superimposed "^" characterizes fields defined at S. In particular, $\hat{\rho}$ is a mass density on S, $\hat{\nabla}_j$ is the tangential gradient at S, f_i is a given force density on S, m is the mass transfer through S and Ω is the mean curvature of S while $\hat{t}_{ij}$ is a 2-dim stress tensor on S (e.g., surface tension). When S has no material properties and $m = 0$, then (3.10) reduces to $[\![t_{ij}]\!] n_j = 0$ at $y = 0$ and this reads $t_{i2} = 0$ if $y < 0$ is a vacuum. But when we consider **(i)** that $S = \{y=0\}$ is a flat ($\Omega = 0$) thermodynamical interface of mass density $\hat{\rho} = \rho_S$ with $\hat{v}_i = v_i = \dot{u}_i$ at $y = 0$ and $\hat{f}_i = f_i = 0$, **(ii)** the problem does not depend on z, eqns. (3.8) and (3.9) for $u = u_3$ reduce to [21]

$$\rho\, \ddot{u}_3 = t_{3\alpha,\alpha} \quad , \alpha = 1,2 \quad , \quad y > 0 \quad ; \quad \rho_S\, \ddot{u}_3 = \hat{t}_{31,1} + t_{32} \tag{3.11}$$

since $\hat{\nabla}_j$ reduces to $\partial/\partial x$. If S, like $S = \{y > 0\}$, has an elastic (but two-dimensional) behavior with shear modulus μ_S, then, (3.11) yield the following problem :

$$\text{(P3)} \quad \boxed{\begin{aligned} \mathcal{D}u \equiv \square_T\, u + u'' = 0 \quad & \text{for} \quad y > 0, \\ u' + \beta\, \square_S\, u = 0 \quad & \text{at} \quad y = 0, \end{aligned}} \tag{3.12}$$

where $\beta = \mu_S/\mu$ has the dimension of a length and $\square_S$ is defined with $c_S^2 = \mu_S/\rho_S$. We see from eqns. (3.12) that introducing the zero-thickness interface has really perturbed the second of (2.2) by adding the β term that includes a new wave operator. Equation $(3.12)_2$ was obtained by MURDOCH [22] who thus deduced a dispersion relation for surface wave solutions $u = u_0 \exp(-\chi_y) \exp[i(\omega t - kx)]$, $\chi > 0$, as

$$D\,(c,k) = (k/\beta)[1 - (\frac{c}{c_S})^2]^2 - [1 - (\frac{c}{c_T})^2] = 0 \,. \tag{3.13}$$

On setting $\bar{\beta} = \bar{\mu}_S/\mu$ and $h = \mu_S/\bar{\mu}_S$ with h finite, eqn. (3.13) yields

$$(kh)^2 = \bar{\beta}^{-2}[1 - (\frac{c}{c_T})^2]/[1 - (\frac{c}{c_S})^2]^2 \quad , \tag{3.14}$$

which is the same as the approximation (3.4) to the first Love mode. The above formulation must be compared to the "thin-plate limit" introduced by TIERSTEN [23], [24], which provides effective boundary conditions - such as $(3.12)_2$- that reproduce the effect of the plate exactly to order (h/λ) for surface waves of wavelength λ -see pp. 448-456 in Ref. 5.

3.4 Curvature, grating and roughness

A "perturbation" in the geometry of the flat limiting plane of the standard problem (2.2) may also favor the existence of pure SH SAWs. This is the case of an appropriately curved limiting surface and an analogy can be drawn with the case of inhomogeneous media bounded by a flat surface : "SH SAWs propagating circumferentially around the stress-free surface of an elastic cylinder that is convex toward vacuum are analogous to Love waves on a planar stress-free surface in which c_T increases exponentially with increasing distance into the medium from the surface", [5], p. 512. Indeed, with a reduced radial variable $y = R \ln(R/r)$ where $0 < r < R$ is the interior of the cylinder, one can write equations analogous to $(3.7)_{1-2}$ where [5]

$$k_T^2(y) = \omega^2/c_T^2(y) \quad \text{with} \quad c_T^2(y) = c_T^2(0)\exp(-2y/R) \quad . \tag{3.15}$$

The resulting waves are obviously dispersive since there exists a characteristic length (radius). These are the celebrated whispering-gallery waves.

Another possibility is provided by a perturbation in the unit normal such as in the second term in (2.5). This occurs when the surface of the substrate presents a regular (periodic) grating or a random roughness. Accounting for the "nonverticality" of the unit normal at the free surface of equation $y = \zeta(x)$ -one-dimensional grating with generators parallel to z- it is easily shown that the boundary condition $(2.2)_2$ for the $u_3 = u$ component is replaced by (Figure 5a)

$$\boxed{u' - (\partial\zeta/\partial x)(\partial u/\partial x) = 0 \quad \text{at} \quad y = \zeta(x) \quad .} \tag{3.16}$$

The second contribution in the left-hand side of (3.16) is the required "perturbation" to favor the existence of SH surface waves. Original works on the matter for SH waves propagating across the grooves of a rectangular grating (Figure 5b) are due to AULD et al. [25] and GULYAEV and PLESSKII [26] who gave approximate solutions and measured the dispersion of the propagating SH SAWs. A nonperturbative solution is due to GLASS and MARADUDIN [27].

In the case of a propagation across a randomly rough surface (Figure 5c) one characteristic length introduces itself as the "transverse correlation length" of the roughness. We refer the reader to original works [28]-[29] for precise results concerning the dispersion and attenuation of the existing SH SAWs which are only weakly bound to the surface, penetrating deeply into the medium (compare to Bleustein-Gulyaev waves below).

To conclude this section concerned with pure mechanical means of favoring the existence of SH SAWs, we note that nonlinear Love waves were considered by KALYANASUNDARAM [30], and the effect of spatial dispersion (due to long-range interaction) by ALLDREDGE [31], and that the right combination

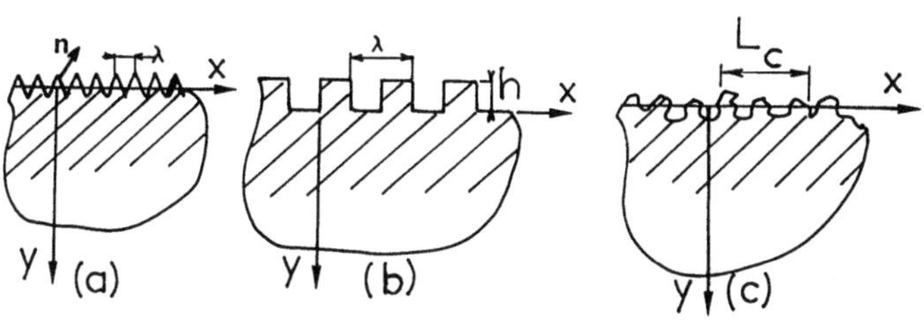

Figure 5 - **(a)** periodic grating ; **(b)** grooves ; **(c)** rough surface

of nonlinearity (in the substrate) and dispersion (related to the superimposed layer) may provide SH SAWs of the type of solitary waves [32], [7] which may not be true solitons (the question is open). Another interesting prospect is provided by SH SAWs on laminated composites (layers parallel to the traction-free limiting plane placed in a slow layer) which exhibit dispersion, a limited frequency range and both band-pass and band stop filter characteristics [33].

4. INTERACTION WITH NONMECHANICAL FIELDS

4.1. Bleustein-Gulyaev SAWs on piezoelectrics

Another way of producing "perturbations" of the type (2.5)-(2.6) while keeping only one elastic component (the SH component $u_3 = u$) is to envisage couplings with nonmechanical fields such as electric and magnetic fields. Since we are concerned with acoustic phenomena, the additional fields may be considered in a quasi-static framework unless there is a real dynamical coupling with acoustics such as resonances. The first known successful example of such a coupling that guarantees the existence of an SH component u_3 was theoretically found by BLEUSTEIN [34] and GULYAEV [35] and its first experimental observation by MAERFELD et al. [36]. Essentially, one has to find a crystal symmetry class for which u_3 uncouples from the elastic component parallel to the sagittal plane but remains coupled to the quasi-static electric potential $\emptyset$. This is achieved in a piezoelectric crystal if the z-axis is a six-fold symmetry axis for a crystal which admits no center of symmetry (e.g., C6v ; Fig. 6a). If we set

$$\bar{\Box}_T \equiv \frac{\partial^2}{\partial x^2} - \frac{1}{\bar{c}_T^2}\frac{\partial^2}{\partial t^2}$$
$$\bar{c}_T^2 = c_T^2(1 + K^2) = \bar{c}_{44}/\rho \quad , \quad K^2 = e_{15}^2/\varepsilon_{11}\, c_{44}^E \quad , \tag{4.1}$$

then instead of (2.2) we have the "mechanical" problem

$$\text{(P4)} \quad \begin{cases} \bar{\mathcal{D}}u \equiv \bar{\Box}_T\, u + u'' = 0 \quad , \quad \text{for } y > 0 \; , \\ u' + \beta\,\psi' = 0 \quad , \quad \text{at } y = 0 \; , \end{cases} \tag{4.2}$$

where $\beta = e_{15}/\bar{c}_{44}$ and $\psi = \emptyset - \alpha u$, $\alpha = e_{15}/\varepsilon_{11}$. Here c_{44}^E, e_{15}, ε_{11} and K are, respectively, an elasticity coefficient, a piezoelectric constant,

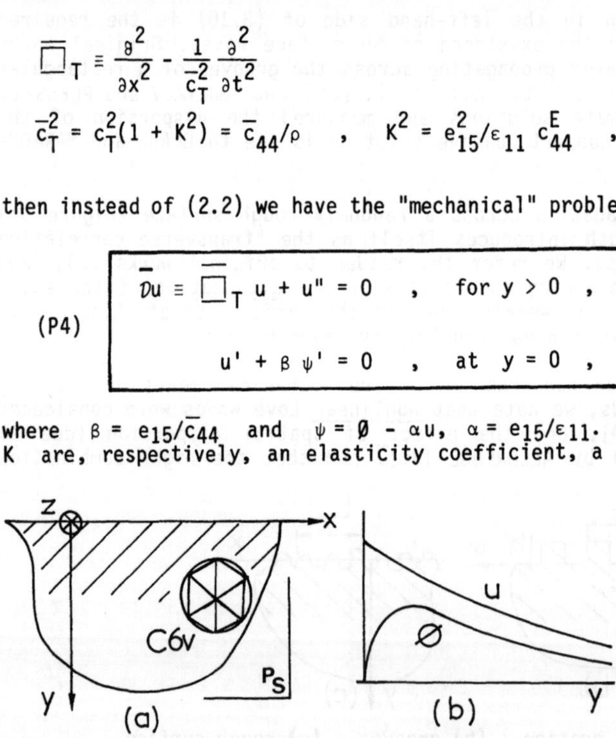

Figure 6 - Bleustein-Gulyaev wave

a dielectric constant, and the electromechanical coupling factor which is responsible for the stiffening of elasticity coefficients. Clearly, the displacement u couples to ψ through the boundary condition $(4.2)_2$ which is therefore perturbed as compared to $(2.2)_2$. But this is not all since the problem now is a two-component wave problem since ψ satisfies also electric conditions (electrically grounded $y = 0$)

$$(P4)_A \left\{ \begin{aligned} &\nabla^2 \psi = 0 \quad \text{for} \quad y \gtrless 0 \ , \\ &\psi + \alpha u = 0 \quad \text{at} \quad y = 0 \ . \end{aligned} \right. \tag{4.3}$$

The resulting SH SAW has a velocity c_{BG} given by

$$c_{BG}^2 = \bar{c}_T^2(1 - \bar{K}^4) \quad , \quad \bar{K}^2 = K^2/(1 + K^2) \quad \text{(grounded boundary)} \ , \tag{4.4}$$

or

$$c_{BG}^2 = \bar{c}_T^2[1 - \bar{K}^4/(1 + \varepsilon_{11})^2] \qquad \text{(free boundary)} \quad . \tag{4.5}$$

As the corrective coefficient is rather small c_{BG} does not differ much from c_T and, correspondingly, Bleustein-Gulyaev (for short BG) waves have a deep penetration and are therefore weakly bound to the bounding surface. In practice, also, BG waves occur in pure form only at sufficiently large distances from the source, e.g., $L \simeq (1/4)(k-k_T)^{-1}$ i.e, $L \simeq 16$cm in CdS at 30MHz, but this distance can be reduced by metallization of the surface. Various direct generalizations of the above BG waves include the use of poled ceramics (poling in the z-direction creating the appropriate induced symmetry -see below), SH waves along an interface between two piezoelectric crystals separated by a narrow gap (GULYAEV and PLESSKII [37]), nonlinear coupling of the electrostrictive type rather than piezoelectricity [38], [39] -then a bias electric field is needed which, in fact, provides a means of controlling the wave characteristics-, BG waves of arbitrary shape $F(\zeta)$ (LARECKI [40] ; pulses are not permitted) and nonlinear BG waves [41]. In the latter case the author notices that nonlinear terms in the field equations are only of even degree in $\nabla \mathbf{u}$ and $\nabla \emptyset$ for crystal symmetries that admit BG waves, allowing thus for a wider range of input-signal levels in linear signal-processing devices using BG waves. Finally, semiconducting rather than dielectric, piezoelectric substrates may be considered. In that case u_3 couples not only with $\emptyset$ but also with charge waves which can be of special interest near to bounding surfaces. In the linearized theory this results in a slight dispersion of the BG wave and an attenuation of characteristic time τ given by (case of a free boundary ; compare (4.5), [42])

$$\begin{aligned} &c_{BG}^2(k) = \bar{c}_T^2 \left[1 - \frac{\bar{K}^4}{(1+\varepsilon_{11})^2} - \frac{g(k)}{\tau_d^2}\right] \ , \\ &\tau^{-1} = 1/2\tau_d \quad , \quad \tau_d = \varepsilon_{11}/\sigma \qquad , \end{aligned} \tag{4.6}$$

where τ_d is the dielectric relaxation time, σ is the electric conductivity and $g(k) \ll \tau_d^2$ for all k's in the long wavelength approximation.

4.2. Mixed BG SAWs on elastic ferroelectrics

In the case of elastic ferroelectric crystals such as barium titanate or sodium nitrite, there exists a possibility of resonance coupling between a dispersive soft ferroelectric mode (oscillations in an ordered system of electric dipoles) and elastic waves. This can be translated to the case of SAWs although the analytical solutions are rather involved. In particular, starting from a nonlinear theory, POUGET and MAUGIN [43] have shown that for the weakest allowed material symmetry, if the electric dipole oscillations take place about a state of initial electric polarization $\mathbf{P}_0$ orthogonal to the sagittal plane, then the general linearized surface wave problem splits into two surface-wave problems, one being a "three-component" Rayleigh problem coupling the elastic displacement parallel to the sagittal plane to the polarization dynamical component orthogonal to P_S, and the other one being a "four-component" BG problem in that it couples the SH elastic displacement to the quasi-static electric potential $\emptyset$ and the two-dimensional component $\mathbf{p}$ in the electric-dipole perturbation parallel to P_S. For the latter problem, noting $\boldsymbol{\nabla}$ the nabla operator in the sagittal plane, it is shown that eqns. (2.2) are replaced by (p_y = y-component of $\mathbf{p}$)

$$\text{(P5)} \quad \begin{array}{ll} \bar{\mathcal{D}}u + \gamma_p \boldsymbol{\nabla}.\mathbf{p} = 0 \quad , & \text{for} \quad y > 0 \quad , \\ u' + \beta_p \, p_y = 0 \quad , & \text{at} \quad y = 0 \quad , \end{array} \tag{4.7}$$

where γ_p and β_p are coupling coefficients that go to zero with $|\mathbf{P}_0|$ and $\bar{\mathcal{D}}$ is defined with an elasticity coefficient stiffened by electromechanical couplings. In addition, the SH displacement u is coupled dynamically to $\mathbf{p}$ and $\emptyset$ by equations of the type

$$(\text{P5})_A \quad \left\{ \begin{array}{l} \ddot{\mathbf{p}} = \mathbf{L}(\nabla^2 u, \nabla\nabla\mathbf{p}, \mathbf{p}, \nabla\emptyset \,|\, \mathbf{P}_0) \quad , \\ \boldsymbol{\nabla}.\mathbf{D} = 0 \quad \text{for} \quad y > 0 \quad , \end{array} \right. \tag{4.8}$$

and boundary conditions, where $\mathbf{L}$ is a vector-valued linear operator acting on the first four arguments and $\mathbf{D}$ is the electric displacement.

Since $\mathbf{L}$ produces a nonhomogeneous polynomial of differentiation, the soft ferroelectric mode is dispersive and through coupling the same holds true of the SH SAW for a grounded boundary $y = 0$. For a free boundary the SH wave is practically not dispersive, a boundary layer effect is exhibited for $\emptyset$, and for wavelengths smaller than that of the resonance point of mixed bulk waves, the mode becomes unstable in time with a marked oscillatory behavior for the amplitudes $\mathbf{p}$ with depth (see graphs in Refs. 43 and 4).

4.3. Mixed magnetoacoustic waves of the SH type

Somewhat analogously to the BG waves in Paragraph 4.1 and 4.2 one can envisage a perturbation of eqns. (2.2), which is due to a coupling with a magnetic field. The outcome is quite different depending on whether the substrate considered is an electricity conductor (e.g., a metal), an elastic paramagnet (e.g., a rare-earth compound which is a good insulator) or an elastic ferromagnet (i.e., a magnetically ordered body which allows for the propagation of spin waves). In the first case the magnetoelastic coupling occurs through a Lorentz force in the equation of motion and a magnetic pressure in the accompanying boundary condition. Then both eqns. (2.2) are perturbed for SH waves which are shown to exist for a bias magnetic field set parallel to the limiting plane $y = 0$ only if this field is tilted on the x-axis (i.e., neither parallel nor orthogonal to the x-axis) [44]. The importance of the resulting SH waves may show up while studying the propagation of cracks in elastic conductors [45].

In the case of elastic paramagnets the magnetoelastic coupling occurs through magnetostriction [46]-[47]. Starting from nonlinear field and constitutive equations the author and HAKMI [48] have shown that for a bias magnetic field set orthogonal to the sagittal plane the linearized surface-wave problem splits into two independent problems, one of which being of the BG type where ϕ now is the quasi-static magnetic scalar potential of the magnetic perturbation. These authors have also accounted for magnetic relaxation effects obtaining thus an attenuation. More recently, ABD-ALLA and MAUGIN [49] have considered the nonlinear case of SH surface waves propagating on a magnetostrictive substrate (e.g., polycrystalline ferromagnetic material) and have constructed a uniformly valid far field solution while exhibiting the anisochronism of the surface-wave solution (change in speed of the fundamental harmonic due to elastic nonlinearities and magnetostrictive couplings -the latter are nonlinear by definition).

To end with, we are left with the most intriguing, as also the most difficult, case, that of surface mixed phonon-magnon as can occur on a substrate of ferromagnetic yttrium-iron-garnet (Y.I.G). Magnons or, in the long-wave approximation, spin waves, are oscillations in the ordered arrangement of magnetic moments that prevails in a ferromagnetic single crystal. DAMON and ESHBACH [50] have shown that surface spin waves could exist. Since spin waves and acoustic waves (phonons) can couple through magnetostriction as they both resort to Bose-Einstein statistics in the quantum description, the question arises of the existence of surface mixed phonon-magnon. It is to PAREKH [51] that goes the merit to have pointed out that SH SAWs coupled to the Damon-Eshbach surface mode could exist, providing thus some kind of magnetoelastic analog to the BG waves of piezoelectrics. But it is only with more thorough studies [52]-[55] that the whole complexity of the problem and its original features were uncovered. For instance, both for pure surface spin waves and magnetoacoustic waves a nonreciprocity of behavior (dispersion) for waves travelling from right to left and left to right is observed (the bias magnetic field, together with the direction of propagation and the unit normal to the limiting plane forming a direct or reciprocal triad depending on the case considered [52]. When they exist, the surface waves are dispersive ; the exchange forces and the allied surface condition concerning magnetic spins create some analytical difficulties in the solution [53], and the mixed SH surface wave is essentially elastic in nature only for one direction of propagation and short wavelengths, with a speed tending to that of SH bulk waves and thus a weak binding to the surface. Finally, whenever the bias magnetic field is set orthogonal to the sagittal plane, starting from a nonlinear theory, using the weakest symmetry requirements and after linearization the author and HAKMI [55] have shown that the surface-wave problem could split in a pure Rayleigh wave problem and a problem of the BG type of dimension four, involving the SH elastic displacement u, the magnetostatic scalar potential ϕ corresponding to perturbations and the two-dimensional perturbation spin component μ parallel to the sagittal plane, resulting thus in a symbolic formulation akin to that in eqns. (4.7)-(4.8) since one then has for u the problem (μ_y = y-component of μ)

$$\text{(P6)} \quad \boxed{\begin{aligned} &\bar{\mathcal{D}}u + \gamma_m \nabla.\mu = 0 \quad , \quad \text{for} \quad y > 0 \; , \\ &u' + \beta_m \mu_y = 0 \quad , \quad \text{at} \quad y = 0 \; , \end{aligned}} \tag{4.9}$$

where γ_m and β_m are coupling coefficients that go to zero with the initial magnetization density and $\bar{\mathcal{D}}$ is defined with an elastic speed altered by magnetomechanical couplings. In addition, μ is governed by equations of the type

$$\begin{aligned} &\dot{\mu} = \gamma\mu_0 \times \mathcal{H} \quad , \quad y > 0 \; , \\ &\nabla.\mathbf{B} = 0 \quad , \quad y \gtrless 0 \; , \end{aligned} \tag{4.10}$$

and boundary (jump) conditions at y = 0. Here γ is the gyromagnetic ratio, $\mathcal{H}$ is an effective magnetic field that includes the effect of u, magnetic anisotropy and exchange forces, and B is the magnetic induction. In both cases (4.7) and (4.9) u is coupled to another dynamical degree of freedom of a nonmechanical nature. However, electric polarization and magnetization are quite different phenomena (think of their microscopic definitions) and the fact that $\gamma^{-1}\mu$ is already an angular momentum means that the first of (4.10) has a structure which radically differs from that of the first of (4.8). The essential result pertaining to the SH surface modes, i.e., the nonreciprocal dispersion relation of mixed modes, is to be found from graphs in original works [52], [55].

5. CONCLUSION

The framed formulas -(2.2), (3.1), (3.5), (3.6), (3.12), (4.2), (4.7) and (4.9)- illustrate in a very concise manner the contents and objective of the present review which fits in Viktorov's original point of view. As compared with our previous review on the same subject matter [4] that was written about 1981, we witness a tendency to account for more complex phenomena (couplings, roughness, nonlinearities, dispersion, attenuation) and, thus, to consider SAWs on real surfaces rather than on ideally simplified models. Of all effects, nonlinearity combined to, if not compensated by, dispersion appears to be the most promising property as also the most difficult one to deal with, but true SH surface solitons on elastic solids are an objective which no longer seems out of reach.

Aknowledgements : The author expresses his debt to A.A. Maradudin who provided very useful, dense preprints which rendered the writing of the present paper much easier than I originally thought. The E.E.C. and the U.S. Army Office in London are also to be thanked for their help in covering local expenses in Nottingham. The paper is dedicated to Acad. Yu. V. Gulyaev who should have been the one to deliver this lecture.

REFERENCES

1. J.D. Achenbach : Wave Propagation in Elastic Solids (North-Holland, Amsterdam, 1973).

2. K.F. Graff : Wave Motion in Elastic Solids (Ohio State University Press, Columbus, Ohio, 1975).
3. I.A. Viktorov : Sov. Phys. Acoustics, 25, 1-9, (1979).
4. G.A. Maugin : In Advances in Applied Mechanics, ed. J.W. Hutchinson, vol. 23 (Academic Press, New York, 1983), pp. 373-434:
5. A.A. Maradudin : In Nonequilibrium Phonon Dynamics, ed. W. Bron (Plenum Press, New York, 1985), pp. 395-599.
6. A.A. Maradudin : In Proc. Intern. Symposium on Surface Waves in Solids and Layered Media (Novosibirsk, July 1-4, 1986), in the press.
7. A.A. Maradudin : "Surface Acoustic Wave on Real Surface", Preprint, 65p., Dept. of Physics, University of California, Irvine (1987).
8. G.W. Farnell : In Acoustic Surface Waves, ed. A.A. Oliner, Topics in Appl. Phys., vol. 24 (Springer, Berlin, Heidelberg, 1978), pp. 13-60.
9. D. Royer and E. Dieulesaint : J. Acoust. Soc. Amer., 76, 1438 (1984).
10. A.C. Eringen and E. Suhubi : Elastodynamics, vol. 2 (Academic Press, New York, 1975).
11. N. Daher and G.A. Maugin : Acta Mechanica, 60, 217 (1986).
12. I.A. Viktorov : In Proc. 10th All-Union Conference on Quantum Acoustics and Acoustoelectronics (Tashkent, USSR, 1978), pp. 101-103.
13. P. Kielczynski : J. Tech. Phys. (Warsaw), 22, 73 (1981).
14. I. Bakirtas and G.A. Maugin : J. Méca. Théor. Appl., 1, 995 (1982).
15. J. Heading : Proc. Cambridge Phil. Soc., 61, 897 (1965).
16. B.S. Westcott : Proc. Cambridge Phil. Soc., 66, 675 (1969).
17. S.N. Bhattacharya : Bull. Seism. Soc. Amer., 60, 1847 (1970).
18. J.E. Gubernatis and A.A. Maradudin : Wave Motion, 9, 111 (1987).
19. P. Kielczynkski and W. Pajewski : NDT International, 18, 25 (1985).
20. R.L. Kovach : Rev. Geophys. and Space Physics, 16, 1 (1978).
21. W. Kosinski : Field Singularities and Wave Analysis in Continuum Mechanics (P.W.N., Warsaw, and Ellis Harwood, Chichester, U.K., 1986).
22. A.I.Murdoch : J. Mech. Phys. Solids, 24, 137 (1976).
23. H.F. Tiersten : J. Appl. Phys., 40, 770 (1969).
24. H.F. Tiersten, B.K. Sinha and T.R. Mecker : J. Appl. Phys., 52, 5614 (1981).
25. B.A. Auld, J.J. Gagnepain and M. Tau : Electronics Lett., 12, 650 (1976).
26. Yu. Y. Gulyaev and V.P. Plesskii : Sov. Phys. Techn. Phys., 23, 266 (1978).
27. N.E. Glass and A.A. Maradudin : Electronics Lett., 17, 728 (1983).
28. A.E. Eguiluz and A.A. Maradudin : Phys. Rev., B28, 711, 728 (1983).
29. O. Hardouin Duparc and A.A. Maradudin : J. Electron. Spectr. and Relat. Phenom., 30, 145 (1983).
30. N. Kalyanasundaram : Int. J. Engng. Sci., 19, 287 (1981).
31. G.P. Alldredge : Phys. Lett., 41A, 281 (1972).
32. K. Bataille and F. Lund : Physica, 6D, 95 (1982).
33. B.A. Auld, G.S. Beaupre and G. Herrmann : In Modern Problems in Elastic Wave Propagation, ed. J. Miklowitz and J.D. Achenbach (J. Wiley-Interscience, New York, 1978), pp. 459-474 ; also Electronics Lett., 13, 525 (1977).
34. J.L. Bleustein : Appl. Phys. Lett., 13, 412 (1968).
35. Yu. V. Gulyaev : Sov. Phys. JETP Lett., 9, 37 (1969).
36. C. Maerfeld, F. Gires and P. Tournois : Appl. Phys. Lett, 18, 269 (1970).

37. Yu. V. Gulyaev and V.P. Plesskii : Sov. Phys. Acoustics, 23, 410 (1977).
38. G.N. Burlak, N.Ya. Kotsarenko and S.V. Koshevaya : Sov. Phys. Solid State, 20, 273 (1975).
39. M.K. Balakirev and A.V. Gorchakov : Sov. Phys. Solid State, 18, 702 (1976).
40. W. Larecki : In these Proceedings (Euromech 226, Nottingham, 1987).
41. N. Kalyanasundaram : J. Sound. Vibr., 96, 411 (1984).
42. N. Daher and G.A. Maugin : Int. J. Engng. Sci., 24, 733 (1986).
43. J. Pouget and G.A. Maugin : J. Acoust. Soc. Amer., 69, 1304 (1981).
44. A.I. Viktorov : Sov. Phys. Doklady, 20, 273 (1975).
45. A. Chattopadhyay and G.A. Maugin : Ind. Jl. Pure Appl. Math., 17, 843 (1986).
46. R.E. Camley and P. Fulde : Phys. Rev., B23, 2614 (1981).
47. C. Lingner and B. Lüthi : Phys. Rev., B23, 256 (1981).
48. G.A. Maugin and A. Hakmi : J. Acoust. Soc. Amer., 76, 826 (1984).
49. A.N. Abd-Alla and G.A. Maugin : In these Proceedings (Euromech 226, Nottingham, 1987).
50. R.W. Damon and J.R. Eshbach : J. Phys. Chem. Solids, 19, 308 (1961).
51. J.P. Parekh : In Proc. IEEE Ultrasonics Symposium, Boston, 1972 (IEEE, New York, 1972), p. 233.
52. R.Q. Scott and D.L. Mills : Phys. Rev., B15, 3545 (1977).
53. R.E. Camley and R.Q. Scott : Phys. Rev., B17, 4327 (1978).
54. R.E. Camley and A.A. Maradudin : Appl. Phys. Lett., 38, 610 (1981).
55. G.A. Maugin and A. Hakmi : J. Acoust. Soc. Amer., 77, 1010 (1985).

Calculation of Interdigital Transducer Interaction with SAWs and BAWs

F. Seifert, A.R. Baghai-Wadji, K.C. Wagner, and G. Kovacs

Technical University, IAEE, Gusshausstr. 27-29,
A-1040 Wien, Austria

I. Introduction

Since the interdigital transducer (IDT) was used by WHITE and VOLTMER /1/ in 1965 to excite surface acoustic waves (SAWs) on plane piezoelectric substrates, signal processing devices based on this effect have evolved to a high degree of maturity and are described comprehensively in the book by MORGAN /2/. These devices are fabricated in a one step planar metal layer technology and consist mainly of launching (L) and receiving (R) - IDTs whose structure determines the electric filter characteristic. Simple physical reasoning shows, that the IDT's SAW impulse response $h(t)$ in a first approximation is given by its structure (i.e. the finger arrangement and mutual finger overlap, called apodisation). Fourier transform (FT^-) yields the transfer function $\overline{H}(\omega)$ of the IDT:

$$\overline{H}(\omega) = FT^-(h(t)). \tag{1}$$

The SAW-filter's transfer function $\overline{H}_F(\omega)$ is given by the product

$$\overline{H}_F(\omega) = \overline{H}_L(\omega) \cdot \overline{H}_R(\omega) \tag{2}$$

of the respective functions of the L- and R-IDT, and the filter's impulse response $h_F(t)$ by the corresponding convolution:

$$h_F(t) = h_L(t) \otimes h_R(t). \tag{3}$$

For nearly all applications the designer of SAW-filters has to meet a given $\overline{H}_F(\omega)$ within stringent specifications. To this goal some iterative computer aided design (CAD) methods have been developed, which are mainly based on the delta function /3/ or impulse /4/ model of the IDT. Here, it is assumed that the source of interaction between IDT and SAW are line charges at the finger edges with amplitudes proportional to the apodisation.
The CAD procedure starts with a splitting up of $\overline{H}_F(\omega)$ into $\overline{H}_L(\omega)$ and $\overline{H}_R(\omega)$ due to Eq.(2). Then the impulse responses $h_L(t)$ and $h_R(t)$ are found by FT^+ and a first approximation IDT geometry is found by fitting the delta functions, (i.e. the finger overlap) into $h_L(t)$ and $h_R(t)$. Since a very crude model of SAW interaction is used up to this point, IDT analysis has to be used now for correction of the transfer function. The accuracy of the models for analysis determines the degree of perfection with which a given $\overline{H}_F(\omega)$ can be met. In the following a short survey of our work to improve the analysis of SAW and bulk wave (BAW) interaction with IDTs is given.

II. The boundary value problem

Since SAW velocity is by a factor of 10^5 slower than light velocity, the time derivatives in Maxwell's equations can be neglected and the problem of SAW excitation by the driving electrical potentials at the IDT's fingers reduces to an electrostatic one. A first approach is purely electrical: Electric field E_i, permittivity tensor ε_{ij}, displacement D_i, potential ϕ and charge density q are related by

$$D_i = \varepsilon_{ij} E_j = -\varepsilon_{ij} \frac{\partial \phi}{\partial x_j}, \tag{4}$$

$$q = \frac{\partial D_i}{\partial x_i} = -\varepsilon_{ij} \frac{\partial^2 \phi}{\partial x_i \partial x_j}. \tag{5}$$

In the above, Einstein's summation convention is used. In Fig.1 the geometry investigated is sketched: At the surface $x_3 = 0$ of a piezoelectric half-space the L-IDT and R-IDT made of infinitely thin, ideally conducting sheets, are placed. Electrical charge is sited exclusively on the finger structure establishing a planar density $\sigma(x_1, x_2)$. This reduces Eq.(5) to

$$\sigma(x_1, x_2) = D_3(x_1, x_2, x_3 = +0) - D_3(x_1, x_2, x_3 = -0). \tag{5a}$$

By the charge distribution $\sigma(x_1, x_2)$, acoustic waves within and on the substrate are excited piezoelectrically. The equation of mechanical motion within the piezoelectric charge free medium reads /2/

$$\rho \frac{\partial^2 u_i}{\partial t^2} = c^E_{ijkl} \frac{\partial^2 u_k}{\partial x_j \partial x_l} + e_{kij} \frac{\partial^2 \phi}{\partial x_j \partial x_k}, \tag{6}$$

$$e_{ijk} \frac{\partial^2 u_j}{\partial x_i \partial x_k} - \varepsilon^S_{ij} \frac{\partial^2 \phi}{\partial x_i \partial x_j} = 0. \tag{7}$$

Here, ρ stands for mass density, c_{ijkl} for the stiffness tensor at $E_i = const$, e_{ijk} for the tensor of piezoelectricity and u_i for the vector of mechanical displacement. The mechanical boundary conditions result from a stress-free surface at $x_3 = 0$. The first step to find the mechanical wave excitation with a given finger potential at the L-IDT using Eqs.(4) to (7) is the determination of the charge distribution $\sigma(x_1, x_2)$ at the metallised regions. To make things simpler we confine our calculations to sinusoidal excitation and to plane waves with the wavenumber vector k_i.

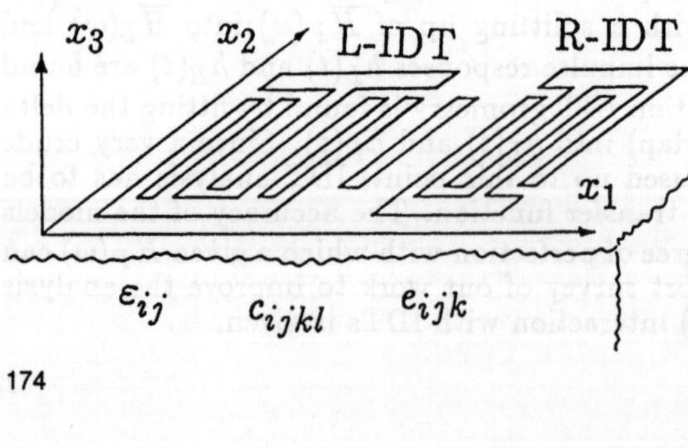

Fig.1 : Geometry of IDTs on piezoelectric half-space

III. Green's function and the method of moments

In 1977, MILSOM et al. /5/ used the concept of Green's function to analyse IDT interaction with acoustic waves. Employing the method of moments HUSSEIN and RISTIC /6/ improved this procedure by calculating surface charge distributions for certain symmetrical IDTs. Our group has extended these calculations to arbitrary two-dimensional IDT geometries and to electrically nonconnected (floating) sheets /7-12/. The linear boundary value problem described above can be solved by means of an appropriate Green's function. In the case of wave excitation by electrically driven IDTs on a piezoelectric half-space in Fig.1, the Green's function G due to an electrical point charge at the surface of the substrate has to be found. Then the electrical potential ϕ is given by

$$\phi(x_1, x_2, x_3) = \int_S\int G(x_1' - x_1, x_2' - x_2, x_3)\sigma(x_1', x_2')dx_1'dx_2'. \tag{8}$$

From the electrical potential ϕ excited by the charge distribution σ at the plane $x_3 = 0$ we can deduce all other physical quantities of interest following Eqs.(6) and (7). In the wavenumber domain (here the respective quantities are marked by a horizontal bar) we find $\overline{\phi}$ by a multiplication of $\overline{G}$ with $\overline{\sigma}$. Then Eq.(8) reads

$$\phi(x_1, x_2, x_3) = \frac{1}{(2\pi)^2}\int_S\int \overline{G}(k_1, k_2, x_3) \cdot \overline{\sigma}(k_1, k_2)e^{-j(k_1x_1+k_2x_2)}dk_1dk_2. \tag{9}$$

We now have to employ the method of moments (MoM) to find the frequency dependent σ from the driving voltage $\phi_L = \phi_{L0} \cdot exp(-j\omega t)$ at the L-IDT fingers. The potential of the floating and R-IDT fingers is not yet known, but additional conditions make our system of equations solvable. In the method of moments all conducting areas at the surface are divided into small subelements. In order to keep computational time low, the subdivision is fine only in areas where a strong change or peaking of charge must be expected (e.g. at the finger edges). For a total of N subelements σ can be written as

$$\sigma(x_1, x_2) = \sum_{n=1}^{N} b_n(x_1, x_2)\sigma_n, \tag{10}$$

where σ_n is the constant charge value at the $n-$th subelement, which is represented by a pulse function b_n. This discrete charge distribution (10) is inserted in (8) to yield the overall potential distribution $\phi(x_1, x_2, x_3, \sigma_1, ..., \sigma_N)$. The potential of each subelement ϕ_k is either given by the actually applied potential ϕ_{Lk} on the L-IDT or it builds up to a value ϕ_{Rk} on floating or R-IDT fingers. In a first step we omit these latter fingers. In order to establish the innerproduct required by the MoM /12/, we assume weighting functions $w_k(x_1, x_2)$, here in the form of pulses for each subelement, to match the potential distribution to the known potentials ϕ_{Lk}:

$$\phi_{Lk} = \int_S\int w_k(x_1, x_2) \cdot \Phi(x_1, x_2, x_3, \sigma_1, ..., \sigma_N)dx_1dx_2. \tag{11}$$

Employing Eqs. (8),(10) and (11) we get a set of N linear equations, each with N terms:

$$\phi_{Lk} = \sum_{l=1}^{N} A_{kl}\sigma_l, \quad k = 1, ..., N, \tag{12}$$

where the elements A_{kl} of the charge-potential-interaction (CPI) matrix are given by

$$A_{kl} = \iint_S w_k(x_1, x_2) \iint_S G(x_1'-x_1, x_2'-x_2, 0) b_l(x_1', x_2') dx_1' dx_2' dx_1 dx_2. \tag{13}$$

In the k−space we get by FT

$$A_{kl} = \frac{1}{(2\pi)^2} \iint_\infty \overline{G}(k_1, k_2, 0) b_l(k_1, k_2) \iint_S w_k(x_1, x_2) e^{j(k_1x_1+k_2x_2)} dx_1 dx_2 . dk_1 dk_2 \tag{14}$$

Since the calculation is easier in k−space, Eqs.(12) and (14) are those applied in our evaluations. The choice of pulse functions for both the basic (b_l) and the weighting (w_k) functions renders the CPI matrix symmetrical. Now, in a second step, we include the R-IDT and floating fingers: If we have altogether M floating sheet structures we get M a priori unknown potentials on the left side of (12). But, from charge neutrality on each such structure we get M additional equations, so that the charge and potential distribution on the whole SAW-filter including L-IDT, R-IDT and floating fingers can be determined. Once $\sigma_1, ..., \sigma_N$ and the resulting potentials on the floating sheets are found, it is relatively easy to analyse all electrical characteristics (i.e. transfer function of IDTs and whole SAW-filter and IDT electrical impedance) with high accuracy. Moreover this procedure helps in the investigation of all kinds of acoustic wave interaction with defined metallic structures at a piezoelectric substrate's plane surface.

IV. SAW and BAW analysis in a sagittal plane

In general, an interdigital transducer (IDT) on the surface of a piezoelectric crystal generates both surface and bulk waves. In this section we present results of analyses taking account of this coupling to bulk modes, which deteriorates the stop bands of many SAW-filters. At present, limited computer resources allow two-dimensional analyses only. Consequently, we consider the x_1x_3-sagittal plane (Fig.1) and assume that all electrodes are parallel to the x_2-direction and have infinite length; i.e. transverse end effects are neglected. In our case the Green's function $G(x_1, x_3)$ is the potential distribution caused by a sinusoidally oscillating line charge situated at the crystal's surface. For an arbitrary material and cut, this function has to be computed numerically using a method described in /6/. Regarding the potential at the surface, we can isolate three contributions to $G(x_1, 0)$: an electrostatic term G^E corresponding to Eq.(5), which shows logarithmic behaviour, a surface wave term G^S, brought about by outgoing undamped SAWs, and a bulk wave term G^B, which decays at least as $r^{-1/2}$. Usually, not $G(x_1, 0)$ but its Fourier-transform $\overline{G}(k_1, 0)$ is used:

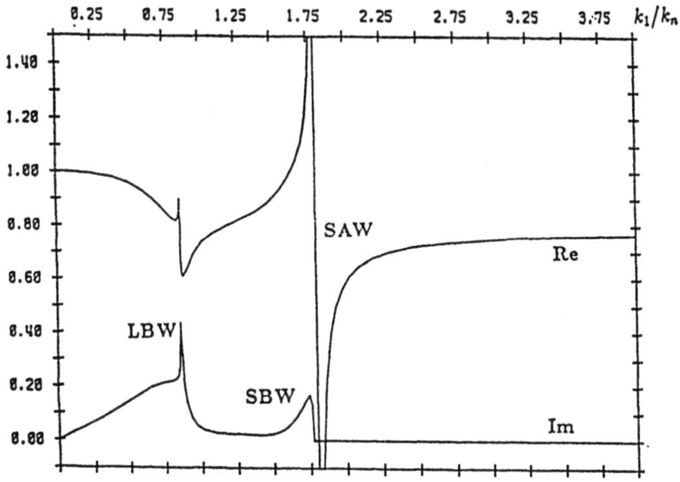

Fig.2 : Normalized plot of $\overline{G}(k_1, 0) \cdot |k_1|$ for YZ-$LiNbO_3$

$$\overline{G}(k_1,0) = \overline{G}^E(k_1,0) + \overline{G}^S(k_1,0) + \overline{G}^B(k_1,0) = \frac{\alpha}{|k_1|} + \frac{\beta}{k_1^2 - k_\infty^2} + \overline{G}^B(k_1,0). \tag{15}$$

As an example, Fig.2 shows $\overline{G}(k_1, 0) \cdot |k_1|$ for Y-cut Z-propagating $LiNbO_3$. Only positive values of the argument are shown, because the function is even. The pole and zero of the function correspond to the Rayleigh wave velocities for free and metallized surfaces, respectively. The discontinuities reflect the cutoff of longitudinal (LBW) and shear (SBW) bulk waves. The third possible bulk wave decouples from the electric field and therefore does not contribute to $\overline{G}(k_1, 0)$. Finally, it is worth noting that the function is purely real above the cutoff of the slowest bulk wave, indicating vanishing energy transport into the bulk.

Figure 3 shows the comparison of a charge distribution, which has been computed taking into account the electrostatic and SAW-interactions only, with the calculation including BAW-excitation. The imaginary part, which is of particular importance, since it determines the amount of radiated acoustic power, has been magnified with respect to the real part. For the L-IDT near its centre frequency, the influence of bulk wave excitation on the charge distribution is, as expected, rather small /11/. However, there are major differences at the ends of the transducer. Therefore bulk wave excitation must not be neglected when end finger effects are analysed. If the transducer is operated at twice its centre frequency as shown in Fig.3, the imaginary part of the charge distribution is clearly dominated by the effects of bulk wave excitation. With the BAW interaction included (Fig.3b), the resulting form of the imaginary part is totally different, indicating a considerable contribution of BAW to IDT interaction at higher frequencies. In Fig.4 angular bulk wave power distributions of a four electrode pair transducer with a metallization ratio $\eta = 0.5$ on YZ-lithium niobate are presented. The modulus of the real part of Poynting's vector P_i has been evaluated for points with a distance of 2 millimeters from the IDT centre, taking account of the quasi-

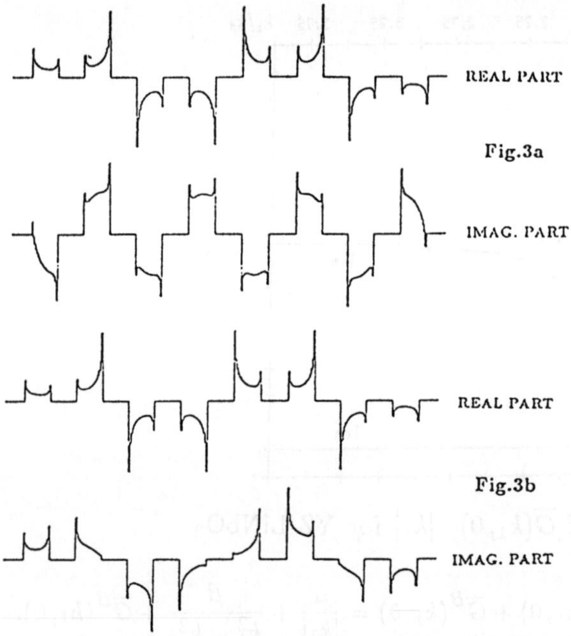

Fig.3 : Charge distribution on split-finger IDT at double center frequency, $\eta = 0.5$, $YZ - LiNbO_3$. a) without BAWs, b) including BAWs

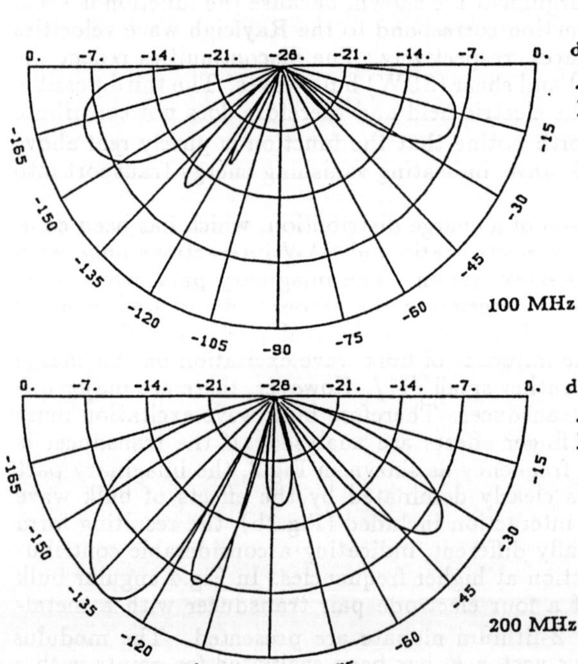

Fig.4 : Radiation pattern of quasilongitudinal wave, four electrode pair IDT on $YZ - LiNbO_3$, center frequency f_0=100 MHz.

longitudinal wave only. Here, Poynting's vector in the quasi-static approximation $P_i = 1/2 \cdot Re[-j\omega(\phi D_i^* + u_j^* T_{ij})]$ was used, where the derivative of the stress tensor T_{ij} stands on the right side of eq.(6). Figure 4 clearly shows that for a certain piezoelectric crystal our method reveals the sharp beams of BAWs which are excited by an IDT structure at a certain frequency. With increasing number of IDT-fingers the beam width decreases.

V. Analysis on the Surface

In the previous section we have analysed wave propagation in one sagittal plane. Now we turn to 2D SAW-effects on the surface of the crystal. No restrictions are imposed on the 2D-geometry of the electrodes (overlap-apodised IDTs, curved-finger IDTs, bus-bars, etc.); bulk wave excitation, however, is neglected.
Using cylindrical coordinates ($x_1 = r \cdot cos\varphi$, $x_2 = r \cdot sin\varphi$, $k_1 = k \cdot cos\psi$, $k_2 = k \cdot sin\psi$) Eq.(11) can be rewritten as

$$\phi(r,\varphi,x_3) = \frac{1}{(2\pi)^2} \int_{\psi_0}^{\psi_0+\pi} \int_{-\infty}^{+\infty} \overline{G}(k,\psi,x_3)\overline{\sigma}(k,\psi) e^{jkr\cos(\psi-\varphi)} |k| dk d\psi. \qquad (16)$$

The Green's function, which can be interpreted here as the electric potential due to a sinusoidally oscillating point charge, is constructed in the wave number space by appropriate rotation of the 1D case. For each angle $\psi = \tilde{\psi}$, specifying a sagittal plane, the Green's function $\overline{G}(k,\tilde{\psi},z)$ can be evaluated using methods similar to those employed for the computation of the effective permittivity $\overline{\epsilon}_{eff}$ /5/,/6/. Particularly, for points at the surface of the substrate ($x_3 = 0$),

$$\overline{G}(k,\tilde{\psi},0) = \frac{1}{|k|\overline{\epsilon}_{eff}(k,\tilde{\psi})}. \qquad (17)$$

The effective permittivity is approximated by $\overline{\epsilon}_{eff}(k,\psi) \simeq \epsilon_\infty(\psi)\frac{k^2-k_0^2(\psi)}{k^2-k_\infty^2(\psi)}$.
Here $k_0(\psi)$ and $k_\infty(\psi)$ are the wave numbers of the plane surface waves propagating in the direction ψ on the free and metallized surfaces, respectively. $\epsilon_\infty(\psi)$ is the effective permittivity for $k \to \infty$. We obtain the Green's function in the real space by a Fourier transform.
The first part of the Green's function, which is determined by the effective permittivity in the limit $|k| \to \infty$, describes the electrostatic interaction:

$$G^E(r,\varphi,0) = \frac{1}{2\pi r} \cdot \frac{1}{\epsilon_\infty(\varphi + \frac{\pi}{2})}. \qquad (18)$$

The electrostatic potential decays with r^{-1}, the anisotropy of the substrate appears in $\epsilon_\infty(\varphi + \frac{\pi}{2})$.
The second or SAW part of the Green's function is determined by the pole and zero distribution of the effective permittivity /13/. Carrying out the k-integration analytically, we obtain

$$G^S(r,\varphi,0) = j\frac{1}{2\pi}\int_{\psi_0}^{\psi_0+\pi} G(\psi)e^{jk_0(\psi)r|cos(\psi-\varphi)|}d\psi. \tag{19}$$

The SAW part of the Green's function is the superposition of plane waves. Each of these waves propagates with a different velocity and is excited with a different amplitude due to the anisotropy of the coupling coefficient

$$G(\psi) = \frac{k_0^2(\psi) - k_\infty^2(\psi)}{2k_0(\psi)\epsilon_\infty(\psi)}.$$

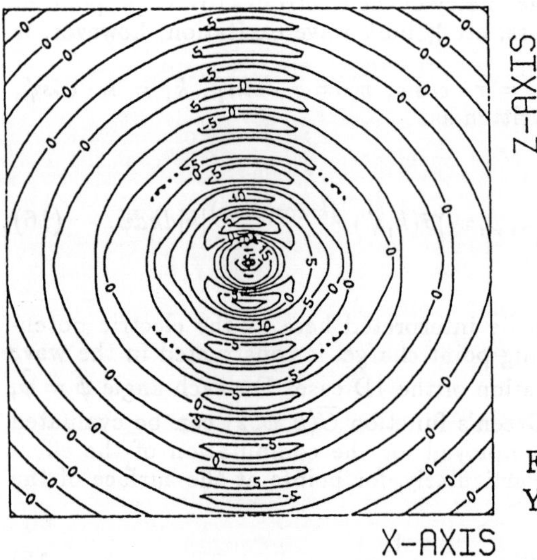

Fig.5a : Real part of G^S on Y-cut $LiNbO_3$

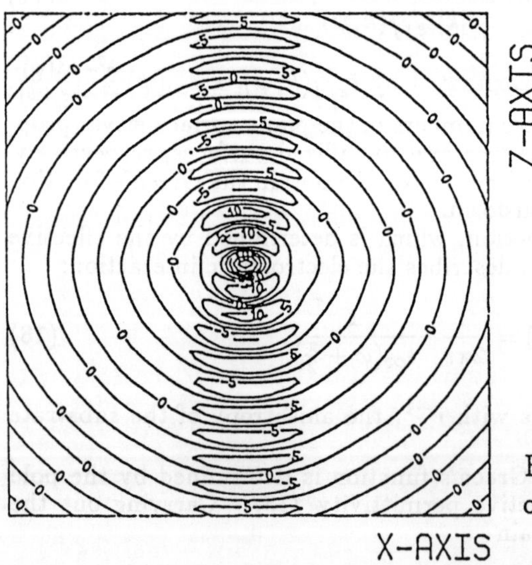

Fig.5b : Imaginary part of G^S on Y-cut $LiNbO_3$

Figure 5 shows the SAW part of the Green's function on the surface of Y-cut $LiNbO_3$. From the centre of these isoplots, where the oscillating point charge is located, waves are excited mainly into the crystalline Z-axis due to the higher coupling coefficient for this direction. The isolines $Re(G^S = 0)$ and $Im(G^S = 0)$, which are constant phase lines too, clearly demonstrate the anisotropy of the surface-wave velocity.
In Fig.6 we show the slowness curve for plane surface waves. This curve has a negative curvature at the crystalline Z-axis. The power flow vector (i.e. group velocity, $\vec{v}_g$) for plane waves near this axis points slightly towards the Z-axis. This explains the self-focussing properties of this crystal cut.
In order to show diffraction effects, in Fig.7 we compare beam profiles perpendicular to the direction of propagation for two different materials, evaluated at various distances from the source. The right column is for a finite line source on an isotropic material having the same properties of slowness and coupling coefficient for all directions as YZ-$LiNbO_3$ has along the crystalline Z-axis. The left column, which is computed for YZ-$LiNbO_3$, demonstrates the self-focussing properties of this cut.

VI. Conclusions

We have shown that the electric charge distribution $\sigma(x_1, x_2)$ on the planar metallic IDT structure on the surface of the piezoelectric substrate is the source of excitation of SAW and BAWs. To determine a satisfactory approximation for $\sigma(x_1, x_2)$ we use the concept of the Green's function G and the method of moments for small subelements of the finger structure. If a certain voltage is applied to the L-IDT our analysis yields the resulting potential for the R-IDT and for floating fingers and the charge distribution on all conducting structures.

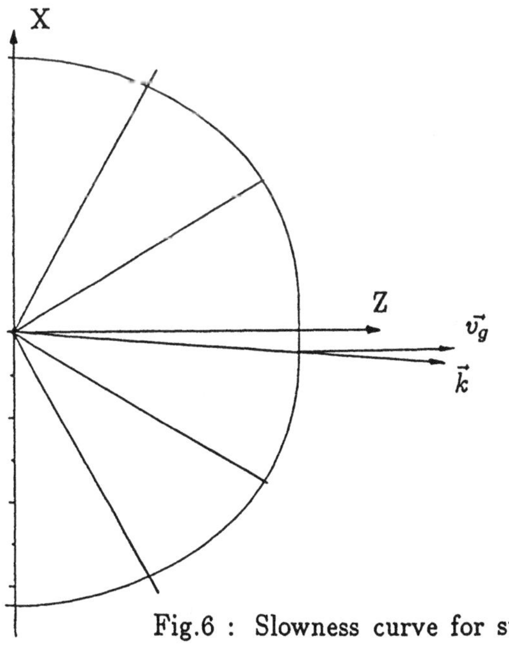

Fig.6 : Slowness curve for surface waves on Y-cut $LiNbO_3$

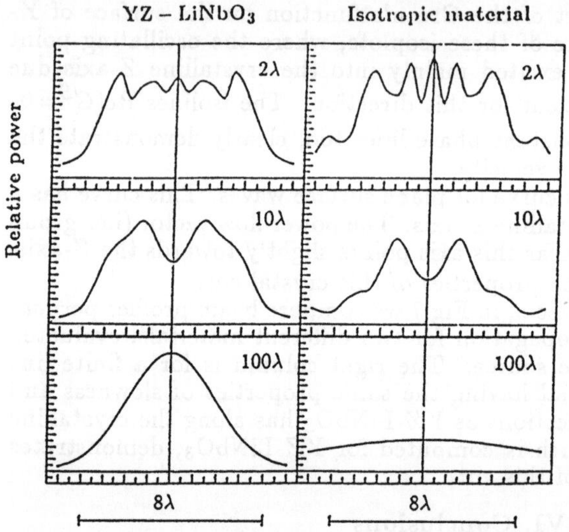

Fig.7 : Beam profiles due to a finite line charge on $YZ - LiNbO_3$ (left column) and isotropic material (right column)

More or less intuitively, floating or unpolarized fingers (e.g. rows of interconnected ones) are often used in SAW devices for matching or reflection purposes. Our evaluation method, for the first time, allows an analysis of these structures in a good approximation to the interaction mechanism; e.g., the unidirectionality of SAW excitation with special finger geometries can be evaluated /14/.
Due to the high precision with which charge distribution can be computed, a good description of special IDT structures in the frequency domain is possible. This leads to the development of devices working up to the seventh harmonic of the fundamental period of the IDT /15/. In this way, with a 0.8 μm technology, which in the fundamental is sufficient for 1 GHz filters, devices up to 2.6 GHz could be fabricated.
In sect. V, an evaluation of the two-dimensional charge distribution on IDT structures was sketched. It takes into account the anisotropy of both SAW slowness and coupling coefficient and yields SAW diffraction patterns for, in principle, arbitrary geometries. Thus this method should replace the Fresnel and angular spectrum of modes techniques for diffraction modelling /16/ which in 1982 initiated high precision applications of SAW-filters.

By calculating the charge and potential distribution $\sigma(x_1, x_2)$ and $\phi(x_1, x_2)$ on the L-IDT and R-IDT simultaneously, the transfer function $\overline{H}_F(\omega)$ can be analysed in one step, rather than splitting it into two parts, following Eq.(2) in conventional design. This new method of analysis takes into account spurious (so-called second-order) effects, namely: capacitive feedthrough, triple transit signal and acoustoelectrical regeneration.

But, finally, we should remark that our model is not at all a comprehensive physical picture of SAW and BAW interaction with real IDTs yet. At least mass loading by IDT structures, temperature and acceleration (i.e. mechanical stress) effects should be considered to more closely approach physical reality, i.e. $\overline{H}_F(\omega)$.

VII. Acknowledgments

This work is supported by Siemens AG, Central Research Laboratories, Munich, West Germany, and the Austrian Science Research Fund, Project P5311. Many thanks are due to Dr. E. Langer for making available his program SABAWA for the calculation of the parameters in (17). Valuable cooperation with Dr. O. Männer, advice by Dr. S. Selberherr and continuous encouragement and support by Dr. G. Riha of Siemens AG are gratefully acknowledged.

VIII. References

/1/ R.M. White and F.W. Voltmer, "*Direct Piezoelectric Coupling to Surface Elastic Waves*", Appl.Phys.Lett. 7, pp.314-316 (1965).

/2/ D.P. Morgan, "*Surface Wave Devices for Signal Processing*", Elsevier 1985.

/3/ R.H. Tancrell and M.G. Holland, "*Acoustic Surface Wave Filters*", Proc. IEEE 59, pp.393-409 (1971).

/4/ C.S. Hartmann et.al., "*Impulse Model Design of Acoustic Surface Wave Filters*", IEEE Trans. MTT-21, pp.162-175 (1973).

/5/ R.F. Milsom, N.H. Reilly AND M. Redwood, "*Analysis of Generation and Detection of Surface and Bulk Acoustic Waves by Interdigital Transducers*", IEEE Trans., SU-24, No.3, May 1977, pp. 147-166.

/6/ A.M. Hussein and V.M. Ristic, "*The Evaluation of the Input Admittance of SAW Interdigital Transducers*", J.Appl.Phys., Vol.50, No.7, July 1979, pp.4794-4801.

/7/ A.R. Baghai-Wadji, "*Berechnung der Quellenverteilung zur Anregung von akus- tischen Wellen in Oberflächenwellenfiltern*", Dr.techn. thesis, Techn. Univ. Wien 1987.

/8/ A.R. Baghai-Wadji, S. Selberherr, F. Seifert, "*On the Calculation of Charge, Electrostatic Potential and Capacitance in Generalized Finite SAW Structures*", IEEE SU-Symp. 1984, Proc. pp.44-48.

/9/ A.R. Baghai-Wadji et al., "*Rigorous 3D Electrostatic Field Analysis of SAW Transducers with Closed-Form Formulae*", IEEE SU-Symp. 1986, Proc. pp.23-27.

/10/ A.R. Baghai-Wadji et al., "*Two Dimensional Green's Function of a Semi-Infinite Anisotropic Dielectric in the Wavenumber Domain*", IEEE Trans. UFFC-3, May 1986, pp.315-317.

/11/ K.C. Wagner et al., "*Interdigital Transducer Interaction with SAWs and BAWs*, IEEE SU-Symp.1987, Proc. to appear spring 1988.

/12/ Liu Yuan, Wang Cheng-Hao and Ying Chong-Fu, "*General Two-dimensional Planar Transducers and Blazed Array-Like Interdigital Transducer*", Proc. IEEE Ultrason. Symp. 1986, Williamsburg, USA, pp.129-132.

/13/ A.R. Baghai-Wadji, O. Männer, S. Selberherr and F. Seifert, "*Analysis and Measurement of Transducer End Radiation in SAW Filters on Strongly Coupling Substrates*", in Proc. 1st European Frequency and Time Forum, Besancon, France, 1987, pp.315-319.

/14/ O. Männer, "*Knotenvariablen Analyse von Oberflächenwellenfilter*", Dr. techn. thesis, Techn. Univ. Wien, 1987.

/15/ B. Fleischmann, W. Ruile and G. Riha, "*Rayleigh-Mode SAW Filters on Quartz for the Timing Recovery at Frequencies above 1 GHz*, IEEE SU-Symp. 1986, Proc. pp.163-167.

/16/ W.R. Mader, C. Ruppel and E. Ehrmann-Falkenau, " *Universal Method for Compensation of SAW Diffraction and other Second Order Effects*, IEEE SU-Symp. 1982, Proc. pp.23-34.

Bleustein-Gulyaev Surface Waves of Arbitrary Form

W. Larecki

Institute of Fundamental Technological Research,
Polish Academy of Sciences, ul. Swiętokrzyska 21,
PL-00-049 Warsaw, Poland

1 Introduction

The aim of this paper is to demonstrate that the theory of surface waves of arbitrary form, developed by FRIEDLANDER /1/, CHADWICK /2/, PARKER /3/, PARKER & TALBOT /4/, CLEMENTS /5/ for surface waves of Rayleigh-type in elastic materials, can be extended to the case of piezoelectric surface waves as well as to show that real Schwartz tempered distributions are admissible as the generalized wave profiles in this theory. For simplicity the analysis is performed for the case of the free-surface Bleustein-Gulyaev surface waves polarized along the sixfold symmetry axis in the 6mm hexagonal piezoelectric crystal. The solution to the governing equations and boundary conditions is assumed in the form of a travelling wave propagating without change of waveform at some subsonic speed in the direction corresponding to the respective space variable. Then we require the theory to predict the same propagation speed for all wave profiles and stipulate the existence, uniqueness and stability of the assumed solution. These assumptions are weaker than involved in /1,2,3,4,5/ and therefore enable us to derive the theory of surface waves of arbitrary form admitting the larger class of wave profiles and less restricted class of solutions. In the framework of this theory we prove that Bleustein-Gulyaev surface waves of $S'(\mathbb{R})$ profiles cannot propagate in the form of impulses.

2 Governing Equations and Boundary Conditions

The classical analysis of the sinusoidal Bleustein-Gulyaev surface waves polarized along the sixfold symmetry axis in the 6mm hexagonal piezoelectric crystal can be found, for example in /6/. Let x_i (i=1,2,3) be rectangular Cartesian coordinates and t denote time. Suppose that the half-space $x_1 > 0$ is occupied by the body and the adjacent region $x_1 < 0$ contains vacuum. The considered surface wave is propagated in the $0x_2$ direction and all fields are independent of x_3. For the Bleustein-Gulyaev surface wave the particle displacement vector $\underline{u}$ is normal to the direction of propagation and parallel to the boundary $x_1=0$. This results in the antiplane stress $\underline{S}$ and the plane electric displacement $\underline{D}$. Let ω and $\tilde{\omega}$ denote the electric potential in the body and in the vacuum, respectively. Let $\tilde{\underline{D}}$ be the electric displacement in the vacuum. Than the following relations describe the fields considered in the subsequent analysis:

$$\underline{u} = (0,0,u(x_1,x_2,t)) \ , \quad \omega = \omega\,(x_1,x_2,t) \quad , \quad \tilde{\omega} = \tilde{\omega}\,(x_1,x_2,t)$$

$$S_{3K} = \frac{\partial}{\partial x_K}\,(c_{44}\,u + e_{15}\,\omega) \quad , \quad v = 4\pi\,e_{15}\,u - \varepsilon_{11}\omega \tag{2.1}$$

$$D_K = \frac{\partial}{\partial x_K}\,v \ , \quad \tilde{D}_K = \frac{\partial}{\partial x_K}\,(-\,\varepsilon_0\,\tilde{\omega}) \quad , \quad K = 1,\,2\,.$$

$c_{44}, e_{15}, \varepsilon_{11}$ are elastic, piezoelectric and dielectric constants of the body, respectively, ε_0 is the electric permeability of the vacuum. After substituting (2.1) into the stress equation of motion and into the charge equation of electrostatics we rearrange the resulting partial differential equations into the following system:

$$\frac{s}{\rho}(u_{,11} + u_{,22}) - \frac{\partial^2 u}{\partial t^2} = 0 \ , \qquad v_{,11} + v_{,22} = 0 \ , \quad x_1 > 0$$
$$\tilde{\omega}_{,11} + \tilde{\omega}_{,22} = 0 \ , \qquad x_1 < 0, \tag{2.2}$$

where

$$s\rho^{-1} := c_{44}\rho^{-1}\,[1 + 4\pi e_{15}^2 (c_{44}\varepsilon_{11})^{-1}] = c_T^2 \ . \tag{2.3}$$

c_T is the transverse body wave speed in the direction Ox_2 and ρ is the density. Vanishing of the tractions and continuity of the electric potential and the normal electric displacement on the boundary $x_1=0$ impose the boundary conditions, which in view of (2.1), (2.3) become

$$s_{31}(0,x_2,t) = s\,u_{,1}(0,x_2,t) - m\,v_{,1}(0,x_2 t) = 0$$
$$D_1(0,x_2,t) - \tilde{D}_1(0,x_2,t) = v_{,1}(0,x_2,t) + \varepsilon_0\,\tilde{\omega}_{,1}(0,x_2,t) = 0 \tag{2.4}$$
$$\omega(0,x_2,t) - \tilde{\omega}(0,x_2,t) = 4\pi m\,u(0,x_2,t) - \varepsilon_{11}^{-1}\,v(0,x_2,t) - \tilde{\omega}(0,x_2,t) = 0,$$

where

$$m := e_{15}\,\varepsilon_{11}^{-1} . \tag{2.5}$$

The boundary conditions (2.1) can be interpreted as the system of coupled Cauchy conditions on $x_1 = 0$:

$$u(0,x_2,t) := k(x_2,t) \qquad \tilde{\omega}(0,x_2 t) := l(x_2,t)$$
$$u_{,1}(0,x_2 t) := g(x_2,t) \qquad \tilde{\omega}_{,1}(0,x_2,t) = -\,s(\varepsilon_0 m)^{-1}\,g(x_2,t)$$
$$v(0,x_2,t) = 4\pi\,e_{15} k(x_2,t) - \varepsilon_{11} l(x_2,t) \tag{2.6}$$
$$v_{,1}(0,x_2,t) = s\,m^{-1}\,g(x_2,t) \ ,$$

where $k(x_2,t), g(x_2,t), l(x_2,t)$ are the independent conditions.

3 Travelling Wave Solution

The travelling wave, propagated without change of wave-form at some subsonic speed w in the Ox_2 direction is assumed to be the form of the solution to the boundary-value problem (2.2), (2.6). According to this assumption, the new independent variables are introduced

$$y := x_1 \ , \quad \Theta := x_2 - w\,t \ , \quad c_T^2 > w^2 > 0 \ , \tag{3.1}$$

and the solution is sought in a form

$$u = U(\Theta,y) \ , \quad v = V(\Theta,y) \ , \quad \tilde{\omega} = W(\Theta,y) . \tag{3.2}$$

After substituting (3.1),(3.2) into (2.2),(2.6) we employ the transformation of the variable y

$$Y := y\, a^{\frac{1}{2}} \;, \qquad r := -y \;, \qquad a = 1 - w^2 c_T^{-2} \;, \tag{3.3}$$

and rearrange the original boundary-value problem as the system of three independent Laplace equations in the upper half-plane subject to the system of three coupled Cauchy conditions

$$\begin{aligned}
&U_{\Theta\Theta} + U_{YY} = 0 \;, \quad U(\Theta,0) = K(\Theta) \;, \quad U_Y(\Theta,0) = a^{-\frac{1}{2}} G(\Theta) \\
&W_{\Theta\Theta} + W_{rr} = 0 \;, \quad W(\Theta,0) = L(\Theta) \;, \quad W_r(\Theta,0) = s\,(\varepsilon_0 m)^{-1} G(\Theta) \\
&V_{\Theta\Theta} + V_{yy} = 0 \;, \quad V(\Theta,0) = 4\pi e_{15} K(\Theta) - \varepsilon_{11} L(\Theta) \;, \\
&\qquad V_y(\Theta,0) = s\, m^{-1} G(\Theta) \;, \\
&\Theta \in \mathbb{R}, \quad Y, y, r \in \mathbb{R}^+ \;,
\end{aligned} \tag{3.4}$$

where, according to (3.1), independent conditions in (2.6) are taken in the following form:

$$k(x_2,t) = K(\Theta) \;, \qquad g(x_2,t) = G(\Theta) \;, \qquad l(x_2,t) = L(\Theta) \;, \tag{3.5}$$

and a suffix notation is employed for the partial derivatives of U,W,V. It follows from (3.1) - (3.4), (2.1) that primitive fields $(u,\omega,\tilde{\omega})$ are expressed in terms of three functions U,W,V, each of them being a solution of the respective Cauchy problem for the Laplace equation. It is well known that the Cauchy problem for the Laplace equation is improperly posed (see, for example /7/). Since the necessary and sufficient condition for the existence of the solution is that the data be continuable as an analytic function throughout the domain of definition of the solution, the solution does not exist globally unless appropriate compatibility relations hold among the data. If the global solution exists, it is unique but it does not depend continuously on the data in the sense of Hadamard. We therefore impose the requirement that *each of the Cauchy problems in (3.4) be well posed* (requirement R1) in order to ensure the existence, uniqueness and stability of the solutions U,W,V. Since the governing equations and boundary conditions involve no natural scale of length or time, the propagation speed w must be the same for all wave profiles, and in fact,it must coincide with the propagation speed of sinusoidal disturbances resulting from the classical theory. Thus we impose in the theory of surface waves of arbitrary form the condition that the unique value of the propagation speed be predicted. It follows from (3.3) that the propagation speed is related to the coefficient a involved in (3.4). The additional requirement that *the boundary-value problem (3.4) be compatible for only one value of* $\sqrt{a}$ (requirement R2) should therefore be imposed, and as a consequence this particular value of a should result from the compatibility condition. It has been pointed out by JOHN /8/ that in the case of the Cauchy problem for the Laplace equation continuous dependence on data can be obtained by suitably restricting the class of solutions. It is also obvious that only for appropriately restricted classes of solution the compatibility condition of the system (3.4) should hold for the unique value of $\sqrt{a}$. We therefore expect that it will be possible to bring about the stability in (3.4) which is necessary for the well-posedness (requirement R1) as well as to satisfy simultaneously the requirement R2 by imposing a certain restrictive condition on the solution to be considered. This condition, however, cannot be inferred from the boundary-value problem alone because of the ill-posedness (see /7/), it should be postulated. Proposing such condition we shall make use of the analytic continuation arguments in order to impose compatibility among the Cauchy data in (3.4), required for the existence of the solution (requirenent R1). It is well known that every solution of the Laplace equation is a smooth (real analytic) function, called harmonic function. Thus the solution of (3.4) consists of three functions $U(\Theta,Y)$, $V(\Theta,y)$, $W(\Theta,r)$ harmonic in the halfplane $\mathbb{R} \times \mathbb{R}^+$. Every half-plane harmonic function can be identified with either real or imaginary part of a half-plane holomorphic function. Let $K = \{z \mid \text{Im } z > 0\}$ be an open upper complex half-plane. We therefore may introduce functions $F_K(z)$, $z = \Theta + i\xi$, $K=1,2,3$ holomorphic in K, such that

$$U(\Theta, Y) = \text{Re } F_1(\Theta + iY) \quad , \quad F_1'(\Theta + iY) = U_\Theta(\Theta,Y) - i\, U_Y(\Theta,Y)$$

$$V(\Theta, y) = \text{Re } F_2(\Theta + iy) \quad , \quad F_2'(\Theta + iy) = V_\Theta(\Theta,y) - i\, V_y(\Theta,y) \tag{3.6}$$

$$W(\Theta, r) = \text{Re } F_3(\Theta + ir) \quad , \quad F_3'(\Theta + ir) = W_\Theta(\Theta,r) - i\, W_r(\Theta,r) \quad ,$$

where F_K' denotes the derivative of F_K. According to the boundary conditions in (3.4), holomorphic functions F_K and F_K' must acquire boundary values on $\text{Im } z = 0$ (limits in a sense of suitable topology as $\text{Im } z \to +0$) denoted by $S_K(\Theta)$ and $T_K(\Theta)$, respectively. It follows from (3.4), (3.6) that the boundary values satisfy the conditions

$$\text{Re } S_1 = K \quad , \quad \text{Re } S_3 = L \quad , \quad \text{Re } S_2 = 4\pi e_{15} K - \varepsilon_{11} L \tag{3.7}$$

$$T_1 = DK - i\, a^{-\frac{1}{2}} G \quad , \quad T_3 = DL - i\, s\, (\varepsilon_0 m)^{-1} G \tag{3.8}$$

$$T_2 = 4\pi e_{15} DK - \varepsilon_{11} DL - i\, s\, m^{-1} G \quad ,$$

where D denotes a distributional derivative. The restrictive condition, which should be postulated in order to satisfy the requirements RI, R2, can be expressed, in view of (3.6) - (3,8), as the condition on holomorphic functions F_K, F_K' and their boundary values S_K and T_K, respectively. It can be done by imposing suitable restrictions on the spaces of functions admissible for F_K and F_K'. Suppose that the functions F_K are of some class $H(K)$ and let $H'(K)$ be the class such that the derivatives of functions of $H(K)$ are in $H'(K)$, where both $H(K)$ and $H'(K)$ are classes of functions holomorphic in K and convergent to boundary values in the sense of respective topologies. Then we propose the condition on the classes $H(K)$ and $H'(K)$ (condition C1) consisting of two postulates.

i) *The following statement holds true. If* $F \in H(K)$ *converges to the boundary value* f *then* $F' \in H'(K)$ *converges to* Df.
ii) $H(K)$ *and* $H'(K)$ *are equipped with the following property. For every* G_1, G_2 *convergent to* g_1, g_2 *, respectively,* $\text{Re } g_2 = a \text{ Re } g_1$ *iff* $\text{Im } g_2 = a \text{ Im } g_1$ *and* $\text{Re } g_2 = a \text{ Re } g_1$ (or $\text{Im } g_2 = a \text{ Im } g_1$) *iff* $G_2 = a\, G_1$, *for all* $a \in \mathbb{R}$.

It should be emphasized that the postulate ii) of C1 displays the intimate links between the real and imaginary components of the boundary values of holomorphic functions as well as the fact that either real or imaginary components of the boundary values uniquely characterize the original holomorphic functions. It follows from (3.8) that

$$\text{Im } T_2 = s\, m^{-1} a^{\frac{1}{2}} \text{ Im } T_1 \quad , \quad \text{Im } T_3 = s\, (\varepsilon_0 m)^{-1} a^{\frac{1}{2}} \text{ Im } T_1 , \tag{3.9}$$

$$\text{Re } T_2 = 4\pi e_{15} \text{ Re } T_1 - \varepsilon_{11} \text{ Re } T_3 . \tag{3.10}$$

Hence (3.9) together with the postulate ii) of C1 yields

$$\mathrm{Re}\, T_2 = s\, m^{-1} a^{\frac{1}{2}}\, \mathrm{Re}\, T_1 \quad , \quad \mathrm{Re}\, T_3 = s\, (\varepsilon_0\, m)^{-1}\, a^{\frac{1}{2}}\, \mathrm{Re}\, T_1 \tag{3.11}$$

and it is consistent with (3.10) for the unique value of $\sqrt{a}$

$$a^{\frac{1}{2}} = p\, \varepsilon_0\, m\, s^{-1} \quad , \quad p := 4\pi e_{15}\, (\varepsilon_0 + \varepsilon_{11})^{-1}. \tag{3.12}$$

As a further consequence of C1, the following relations result:

$$\begin{aligned} &DS_2 = T_2 = \varepsilon_0\, p\, DS_1 = \varepsilon_0\, p\, T_1 \quad , \quad DS_3 = T_3 = p\, T_1 = p\, DS_1 \\ &F_2' = \varepsilon_0\, p\, F_1' \quad , \quad F_3' = p\, F_1' \ . \end{aligned} \tag{3.13}$$

Solving (3.13) with respect to S_2, S_3, F_2, F_3 and taking the real part of the solution we obtain

$$\mathrm{Re}\, S_2 = \varepsilon_0\, p\, \mathrm{Re}\, S_1 + \varepsilon_0\, p\, C_1 \quad , \quad \mathrm{Re}\, S_3 = p\, \mathrm{Re}\, S_1 + p\, C_2 \tag{3.14}$$

$$\mathrm{Re}\, F_2 = \varepsilon_0\, p\, \mathrm{Re}\, F_1 + \varepsilon_0\, p\, C_1 \quad , \quad \mathrm{Re}\, F_3 = p\, \mathrm{Re}\, F_1 + p\, C_2 . \tag{3.15}$$

Substituting (3.14) and (3.12) into (3.7) results in the condition

$$\varepsilon_0\, C_1 + \varepsilon_{11}\, C_2 = 0 . \tag{3.16}$$

Without loss of generality we may put $C_1=0$ since the constants C_1, C_2, related by (3.16), represent and arbitrary constant added to the electrostatic potential. Then it follows from (3.15), (3.6), (3.4) that three unknowns U,V and W can be expressed in terms of one harmonic function $\Phi(\Theta,\xi)$

$$U(\Theta, Y) = \Phi(\Theta, Y), \quad V(\Theta, y) = \varepsilon_0\, p\, \Phi(\Theta, y) \ , \ W(\Theta, r) = p\, \Phi(\Theta, r), \tag{3.17}$$

which is the solution of the Cauchy problem for the Laplace equation in the half-plane $\Theta \varepsilon \mathbb{R}, \xi \varepsilon \mathbb{R}^+$

$$\Phi_{\Theta\Theta} + \Phi_{\xi\xi} = 0 \ , \quad \Phi(\Theta, 0) = K(\Theta) \ , \ \Phi_\xi(\Theta, 0) = a^{-\frac{1}{2}}\, G(\Theta) := M(\Theta), \tag{3.18}$$

and the functions F_K can be written in the form

$$F_1 = F \ , \ F_2 = \varepsilon_0\, p\, F \ , \quad F_3 = p\, F \ , \tag{3.19}$$

where

$$\Phi(\Theta, \xi) = \mathrm{Re}\, F\, (\Theta + i\xi) \quad , \quad F'(\Theta + i\xi) = \Phi_\Theta(\Theta, \xi) - i\Phi_\xi(\Theta, \xi). \tag{3.20}$$

It follows from (3.19) that $F \varepsilon H(K)$ and $F' \varepsilon H'(K)$. Equations (3.18),(3.20) together with the postulate i) of C1 yield the boundary value T of F′

$$T = DK - i\, M = DS \quad , \quad K = \mathrm{Re}\, S \ , \tag{3.21}$$

where S is the boundary value of F. In (3.21) DK and M are uniquely related to each other since the postulate ii) of C1 provides the links between the real and imaginary boundary-value components of $H'(K)$ holomorphic functions. Hence for given K and for the class of solutions implicitly restricted by the condition C1, the Cauchy problem for the Laplace equation (3.18) admits only one M (uniquely determined by DK) and it therefore reduces to the properly posed Dirichlet problem. Thus the condition C1 is the sufficient condition for the well-posedness of the Cauchy problems in (3.4) (requirement R1).

As follows from (3.9) - (3.12), C1 is also the sufficient condition that the boundary-value problem (3.4) be consistent for only one value of the parameter a (requirement R2). Substituting the value of a given by (3.12) into (3.3) we obtain the propagation speed w which equals to the propagation speed of sinusoidal disturbances resulting from the classical theory /6, Equation (III.1.18)/. Substituting (3.2) into (2.1) and employing (3.1), (3.3), (3.12), (3.17) we finally obtain the representation of all fields considered in this theory

$$u_3 = u(\Theta,y) = \Phi(\Theta, y\, a^{\frac{1}{2}}) \quad , \tilde{\omega} = \tilde{\omega}(\Theta,y) = p\,\Phi(\Theta,-y)$$
$$\omega = \omega(\Theta,y) = 4\pi e_{15}\, \varepsilon_{11}^{-1}\, \Phi(\Theta, y\, a^{\frac{1}{2}}) - \varepsilon_0\, \varepsilon_{11}^{-1}\, p\,\Phi(\Theta,y)$$
$$S_{31} = S_{31}(\Theta,y) = \varepsilon_0\, \varepsilon_{11}^{-1}\, p\, e_{15}\, [\Phi_\xi(\Theta, y\, a^{\frac{1}{2}}) - \Phi_\xi(\Theta,y)]$$
$$S_{32} = S_{32}(\Theta,y) = (c_{44} + 4\pi e_{15}^2 \varepsilon_{11}^{-1})\, \Phi_\Theta(\Theta, y\, a^{\frac{1}{2}}) - \varepsilon_0\, \varepsilon_{11}^{-1} p\, \Phi_\Theta(\Theta,y)$$
$$D_1 = D_1(\Theta,y) = \varepsilon_0\, p\, \Phi_\xi(\Theta,y) \quad , \quad D_2 = D_2(\Theta,y) = \varepsilon_0\, p\Phi_\Theta(\Theta,y) \tag{3.22}$$
$$\tilde{D}_1 = \tilde{D}_1(\Theta,y) = \varepsilon_0\, p\, \Phi_\xi(\Theta,-y) \quad , \quad \tilde{D}_2 = \tilde{D}_2(\Theta,y) = -\varepsilon_0\, p\Phi_\Theta(\Theta,-y) \; ,$$

where $y = x_1$, $\Theta = x_2 - wt$, $w^2 = (1-a)\, c_T^2$ and the parameters $\sqrt{a}$ and p, given by (3.12), (2.3), are the combinations of material constants. The representation (3.22) yields the boundary values of these fields on $x_1 = 0$

$$u_3(x_1=0) = u(\Theta,0) = K(\Theta) \quad , \quad \omega(x_1=0) = \omega(\Theta,0) = p\, K(\Theta) = \tilde{\omega}(\Theta,0) = \tilde{\omega}(x_1=0)$$
$$S_{31}(x_1=0) = S_{31}(\Theta,0) = 0 \quad , \quad S_{32}(x_1=0) = S_{32}(\Theta,0) = (c_{44} + p\, e_{15})\, DK(\Theta)$$
$$D_1(x_1=0) = D_1(\Theta,0) = \varepsilon_0\, p\, M(\Theta) = \tilde{D}_1(\Theta,0) = \tilde{D}_1(x_1=0) \tag{3.23}$$
$$D_2(x_1=0) = D_2(\Theta,0) = \varepsilon_0\, p\, DK(\Theta) = -\tilde{D}_2(\Theta,0) = -\tilde{D}_2(x_1=0) \quad ,$$

which are, in turn, related to the boundary value T of the holomorphic function $F'\varepsilon H'(K)$, given by (3.21)

$$\begin{aligned} \mathrm{Re}\, T(\Theta) &= DK(\Theta) = (\varepsilon_0\, p)^{-1}\, D_2(\Theta,0) = -(\varepsilon_0\, p)^{-1}\, \tilde{D}_2(\Theta,0) \\ \mathrm{Im}\, T(\Theta) &= -M(\Theta) = -(\varepsilon_0\, p)^{-1}\, D_1(\Theta,0) = -(\varepsilon_0\, p)^{-1}\, \tilde{D}_1(\Theta,0) \quad , \end{aligned} \tag{3.24}$$

where the condition C1 provides the unique bond between Re T and Im T.

4 Discussion of the Theory and the Properties of the Solution

Equations (3.17) and the representation of the fields (3.22) show that the present theory of Bleustein-Gulyaev surface waves fulfils the condition (denoted as C2) that *all primitive fields of the theory* $(u_3, \omega, \tilde{\omega})$ *be expressed in terms of a single half-plane harmonic function* $(\Phi(\Theta,\xi))$. It can be shown that the same results can be recovered if the condition C2 is postulated instead of the condition C1. When C2 is assumed to hold, then the solution of (3.4) can be taken in the form

$$U(\Theta,Y) = \Phi(\Theta,Y) \quad , \quad V(\Theta,y) = \alpha\, \Phi(\Theta,y) + C_1 \quad , \quad W(\Theta,r) = \beta\Phi(\Theta,r) + C_2 . \tag{4.1}$$

Inserting (4.1) into the boundary-value problem (3.4) and stipulating the consistency of the system of boundary conditions we obtain the set of relations determining constants α, β, C_1, C_2 and the parameter $\sqrt{a}$, which lead to (3.12),(3.16),(3.17) and (3.18). Assuming (3.20),(3.21) and employing (3.16),(3.17) we arrive at (3.19),(3.13),(3.14),(3.15). Then, by comparison with (3.6),(3.8), we deduce the properties of holomorphic functions F_K, F'_K, which are the same as postulated by the condition C1. Hence the condition C2 is equivalent to the condition C1. It should be emphasized that the sufficient condition ensuring the fulfilment of the requirements R1, R2, formulated equivalently as either C1 or C2, plays the role of fundamental assumption in the derivation of this theory. This fact, i.e. sufficiency but not necessity of the assumed condition, raises the question whether this condition can be weakened. Taking into account (3.8) we are able to prove that the necessary and sufficient condition for the uniqueness of $\sqrt{a}$ (requirement R2) is that the *boundary values* T_K *of* F'_K *be linearly dependent* (denoted as condition C3). This condition yields

$$T_2 = 4\pi e_{15} T_1 - \varepsilon_{11} T_3 \tag{4.2}$$

and is fulfilled by the relations (3.13) implied by the condition C1. It can be proved that the condition C3 does not ensure the well-posedness of the Cauchy problems in (3.4) - requirement R1. For the proof, the solution of (3.4) satisfying (4.2) and exhibiting lack of the continuous dependence on data can be constructed by using the well-known Hadamard example /9/. Unfortunately, we are not able to provide the necessary and sufficient condition that the Cauchy problem for the Laplace equation be well posed (see/7/). The question raised therefore cannot be definitely answered, but it can only be stated that it is not possible to replace the assumed sufficient condition C1, C2 by the weaker condition C3 and, as a consequence, to derive the theory from the uniqueness of the propagation speed alone.

TAUWERTER /10/ introduced the class H^+ of "upper regular functions" by the following definition: $F(z)$, $z=\Theta + i\xi$, is said to be of class H^+ if it is holomorphic for $\mathrm{Im}\, z > 0$ and if it is bounded by a polynomial uniformly in every closed half plane $\mathrm{Im}\, z \geq \delta > 0$. Let $\hat{H}^+$ be the class of Laplace transforms of distributions in $\mathcal{S}'(\mathbb{R}) \cap \mathcal{D}'_+(\mathbb{R})$. BELTRAMI & WOHLERS /11,12,13/ showed that $\hat{H}^+$ is the subspace of H^+ comprising all those H^+ functions which attain boundary values in the $\mathcal{S}'$ topology as $\mathrm{Im}\, z \to +0$. It is proved /11,12,13/ that every half-plane holomorphic function $F \in \hat{H}^+$ converges in the $\mathcal{S}'$ sense as $\mathrm{Im}\, z \to +0$ to some S such that $S = \mathcal{F}f$ iff $F = \mathcal{L}f$ and $f \in \mathcal{S}'(\mathbb{R}) \cap \mathcal{D}'_+(\mathbb{R})$, where $\mathcal{F}$ denotes the Fourier transform (in notation of GELFAND & SHILOV /14/) and $\mathcal{L}$ denotes the respective Laplace transform. The boundary values of $\hat{H}^+$ functions are those $\mathcal{S}'(\mathbb{R})$ distributions, which are the Fourier transforms of $\mathcal{S}'(\mathbb{R}) \cap \mathcal{D}'_+(\mathbb{R})$ distributions, and real and imaginary components of these boundary values are uniquely related to each other by the generalised Hilbert reciprocal transform /12, Theorem 1/,/13, Theorems 3.10,3.13/. It can be proved that if $F \in \hat{H}^+$ and converges to some $S \in \mathcal{S}'(\mathbb{R})$, then $F' \in \hat{H}^+$ and converges to $T = DS$. Hence the class $\hat{H}^+$ is closed with respect to differentiation. These properties make the class $\hat{H}^+$ allowable as $H(K)$ and $H'(K)$ in our theory and, in fact, $\hat{H}^+$ is the largest class of half-plane holomorphic functions with the properties stipulated by the postulate ii) of C1. For the most general setting of this theory we take $H(K) = \hat{H}^+$. Then an arbitrary real $\mathcal{S}'(\mathbb{R})$ distribution is allowed to be the wave profile K in (3.23),(3.21), (3.18) since it is always possible to associate the imaginary $\mathcal{S}'(\mathbb{R})$ distribution P such that $K + P = \mathcal{F}f$, $\mathrm{supp}\, f \subseteq [0, +\infty)$ and K and P are uniquely related to each other by the generalized Hilbert reciprocal transform. The half-plane harmonic function $\Phi(\Theta, \xi)$, which determines the fields (3.22) corresponding to the wave profile K, can be represented either as $\Phi = \mathrm{Re}\, \mathcal{L}f$ or by means of the generalized Cauchy integral formula given in /13, § 3.2/. The function $F = \mathcal{L}f$ and also its derivatives are of the class H^+ and therefore are bounded by polynomials. Hence the harmonic function Φ , together with its derivatives, is also bounded by a polynomial in Θ and ξ . This fact leads to the conclusion that in this S'setting, when all real $S'(\mathbb{R})$ distributions are allowed to be the wave profiles, the theory admits the fields (3.22) of polynomial growth to infinity as $|x_1| \to +\infty$. In typical surface wave theories, the fields of definite physical meaning like **particle displacement**, electric potential, strain, stress, electric displacement, etc. are required to tend to zero as the vertical space variable tends to infinity. This property can be assured in this theory by restricting the space of allowable wave profiles to appropriate subspaces of $S'(\mathbb{R})$ and, correspondingly, by assuming for $H(K)$ and $H'(K)$ respective classes smaller than $\hat{H}^+$. For example, if we restrict the space of allowable wave profiles to the Sobolev space $W^{1,2}(\mathbb{R})$, then the classes $H(K)$, $H'(K)$ coincide with the half-plane Hardy Class H^2 /15/ . In this case the fields (3.22) attain the respective boundary values (3.23) in the L^2 norm as well as pointwise a.e. and tend towards zero not slower than $|x_1|^{-\frac{1}{2}}$as $|x_1| \to +\infty$. It is easy to verify that the standard results for the sinusoidal Bleustein-Gulyaev surface wave of angular frequency ω are recovered on setting in (3.19), (3.20) $F(z) = i\,A \exp(i\,k\,z)$, where $\omega = k\,w$ and A is an arbitrary real constant. The general formulation of the theory reveals one very important feature of the Bleustein-Gulyaev surface waves. The distribution T in (3.21) is the boundary value of $F' \varepsilon \hat{H}^+$ and, in virtue of /12, Theorem 4/,/13,Corollary 3.4/, Re T and Im T cannot both vanish on any open set. It then follows from (3.24) that $D_1(\Theta,0)$ and $D_2(\Theta,0)$ cannot simultaneously vanish on the set of positive measure and Bleustein-Gulyaev surface waves of the $S'(\mathbb{R})$ wave profiles therefore cannot propagate in the form of impulses with well defined wave fronts.

References

1. F.G.Friedlander: Quart.J. Mech.Appl.Math. 1 376 (1948)
2. P.Chadwick: J.Elasticity, 6. 73 (1976)
3. D.F.Parker: Physica. 16D, 385 (1985)
4. D.F.Parker, F.M. Talbot: J.Elasticity. 15, 389 (1985)
5. D.L.Clements: Acta Mechanica. 55, 31 (1985)
6. M.K.Balakirev, I.A.Gilinski: Waves in Piezocrystals, Nauka, Novosibirsk 1982 (in Russian)
7. L.E.Payne: In Symposium on Non-Well Posed Problems and Logarithmic Convexity, ed . by R.J.Knops, Lecture Notes in Mathematics, vol.316 (Springer-Verlag, Berlin, Heidelberg, New York 1973) p.1
8. F.John: Comm. Pure Appl.Math. 12, 551 (1960)
9. J.Hadamard: Lectures on Cauchy's Problem in Linear Partial Differential Equations , Dover Publications, New York 1952
10. H.A.Lauwerier: Arch.Rat.Mech.Anal.13, 157 (1963)
11. E.J.Beltrami, M.R.Wohlers: Arch.Rat.Mech.Anal. 18, 304 (1965)
12. E.J.Beltrami, M.R. Wohlers: J.Math.Mech. 15, 137 (1966)
13. E.J.Beltrami, M.R. Wohlers: Distributions and the Boundary Values of Analytic Functions, Academic Press, New York and London 1966
14. I.M.Gelfand, G.E.Shilov: Generalized Functions, vols.1.2.3, Moscow 1958 (in Russian)
15. P.Koosis: Introduction to H^p Spaces, Cambridge University Press, Cambridge 1980.

A Dynamic Problem for a Piezoelectric Medium

J.P. Nowacki[1] *and C. Trimarco*[2]

[1]Institute of Fundamental Technological Research,
Polish Academy of Sciences, ul. Swiętokrzyska 21,
PL-00-049 Warsaw, Poland
[2]Instituto di Matematica Applicata "U. Dini",
Facolta di Ingegneria, Pisa, Italy

The axisymmetric problem for an elastic dielectric with polarization gradient effects is considered, in which an axisymmetric, time-varying loading is prescribed on the boundary of the half space. Wave propagation in half space is examined in the monochromatic case. Surface waves are discussed, assuming the specific form of loading and the closed form solution is given.

1. INTRODUCTION

Elastic dielectric materials, in which the linear piezoelectric effect occurs, have become important in modern technology. The classical linear theory of piezoelectricity created by Voigt [1] is concerned with the interaction between the mechanical strain and the electric field and the electric polarization.
According to this theory, there is no direct coupling between the mechanical displacement and the polarization in centrosymmetric crystals. However Mindlin [2] has shown that if the stored energy function depends not only on the strain tensor and polarization vector but also on the polarization gradient tensor, such a coupling effect exists. This extension accommodates observed phenomena, such as electromechanical interaction in symmetric and non-symmetric materials and the capacitance phenomenon of a thin dielectric film. It also includes a surface energy of deformation and polarization, which is absent from the previous theories but which has been observed in the laboratory and calculated from atomic considerations.
Recent years have witnessed an increasing interest in the various problems of dielectrics with polarization gradient. Glockner and his co-workers considered the propagation of waves in thermoelastic centrosymmetric dielectrics [3,4,5]. Love waves, longitudinal and transverse waves in a dielectric layer, torsional waves in a cylinder and waves in layered media have been investigated [6,7].
The paper is devoted to the axially symmetric Lamb problem in a semi-infinite dielectric elastic solid. In the case of an elastic dielectric the generalized Lamb problem is more complicated than in classical elasticity, since the dynamic state of the body is now described by seven independent functions.

2. GOVERNING EQUATIONS

The system of basic linear equations for an elastic dielectric including polarization gradient effects occupying the domain V and bounded by the surface S is given by

(i) The field equations

$$\sigma_{ij,i} + \rho f_j = \rho \frac{\partial^2 u_j}{\partial t^2}, \qquad \sigma_{ij} = \sigma_{ji},$$

$$E_{ij,i} + {}_LE_j + E_j^{MS} = E_j^o,$$

$$-\varepsilon_0 \varphi_{,ii} + P_{i,i} = 0 \qquad \text{in } V,$$

$$\varphi_{,ii} = 0 \qquad \text{in } V'. \tag{2.1}$$

(ii) The boundary conditions

$$\sigma_{ij} n_i = k_j,$$

$$E_{ij} n_i = S_j$$

$$(P_i - \varepsilon_o \| \varphi_{,i} \|) = 0 \qquad \text{on } S. \tag{2.2}$$

(iii) The constitutive equations

$$\sigma_{ij} = d_{12}\delta_{ij}P_{nn} + d_{44}(P_{ji} + P_{ij}) + c_{12}\,\delta_{ij}\,S_{kk} + 2c_{44}S_{ij},$$

$$-{}_LE_k = aP_k,$$

$$E_{ij} = b_{12}\delta_{ij}P_{nn} + b_{44}(P_{ij} + P_{ji}) + b_{77}(P_{ij} - P_{ji}) + 2d_{44}\,S_{ij} + b_o\delta_{ij}. \tag{2.3}$$

(iv) The kinematic relations

$$E_i^{MS} = -\varphi_{,i}, \qquad S_{ij} = \frac{1}{2}(u_{i,j} + u_{j,i}), \qquad P_{ij} = P_{j,i}, \tag{2.4}$$

where σ_{ij}, E_{ij} and S_{ij} designate respectively the components of the stress tensor, the electric tensor and the strain tensor, respectively; u_i, P_i, ${}_LE_i$, E_i^{MS}, E_j^o, f_i and n_i are respectively the components of displacement vector, polarization vector, effective local electric force vector, Maxwell self-field vector, body force vector, external electric field vector and the exterior unit normal vector; φ, $\|\varphi_{,i}\|$, denote Maxwell potential, jump in $\varphi_{,i}$ across S; k_i and S_i are surface traction and surface electric force, while V' stands for the external vacuum and ε_0 its permittivity.

Substituting (2.3) and (2.4) into (2.1) yields the Navier equations of elastic dielectrics as

$$c_{44}\nabla^2\mathbf{u} + (c_{12} + c_{44})\nabla\nabla\cdot\mathbf{u} + d_{44}\nabla^2\mathbf{P} + (d_{12} + d_{44})\nabla\nabla\cdot\mathbf{P} + \rho\mathbf{f} = \rho\frac{\partial^2\mathbf{u}}{\partial t^2},$$

$$d_{44}\nabla^2\mathbf{u} + (d_{12} + d_{44})\nabla\nabla\cdot\mathbf{u} + (b_{44} + b_{77})\nabla^2\mathbf{P} + (b_{12} + b_{44} - b_{77})\nabla\nabla\cdot\mathbf{P}$$

$$-a\mathbf{P} - \nabla\varphi + \mathbf{E}^o = 0,$$

$$\nabla\cdot\mathbf{P} - \varepsilon_0\nabla^2\varphi = 0. \tag{2.5}$$

3. THE GENERALIZED LAMB PROBLEM

Consider now the axisymmetric problem of dynamic piezoelectricity, when on the boundary of the half-space the prescribed loadings are axisymmetric and vary in time. In view of the assumed axial symmetry of the loadings we shall work in the cylindrical coordinate system (r,θ,z) and we assume that all quantities are independent on the angle θ. In this case the system of equations (2.5) describing the motion of the medium splits up into two independent systems (for $\mathbf{f} = 0$, $\mathbf{E}^o = 0$):

$$c_{44}(\nabla_1^2 u_r - \frac{u_r}{r^2}) + (c_{12}+c_{44})\frac{\partial e}{\partial r} + d_{44}(\nabla_1^2 P_r - \frac{P_r}{r^2}) + (d_{12}+d_{44})\frac{\partial \upsilon}{\partial r} = \rho\frac{\partial^2 u_r}{\partial t^2},$$

$$c_{44}\nabla_1^2 u_z + (c_{12}+c_{44})\frac{\partial e}{\partial z} + d_{44}\nabla_1^2 P_z + (d_{12}+d_{44})\frac{\partial \upsilon}{\partial z} = \rho\frac{\partial^2 u_z}{\partial t^2},$$

$$d_{44}(\nabla_1^2 u_r - \frac{u_r}{r^2}) + (d_{12}+d_{44})\frac{\partial e}{\partial r} + b_{44}(\nabla_1^2 P_r - \frac{P_r}{r^2}) + (b_{12}+b_{44}-b_{77})\frac{\partial \upsilon}{\partial r}$$

$$-aP_r - \frac{\partial \varphi}{\partial r} = 0,$$

$$d_{44}\nabla_1^2 u_z + (d_{12}+d_{44})\frac{\partial e}{\partial z} + b_{44}\nabla_1^2 P_z + (b_{12}+b_{44}-b_{77})\frac{\partial \upsilon}{\partial z} - aP_z - \frac{\partial \varphi}{\partial z} = 0,$$

$$\upsilon - \varepsilon_0\nabla_1^2\varphi = 0,$$

and

$$c_{44}(\nabla_1^2 - \frac{1}{r^2})\, u_\theta + d_{44}(\nabla_1^2 - \frac{1}{r^2})\, P_\theta = \rho\frac{\partial^2 u_\theta}{\partial t^2},$$

$$d_{44}(\nabla_1^2 - \frac{1}{r^2})\, u_\theta + (b_{44}+b_{77})(\nabla_1^2 - \frac{1}{r^2})\, P_\theta - aP_\theta = 0\,, \qquad (3.1)$$

where

$$e = \frac{1}{r}\frac{\partial}{\partial r}(ru_r) + \frac{\partial u_z}{\partial z}, \qquad \upsilon = \frac{1}{r}\frac{\partial}{\partial r}(rP_r) + \frac{\partial P_z}{\partial z},$$

$$\nabla_1^2 \equiv \frac{\partial^2}{\partial r^2} + \frac{1}{r}\frac{\partial}{\partial r} + \frac{\partial^2}{\partial z^2}.$$

A. At first we shall be concerned with the system $(3.1)_1$. Let us express the displacement u_r, u_z and the polarization P_r, P_z in terms of potentials ψ, χ, L, R:

$$u_r = \frac{\partial \psi}{\partial r} - \frac{\partial L}{\partial z}, \qquad P_r = \frac{\partial \chi}{\partial r} - \frac{\partial R}{\partial z},$$

$$u_z = \frac{\partial \psi}{\partial z} + \frac{\partial}{r\partial r}(rL), \qquad P_z = \frac{\partial \chi}{\partial z} + \frac{\partial}{r\partial r}(rR). \qquad (3.2)$$

Substituting the above into (3.1) we obtain two independent systems:

$$c_{11}\nabla_1^2\psi + d_{11}\nabla_1^2\chi = \rho\frac{\partial^2\psi}{\partial t^2},$$

$$d_{11}\nabla_1^2\psi + (b_{11}\nabla_1^2 - a)\chi - \varphi = 0,$$

$$\nabla_1^2\chi - \varepsilon_0\nabla_1^2\varphi = 0$$

and

$$c_{44}\nabla_1^2 L + d_{44}\nabla_1^2 R = \rho\frac{\partial^2 L}{\partial t^2},$$

$$d_{44}\nabla_1^2 L + [(b_{44} + b_{77})\nabla_1^2 - a\,]\,R = 0, \tag{3.3}$$

where

$c_{11} = c_{12} + 2c_{44}$, $\quad b_{11} = b_{12} + 2b_{44}$, $\quad d_{11} = d_{12} + 2d_{44}$.

Both these systems can be transformed into

$$\nabla_1^2\,[l_1^2\,\nabla_1^2\,\nabla_1^2 - \nabla_1^2\,(1 + \frac{b_{11}}{(a + \varepsilon_0^{-1}c_1^2)}\frac{\partial^2}{\partial t^2}) + \frac{\partial^2}{c_1^2\partial t^2}](\,\psi, \chi, \varphi\,) = 0$$

$$[l_2^2\,\nabla_1^2\,\nabla_1^2 - \nabla_1^2\,(1 + \frac{b_{44} + b_{77}}{ac_2^2}\frac{\partial^2}{\partial t^2}) + \frac{\partial^2}{c_2^2\partial t^2}](\,R, L\,) = 0, \tag{3.4}$$

where the following notation has been introduced:

$$c_1^2 = \frac{c_{11}}{\rho}, \qquad l_1^2 = \frac{c_{11}b_{11} - d_{11}^2}{(a + \varepsilon_0^{-1})} > 0,$$

$$c_2^2 = \frac{c_{44}}{\rho}, \qquad l_2^2 = \frac{c_{44}(b_{44}+b_{77}) - d_{44}^2}{ac_{44}} > 0.$$

Assuming disturbances of the monochromatic form $f(r,t) = f^*(r)e^{-i\omega t}$
and performing on (3.4) the Hankel transform, defined by

$$\mathcal{H}_\nu\{f^*(r)\}(\alpha) = \int_0^\infty dr f^*(r) r J_\nu(\alpha r),$$

in the case $\nu = 0$ gives

$$(\frac{\partial^2}{\partial z^2} - \alpha^2)[l_1^2\,(\frac{\partial^2}{\partial z^2} - \alpha^2)^2 - (\frac{\partial^2}{\partial z^2} - \alpha^2)(1 - \eta_1) - \sigma_1^2]\,(\mathcal{H}_0\{\psi^*\}, \mathcal{H}_0\{\chi^*\}, \mathcal{H}_0\{\varphi^*\}) = 0,$$

$$[l_2^2\,(\frac{\partial^2}{\partial z^2} - \alpha^2)^2 - (\frac{\partial^2}{\partial z^2} - \alpha^2)(1 - \eta_2) - \sigma_2^2]\,(\mathcal{H}_0\{R^*\}, \mathcal{H}_0\{L^*\}) = 0, \tag{3.5}$$

where

$$\eta_1=\frac{b_{11}\omega^2}{(a+\varepsilon_o^{-1})c_1^2},\qquad \eta_2=\frac{(b_{44}+b_{77})\omega^2}{ac_2^2},\qquad \sigma_1=\frac{\omega}{c_1},\qquad \sigma_2=\frac{\omega}{c_2}.$$

The solution of these equations, which vanishes at infinity, takes the form

$$\mathcal{H}_0\{\psi^*\} = A_1\exp(-\beta_1 z) + A_2\exp(-\beta_2 z),$$
$$\mathcal{H}_0\{\chi^*\} = \kappa_1 A_1\exp(-\beta_1 z) + \kappa_2 A_2\exp(-\beta_2 z) + A_3\exp(-\alpha z),$$
$$\mathcal{H}_0\{\varphi^*\} = \varepsilon_0^{-1}\kappa_1 A_1\exp(-\beta_1 z) + \varepsilon_0^{-1}\kappa_2 A_2\exp(-\beta_2 z) + aA_3\exp(-\alpha z),$$

$$\mathcal{H}_0\{R^*\} = B_1\exp(-\gamma_1 z) + B_2\exp(-\gamma_2 z),$$
$$\mathcal{H}_0\{L^*\} = \tau_1 B_1\exp(-\gamma_1 z) + \tau_2 B_2\exp(-\gamma_2 z), \tag{3.6}$$

where

$$\beta_{1,2} = [\,\alpha^2 + \frac{1}{2l_1^2}\{\,(1-\eta_1) + [(1-\eta_1)^2 \pm 4\sigma_1^2 l_1^2\,]^{1/2}\,\}\,]^{1/2},$$

$$\gamma_{1,2} = [\,\alpha^2 + \frac{1}{2l_2^2}\{\,(1-\eta_2) + [(1-\eta_2)^2 \pm 4\sigma_2^2 l_2^2\,]^{1/2}\,\}\,]^{1/2},$$

$$\kappa_\xi = -\frac{c_{11}}{d_{11}}\,\frac{\beta_\xi^2-\alpha^2+\sigma_1^2}{\beta_\xi^2-\alpha^2},\qquad \tau_\xi = -\frac{c_{44}}{d_{44}}\,\frac{\gamma_\xi^2-\alpha^2+\sigma_2^2}{\gamma_\xi^2-\alpha^2},\qquad \xi = 1, 2.$$

and the following inequalities are satisfied:

$$\alpha^2 l_\xi^2 > \frac{1}{2}\{(1-\eta_\xi) + [(1-\eta_\xi)^2 + 4\sigma_\xi^2 l_\xi^2\,]^{1/2}\,\}.$$

The quantities A_1, A_2, A_3, B_1, B_2 involved in equations (3.6) will be determined from the boundary conditions, which can be written in the form

$$\sigma_{zz}\Big|_{z=0} = 2c_{44}\frac{\partial u_z}{\partial z} + c_{12}e + 2d_{44}\frac{\partial P_z}{\partial z} + d_{12}\upsilon\Big|_{z=0} = h_1(r,0,t),$$

$$\sigma_{zr}\Big|_{z=0} = c_{44}\Big(\frac{\partial u_r}{\partial z}+\frac{\partial u_z}{\partial r}\Big) + d_{44}\Big(\frac{\partial P_r}{\partial z}+\frac{\partial P_z}{\partial r}\Big)\Big|_{z=0} = h_2(r,0,t),$$

$$E_{zz}\Big|_{z=0} = 2d_{44}\frac{\partial u_z}{\partial z} + d_{12}e + 2b_{44}\frac{\partial P_z}{\partial z} + b_{12}\upsilon\Big|_{z=0} = h_3(r,0,t),$$

$$E_{zr}\Big|_{z=0} = d_{44}\Big(\frac{\partial u_r}{\partial z}+\frac{\partial u_z}{\partial r}\Big) + b_{44}\Big(\frac{\partial P_r}{\partial z}+\frac{\partial P_z}{\partial r}\Big) + b_{77}\Big(\frac{\partial P_r}{\partial z}-\frac{\partial P_z}{\partial r}\Big)\Big|_{z=0} = h_4(r,0,t),$$

$$-\varepsilon_o\,\Big\|\frac{\partial\varphi^+}{\partial z}-\frac{\partial\varphi}{\partial z}\Big\| + P_z\Big|_{z=0} = 0. \tag{3.7}$$

In vacuum, Laplace's equation must be satisfied, so that

$$\nabla_1^2\varphi^+ = 0.$$

The solution in our case has the form

$\mathcal{H}_0\{\varphi^+(r, z, t)\} = A_0 e^{-\alpha z} e^{-i\omega t}$

and the electric potential must be continuous at the plane $z = 0$, i.e.,

$\varphi^+ - \varphi \Big|_{z=0} = 0.$

As a special case we can consider the influence of a concentrated harmonic force at the origin of the coordinate system:

$h_1(r, 0, t) = h_1(r)e^{-i\omega t} = \dfrac{h_0\delta(r)}{2\pi r} e^{-i\omega t}$ and $\mathcal{H}_0\{h_0^*(r)\} = \dfrac{h_1}{2\pi}.$

Expressing the displacement u_r, u_z and polarization P_r, P_z in terms of potentials ψ, χ, R, L and applying the Hankel transform we obtain a system of linear homogeneous equations from which we can determinate the quantities $A_{0,1,2,3}$ and $B_{1,2}$. Knowing these we can perform the inverse Hankel transform in order to find the displacement and polarization field. However, our solution is formal only, since the computation of the inverse transform in our case is mathematically difficult. Nevertheless our results should be useful in investigating surface waves in a dielectric half space.

B. Let us now consider the second system $(3.1)_2$ which can be written in the following form:

$$c_{44}\Big(D - \frac{\rho}{c_{44}}\frac{\partial^2}{\partial t^2}\Big) u_\theta + d_{44}\, D\, P_\theta = 0,$$

$$d_{44}\, D\, u_\theta + (b_{44} + b_{77})\Big(D - \frac{a}{b_{44} + b_{77}}\Big)P_\theta = 0 \tag{3.8}$$

or, after manipulations, for monochromatic waves

$$[\, DD + (\eta_2 - 1) l_2^{-2} D + \sigma_2^2\, l_2^{-2}\,](u_\theta^*, P_\theta^*) = 0,$$

where the following notation has been introduced:

$$D = \nabla_1^2 - \frac{1}{r^2}.$$

Finally it takes the form

$$(D - \mu_1)(D - \mu_2)(u_\theta^*, P_\theta^*) = 0, \tag{3.9}$$

where

$$\mu_{1,2} = \frac{1}{2l_2^2}\{ (1 - \eta_2) + [(1 - \eta_2)^2 \pm 4\sigma_2^2 l_2^2]^{1/2} \}$$

and let us observe that μ_i, like τ_i and κ_i, does not depend on α. Introducing the potentials U_θ^* and Π_θ^* given by the relations

$$u_\theta^* = \frac{\partial}{\partial r} U_\theta^*, \qquad P_\theta^* = \frac{\partial}{\partial r}\Pi_\theta^*, \tag{3.10}$$

the equations (3.9) can be written in the form

$$\frac{\partial}{\partial r}(\nabla_1^2 - \mu_1)(\nabla_1^2 - \mu_2)(U_\theta^*, \Pi_\theta^*) = 0, \tag{3.11}$$

where we have used the relation $\frac{\partial}{\partial r}(\nabla_1^2) = D\frac{\partial}{\partial r}$.

We must first perform the Hankel transform of order unity $\mathcal{H}_1$ and using the relation

$$\int_0^\infty dr\ \{\frac{\partial}{\partial r} f^*(r)\}\ rJ_1(\alpha r) = -\alpha \int_0^\infty dr\ \{\ f^*(r)\ rJ_0(\alpha r)\}$$

we obtain

$$-\alpha\ [(\frac{\partial^2}{\partial z^2} - \alpha^2 - \mu_1^2)(\frac{\partial^2}{\partial z^2} - \alpha^2 - \mu_2^2)]\ \mathcal{H}_0\{U_\theta^*, \Pi_\theta^*\} = 0.$$

The solution is given by the functions

$$\mathcal{H}_0\{U_\theta^*\} = C_1\exp(-\gamma_1 z) + C_2\exp(-\gamma_2 z),$$
$$\mathcal{H}_0\{\Pi_\theta^*\} = \tau_1 C_1\exp(-\gamma_1 z) + \tau_2 C_2\exp(-\gamma_2 z). \qquad (3.12)$$

The boundary conditions will have the form

$$\sigma^*_{z\theta}\Big|_{z=0} = \frac{\partial}{\partial r}[c_{44}\frac{\partial U_\theta^*}{\partial z} + d_{44}\frac{\partial \Pi_\theta^*}{\partial z}]\Big|_{z=0} = f^*(r),$$
$$E^*_{z\theta}\Big|_{z=0} = \frac{\partial}{\partial r}[d_{44}\frac{\partial U_\theta^*}{\partial z} + (b_{44} + b_{77})\frac{\partial \Pi_\theta^*}{\partial z}]\Big|_{z=0} = 0. \qquad (3.13)$$

Substituting the solution into the transformed boundary conditions, we obtain, after simple calculations,

$$C_1 = \frac{\mathcal{H}_1\{f^*(r)\}}{\alpha\gamma_1}C_3, \qquad C_3 = -\frac{d_{44} + \tau_1(b_{44} + b_{77})}{\Delta},$$
$$C_2 = \frac{\mathcal{H}_1\{f^*(r)\}}{\alpha\gamma_2}C_4, \qquad C_4 = -\frac{d_{44} + \tau_2(b_{44} + b_{77})}{\Delta},$$

where

$$\Delta = (c_{44} + \tau_1 d_{44})[d_{44} + \tau_2(b_{44} + b_{77})] - (c_{44} + \tau_2 d_{44})[d_{44} + \tau_1(b_{44} + b_{77})].$$

Inverting the Hankel transform of the functions U_θ^* and Π_θ^* we arrive at the final formulae

$$U_\theta(r, z, t) = e^{-i\omega t}\int_0^\infty d\alpha\ \frac{\mathcal{H}_1\{f^*(r)\}}{-\alpha}\ [\ C_3\frac{\exp(-z(\alpha^2 + \mu_1^2)^{1/2})}{(\alpha^2 + \mu_1^2)^{1/2}}$$
$$-\ C_4\frac{\exp(-z(\alpha^2 + \mu_2^2)^{1/2})}{(\alpha^2 + \mu_2^2)^{1/2}}\ ]\ \alpha J_0(\alpha r),$$

$$\Pi_\theta(r, z, t) = e^{-i\omega t} \int_0^\infty d\alpha \frac{\mathcal{H}_1\{ f^*(r)\}}{-\alpha} [\tau_1 C_3 \frac{\exp(-z(\alpha^2 + \mu_1^2)^{1/2})}{(\alpha^2 + \mu_1^2)^{1/2}}$$

$$- \tau_2 C_4 \frac{\exp(-z(\alpha^2 + \mu_2^2)^{1/2})}{(\alpha^2 + \mu_2^2)^{1/2}}] \alpha J_0(\alpha r). \qquad (3.14)$$

Now, substituting particular forms of loadings we can calculate the corresponding displacement u_θ and polarization P_θ. Let us consider for example the following type of loading: $f(r, t) = \frac{\partial}{\partial r}(\frac{h_0\delta(r)}{2\pi r}) e^{-i\omega t}$. Then its Hankel transform is $\mathcal{H}_1\{ f(r,t)\} = -\frac{h_0\alpha}{2\pi} e^{-i\omega t}$.

Substituting into the above formula (3.14) and performing the integration [8], we arrive at

$$U_\theta(r, z, t) = \frac{h_0 e^{-i\omega t}}{2\pi} [C_3 \frac{\exp(-\mu_1(r^2 + z^2)^{1/2})}{(r^2 + z^2)^{1/2}} - C_4 \frac{\exp(-\mu_2(r^2 + z^2)^{1/2})}{(r^2 + z^2)^{1/2}}]$$

$$\Pi_\theta(r,z,t) = \frac{h_0 e^{-i\omega t}}{2\pi} [\tau_1 C_3 \frac{\exp(-\mu_1(r^2 + z^2)^{1/2})}{(r^2 + z^2)^{1/2}} - \tau_2 C_4 \frac{\exp(-\mu_2(r^2 + z^2)^{1/2})}{(r^2 + z^2)^{1/2}}].$$

Now, using the formula (3.10) we are able to write the expressions for displacement and polarization in the analytical form.

ACKNOWLEGEMENTS

The authors wish to thank Professor Tristano Manacorda of the University of Pisa for several helpful and informative discussions.

REFERENCES

1. W. Voigt, *Lehrbuch der Kristall-physik*, Teubner, Leipzig, (1910).
2. R. D. Mindlin, Journal of Elasticity, **2**, 217, (1972).
3. J. P. Nowacki and P. G. Glockner, Int. J. Solids and Structures, **15**, 183, (1979).
4. J. P. Nowacki and P. G. Glockner, Int. J. Engng. Sci. **19**, 603, (1981).
5. S. Dost, M. Epstein, S. Gözde, in *The Mechanical Behavior of Electromagnetic Solid Continua*, p.223, G. A. Maugin (ed), North-Holland, Amsterdam, (1984).
6. J. P. Nowacki, in *The Mechanical Behavior of Electromagnetic Solid Continua*, p. 211, G. A. Maugin (ed), North-Holland, Amsterdam,(1984).
7. K. Majorkowska-Knapp, ZAAM **66**, 4, T66-67, (1986).
8. A. Erdelyi, (ed), *Tables of Integral Transforms, Vol.II*, McGraw-Hill Book Company, (1954).

Electromechanical Surface Waves in Layered Elastic Dielectrics with Polarization Gradient

K. Majorkowska-Knap[1] *and J. Lenz*[2]

[1]Institute of Mechanics of Engineering Constructions,
Warsaw Technical University, Lukasiewicza 17, PL-09-400 Plock, Poland

[2]Institut für Theoretische Mechanik, Universität Karlsruhe,
Kaiserstraße 12, D-7500 Karlsruhe, Fed. Rep. of Germany

1. Introduction

In recent years a continually increasing number of publications, in theoretical as well as experimental research, has been devoted to the coupled effects of various fields such as mechanical, thermal, electromagnetic, diffusion, radiation and other fields.

Surface acoustic waves have been studied for about a century now, but there has again been an enormous increase in the interest shown in these waves during the past twenty years [1]. Such waves can also exist in piezoelectric media, in which a deformation produces an electric field, thus giving rise to a coupling of mechanical and electric phenomena. Important technical applications of electromechanical surface waves [2,3] include for instance signal processing (with applications to radar, sonar and telecommunications), non-destructive evaluation and flaw detection and various aspects of seismology.

Because of their extremely low velocity and wavelengths, when compared with electromagnetic waves of the same frequency, acoustic wave devices offer a dramatic reduction in size and weight in comparison with electromagnetic devices. Some of these acoustic surface wave devices employ a thin layer of piezoelectric material on a distinct piezoelectric substrate. This layer structure permits, besides the propagation of Rayleigh surface modes, the propagation of Love surface waves, representing transverse modes with horizontal polarization [4]. Within the framework of the classical theory of piezoelectricity [5], Love waves in layered anisotropic media were studied, among others, in [6 – 10].

In this contribution the propagation of electromechanical (piezoelectrically stiffened) Love waves in layered elastic dielectrics is treated on the basis of MINDLIN's linear model of piezoelectricity [11,12]. This theory incorporates, besides strain and polarization, the polarization gradient into the internal energy density and hence reduces the "gap" between the continuum and the lattice theory, i.e. between the classical theory of piezoelectricity and the dynamical theory of crystal lattices. Besides taking into account more precisely the microscopic structure of matter, Mindlin's model is also able to accommodate the piezoelectric effect in centrosymmetric (including isotropic) materials, too, and to define a surface energy of deformation and polarization.

Our investigation is confined to the study of small amplitude waves using the quasistatic approximation for the electromagnetic field. Special attention is directed towards the conditions for existence of surface waves, the (numerical) analysis of the dispersion relation and the inspection of displacement and polarization.

2. Outline of Mindlin's Theory

As an extension of TOUPIN's [13] classical variational principle for the equilibrium equations of piezoelectricity, MINDLIN [11] formulated Hamilton's principle for elastic, dielectric continua in the following (linear) version:

$$\delta \int_{t_o}^{t_1} dt \int_{V^*} \left(\frac{1}{2}\varrho \dot{\mathbf{u}} \circ \dot{\mathbf{u}} - H \right) dV$$

$$+ \int_{t_0}^{t_1} dt \left\{ \int_V \left(\mathbf{f} \circ \delta \mathbf{u} + \mathbf{E}^0 \circ \delta \mathbf{P} \right) dV + \int_S \mathbf{t} \circ \delta \mathbf{u} \, dS \right\} = 0, \tag{1}$$

where ϱ is the mass density, H is the electric enthalpy density, $\mathbf{u}$ and $\mathbf{P}$ are the displacement and polarization, $\mathbf{f}$ and $\mathbf{t}$ are the external body force and surface traction, and $\mathbf{E}^0$ is the external electric field. In (1) V denotes the volume occupied by the body, V^* the entire space and S the surface separating V from vacuum V'.

In addition to including the kinetic energy, Mindlin takes into account the polarization gradient $\operatorname{Grad}\mathbf{P}$ in the stored energy density W which leads to the following expression for the electric enthalpy density:

$$H = W(\boldsymbol{\varepsilon}, \mathbf{P}, \operatorname{Grad}\mathbf{P}) - \frac{1}{2}\varepsilon_0 \operatorname{grad}\varphi \circ \operatorname{grad}\varphi + \operatorname{grad}\varphi \circ \mathbf{P}, \tag{2}$$

where $\boldsymbol{\varepsilon} = \frac{1}{2}[\operatorname{Grad}\mathbf{u} + (\operatorname{Grad}\mathbf{u})^T]$ is the linear deformation tensor, φ is the potential of the Maxwell self field $\mathbf{E}^{MS} : \mathbf{E}^{MS} = -\operatorname{grad}\varphi$, and ε_0 is the permittivity of vacuum.

The variational principle (1) yields the following *field equations*:

$$\left.\begin{aligned} \operatorname{Div}\boldsymbol{\sigma} + \mathbf{f} &= \varrho\ddot{\mathbf{u}} \\ \mathbf{E}^L + \operatorname{Div}\boldsymbol{E}^L - \operatorname{grad}\varphi + \mathbf{E}^0 &= \mathbf{0} \\ -\varepsilon_0\Delta\varphi + \operatorname{div}\mathbf{P} &= 0 \end{aligned}\right\} \text{ in } V$$
$$\Delta\varphi = 0 \qquad \text{in } V' \tag{3}$$

and *boundary conditions* ($\mathbf{n}$: outward unit normal vector):

$$\boldsymbol{\sigma}\mathbf{n} = \mathbf{t} \; ; \; \boldsymbol{E}^L\mathbf{n} = \mathbf{0} \; ; \; \mathbf{n} \circ \{\mathbf{P} - \varepsilon_0 [\![\operatorname{grad}\varphi]\!]\} = 0 \qquad \text{on } S \tag{4}$$

with $[\![\;]\!]$ denoting the jump across S, and where the local electric field $\mathbf{E}^L$, local electric tensor $\boldsymbol{E}^L$ and the stress tensor $\boldsymbol{\sigma}$ are given by

$$\mathbf{E}^L = -\frac{\partial W}{\partial \mathbf{P}} \; ; \; \boldsymbol{E}^L = \frac{\partial W}{\partial (\operatorname{Grad}\mathbf{P})} \; ; \; \boldsymbol{\sigma} = \frac{\partial W}{\partial \boldsymbol{\varepsilon}} \; . \tag{5}$$

For the stored energy density Mindlin takes the polynomial approximation

$$W = \frac{1}{2}c_{ijkl}\varepsilon_{ij}\varepsilon_{kl} + \frac{1}{2}a_{ij}P_iP_j + f_{ijk}\varepsilon_{ij}P_k + b^0_{ij}P_{j,i}$$

$$+ \frac{1}{2}b_{ijkl}P_{j,i}P_{l,k} + g_{ijk}P_iP_{k,j} + d_{ijkl}P_{j,i}\varepsilon_{kl} \tag{6}$$

from which, with the aid of (5), $\mathbf{E}^L, \boldsymbol{E}^L$ and σ can be determined. For a centrosymmetric isotropic material these constitutive equations have the form

$$
\begin{aligned}
\mathbf{E}^L &= -\, a\mathbf{P} \\
\boldsymbol{E}^L &= d_{12}(\operatorname{tr}\boldsymbol{\varepsilon})\boldsymbol{1} + 2d_{44}\boldsymbol{\varepsilon} + b_{12}(\operatorname{div}\mathbf{P})\boldsymbol{1} + b^0\boldsymbol{1} \\
&\quad + b_{44}\left\{\operatorname{Grad}\mathbf{P} + (\operatorname{Grad}\mathbf{P})^T\right\} + b_{77}\left\{\operatorname{Grad}\mathbf{P} - (\operatorname{Grad}\mathbf{P})^T\right\} \\
\boldsymbol{\sigma} &= c_{12}(\operatorname{tr}\boldsymbol{\varepsilon})\boldsymbol{1} + 2c_{44}\boldsymbol{\varepsilon} + d_{12}(\operatorname{div}\mathbf{P})\boldsymbol{1} \\
&\quad + d_{44}\left\{\operatorname{Grad}\mathbf{P} + (\operatorname{Grad}\mathbf{P})^T\right\} \quad .
\end{aligned} \tag{7}
$$

3. Formulation of Problem

As pictured in Fig. 1, we consider a body consisting of the homogeneous layer $-h \leq x_1 \leq 0$ of a centrosymmetric, isotropic, elastic dielectric, overlaying the homogeneous, isotropic, purely elastic half-space $x_1 \geq 0$ (i.e. we neglect the dielectric properties of the half-space). This model serves as an idealization of a real surface wave delay line where a thin layer of one material is attached to a comparatively thick substrate of a different material (see e.g. [14]).

We are concerned with the propagation of electromechanical (piezoelectrically stiffened) Love surface waves in this medium, i.e. waves polarized in x_3-direction (c.f. vibration direction of displacement and polarization in Fig. 1) and propagating in x_2-direction. The material coefficients of the dielectric layer (thickness h) are given by $\varrho, c_{ij}, d_{ij}, b_{ij}, a$, those of the elastic half-space by ϱ', c'_{ij}. We are interested in the existence conditions for such waves, in the dispersion relations of their various modes as well as in the accompanying displacement and polarization profiles.

In the absence of an external electric field ($\mathbf{E}^0 = \mathbf{0}$), body forces ($\mathbf{f} = \mathbf{0}$) and surface tractions ($\mathbf{t} = \mathbf{0}$), Mindlin's set of equations (3) – (5) together with the constitutive laws (7) yields the *field equations*

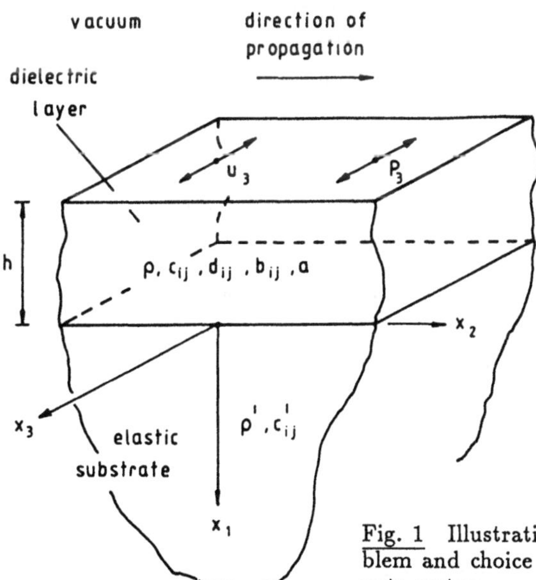

Fig. 1 Illustration of problem and choice of coordinate system

$$
\left.\begin{aligned}
c_{44}\Delta\mathbf{u} + (c_{12}+c_{44})\,\mathrm{grad\,div}\,\mathbf{u} + d_{44}\Delta\mathbf{P} & \\
+(d_{12}+d_{44})\mathrm{grad\,div}\,\mathbf{P} &= \varrho\ddot{\mathbf{u}} \\
d_{44}\Delta\mathbf{u} + (d_{12}+d_{44})\mathrm{grad\,div}\,\mathbf{u} + (b_{44}+b_{77})\Delta\mathbf{P} & \\
+(b_{12}+b_{44}-b_{77})\mathrm{grad\,div}\,\mathbf{P} - a\mathbf{P} - \mathrm{grad}\,\varphi &= \mathbf{0} \\
-\varepsilon_0\Delta\varphi + \mathrm{div}\,\mathbf{P} &= 0
\end{aligned}\right\} \quad \text{for } x_1 \geq -h \quad (8)
$$

$$
\Delta\varphi = 0 \qquad \text{for } x_1 < -h \quad (9)
$$

and the *boundary conditions* ($\mathbf{n} = -\mathbf{e}_1$)

$$
\sigma_{1i} = 0 \; ; \; E^L_{1i} = 0 \; ; \; P_1 - \varepsilon_0 [\![\varphi_{,1}]\!] = 0 \qquad \text{on } x_1 = -h \qquad (10)
$$

which have to be augmented by the *continuity conditions*

$$
\mathbf{u} = \mathbf{u}' \; ; \; \boldsymbol{\sigma}\mathbf{e}_1 = -\boldsymbol{\sigma}'(-\mathbf{e}_1) \; ; \; \boldsymbol{E}^L\mathbf{e}_1 = -\boldsymbol{E}^{L'}(-\mathbf{e}_1) \qquad \text{on } x_1 = 0 \qquad (11)
$$

(where the unprimed quantities refer to the dielectric layer and the primed ones to the half-space).

Confining ourselves to the propagation of *plane* surface waves of the Love-type with constant phase velocity p, means that we are looking for solutions of (8) – (11) of the form

$$
\mathbf{u} = u_3(x_1,x_2,t)\mathbf{e}_3 \; ; \; \mathbf{P} = P_3(x_1,x_2,t)\mathbf{e}_3 \; ; \; \varphi = \varphi(x_1,x_2,t) \; . \qquad (12)
$$

4. Analytical Solution

Under the aforementioned assumption (12) the system (8) – (11) reduces to the boundary value problem

$$
\left.\begin{aligned}
c_{44}\left(\Delta - \frac{\varrho}{c_{44}}\partial_t^2\right)u_3 + d_{44}\Delta P_3 &= 0 \\
d_{44}\Delta u_3 + \{(b_{44}+b_{77})\Delta - a\}P_3 &= 0
\end{aligned}\right\} \quad \text{for } -h \leq x_1 \leq 0 \qquad (13)
$$

$$
c'_{44}\left(\Delta - \frac{\varrho'}{c'_{44}}\partial_t^2\right)u'_3 = 0 \qquad \text{for } x_1 \geq 0 \qquad (14)
$$

$$
\sigma_{13}(x_1 = -h, x_2, t) = 0 \; ; \; E^L_{13}(x_1 = -h, x_2, t) = 0 \qquad (15)
$$

$$
u_3(x_1 = 0, x_2, t) = u'_3(x_1 = 0, x_2, t) \; ; \; E^L_{13}(x_1 = 0, x_2, t) = 0
$$

$$
\sigma_{13}(x_1 = 0, x_2, t) = \sigma'_{13}(x_1 = 0, x_2, t) \; . \qquad (16)
$$

We now seek solutions of (13) – (16) in the form of plane *harmonic* waves:

$$
\begin{aligned}
u_3(x_1,x_2,t) &= u^0_3(x_1)\exp\{iK(x_2 - pt)\} \\
P_3(x_1,x_2,t) &= P^0_3(x_1)\exp\{iK(x_2 - pt)\} \\
u'_3(x_1,x_2,t) &= u^{0'}_3(x_1)\exp\{iK(x_2 - pt)\} \; .
\end{aligned} \qquad (17)
$$

We express the phase velocity $p = \omega/K$ (ω: angular frequency, K : wave number) as

$$
p = c\sqrt{\frac{c_{44}}{\varrho}} = c'\sqrt{\frac{c'_{44}}{\varrho'}} \, , \qquad (18)
$$

where c and c' denote relative phase velocities and where the roots characterize the propagation velocities of transverse bulk waves in the layer and in the substrate, respectively.

For the sake of brevity, we omit the straightforward, yet lengthy and tedious calculations and rather present the final results. The *dispersion relation*

$$c = c(\kappa) \quad ; \quad \text{where } \kappa = Kh \tag{19}$$

is given by the following complicated transcendental equation:

$$s_1 \cdot \tan(s_1 h) = \frac{c'_{44} s' \frac{\kappa}{h} \left\{ \hat{b}_{44} \left(s_1^2 + \frac{\kappa^2}{h^2} \right) + a \right\} s_2 \cdot \tanh(s_2 h)}{c_{44} a l^2 (s_1^2 + s_2^2) s_2 \cdot \tanh(s_2 h) + c'_{44} s' \frac{\kappa}{h} \left\{ \hat{b}_{44} \left(s_2^2 - \frac{\kappa^2}{h^2} \right) - a \right\}} \tag{20}$$

subject to the restriction

$$\frac{c_{44}}{\varrho} \left\{ 1 - \frac{\kappa^2 d_{44}^2}{c_{44} \left(a h^2 + \kappa^2 \hat{b}_{44} \right)} \right\} < p^2 < \frac{c'_{44}}{\varrho'} \ . \tag{21}$$

In these formulae the abbreviations

$$s_1 = \sqrt{\frac{-B + \sqrt{B^2 - 4AD}}{2A}} \quad ; \quad s_2 = \sqrt{\frac{B + \sqrt{B^2 - 4AD}}{2A}} \quad ; \quad s' = \sqrt{1 - \frac{\varrho' c_{44}}{\varrho c'_{44}} c^2}$$

and

$$\hat{b}_{44} = b_{44} + b_{77} \quad , \quad A = l^2 = \frac{c_{44} \hat{b}_{44} - d_{44}^2}{a c_{44}} > 0$$

$$B = 1 + \frac{\kappa^2}{h^2} \left\{ 2l^2 - \frac{\hat{b}_{44}}{a} c^2 \right\} \quad , \quad D = \frac{\kappa^2}{h^2} \left\{ 1 - c^2 + \frac{\kappa^2}{h^2} \left(l^2 - \frac{\hat{b}_{44}}{a} c^2 \right) \right\}$$

have been used. It may easily be shown that for vanishing dielectric constant d_{44}, i.e. $d_{44} = 0$, (20) reduces to the well-known dispersion relation

$$\hat{s}_1 \cdot \tan(\hat{s}_1 h) = \frac{c'_{44}}{c_{44}} s' \frac{\kappa}{h} \quad ; \quad \text{where } \hat{s}_1 = \frac{\kappa}{h} \sqrt{c^2 - 1} \tag{22}$$

for Love waves in the purely elastic case, and (21) delivers the "classical" existence condition

$$\frac{c_{44}}{\varrho} < p^2 < \frac{c'_{44}}{\varrho'} \ . \tag{23}$$

As in the classical case, the roots of (20) determine a *series of modes of propagation*. However, the restriction (21) allows for *two distinct existence conditions:*

$$\text{I \quad "Classical" case [15]:} \quad \frac{c_{44}}{\varrho} < \frac{c'_{44}}{\varrho'} \quad \text{for } \underline{\text{every}}\ \omega \tag{24}$$

and

II "Non-classical" case: $$\frac{c_{44}}{\varrho} > \frac{c'_{44}}{\varrho'} > \frac{c_{44}}{\varrho} - \frac{d_{44}^2}{\varrho \hat{b}_{44}} \qquad (25)$$

for *cut-off frequencies* $$\omega^2 > \frac{a\frac{c'_{44}}{\varrho'}\left(\frac{c_{44}}{\varrho} - \frac{c'_{44}}{\varrho'}\right)}{\frac{d_{44}^2}{\varrho} + \hat{b}_{44}\left(\frac{c'_{44}}{\varrho'} - \frac{c_{44}}{\varrho}\right)} \; . \qquad (26)$$

Finally, we present the solutions for the displacement and polarization fields:

$$\left.\begin{aligned} u_3(x_1,x_2,t) &= -\alpha\hat{u}_3(x_1)\exp\{iK(x_2-pt)\} \\ P_3(x_1,x_2,t) &= -\alpha\hat{P}_3(x_1)\exp\{iK(x_2-pt)\} \end{aligned}\right\} \quad \text{for } -h \le x_1 \le 0 \qquad (27)$$

$$u'_3(x_1,x_2,t) = -\alpha\hat{u}'_3(x_1)\exp\{iK(x_2-pt)\} \quad \text{for } x_1 \ge 0 \qquad (28)$$

where α is an arbitrary constant and the amplitude functions are given by

$$\hat{u}_3(x_1) = \frac{\cos\{s_1(x_1+h)\}}{\sin(s_1 h)} + \frac{s_1(d_{44}+\hat{b}_{44}\kappa_1)\cdot\cosh\{s_2(x_1+h)\}}{s_2(d_{44}+\hat{b}_{44}\kappa_2)\cdot\sinh(s_2 h)}$$

$$\hat{P}_3(x_1) = \kappa_1\frac{\cos\{s_1(x_1+h)\}}{\sin(s_1 h)} + \kappa_2\frac{s_1(d_{44}+\hat{b}_{44}\kappa_1)\cdot\cosh\{s_2(x_1+h)\}}{s_2(d_{44}+\hat{b}_{44}\kappa_2)\cdot\sinh(s_2 h)}$$

$$\hat{u}'_3(x_1) = \left\{\cot(s_1 h) + \frac{s_1}{s_2}\frac{d_{44}+\hat{b}_{44}\kappa_1}{d_{44}+\hat{b}_{44}\kappa_2}\coth(s_2 h)\right\}\exp(-Ks'x_1) \qquad (29)$$

with

$$\kappa_1 = -\frac{d_{44}(s_1^2+K^2)}{\hat{b}_{44}(s_1^2+K^2)+a} \quad ; \quad \kappa_2 = -\frac{d_{44}(s_2^2-K^2)}{\hat{b}_{44}(s_2^2-K^2)-a} \; .$$

5. Numerical Results and Discussion

The following numerical results were obtained on a SIEMENS 7780-computer of the "Rechenzentrum der Universität Karlsruhe".

In these calculations the layer thickness was chosen throughout as $h = 10^{-3}$ cm. For the analysis of the dispersion relation (20) in the *"classical" case* (24) we took *sodium chloride (NaCl)* as material for the layer and *silicon* as material for the half-space. Both materials belong to the cubic crystal system (class $m3m$), but it can be shown that for such materials the field equations and boundary conditions, presented above, remain unchanged.

The following values were assigned to the material constants: $c_{44} = 1.266 \cdot 10^6$ N/cm^2; $\varrho = 2.164$ g/cm^3 [16]; $b_{44} = b_{77} = 3.44 \cdot 10^{-2}$ Ncm^4/C^2; $d_{44} = -1.7 \cdot 10$ Ncm/C; $a = 1.74 \cdot 10^4$ Ncm^2/C^2 [17]; $c'_{44} = 7.96 \cdot 10^6$ N/cm^2; $\varrho' = 2.33$ g/cm^3 [18].

Figure 2 shows the dispersion curves for this combination of materials. The maximum value of the relative phase velocity c is given by $c_{max} = c_T^S$ (relative propagation velocity of transverse bulk waves in the substrate), the minimum value by $c_{min} = c_T^L$ (relative propagation velocity of transverse bulk waves in the layer). These curves resemble much those which are obtained for vanishing values of the dielectric coefficients (purely elastic layer). Indeed, a comparison of the computed values of phase velocities

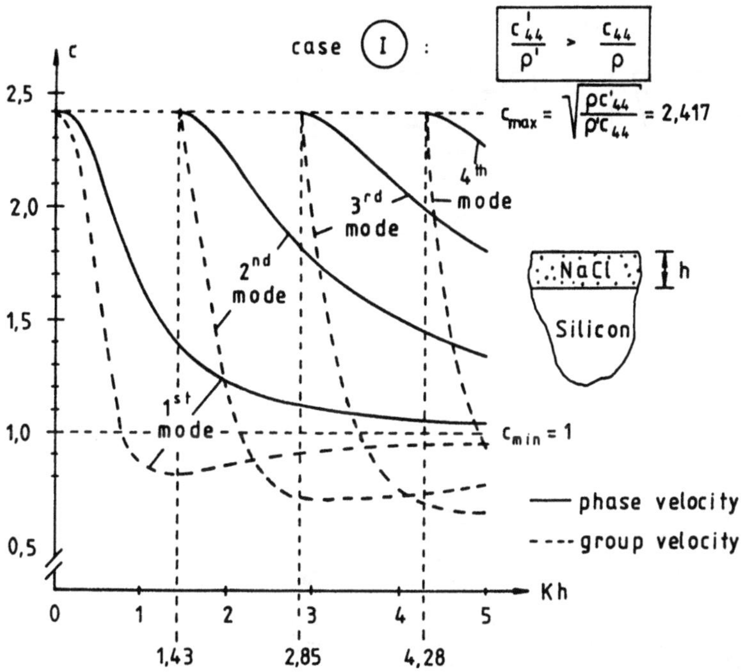

Fig. 2 Dispersion curves for the first four modes (relative phase velocity and relative group velocity as function of $\kappa = Kh$) in the *"classical" case* I

shows, that the dielectric property of the layer *lowers* the phase velocities only in the magnitude of a few ‰ in contrast to the purely elastic case.

For the value $\kappa = Kh = 5$ Fig. 3 presents the displacement and polarization profiles of the first four modes as computed from (29). For all modes displacement and polarization deflect in the same direction. Fig. 3 also illustrates the general result that for a given value of Kh the wave penetrates the deeper into the half-space the higher the mode-number.

Figure 4 shows the ratios δ (for displacement) and ψ (for polarization) of the amplitudes at the material interface and at the free surface as function of $\kappa = Kh$. It is evident that the amplitude ratio of polarization of the first mode differs from that of the higher modes.

We finally draw attention to the dispersion curves in the *"non-classical" case*. With the values for the material coefficients ϱ, d_{44} and $\hat{b}_{44} = b_{44} + b_{77}$, as noted above for sodium chloride, we obtain from (25)

$$5.850 > \frac{c'_{44}}{\varrho'} \; [10^{10} \text{ cm}^2\text{s}^{-2}] > 5.831 \; . \tag{30}$$

Unfortunately, no real material with a value c'_{44}/ϱ' falling into this narrow range, could be found in the literature. We were therefore forced to choose a fictitious material for the simulation of the *"non-classical" case*, taking $\varrho' = 2$ g/cm^3 and $c'_{44} = 1.168 \cdot 10^6$

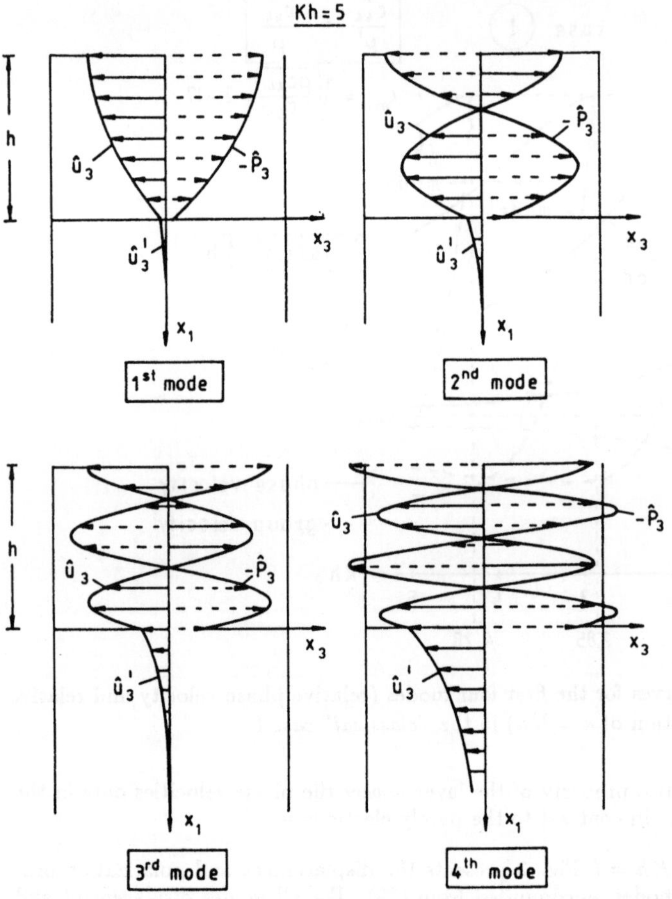

Fig. 3 Displacement and polarization profiles of the first four modes for $Kh = 5$

N/cm^2 which delivers $c'_{44}/\varrho' = 5.84 \cdot 10^{10}$ cm^2s^{-2}. The accompanying dispersion curves are shown in Fig. 5 where a logarithmic scale has been used for the abscissa. The cut-off frequency which can be determined from (26) or from the cut-off value $Kh \approx 0.56$ in Fig. 5, is given by $\omega \approx 1.23 \cdot 10^8$ s^{-1}.

An experimental determination of the cut-off frequency (26) in the *"non-classical" case* II could give further evidence of the necessity to introduce the polarization gradient into the theory as done in Mindlin's model. It should be noted, however, that the values of the material coefficients d_{44}, b_{44}, b_{77} for *NaCl* which we gathered from [17] in order to derive (30), were actually calculated by the same authors in a preceding paper [19]. But there they present the value $d_{44} = -2.44 \cdot 10^2$ Ncm/C (instead of $d_{44} = -1.7 \cdot 10$ Ncm/C, used in the calculations above) which would, instead of (30), lead to the margin

$$5.850 > \frac{c'_{44}}{\varrho'} \; [10^{10} \; cm^2s^{-2}] > 1.850$$

thus giving way to a number of materials admissable as substrate. Unfortunately, no hint can be found in [17] why the original values of [19] have been corrected.

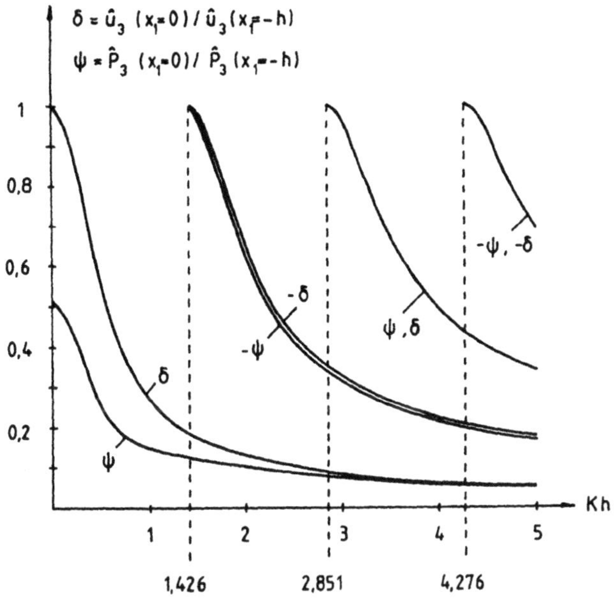

Fig. 4 Amplitude ratios of displacement and polarization as function of $\kappa = Kh$

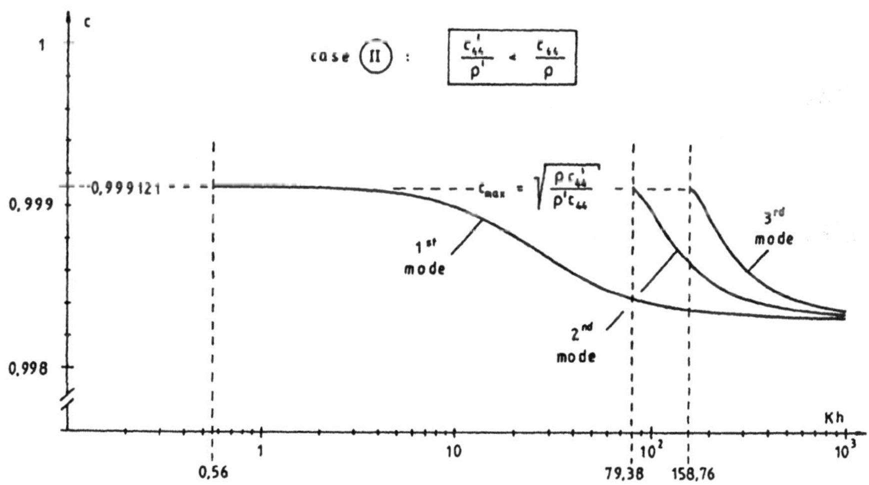

Fig. 5 Dispersion curves for the first three modes in the *"non-classical" case* II

Acknowledgement

The work of the first-named author was sponsored by C.P.B.P. 01.08.

References

1. A.A. Maradudin: In *Festkörperprobleme XXI, Advances in Solid State Physics*, ed. by J. Treusch (Vieweg, Braunschweig 1981), p. 25–116
2. W.P. Mason: J. Acoust. Soc. Am. *70* (1981)
3. E. Dieulesaint, D. Royer: In *The Mechanical Behavior of Electromagnetic Solid Continua*, ed. by G.A. Maugin (North-Holland, Amsterdam, New York, Oxford 1984), p. 3–15
4. G.A. Maugin: Adv. in Appl. Mech. *23* (1983)
5. W. Voigt: *Lehrbuch der Kristallphysik* (Johnson Reprint Co., New York, Teubner, Stuttgart 1966)
6. E. Dieulesaint, D. Royer: *Ondes élastiques dans les solides* (Masson, Paris 1974), see ref. p. 295
7. V.R. Velasco, F. Garcia-Moliner: Surface Sci. *143* (1984)
8. G.W. Farnell: In *Topics in Applied Physics 24: Acoustic Surface Waves*, ed. by A.A Oliner (Springer, Berlin, Heidelberg, New York 1978), p. 13–60
9. B.A. Auld: J. Acoust. Soc. Am. *70* (1981)
10. G.G. Kessenikh, L.A. Shuvalov: Solid State Phys. *24* (1982)
11. R.D. Mindlin: Int. J. Solids Structures *4* (1968)
12. R.D. Mindlin: J. of Elasticity *2* (1972)
13. R.A. Toupin: J. Rat. Mech. Anal. *5* (1968)
14. C. Lardat, C. Maerfeld, P. Tournois: Proc. of the IEEE *59* (1971)
15. K. Majorkowska-Knap: ZAMM *66*, 4 (1986)
16. Landolt-Börnstein: *Zahlenwerte und Funktionen aus Naturwissenschaften und Technik*, Neue Serie, Vol. III, 11, ed. by K.-H. Hellwege (Springer, Berlin, Heidelberg, New York 1979)
17. A. Askar, P.C.Y. Lee, A.S. Cakmak: Int. J. Solids Structures *7* (1971)
18. A.J. Slobodnik, E.D. Conway, R.T. Delmonico: *Microwave Acoustics Handbook, Vol. 1A, Surface Wave Velocities*, AFCRL-TR-73-0597 (Air Force Cambridge Research Laboratories, Hanscom AFB, Mass. 1973)
19. A. Askar, P.C.Y. Lee, A.S. Cakmak: Phys. Rev. B, *1* (1970)

Surface Waves in a Semiconductor with Relaxation of Charge Carrier Field

B. Maruszewski

Technical University of Poznań, Institute of Applied Mechanics, ul. Piotrowo 3, PL-60-965 Poznań, Poland

The propagation of coupled surface waves in a semiconducting half-space with relaxation of charge carrier field is considered. Based on general equations, a case of elasto-electronic surface waves is carefully investigated. The dispersion and distribution of amplitudes in germanium is presented.

1. Basic Equations

Classical description of the acoustic surface waves with their fundamental applications in practice has been very well presented by OLINER /1/. It is well known that they are not only of a pure acoustic or elastic character. These waves can also be coupled with other physical fields. MAUGIN /2/,/3/ and RYMARZ /4/ deal carefully with many cases concerning propagation of them in materials of various physical properties.

The paper aims at investigating the propagation of the elasto- electronic surface waves. The theory concerning the reciprocal couplings of elastic, thermal, electromagnetic and charge carrier fields in semiconductors within a model of a hyperbolic system of field equations in the presence of heat and electricity conduction has been given by MARUSZEWSKI /5/ and MARUSZEWSKI and LEBON /6/. Based on that, the linearized coupled equations take the form

$$\mu u_{i,jj} + (\lambda+\mu) u_{j,ij} - \varrho \ddot{u}_i - \lambda^n N_{,i} - \lambda^p P_{,i} - \lambda^T \Theta_{,i} = 0$$

$$\varkappa \Theta_{,ii} + m^{nq} N_{,ii} + m^{pq} P_{,ii} - \left(1 + \tau^Q \frac{\partial}{\partial t}\right)\left(\varrho c \dot{\Theta} + \varrho T_o \alpha^n \dot{N}\right.$$

$$\left. + \varrho T_o \alpha^p \dot{P} + T_o \lambda^T \dot{u}_{k,k}\right) - \varrho\left(a_1^n \dot{N} + a_1^p \dot{P}\right) = -\left(a_1^n g^n + a_1^p g^p\right)$$

$$\varrho D^n N_{,ii} + m^{qn} \Theta_{,ii} - \varrho\left(1 - a_2^n T_o \alpha^n + \tau^n \frac{\partial}{\partial t}\right) \dot{N} - a_2^n\left(\varrho c \dot{\Theta}\right.$$

$$\left. + \varrho T_o \alpha^p \dot{P} + T_o \lambda^T \dot{u}_{k,k}\right) = -\left(1 + \tau^n \frac{\partial}{\partial t}\right) g^n$$

$$\varrho D^p P_{,ii} + m^{qp} \Theta_{,ii} - \varrho\left(1 - a_2^p T_o \alpha^p + \tau^p \frac{\partial}{\partial t}\right) \dot{P} - a_2^p\left(\varrho c \dot{\Theta}\right.$$

$$\left. + \varrho T_o \alpha^n \dot{N} + T_o \lambda^T \dot{u}_{k,k}\right) = -\left(1 + \tau^p \frac{\partial}{\partial t}\right) g^p , \qquad (1.1)$$

where $\theta = T - T_0$, $N = n - n_0$ and $P = p - p_0$ denote the relative temperature and concentrations of electrons and holes, respectively, then τ^q, τ^n and τ^p denote the relaxation times of heat and charge carrier fields, g^n, g^p - the recombination functions and λ^n, λ^p, λ^T, m^{nq}, m^{pq}, m^{qn}, m^{qp}, α^n, α^p, a_1^n, a_1^p, a_2^n, a_2^p - the coupling constants.

2. Elasto-Electronic Surface Waves

The subject of our considerations is the propagation of elasto-electronic surface waves in the x_1 direction at the surface $x_3=0$ of the semiconducting halfspace $x_3 \geqslant 0$. The influence of thermal and hole fields as well as the electric field resulting from heterogenity of the electron field are omitted. From (1.1) there results the following system of wave equations:

$$\mu u_{i,jj} + (\lambda + \mu) u_{j,ij} - \rho \ddot{u}_i - \lambda^n N_{,i} = 0$$

$$\rho D^n N_{,ii} - \rho\left(1 + \tau^n \frac{\partial}{\partial t}\right) \dot{N} - \frac{\rho}{\tau_n^+}\left(1 + \tau^n \frac{\partial}{\partial t}\right) N = 0 . \qquad (2.1)$$

The recombination function for electrons is given as follows /7/:

$$g^n(x_1, x_3, t) = \frac{\rho}{\tau_n^+} N(x_1, x_3, t) , \qquad (2.2)$$

where τ_n^+ is the life-time of electrons. Assuming that the plane $x_3 = 0$ is free of mechanical loadings the boundary conditions for (2.1) we get /8/,/9/

$$u_{3,1} + u_{1,3} = 0$$

$$\lambda u_{1,1} + (\lambda + 2\mu) u_{3,3} - \lambda^n N = 0$$

$$\rho D^n N_{,3} + \rho s^n\left(\tau^n \frac{\partial}{\partial t} + 1\right) N = 0 , \qquad (2.3)$$

where s^n denotes the velocity of the surface recombination. Decomposing the displacement vector u_i into the potentials ϕ and $\Psi_i = [0, \Psi, 0]$ according to

$$u_i = \phi_{,i} + \varepsilon_{ijk} \Psi_{k,j} , \qquad \Psi_{i,i} = 0 \qquad (2.4)$$

(2.1) and (2.3) take the form

$$(\lambda + 2\mu)\nabla_1^2 \phi - \rho \ddot{\phi} - \lambda^n N = 0$$

$$\mu \nabla_1^2 \Psi - \rho \ddot{\Psi} = 0$$

$$\rho D^n \nabla_1^2 N - \rho \tau^n \ddot{N} - \rho\left(1 - \frac{\tau^n}{\tau_n^+}\right) \dot{N} - \frac{\rho}{\tau_n^+} N = 0 , \qquad (2.5)$$

$$(2\mu + \lambda)\nabla_1^2\phi + 2\mu(\Psi_{,13} - \phi_{,11}) - \lambda^n N = 0$$

$$2\phi_{,13} + \nabla_2^2\Psi = 0 \qquad \nabla_{1;2}^2 = \frac{\partial^2}{\partial x_1^2} \pm \frac{\partial^2}{\partial x_3^2}$$

$$D^n N_{,3} + s^n \tau^n \dot{N} + s^n N = 0 . \tag{2.6}$$

The solution of our problem is looked for in such a form that the partial waves should be of the type

$$\phi = \phi^o \exp(-\beta k x_3)\exp[ik(x_1 - vt)]$$

$$\Psi = \Psi^o \exp(-\bar{\beta} k x_3)\exp[ik(x_1 - vt)]$$

$$N = N^o \exp(-\beta k x_3)\exp[ik(x_1 - vt)]. \tag{2.7}$$

In this way from (2.5) - (2.7) assuming that for the considered semiconductor of the relaxation type $\tau_n^+ = \tau^n$, we find for fixed v and k the depths of penetration of those waves

$$\vartheta_{1;2} = -\frac{1}{2D^n}\left[\left(D^n + \tau^n c_L^2\right)\frac{v^2}{c_L^2} - \frac{1}{k^2\tau_n^+}\right] \pm \frac{1}{2}\sqrt{\Delta} ,$$

$$\Delta = \frac{1}{(D^n)^2}\left[\left(D^n + \tau^n c_L^2\right)\frac{v^2}{c_L^2} - \frac{1}{k^2\tau_n^+}\right]^2 - 4\frac{v^2}{D^n c_L^2}\left(v^2\tau^n - \frac{1}{k^2\tau_n^+}\right) , \tag{2.8}$$

$$\bar{\beta} = \sqrt{1 - \frac{v^2}{c_T^2}} , \tag{2.9}$$

where $c_L^2 = (\lambda + 2\mu)/\varrho$, $c_T^2 = \mu/\varrho$, $\vartheta = \beta^2 - 1$.

The solutions $(2.7)_{1;3}$ we can now rewrite into the forms

$$\phi = \left[\phi_1^o \exp(-\beta_1 k x_3) + \phi_2^o \exp(-\beta_2 k x_3)\right]\exp[ik(x_1 - vt)]$$

$$N = \frac{\varrho k^2}{\lambda^n}\left[\left(v^2 + c_L^2\vartheta_1\right)\phi_1^o \exp(-\beta_1 k x_3) + \left(v^2 + c_L^2\vartheta_2\right)\phi_2^o \exp(-\beta_2 k x_3)\right]\exp[ik(x_1 - vt)]. \tag{2.10}$$

Putting $(2.7)_2$ and (2.10) into (2.6) we get the equation for v (for fixed k)

$$\Lambda(v,k) = \left[D^n\beta_1 k + s^n(ik\tau^n v - 1)\right]\left(\frac{v^2}{c_L^2} + \vartheta_1\right)\left[\left(2 - \frac{v^2}{c_T^2}\right)^2\right.$$

$$- 4 \beta_2 \sqrt{1 - \frac{v^2}{c_T^2}}\,\Big] - \Big[D^n \beta_2 k + s^n\big(ik\,\tau^n v - 1\big)\Big]\Big(\frac{v^2}{c_L^2} + \vartheta_2\Big)\Big[\Big(2 - \frac{v^2}{c_T^2}\Big)^2 - 4\beta_1\sqrt{1 - \frac{v^2}{c_T^2}}\,\Big] = 0 . \qquad (2.11)$$

As a result we obtain that $\beta_1 = \beta_2 = \beta$ and

$$v = \frac{c_L}{k}\sqrt{\frac{\tau^n c_L^2 - D^n}{\tau_n^+\left[(D^n + \tau^n c_L^2)^2 - 4\,\tau^n c_L^2 D^n\right]}} . \qquad (2.12)$$

Finally the considered surface waves we can present in the form

$$u_1 = u_1^o(x_3)\,\exp\big[ik(x_1 - vt)\big]$$
$$= i\left[k\,\phi^o \exp(-\beta k x_3) - i\bar{\beta}\,k\,\Psi^o \exp(-\bar{\beta} k x_3)\right]\exp\big[ik(x_1 - vt)\big]$$
$$u_3 = u_3^o(x_3)\,\exp\big[ik(x_1 - vt)\big]$$
$$= \left[-\beta k\,\phi^o \exp(-\beta k x_3) + ik\,\Psi^o \exp(-\bar{\beta} k x_3)\right]\exp\big[ik(x_1 - vt)\big],$$

$$N = N^o(x_3)\exp\big[ik(x_1 - vt)\big]$$
$$= \frac{\varrho k^2}{\lambda^n}\left[v^2 + c_L^2(\beta - 1)\right]\phi^o \exp(-\beta k x_3)\exp\big[ik(x_1 - vt)\big] . \qquad (2.13)$$

The components u_1 and u_3 are phase-shifted by $\pi/2$. Thus the elastic wave is elliptically polarized in the sagittal plane and the electron wave is linearly polarized.

The numerical results obtained for germanium are presented on Figures 1 - 3.

From the dispersion curves (Fig.1), it follows that the region of existence of the elasto-electronic surface waves depends not only on k, but also on the admissible values of τ_n^+ . Fig.2 shows that the longitudinal component penetrates the body deeper than the transverse one. The analysis of Fig.3 shows that if in a classical case of Rayleigh waves the inequality $u_1^o < u_3^o$ is always satisfied, then in the considered case the u_1^o and u_3^o curves cross each other. Such situation is caused by the elasto-electronic coupling. A similar phenomenon is observed when an elastic body is covered with an infinitely thin conducting film /10/ with the difference that, in our case, the SH-surface wave disappears.

Fig. 1. Dispersion curves and area of existence of the elasto-electronic surface waves

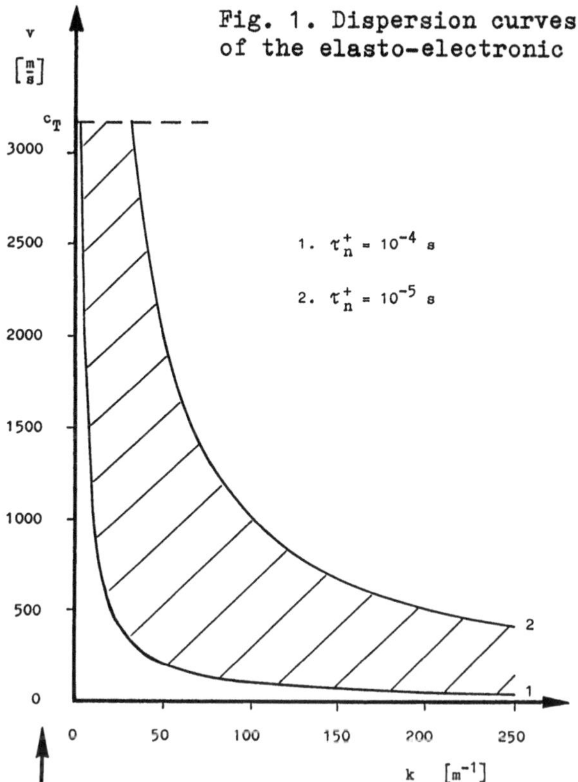

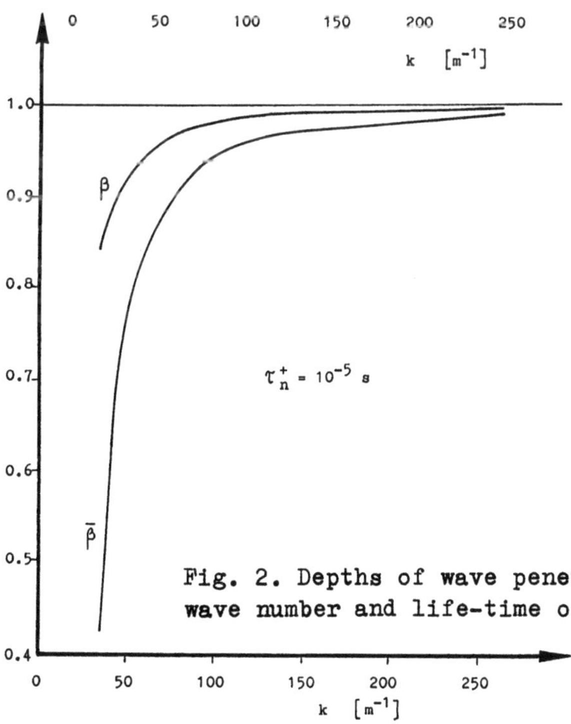

Fig. 2. Depths of wave penetration as functions of wave number and life-time of electrons

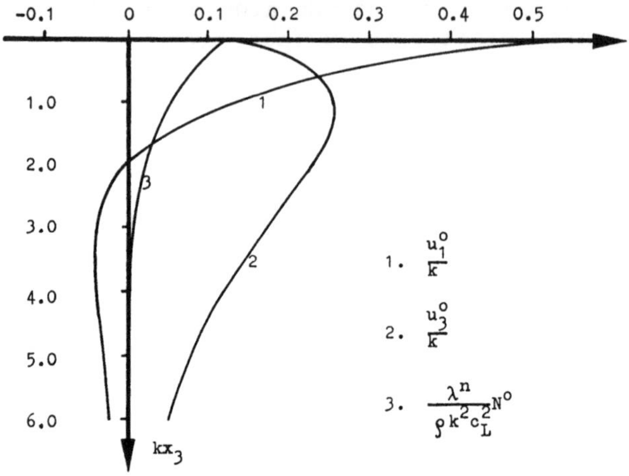

Fig. 3. Distribution of amplitudes in the elasto-electronic surface waves

Acknowledgment

The research reported here has been supported by CPBP 01.08.

References

1. Acoustic Surface Waves, ed. by A.A.Oliner, Topics Appl. Phys. Vol.24 /Springer, Berlin, Heidelberg 1978/
2. G.A.Maugin: Nonlinear Electromechanical Effects and Applications, Ser.Theor.Appl.Mech., Vol. 1, ed. by R.K.T.Hsieh, World Scientific, Singapore 1985
3. G.A.Maugin: In Applied Mechanics, ed. by J.W.Hutchinson, Vol. 23, Academic Press, New York 1983
4. S.Kaliski,Cz.Rymarz:In Drgania i Fale, ed. by S.Kaliski, Mechanika Techniczna, Vol. 3, PWN, Warszawa 1986
5. B.Maruszewski: Arch.Mech. 38 /1986/ p.83
6. B.Maruszewski,G.Lebon: Int.J.Engn.Sci. 24 /1986/ p.583
7. K.Seeger: Semiconductor Physics, Springer Ser.Solid-State Sci., Vol.40 /Springer, Berlin, Heidelberg 1973/
8. B.Maruszewski: In Electromagnetomechanical Interactions in Deformable Solids and Structures, ed. by Y.Yamamoto, K.Miya, North-Holland, Amsterdam 1987
9. A.Many,J.Goldstein,N.B.Grover: Semiconductor Surfaces, North-Holland, Amsterdam 1965
10. I.B.Jakovkin,D.W.Petrov: Difrakcja Sveta na Akusticzeskich Poverchnostnych Volnach, Nauka, Novosibirsk 1979

Part V

Surface Waves on Composites

Acoustic Properties of Periodic Layers

G.W. Farnell

Department of Electrical Engineering, McGill University,
Montreal, Canada H3A 2A7

1. INTRODUCTION

There is a renewed interest in the acoustic properties of periodically layered media, in part because of the need to know more details of the long-wavelength limit of the phonon spectrum encountered in the scattering of conduction electrons in super lattices, and in part because of the possibility of fabricating new materials for device applications. This paper is a pictorial summary, first of the propagation of acoustic waves in the bulk of an infinite periodically layered medium when the layers may be anisotropic, the case encountered in the super lattice where the individual layers are separately single crystal. The summary is then extended to cover some of the properties of the surface waves which will propagate on the free surface of a half-space consisting of the periodic medium. Finally the effects of having but a finite number of layers on top of a homogeneous substrate are outlined. The background work with isotropic layers is covered in References 1, 2 and 3.

The geometry of the problem is shown in Fig 1 for the case of two layers of different materials, of different crystal orientations and of respective thickness h_2 and h_3. The designation "material 1" is reserved for a substrate to be considered later. The medium is assumed for now to be infinite in all directions with the X_3 axis perpendicular to the layers as shown. The mechanical displacements of all the acoustic waves to be discussed are assumed to be of the form

$$\bar{U}(X_3)e^{jkX_1} \tag{1}$$

in which the wave number k along the X_1 axis is taken to be real and the "velocity" v is taken as the component along this axis and defined by $v = \omega/k$. Since the medium in Fig 1 is periodic in the X_3 direction, the displacement must be of the form

$$\bar{U}(X_3 + d) = \bar{U}(X_3)e^{j\beta d} \tag{2}$$

where the quantity β gives the phase shift between equivalent points in adjacent layers. At sufficiently low frequencies where β is purely real it can be thought of as a transverse wave number giving the phase shift per unit length in the X_3 direction as indicated on the sketch of axes in Fig 1.

In order to present a consistent and developing picture of the dominant features of propagation in such a periodic medium, in every example the materials will be taken as Silicon and Germanium with their crystalline axes aligned and with equal layer thickness $h_2 = h_3 = 100$. The unit of thickness is arbitrary but if it is taken as microns then the

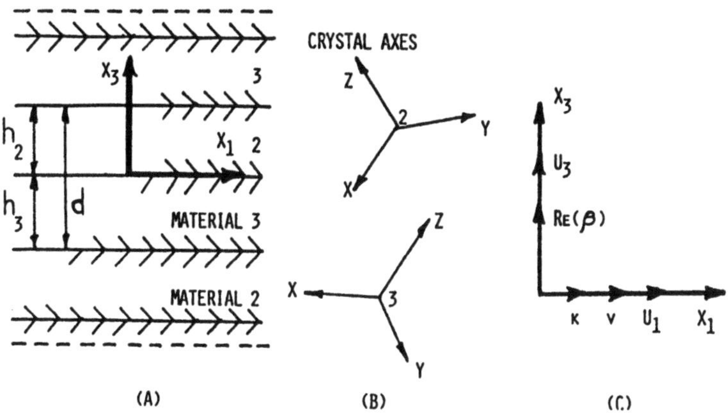

Fig. 1: Geometry of the problem.

frequency will be measured in megahertz. The case of these two aligned cubic crystals illustrates many of the effects of anisotropy since the common alignment direction is yet to be specified with respect to the X_1 and X_3 axes, but of course the effects of piezoelectricity have to be ignored in such an example.

All of the numerical results to be shown were calculated using a propagation-matrix type of computer program [4] developed for general layered, anisotropic, piezoelectric, surface-wave problems but modified to take account of the periodic nature of the medium under consideration.

2. SAGITTAL PLANE SLOWNESS

If the bulk medium consisted of identical layers of Silicon alone then for the crystalline axes aligned parallel to X_1 and X_3, that is propagation along the [100] direction in the basal plane, the sagittal plane slowness surfaces for a frequency f=9 would be as shown by the curves marked Si in Fig 2(a). The inner curve is for the quasi-longitudinal wave and the outer for the quasi-transverse wave. Here the curves are symmetric with respect to the k-axis. If the layers were Germanium alone, the slowness curves would be the two marked Ge. Note that while the material is homogeneous, the real values of β are restricted for comparison with later cases to lie in the range $-\pi$ to $+\pi$, thus the inverted branch of the quasi-transverse curve shown is translated by a reciprocal lattice vector from the corresponding branch below the axis. The two corresponding curves for the periodic combination of Si and Ge ($h_2 = h_3 = 100$) are shown by the dotted curves marked T_0 and L_0 in Fig 2(a). As in the isotropic case for such a low frequency, the slowness curves for the periodic material are of the form of a weighted average between those corresponding to the two materials separately.

In this cubic problem there are four values of β at each value of k corresponding to sagittal plane displacements, as well as the two for the SH displacements to be considered later. When the β values are real, then when plotted against k as in Fig 2(a) they give the slowness surfaces which have been reproduced in Fig 2(b) at the same frequency. Recall that these curves have mirror symmetry about the horizontal axis. For certain

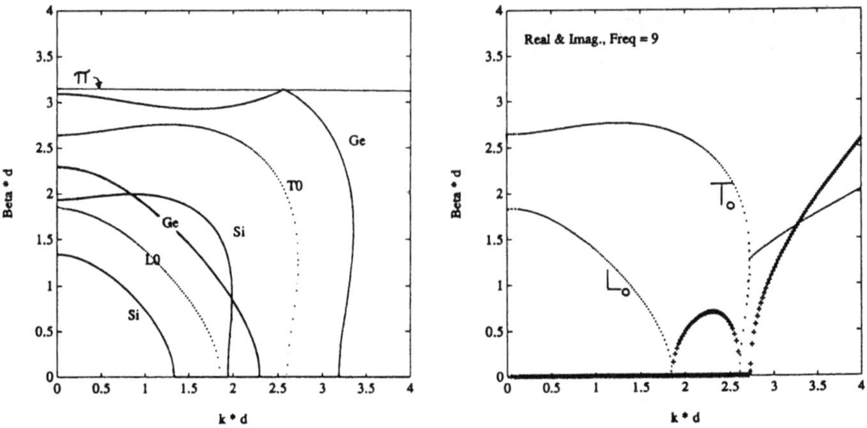

Fig. 2: Sagittal-plane roots β as a function of real wave number k for layered periodic material, Si and Ge, $h_2 = h_3 = 100$, basal plane, k along X_1, f=9. (a) left: Solid lines are slowness curves for homogeneous Si or Ge, dotted curves for composite material. (b) right: Real (dots) and imaginary (crosses) parts of β for composite material at same frequency, f=9.

regions of k, the "roots" β which satisfy the wave equation in each medium and the boundary conditions become complex much as for the four roots in the homogeneous case. In Fig 2(b) the real parts of the values of β are shown by the dotted curves and the imaginary parts by the crosses. For small values of k the four roots are all real, for the next range two are real and two are imaginary, for the next small range all four are real again, and finally for large values of k the roots are complex, here being conjugate and negative- conjugate pairs. The imaginary parts of these roots become important in the surface-wave problem to be considered later. Thus Fig 2(b) is a representative plot of the low frequency behaviour of the roots and is similar to that of a homogeneous medium.

Let us now see what happens when the frequency is increased progressively so that the wavelengths involved become comparable to the period of the layers. In Fig 3(a), the branches are labelled as they might be in a perturbation calculation of the dispersion though this is not a perturbation calculation and such designations become blurred as the frequency is increased. The branch corresponding to the quasi-longitudinal wave is merely expanded from its form in Fig 2(a). On the other hand the quasi-transverse branch labelled T_0 now interacts with its counterpart emanating from the next reciprocal lattice point to create the gaps on the Brillouin zone boundary, $Re\{\beta d\} = \pi$ In these gap regions of k, β has an imaginary part and thus the wave increases or decreases with X_3. The existence of the two gap regions on the boundary is associated with the bulge in the slowness surface and is not present when the materials are isotropic. The particle displacements associated with representative points on these "slowness" curves will be considered below.

As the frequency is increased further to f=13 there is a mixed-mode type of interaction which takes place within the Brillouin zone as illustrated in Fig 3(b). The

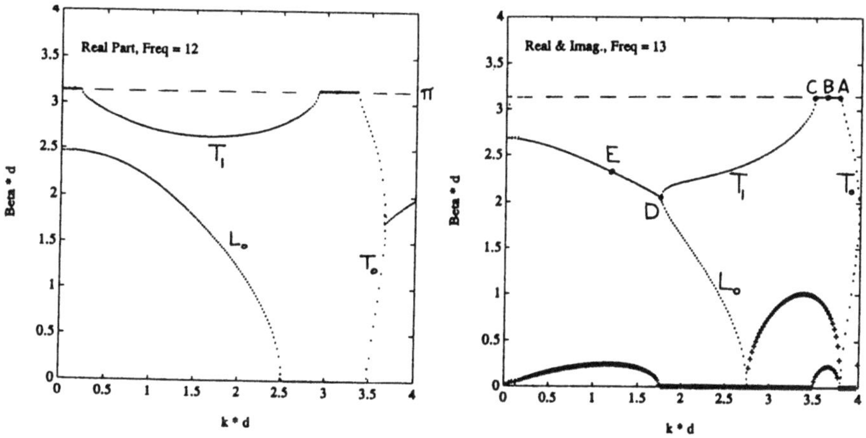

Fig. 3: Sagittal-plane roots for same geometry as in Fig 2. (a) left: Real part for f=12. (b) right: Real (dots) and imaginary (crosses) parts for f=13.

imaginary part of βd which gives the transverse attenuation per cell is shown again by the crosses, while the real part is given by the dotted curves. The quasi-transverse gap on the zone boundary covers a smaller range of k as the frequency is increased, disappears about f=16, and reappears at higher frequencies as shown in Fig 4. This vanishing of the gap is related to the so-called Lamé-wave condition which is met in the propagation of sagittal waves in isotropic plates. Two other points should be noted with respect to the f=18 curves of Fig 4: first, the gap on the Brillouin zone boundary which would correspond in a perturbation calculation to the interaction of the quasi-longitudinal wave of the central zone with its counterpart translated by one reciprocal lattice vector; and second, the rather complicated mixture of modes developing for axial propagation.

The displacement associated with axial propagation of the quasi-longitudinal mode at the low frequency represented by Fig 2(b) is illustrated in Fig 5. Here U_1, the

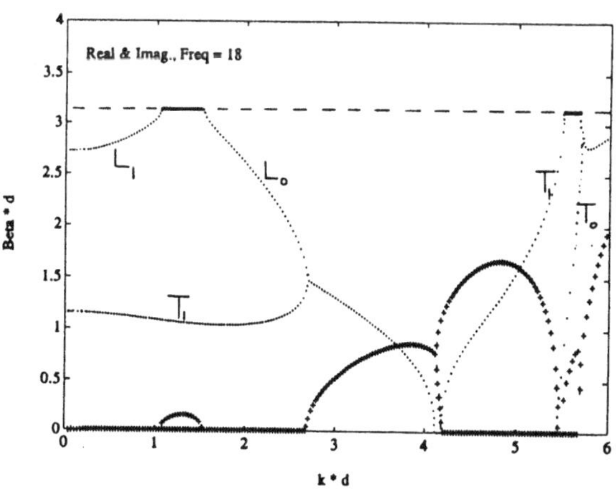

Fig. 4: Real (dots) and imaginary (crosses) parts of sagittal-plane roots for same geometry as in Fig 2, f=18.

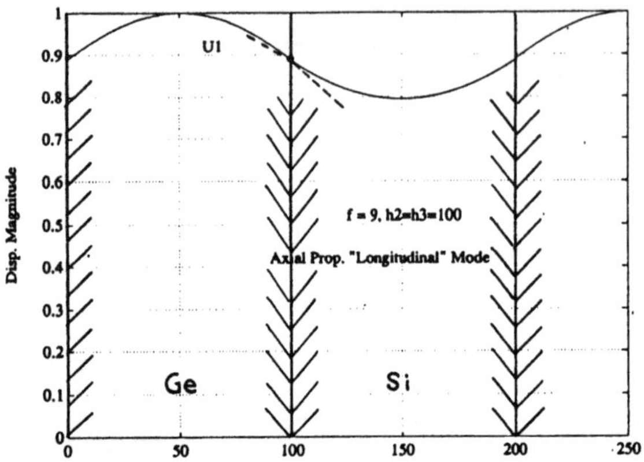

Fig. 5: Longitudinal displacement (U_1) as a function of depth (X_3) for propagation parallel to the layers for mode L_0 in Fig 2(a).

component along the direction of propagation, is even with respect to the centre lines in both of the layers and is the dominant component. In this direction U_3 is odd in both layers and is negligible to the scale of Fig 5. The boundary conditions contain the continuity of displacement and normal components of stress across each interface, the latter gives here the discontinuity in the derivative of displacement seen across the interfaces. At low frequencies the waves in the periodic medium are grossly of the form of the corresponding ones in a homogeneous material (e.g. with this symmetry U_1 would be constant with depth while U_3 would be zero) but there are periodic fluctuations of the amplitude and the phase about the values appropriate to a homogeneous "average" material. The main concern in the remainder of this paper will be with the range of frequency where the relevant wavelengths are of the order of the period of the layers.

Let us consider now the displacements associated with two forms of the interaction gaps, in particular for the points marked A to E on the slowness curve for f=13 in Fig 3(b). Figure 6 shows the two components of displacement for the point A at the right edge of the gap on the zone boundary. There is no phase shift with depth for U_1 or U_3 except for the cell to cell change of sign associated with the real part of β being equal to π/d and there is no decay or growth with depth because the imaginary part of β is zero. It is seen that U_3 is odd in the Germanium layer and thus the polarization is purely

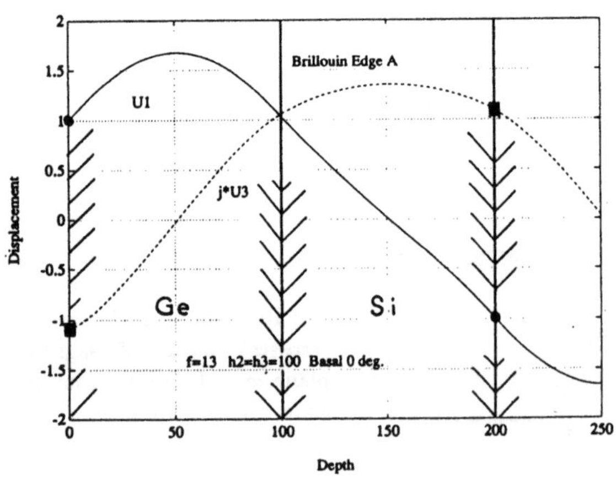

Fig. 6: Displacement components for propagation at the point on the Brillouin-zone edge marked A on Fig 3(a).

longitudinal on this centre line while it is purely transverse on each centre line of the Silicon, at other depths the polarization is elliptical with the major axes of the ellipse parallel and perpendicular to the layers. At the other extremity of the gap, point C in Fig 3(b), the displacements are similar in form to Fig 6 except for an interchange of U_1 with U_3 so that U_1 is odd in the Germanium and even in the Silicon. At the intervening point B the displacements are intermediate between those for the extremities, A and C, but in addition, because of the imaginary part to β, there is a cell-to-cell growth or decay of the amplitude. Again there is no phase shift with depth except for the zero crossings.

At the edge of the mixed-mode gap marked as D in Fig 3(b), the magnitude of the displacements are as shown in Fig 7(a). These magnitudes are seen to be comparable at each depth, to be non-zero everywhere, to be symmetrical about each centre line, and to be periodic with depth since the imaginary part of β is zero at this point. The behaviour with depth of the phase of the two components is illustrated in Fig 7(b). It is seen that the phase of a component at a given point in one cell differs from that in the adjacent cell by βd, or by βd plus or minus some multiple of 2π, as demanded by the Floquet condition of Eq 2. The broken lines in Fig 7(b) have slopes corresponding to β and $\beta - 2\pi/d$. The point-to-point phase shift within a cell departs markedly from the average cell-to-cell phase shift. The displacements at point E on Fig 3(b) are similar to those of point D except that they decay with depth because of the imaginary part of β.

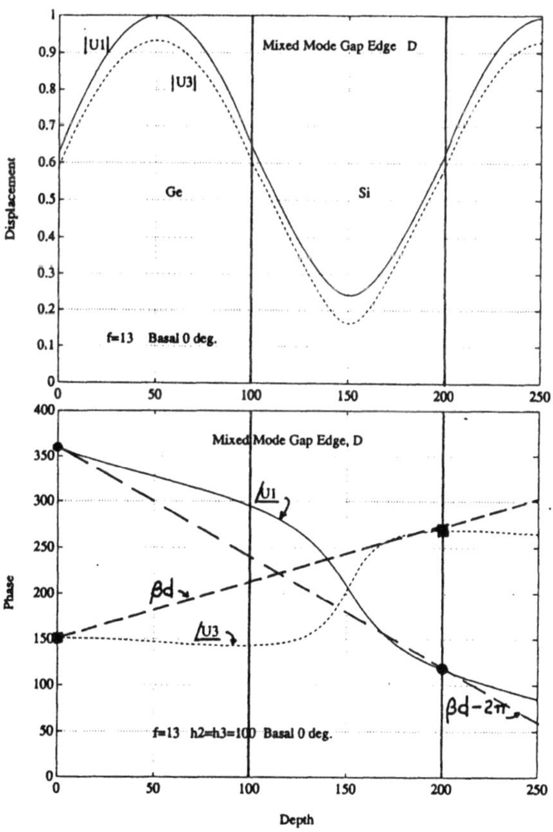

Fig. 7: (a) Magnitude and (b) phase of displacement components for propagation at the gap-edge marked D in Fig 3(a).

Figures 2 to 4 have given the slowness curves at different values of frequency. It is informative to look at similar information but plotted differently, namely as dispersion curves giving the frequency of the different modes as a function of the wave number in different directions. Such a dispersion curve is shown in Fig 8 for propagation along the X_1 axis in this case where it is a cubic axis and the layers are in the basal plane. At low frequencies the slope of each curve which gives the phase velocity is constant for the two modes, one grossly longitudinal and one grossly transverse (SV), corresponding to the homogeneous "average" medium. As the frequency is increased other branches occur with different displacement distributions. For example at f=25 in Fig 8 the four modes all have displacement distributions which are very approximately of the form $(1 + cos2\pi X_3/d)$ or $(sin2\pi X_3/d)$ across the unit cell, the function being centred on either the Silicon or the Germanium and applying either to the U_1 or the U_3 component. As a result the four modes from left to right could be described as being respectively: "mostly transverse, most energy in the Ge"; "mostly longitudinal, most energy in the Ge"; "mostly transverse, most energy in the Si"; and "mostly longitudinal, most energy in the Si".

The dispersion curves for propagation perpendicular to the layers when again the layers are parallel to the basal plane is shown in Fig 9. Thus for propagation directly

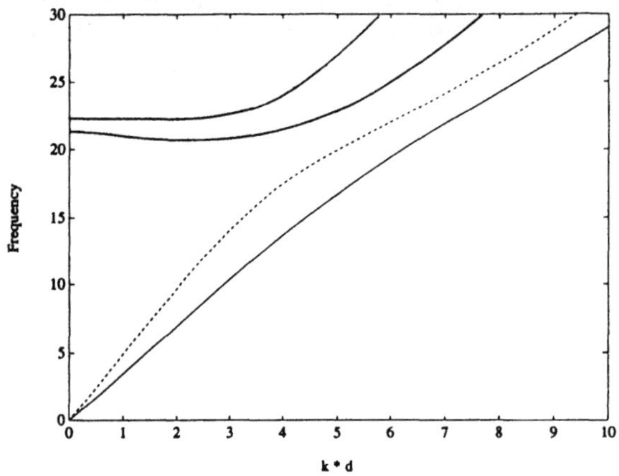

Fig. 8: Sagittal-plane dispersion curves for propagation parallel to the layers for the geometry of Fig 2(a).

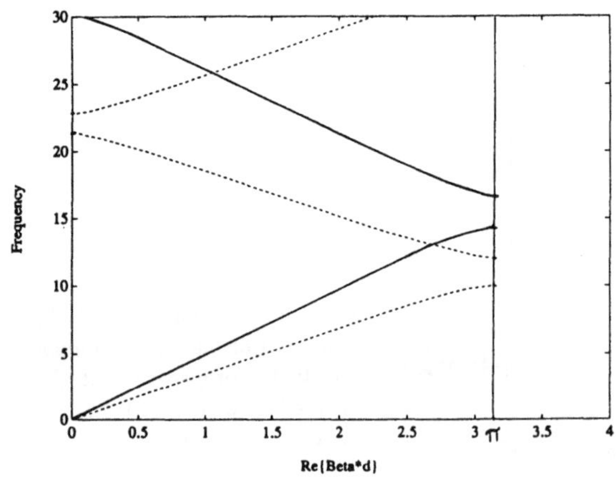

Fig. 9: Sagittal-plane dispersion curves for propagation perpendicular to the layers for the geometry of Fig 2(a).

along this axis where U_1 and U_3 are uncoupled, the curves have the appearance which would be expected of two uncoupled one-dimensional periodic systems plotted in the usual reduced zone scheme. For example, the shear and the longitudinal curves cross at about f=13 this corresponds to the k=0 condition in the slowness curve of Fig 3(b) where the imaginary part of β is zero. But note from the latter drawing that for a wide range of angles away from the perpendicular the transverse and the longitudinal modes couple and there is no unattenuated propagation.

Before turning to the surface-wave problem it is interesting to see some of the changes in propagation which occur if the plane of the interfaces is not a mirror plane of the crystals as has been the case in Figures 2 to 9. In Fig 10(a) the [111] axes of both the Germanium and the Silicon are normal to the layers and propagation is at 30 degrees to the [110]-axis, a combination for which the U_2 displacement again separates from the sagittal displacements. Here at this relatively low frequency we again have the quasi-longitudinal and quasi-transverse slowness branches, real β, as in the mirror-plane case of Fig 2(b) but now the curves are not symmetric. Note that the greater apparent departure from circles is due to the difference between the horizontal and the vertical scales, and that the roots occur in conjugate pairs. Here when two of the roots become complex between the intercepts of the L_0 and T_0 curves the imaginary part of β is not zero, moreover at higher values of k the roots are in conjugate pairs but the magnitude

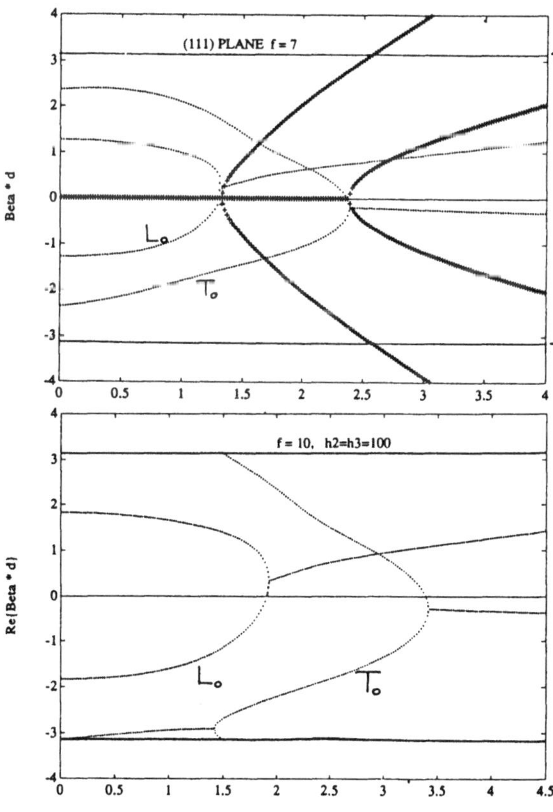

Fig. 10: Sagittal-plane roots when X_1 is along the [111] axis and X_3 is 30° from [110]. (a) Real (dots) and imaginary (crosses) parts for f=7. (b) Real part for f=10.

of the real parts are seen to be unequal. Similar behaviour is observed for homogeneous materials of the same symmetry. The effects of increasing the frequency are illustrated by Figures 10 and 11. In Fig 10(b) where f=10 the gap associated with the interaction of the branches T_0 and T_{-1} is seen in the lower part of the picture. It should be noted that this gap, even though it results from the interaction of similar waves, does not occur on the zone boundary and that there is but one gap of this type within the range of k where it exists. Directly along the negative trigonal [111] axis the real part of βd is equal to $-\pi$ and the imaginary part is zero. Moving to the higher frequency f=13 of Fig 11(a), the similar-mode gap has moved to the right but is still within the first zone, and now there are the mixed-mode interactions at two unsymmetric locations across the zone. As before in each of these gaps or interaction regions the imaginary part of β is non-zero but is not shown here. At the still higher frequency of Fig 11(b) are seen the similar-mode interaction gaps for quasi-longitudinal and for quasi-transverse waves, one each and within the zone. Also seen is the rather complicated interaction near the centre of this picture.

A discussion of the propagation of the waves polarized along X_2, SH waves, in these cases where such waves are separated from the sagittally polarized waves will be omitted here though it will be touched upon in the consideration of surface waves to follow.

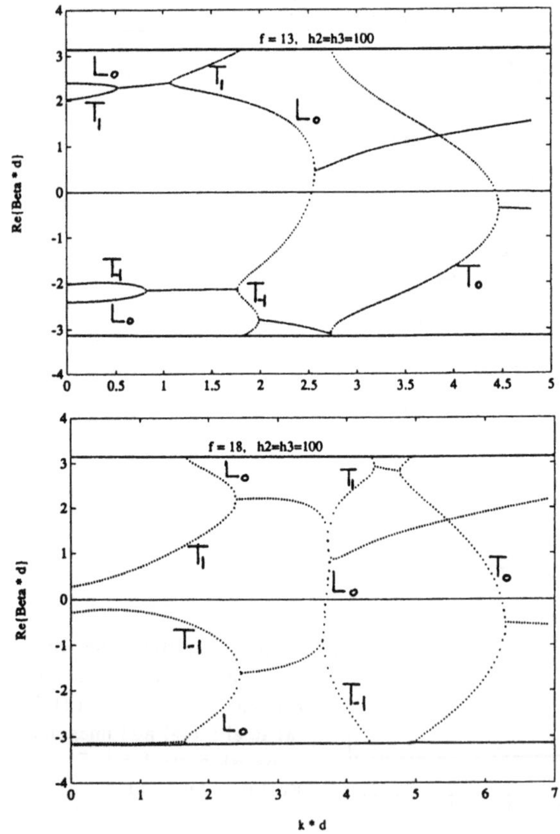

Fig. 11: Real part of the sagittal plane roots for the geometry of Fig 10. (a) f=13, (b) f=18.

3. SURFACE WAVES

Let us now turn to consideration of the propagation of surface waves on a periodic substrate occupying the half-space $X_3 < 0$. To be called a "surface wave", here as in homogeneous substrate case, the wave must propagate unattenuated in the X_1 direction (that is have real k and integrated energy flow parallel to the surface), must satisfy the free-surface boundary conditions at $X_3 = 0$, must vanish at infinite depths, must satisfy the wave equation in the appropriate material at each depth and must satisfy the interfacial boundary conditions between each layer. Stated in another way the surface-wave solution must be a combination of partial wave terms corresponding to roots at equal k in the pictures such as Figures 2 to 4 or 10 and 11 with negative imaginary parts, the combination being chosen to satisfy the free-surface boundary conditions.

Continuing one of the previous examples, the medium will consist of alternating layers of Germanium and Silicon with the basal planes of each parallel to the layers, but now the medium only fills the half-space $X_3 < 0$ and here the topmost layer is taken to be a full layer, h=100, of either Germanium or Silicon as the case may be. Again the first case to be considered is for propagation along the crystalline axis and we start with the separated sagittal-plane displacements. At very low frequencies, in view on the one hand of the "average" nature of the bulk waves at such frequencies as discussed previously, and on the other hand the asymptotic behaviour of surface waves propagating in a single layer on a homogeneous substrate, it is not surprising to find that the low-frequency velocity of the sagittal or Rayleigh-type surface wave is somewhat less than that of the vertically polarized axial shear wave of the composite material. This surface-wave velocity is shown at f=0 in Fig 12(a) but the latter drawing is rather cluttered and must be considered curve by curve.

The curves marked T and L at low frequency just convey the same information as contained in the other type of axial dispersion curves in Fig 8 but here plotted with velocity as a function of frequency. From the point of view of surface waves, in the curves of β versus k as in Figures 2 to 4 any purely real values of β represent waves radiating energy at an angle to the free surface. Thus in general it is not possible to have surface waves at a value of k at which there are purely real values of β, because the waves corresponding to the latter would radiate into the substrate or require sources at minus infinity. As a result there are areas such as the cross-hatched ones of the velocity-frequency plot of Fig 12(a) in which surface waves satisfying the definition above are not possible. There is always the low velocity region corresponding to the extreme right side of the slowness plots in which the roots all have an imaginary part and surface waves are possible. Similarly it is seen from such plots as Fig 3(b) that in the periodic case there are ranges at lower values of k (higher velocities) at which there are gaps within which there are no real values of β. Such gaps are plotted as the unshaded regions in Fig 12(a). Thus in searching for surface-wave solutions it is only necessary to consider the latter regions. The straight line for the condition $\lambda = 2\pi/k = d$ is shown as a reference line dividing "high" from "low" frequency regions.

The two curves marked Ge are the dispersion curves for the only surface-wave modes in the range shown for the case in which the Germanium is the topmost layer. Similarly the curves marked Si have a Silicon layer at the free surface. The lower Ge curve starts at the Rayleigh-wave velocity corresponding to the "average" material, descends as the frequency is increased because of the slower layer at the top and asymptotes at high frequency to the Rayleigh velocity of the topmost layer. On the other hand with Silicon

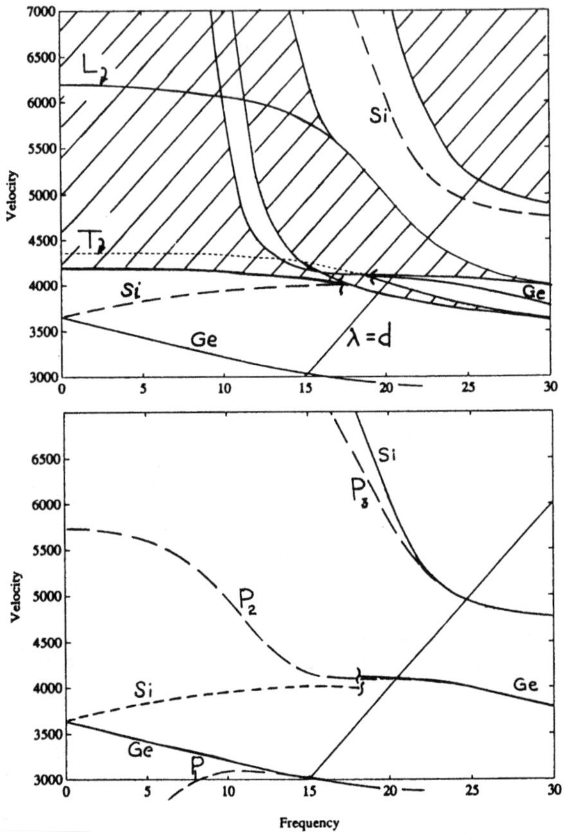

Fig. 12: (a) Dispersion curves for sagittal surface-wave modes, free surface, crystal geometry as in Fig 2. Top full layer: Germanium, marked Ge; Silicon, marked Si. Shaded regions are ones in which coupling to radiating waves would take place. (b) Surface-wave dispersion curves as in (a) and the first three plate modes of a composite Ge-Si plate also with $h_2 = h_3 = 100$ and crystalline orientation as in Fig 2.

on top the upper layer is faster than the "average" and the curve for the first mode rises with frequency, but much as with single layer propagation, the velocity approaches a radiation region, the penetration of the wave becomes deeper and deeper until finally it begins to radiate into the substrate and no longer meets our definition of a surface wave. The other two curves marked Ge and Si are for the higher mode when the Germanium or the Silicon respectively is the top layer.

In Fig 12(b) the four surface-wave dispersion curves of Fig 12(a) are reproduced. Also shown on Fig 12(b) are the dispersion curves for a composite plate of Silicon and Germanium layers each with h=100 and with the same crystal orientation as in the periodic medium. The similarity of the higher mode curves of the periodic case to those of the plate when the latter lie within a gap is noteworthy, this and the displacements now to be considered make it tempting to think of the higher surface-wave modes being related to a composite plate "floating" on a cut-off substrate of the remainder of the periodic medium.

4. SURFACE WAVE DISPLACEMENTS

Turning now to typical displacements for the surface-wave modes of the periodic substrate, we see that at low frequencies as represented in Fig 13(a) the displacement

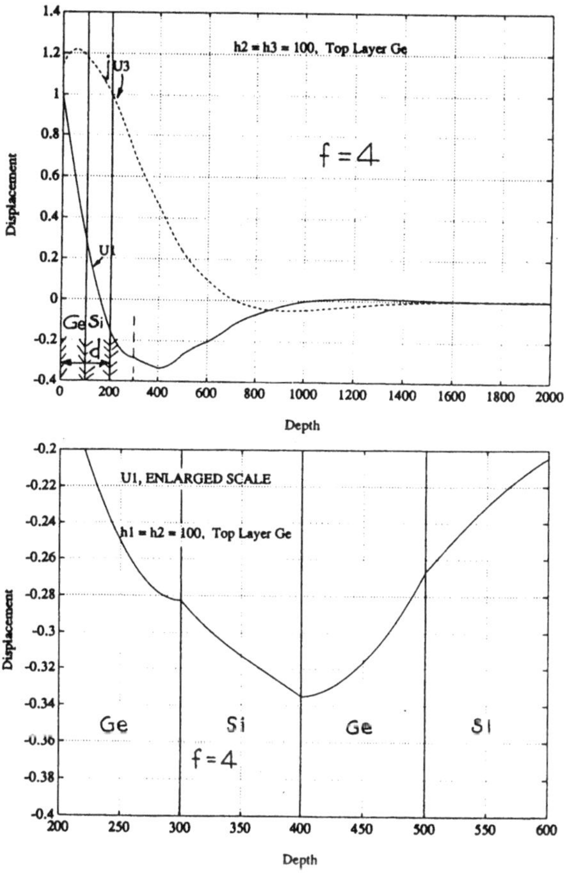

Fig. 13: (a) Depth dependence of the displacement components for the Rayleigh wave at a low frequency, f=4 in Fig 12(a). (b) Enlarged view of a part of the U1 curve in (a) showing the periodic details.

variation with depth is grossly what would be expected of an "average" homogeneous medium of the same symmetry, moreover, the form of the low frequency curves is about the same regardless of the choice of top layer. However when a region of a curve of Fig 13(a) is enlarged, as in Fig 13(b), the periodic structure becomes evident even at this low frequency. The same type of microstructure is present throughout the curve and is similar cell-to-cell with either Germanium or Silicon on top.

Figure 14(a) gives the displacement for the second mode of Fig 12(a) at f=22 with Germanium as the top layer. Other than the zero-crossings there is no progressive phase shift with depth, and the two components are 90° out of phase. Here the decay is oscillatory and associated with this type of gap, $Re\{\beta\} = \pi$, alternate cells are 180° out of phase. The broken curves of Fig 14(b) are an enlargement of the topmost cell, pair of layers, from Fig 14(a). The solid curves are the displacements of a composite plate, free on both surfaces, with the same geometry and at the same frequency. The similarity of the two pairs of curves is evident.

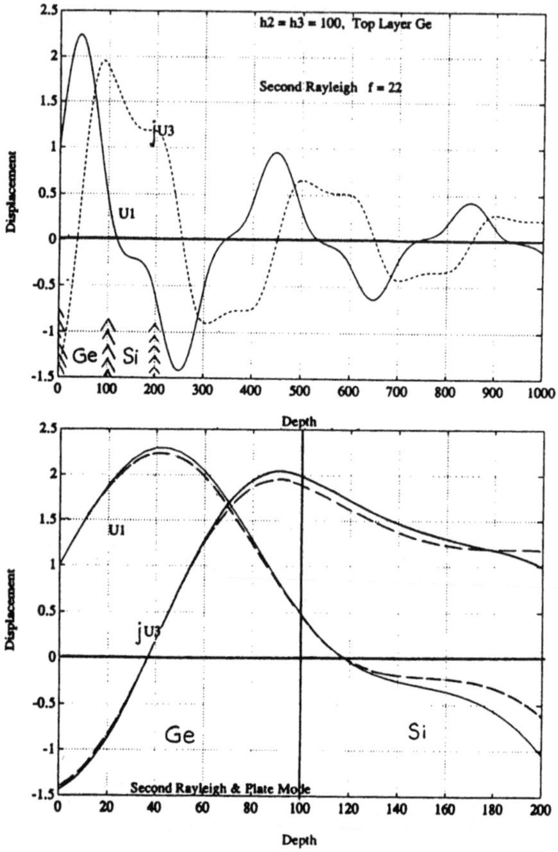

Fig. 14: (a) Depth dependence of the displacement components for the second Rayleigh mode at f=22 in Fig 12(a), Ge as top layer. (b)Displacement components in the first cell of the periodic medium for the mode shown in (a) (broken curves). Solid curves are displacement components of the second composite-plate mode, Fig 12(b), at the same frequency and with the same geometry.

5. SH WAVES

Up to this point we have been concerned with sagittal plane displacements and only in geometries where these displacements are separated from the SH component U_2. For separated cases, slowness curves of the form considered above can be drawn for the SH waves but they are somewhat simpler because there are no mixed-mode interactions. For example for the Ge-Si basal plane case considered above the forbidden regions for surface waves having only U_2 components would be as shown cross-hatched in Fig 15. The lower limit of the lower region is the dispersion curve for axial propagation of the bulk mode in this medium. In a homogeneous medium of this geometry, the SH bulk mode satisfies the free surface boundary conditions at all frequencies but it probably should not be called a surface wave because it does not decay with depth. Here the same condition applies as indicated by the f=0 point on Fig 15, but now in the periodic case as the frequency is increased with the slow layer on top, there is a mode which might be called the first Love mode in analogy with the case of a single layer on a homogeneous substrate. This mode does not exist if the top layer is faster than that of the "average" material. The curve asymptotes to the SH wave velocity of the topmost layer, here Germanium. In this configuration a second Love mode appears for a range of frequencies and is marked L_2 on Fig 15. In the range of this illustration there are no Love-type modes when the top layer is Silicon. Also shown on the graph are the

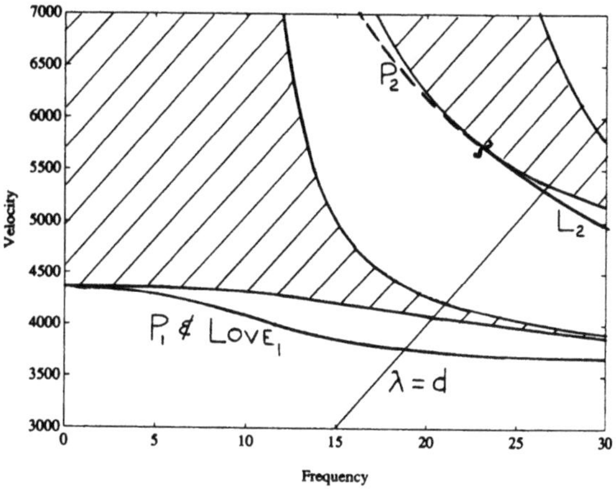

Fig. 15: Dispersion curves for transverse modes corresponding to the sagittal modes of Fig 12. "Love" modes for the semi-infinite periodic medium with Ge as the top layer, and the transverse plate modes for the composite plate.

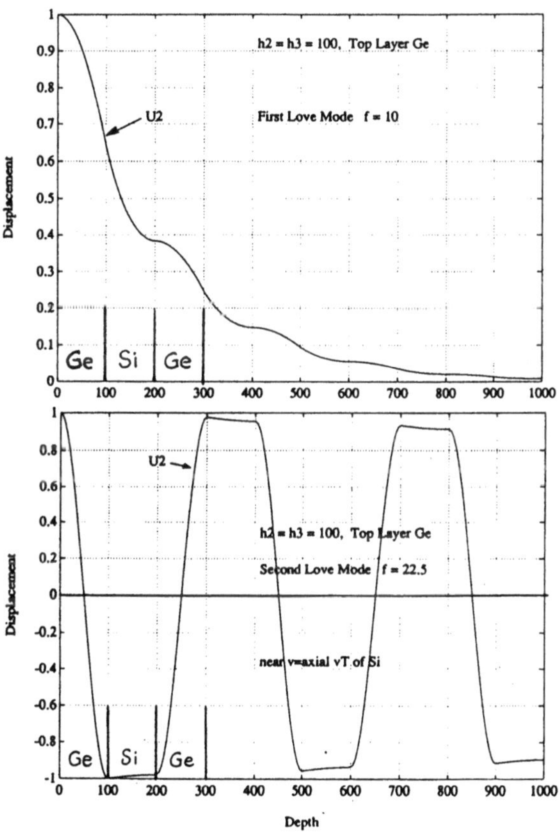

Fig. 16: Displacement as a function of depth for Love-type modes of Fig 15. (a) First Love mode at f=10. (b) Second Love mode at f=22.5.

dispersion curves of the first two modes of a composite plate of the thickness of the unit cell and of the same crystal orientation.

The form of the displacement for the first Love mode at an intermediate value of frequency, f=10, is shown in Fig 16(a) and has the gross form seen in single-layer geometries. U_2 is real at all depths. Here however there is again periodic structure to the curves. If the displacement curve for the composite plate at the same frequency is superimposed on the curve of Fig 16(a) over the first cell, the two curves are indistinguishable. Figure 16(b) shows the displacement of the second Love mode of Fig 15 at f=22.5 where it is approaching the radiation condition and where it is also near the SH velocity of the Silicon. The former results in the slow decay of the oscillatory behaviour while the latter results in the almost constant value of U_2 across each Silicon layer. In the type of gap where this mode exists, $Re\{\beta d\} = \pi$, and thus in the oscillations of Fig 16(b) alternate cells are 180° out of phase.

6. ANGULAR DEPENDENCE

When our $X_1 - X_2$ plane is a basal plane of the two crystals and the propagation direction X_1 is either along the crystal axis or at 45° to it, the sagittal and the SV components separate. However for propagation angles between these values the components are all intercoupled by the wave equation and the boundary conditions. Figure 17 gives some features of the angular dependence of surface waves in the basal plane. The cross-hatched region again represents the radiation region and for the frequencies of concern here is not very frequency dependent. The curve marked 0 gives the angular dependence of the surface wave at very low frequencies. It has the same form as the corresponding curve of a homogeneous substrate of the same symmetry and anisotropy. It starts along the axis, 0°, at the separated Rayleigh velocity of the "average" material as seen in Fig 12(a). As the angle is increased, it becomes a generalized Rayleigh wave with all three components of displacement and gradually degenerates into the pure SH wave at 45° which satisfies the boundary conditions. Also at 45° and marked by a cross is the end point of the so-called pseudo-surface-wave branch met also with homogeneous substrates of this symmetry. This pure Rayleigh wave with its sagittal displacements exists only along the uncoupled direction, for other directions it is coupled to bulk modes, represented by the cross-hatching, and radiates.

If the top layer is Silicon and the frequency is increased to f=10 then the angular dependence of the surface-wave velocity is as shown by the curve marked 10Si in Fig 17. Here again along the axis there is a pure Rayleigh mode as marked by the cross and indicated also as part of the dispersion curve of Fig 12(a). As the angle is increased at this frequency the dispersion curve approaches the radiation region, the penetration

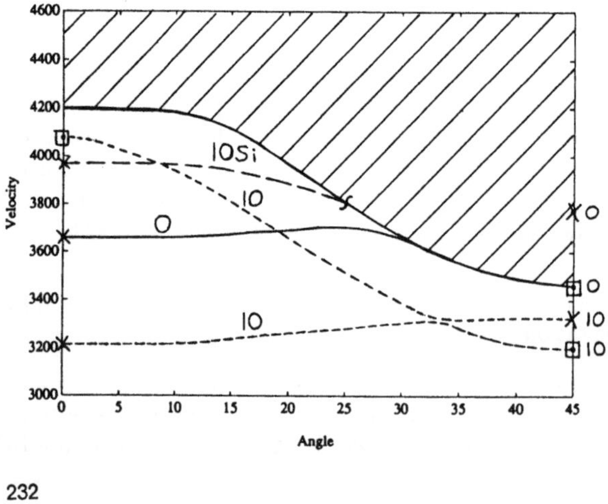

Fig. 17: Angular dependence of the surface-wave velocity in the basal plane of the Ge-Si combination with f=0 (marked 0) and at f=10 either with Silicon as the top layer (10Si) or with Ge (10).

becomes deeper, and eventually it radiates. If the top layer is Germanium at the frequency f=10, we have two modes to consider for propagation along the axis, the pure Rayleigh mode of Fig 12(a) and the pure Love mode of Fig 15. The angular dependence of these two modes is indicated by the two curves marked 10 in Fig 17. While both are mixed, one is predominantly sagittal and the other is predominantly SH; at 45° they are again separate, Rayleigh and Love.

For intermediate angles we can again consider the dispersion curves, for example at 40° in the basal plane as shown in Fig 18. Now because of the interaction of the three displacement components the radiation region is more or less the combination of two sets of regions such as in Figures 12(a) and 15. Thus the gap regions tend to be smaller than in the separated cases. The curves in the gaps are the dispersion curves with either Germanium or Silicon, as marked, for the top layer. In particular the two modes for f=10 at this angle in Fig 17 are seen here again.

It is interesting to return for a moment to a separated case in which the sagittal displacements are again independent of the SH components, but now to choose the geometry of Figures 10 and 11, that is the (111)-plane case where the crystal plane parallel to the layers is no longer a mirror plane. Here the dispersion curves are much as for the basal plane case shown in Fig 12(a) except that the numerical values are somewhat different. Now it should be recalled that the slowness curves are not symmetrical in the k-axis; as a result there are phase shifts in the displacements as a function of depth. For example at the relatively low frequency of f=4, the magnitude and phase of the displacement components are as shown in Fig 19. It is seen that now not only does the ratio of the major to the minor axis of the displacement ellipse change with depth, but the orientation of these axes changes also.

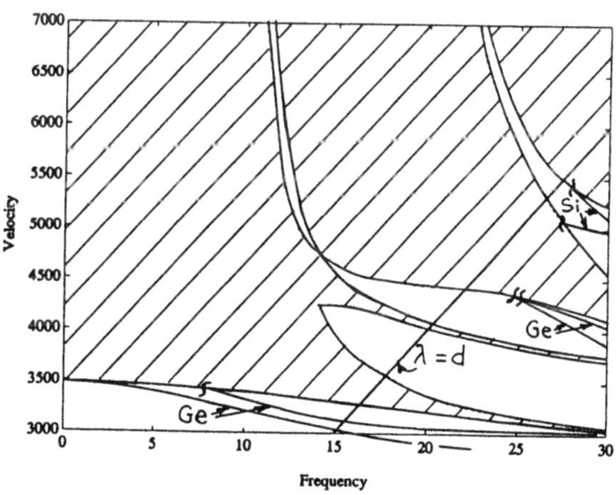

Fig. 18: Dispersion curves in the basal plane at 40 degrees (see Fig 17) with Germanium as the top layer (marked Ge) or with Silicon (marked Si).

7. FINITE NUMBER OF LAYERS

Let us now turn briefly to a related problem. In practice periodic layers of superlattices do not exist as infinite half-spaces, but rather as a large but finite number of periodic layers on an underlying substrate material. Questions arise about the transition between such a finite case and the infinite case considered above. For simplicity let us consider the sagittal modes in an isotropic combination of a fast material, SiO_2 and a slow one, gold, with the slow material as the topmost and with the fast material as the substrate. Here

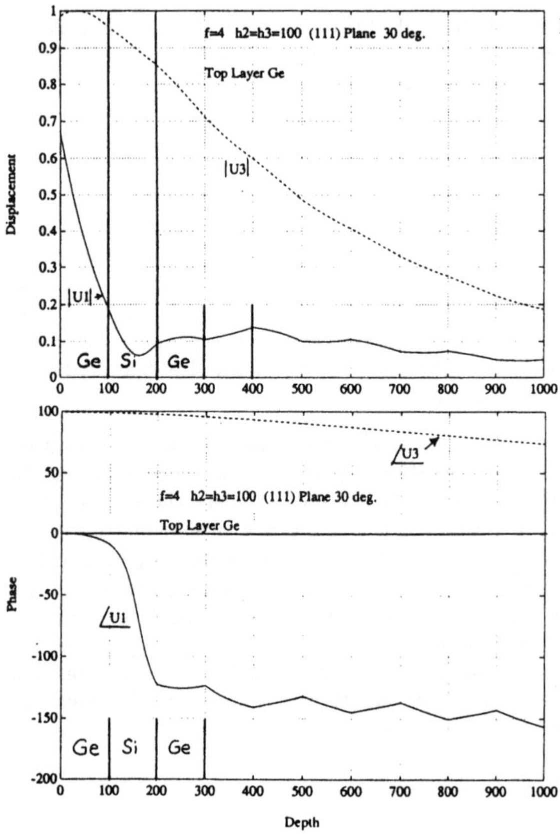

Fig. 19: (a) Magnitude and (b) phase of displacement components as a function of depth for the (111)-plane geometry of Figures 10 and 11. Here the top layer is Ge and f=4.

the "average" periodic material is slower than the substrate. The transverse velocity of the substrate forms an upper limit to non-radiating surface-wave modes and is shown by the cross-hatched strip across the top of Fig 20. If we had a single homogeneous layer on the homogeneous substrate, there would be a succession of modes more or less of the form shown by the curves 1, 2 and 3 of Fig 20. The first of these modes begins at the substrate Rayleigh velocity and asymptotes at high frequencies to the Rayleigh velocity of the top layer. The general shape of this curve is the same when the "layer" consists of many periodic layers, the initial slope being roughly proportional to the number of such periodic layers, that is, to the gross "layer" thickness. Also shown on Fig 20 is the dispersion curve of the first Rayleigh mode when the periodic medium becomes thick enough to be regarded as a half-space. When there are layers on a substrate and the substrate has the highest velocity as in the case here, the higher modes in the layer involve propagating waves, real values of β. Thus here where the "layer" is a periodic combination, we can have the higher surface-wave modes only in the regions where β of the composite material is real. Thus in this case we can have modes only in the regions which were forbidden in the earlier discussion. For example, the previous "gaps" where β has an imaginary part are now regions within which the dispersion curves do not exist and are shown cross-hatched in Fig 20. The allowed dispersion curves "avoid" these regions as sketched in this graph. As the number of periodic layers of a given periodicity is increased, the density of the modes in each of

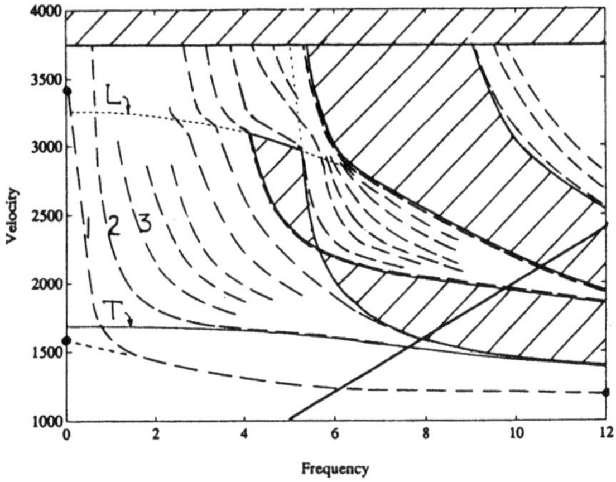

Fig. 20: Sketch of dispersion curves for a finite number of periodic layers on an infinite substrate. Numerical values correspond to about eight periods of a gold and fused quartz combination (h_2 = h_3 = 100) on a fused quartz substrate with gold as the top layer.

the allowed regions increases; the density in the sketch is for eight periods. The mode mixing evident in this curve and the asymptotic modes, for example the one passing through the crossover of the Brillouin zone gap will not be considered in detail here.

ACKNOWLEGEMENT

This research is supported by the Natural Sciences and Engineering Research Council of Canada.

REFERENCES

1. S.M. Rytov, "Acoustical Properties of a Thinly Laminated Medium", Soviet Phys. Acoustics 2, (1956), 68-80.
2. B.A. Auld, Y.H. Shui and Y. Wang, "Elastic Wave Propagation in Three-dimensional Periodic Composite Materials", J. de Physique, 45, (1984), 159-163.
3. B. Djafari, L. Dobrzynski, O.H. Duparc, R.E. Camley and A.A. Maradudin, "Sagittal Elastic Waves in Infinite and Semi-infinite Superlattices, Phys. Rev. B, 28, (1983), 1711-1720.
4. A.H. Fahmy and E.L. Adler, "Multilayer Acoustic Surface-Wave Program", Proc. IEEE, 122, (1975) 470-472.

Surface Waves on a Laminated Half-Space

W.A. Green

Department of Theoretical Mechanics, University of Nottingham, Nottingham, NG72RD, U.K.

1. Introduction

For a half-space formed by a periodic repetition of a basic laminated cell, AULD, BEAUPRE and HERRMANN [1],[2],[3] have demonstrated the existence of horizontal shear waves (anti-plane strain) travelling along the free surface and decaying with depth. The occurrence of these surface waves is associated with the presence of stopping bands in the infinite laminated medium, these stop bands being characterized by complex values of the Floquet wavenumber. Like the closely related Love waves, which propagate in a layer on a homogeneous half space, these surface waves exhibit dispersion. They differ from Love waves, firstly in that some modes propagate only over a finite range of frequencies and, secondly in that it is not necessary for the surface layer to have a shear wave speed lower than that in the underlying layer. The equivalent problem for sagittal surface waves (plane strain) has been examined by DJAFARI-ROUHANI et al. who consider the case of a free surface parallel to the layering [4] and a free surface perpendicular to the layering [5].

In the work outlined in the preceding paragraph the authors are concerned with laminates formed of alternate layers of two different elastic media. Here we direct our attention to a laminate formed from layers of a single fibre reinforced material in which the reinforcement is provided by a family of straight parallel fibres lying in the plane of each layer. We consider a half-space consisting of an infinite sequence of layers each of uniform thickness 2h, of this material, assembled so that the fibre directions in alternate layers are orthogonal to each other. Thus, the basic unit cell consists of a two-ply lay-up, of overall thickness 4h, the plies having fibre directions at 0^0 and 90^0 to some reference direction parallel to the plane of the laminae. Alternatively we may regard the unit cell as consisting of a layer of thickness 2h of the material with one of the fibre orientations, bounded on either side by layers each of thickness h of the same material but with the other fibre orientation. It is this latter interpretation which we find most convenient for analysis and which we employ in the rest of this paper. Despite the fact that the layers are formed from identical material it is convenient to refer to layers of one orientation as material 1 and layers of the other orientation as material 2. The fibre reinforced material is modelled as a transversely isotropic elastic continuum with the axis of transverse isotropy along the fibre direction. Adopting this continuum model to examine wave propagation is equivalent to assuming that the wavelengths are large compared with the fibre diameter and the inter-fibre spacing, so that scattering effects due to individual fibres are negligible. A typical sheet of carbon fibre reinforced prepeg will have thickness of the order of 124 μm and contain some 60% by volume of fibre with the fibre diameter and inter-fibre spacing each of the order of 6 μm. For such a material the continuum approach might be expected to be valid for wavelengths of the order of 1/3 the layer thickness or greater, corresponding to frequencies ranging from some 50 MHz down to zero.

We may simplify the equations governing wave propagation in the layered medium by making the additional assumption that the transversely isotropic continuum is inextensible in the direction of transverse isotropy. This is an idealization of the fact that the extensional modulus of the composite along the fibre direction can be of the order of 100 times the corresponding modulus in the cross-fibre direction. A detailed consideration of the effect of this assumption is contained in the papers by BAYLIS and GREEN [6],[7]. Since the composite is modelled as an anisotropic material, a plane wave propagating in the plane of the layers will in general involve all three components of displacement. For the particular case of propagation along one of the fibre directions the disturbance separates out into anti-plane strain motion, involving the horizontal shear displacement only and plane-strain motion, involving sagittal plane displacements only. For the anti-plane motion, the laminate characteristics are the same as those of the two ply lay up of two isotropic materials considered in [1-3]. For the plane strain motion the laminate characteristics differ from those considered in [4] by virtue of the inextensibility constraint associated with the motion in one set of plies. The effect of this is to reduce the order of the transfer matrix, which expresses conditions at the top of a unit cell in terms of those at the bottom, from order 4 to order 2 and consequently it is possible to obtain explicit expressions for the Floquet wavenumber associated with waves in a laminate of infinite extent. The derivation of this transfer matrix is given in Section 2.

The study of Floquet waves in a laminate formed of an infinite sequence of these unit cells has been carried out by BAYLIS [8], [9] who has, in particular, established the boundaries between the passing and stopping bands for waves travelling parallel to one of the fibre directions. In Section 3 we make use of these results to identify regions of the frequency/wavenumber space associated with the existence of waves decaying with depth in a half space of the laminated medium. For these waves, the normal component of traction and the normal displacement at the surface of the half space is given in terms of an eigenvector of the transfer matrix. We obtain an equation for the thickness of a superimposed layer of material which, when bonded to the surface of the half-space, will transform this eigenvector into the condition of zero traction at the free surface. Results are obtained for both an overlayer of inextensible material and of isotropic material.

2. Propagator Solution

We choose a cartesian system of axes $Ox_1x_2x_3$ with Ox_1 normal to the plane of the layers and Ox_3 as the reference direction. Then the fibres in the two outer layers of the unit cell (material 1) are parallel to the x_3-axis and those in the two inner layers (material 2) are parallel to the x_2-axis.

We consider displacements of the form

$$u_1 = U_\alpha(x_1)\sin(kx_3-\omega t),\quad u_2 = 0,\quad u_3 = W_\alpha(x_1)\cos(kx_3-\omega t),\quad \alpha=1,2 \quad , \tag{1}$$

where the suffix α refers to the components in material α. These displacements correspond to plane waves of angular frequency ω and wave number k propagating along the x_3-direction. For such waves the condition of inextensibility in material 1 ($W_1=0$) reduces the system of governing equations to a single second order equation in the normal displacement $U_1(x_1)$ from which all the non-zero stresses may be determined. Introducing the normal stress amplitude $T_\alpha(x_1)$ such that

$$t_{11}^{(\alpha)} = \rho c_2^2 k T_\alpha(x_1)\cos(kx_3-\omega t) \quad , \tag{2}$$

the solution to the governing equation may be written in the form

$$\underset{\sim}{X}_1(x_1) = \underset{\sim}{M}(x_1)\,\underset{\sim}{X}_1(0) \quad , \tag{3}$$

where

$$\underset{\sim}{X}_1(x_1) = \begin{bmatrix} T_1(x_1) \\ U_1(x_1) \end{bmatrix} \quad \text{and} \quad \underset{\sim}{M}(x_1) = \begin{bmatrix} C & pS \\ S/p & C \end{bmatrix} . \tag{4}$$

Here $p^2 = (1-v^2/c_3^2)$, $v^2 = \omega^2/k^2$, $C = \cosh kpx_1$, $S = \sinh kpx_1$, $c_1^2 = (\lambda+2\mu_T)/\rho$, $c_2^2 = \mu_T/\rho$, $c_3^2 = \mu_L/\rho$, ρ is the density and λ, μ_L and μ_T are the elastic moduli of the material.

The equations of motion in material 2 are identical with those for an isotropic material with Lamé constants λ and μ_T and density ρ and the solution has the form

$$\underset{\sim}{Y}_2(x_1) = \underset{\sim}{P}(x_1)\, \underset{\sim}{Y}_2(0), \tag{5}$$

where $\underset{\sim}{Y}_2(x_1) = (T_2(x_1), R_2(x_1), U_2(x_1), W_2(x_1))^T$ and T denotes the transpose. The function $R_2(x_1)$ is related to the shear stress $t^{(2)}_{13}$ by the expression

$$t^{(2)}_{13} = \rho c_2^2 k\, R_2(x_1)\, \sin(kx_3-\omega t), \tag{6}$$

and the propagator matrix $\underset{\sim}{P}(x_1)$ is given by

$$\begin{pmatrix} \alpha C_1+(1-\alpha)C_2 & \dfrac{\alpha S_1}{q_1}+(1-\alpha)q_2 S_2 & -\dfrac{2\gamma\alpha^2}{(1-\alpha)}\dfrac{S_1}{q_1}+2\gamma(1-\alpha)q_2 S_2 & 2\gamma\alpha(C_1-C_2) \\ (1-\alpha)q_1 S_1+\dfrac{\alpha S_2}{q_2} & (1-\alpha)C_1+\alpha C_2 & -2\gamma\alpha(C_1-C_2) & 2\gamma(1-\alpha)q_1 S_1-\dfrac{2\gamma\alpha^2}{(1-\alpha)}\dfrac{S_2}{q_2} \\ -\dfrac{(1-\alpha)}{2\gamma}\left(q_1 S_1-\dfrac{S_2}{q_2}\right) & -\dfrac{(1-\alpha)}{2\gamma}(C_1-C_2) & \alpha C_1+(1-\alpha)C_2 & -(1-\alpha)q_1 S_1-\alpha\dfrac{S_2}{q_2} \\ \dfrac{(1-\alpha)}{2\gamma}(C_1-C_2) & \dfrac{(1-\alpha)}{2\gamma}\left(\dfrac{S_1}{q_1}-q_2 S_2\right) & -\alpha\dfrac{S_1}{q_1}-(1-\alpha)q_2 S_2 & (1-\alpha)C_1+\alpha C_2 \end{pmatrix}. \tag{7}$$

The terms appearing in the matrix in (7) are defined by

$$S_1 = \sinh kq_1x_1,\quad S_2 = \sinh kq_2x_1,\quad C_1 = \cosh kq_1x_1,\quad C_2 = \cosh kq_2x_1,$$

$$\alpha = 1-\frac{2c_2^2}{v^2},\quad \gamma = \frac{c_2^2}{c_1^2},\quad q_1 = \left[1-\frac{v^2}{c_1^2}\right]^{\frac{1}{2}},\quad q_2 = \left[1-\frac{v^2}{c_2^2}\right]^{\frac{1}{2}}.$$

Consider now the unit cell of the semi-infinite layered medium made up of two layers (each of thickness h), of material 2 sandwiched by a layer of thickness h of material 1 on either side. In general, the continuity conditions at the interface between materials 1 and 2 require both displacement components U, W, and the traction components t_{11}, t_{13}, to be continuous. However, the constraint of inextensibility involves the existence of a reaction stress in the fibre direction which can become singular, allowing a discontinuity in shear stress across the interface. The conditions to be satisfied at the interface are then the continuity of U and T (or t_{11}) and the requirement that $W_2 = 0$ to ensure continuity of W. Using these conditions it is possible to relate the vector $\underset{\sim}{X}^B$ at the bottom of the unit cell to the vector $\underset{\sim}{X}^A$ at the top in the form

$$\underset{\sim}{X}^A = \frac{\underset{\sim}{M}\,\underset{\sim}{R}\,\hat{\underset{\sim}{R}}\,\underset{\sim}{M}}{|\det \underset{\sim}{R}|}\underset{\sim}{X}^B = \frac{\underset{\sim}{Q}\,\hat{\underset{\sim}{Q}}}{|\det \underset{\sim}{Q}|}\underset{\sim}{X}^B = \underset{\sim}{D}\,\underset{\sim}{X}^B, \tag{8}$$

where

$$\underset{\sim}{R} = \begin{bmatrix} r_{11} & r_{12} \\ r_{21} & r_{22} \end{bmatrix},\quad \hat{\underset{\sim}{R}} = \begin{bmatrix} r_{22} & r_{12} \\ r_{21} & r_{11} \end{bmatrix},$$

$$\underset{\sim}{Q} = \underset{\sim}{M}\,\underset{\sim}{R} = \begin{bmatrix} q_{11} & q_{12} \\ q_{21} & q_{22} \end{bmatrix},\quad \hat{\underset{\sim}{Q}} = \hat{\underset{\sim}{R}}\,\underset{\sim}{M} = \begin{bmatrix} q_{22} & q_{12} \\ q_{21} & q_{11} \end{bmatrix}, \tag{9}$$

and the elements r_{ij} of $\underset{\sim}{R}$ are given in terms of the elements $p_{\ell m}$ of $\underset{\sim}{P}$ by the relations

$$r_{11} = p_{11} - \frac{p_{14}p_{41}}{p_{44}},\quad r_{12} = p_{13} - \frac{p_{12}p_{43}}{p_{42}},$$

$$r_{21} = p_{31} - \frac{p_{41}p_{34}}{p_{44}},\quad r_{12} = p_{33} - \frac{p_{32}p_{43}}{p_{42}}. \tag{10}$$

In (8), (9) and (10) the elements of $\underset{\sim}{M}$ and $\underset{\sim}{P}$ are evaluated at $x_1 = h$. Equation (8) allows us to determine the Floquet wave number for motion in an infinite laminated material and will be employed in the following section to identify the stopping bands.

3. Surface Waves

We now consider a laminated half space $x_1 < 0$ formed of a repeating sequence of unit cells, overlaid by a layer of depth d of some other elastic material, bonded to the half space at $x_1 = 0$. Because of the periodicity associated with the unit cell of depth $\ell = 4h$, the vector $\underset{\sim}{X}^A$ at the top of each unit cell is related to the vector $\underset{\sim}{X}^B$ at the bottom of the cell through the Bloch theorem, in the form

$$\underset{\sim}{X}^A = e^{ir\ell} \underset{\sim}{X}^B . \tag{11}$$

Equations (8) and (11) taken together imply that the Floquet wavenumber r is given in terms of ω,k and the cell depth ℓ by the condition

$$|\underset{\sim}{D} - e^{ir\ell} \underset{\sim}{I}| = 0 , \tag{12}$$

so that $e^{ir\ell}$ is an eigenvalue of $\underset{\sim}{D}$. Denoting the eigenvalues of $\underset{\sim}{D}$ by λ_1 and λ_2 we can express these in terms of the elements of $\underset{\sim}{Q}$, in the form

$$\lambda_1 = \frac{(\sqrt{q_{11}q_{22}} + \sqrt{q_{12}q_{21}})}{(\sqrt{q_{11}q_{22}} - \sqrt{q_{12}q_{21}})} , \qquad \lambda_2 = \frac{1}{\lambda_1} , \tag{13}$$

and the associated eigenvectors are

$$\underset{\sim}{X}_1 = \begin{bmatrix} \sqrt{q_{11}q_{12}} \\ \sqrt{q_{21}q_{22}} \end{bmatrix} \quad \text{and } \underset{\sim}{X}_2 = \begin{bmatrix} \sqrt{q_{11}q_{12}} \\ -\sqrt{q_{21}q_{22}} \end{bmatrix} . \tag{14}$$

If $|\lambda_1| > 1$ the eigenvector $\underset{\sim}{X}_1$ decays in moving down through the material from the interface $x_1 = 0$, and the disturbance associated with it is therefore a surface wave. The condition for this to be satisfied is that

$$q_{11}q_{22}q_{12}q_{21} > 0 \tag{15}$$

and it follows that the eigenvector is then also real. When the inequality (15) is reversed the eigenvalues λ_1 and λ_2 are complex conjugates with unit modulus and correspond to a disturbance which is periodic with depth in the half space. Thus the stopping and passing bands are separated by the roots of the equations

$$q_{11} = 0, \quad q_{22} = 0, \quad q_{12} = 0 , \quad q_{21} = 0 . \tag{16}$$

Each of (16) corresponds to an infinite set of dispersion curves relating ω to k. These equations have been studied in [8] and [9] and the first four harmonics of each set of curves is shown in Figure 1 for a particular carbon fibre epoxy resin composite for which $c_1^2/c_2^2 = 4.297$ and $c_3^2/c_2^2 = 2.301$. These curves (taken from [9]) correspond to the edges of the dispersion surface in ω,k,r space plotted on an extended zone scheme. The curves $q_{12} = 0$, $q_{21} = 0$, correspond to $r\ell = 2k\pi$ (k=0,1,...) whilst $q_{11} = 0$, $q_{22} = 0$ correspond to $r\ell = (2k+1)\pi$ (k=0,1,...). The surface has two sheets so that the fundamental of $q_{12} = 0$ and $q_{21} = 0$ both occur at $r = 0$ whilst the fundamentals of $q_{11} = 0$ and $q_{22} = 0$ come at $r\ell = \pi$. The regions shown hatched in Figure 1 correspond to complex values of r and are associated with eigenvalues $|\lambda_1| > 1$ and the region of surface disturbances. The curves themselves correspond to $\lambda_1 = \lambda_2 = 1$ and $q_{11} = 0$ and $q_{12} = 0$ both give traction free conditions on the surface $x_1 = 0$ (and therefore $x_1 = -s\ell$, (s = 1,2,...). The latter corresponds to a disturbance which is antisymmetric relative to the middle surface of each layer whereas the former is symmetric relative to the middle surface of the layers of material 2 and antisymmetric relative to the middle surface of the layers of material 1. The curves $q_{22} = 0$ and $q_{21} = 0$ both give zero normal displacement at $x_1 = 0$ (and $x_1 = -s\ell$, s =1,2,...), the

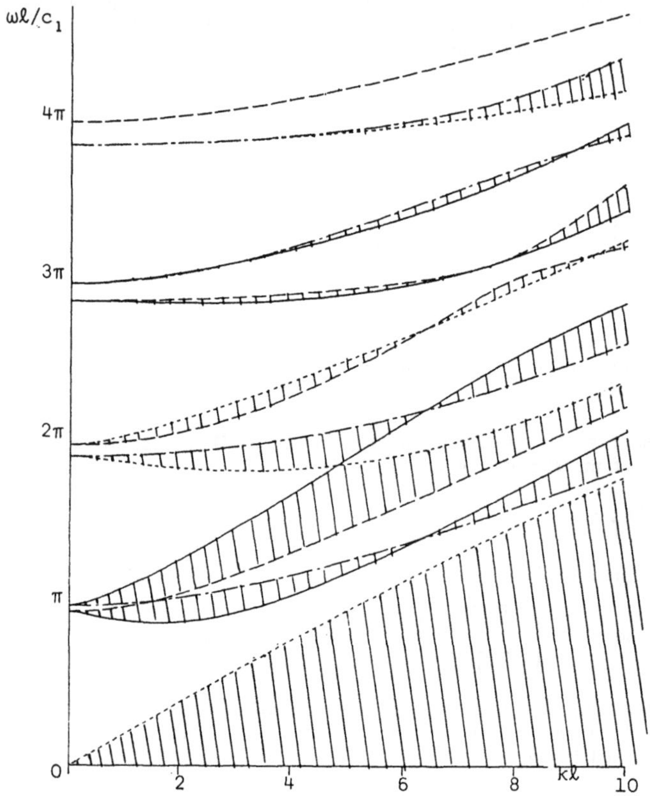

Figure 1 First four harmonics of the curves $q_{11} = 0$ (solid), $q_{12} = 0$ (dot), $q_{21} = 0$ (dash), $q_{22} = 0$ (dot/dash)

latter being associated with a disturbance which is symmetric relative to the middle surface of each layer and the former associated with a symmetric motion relative to the middle surface of the layers of material 1 but antisymmetric relative to the middle surface of the layers of material 2. These results have a strong similarity with those for S.H. waves in a laminate, which are detailed in [2] and it is possible to make use of the geometric argument put forward in [2] to show the existence of surface waves when the overlayer is made of material 1. This is possible because the traction free boundary condition in material 1 involves the normal stress $t^{(1)}_{11}$ only, since a discontinuity is possible in the shear stress component $t^{(1)}_{13}$ and the situation is then analagous to that for S.H. motion which involves only one non-zero traction component. When the overlayer consists of a general material however, it is necessary to impose the condition that both components of traction vanish at the free surface and the geometric argument is no longer applicable. We accordingly adopt an algebraic approach in our consideration of a variety of overlayers.

We begin with an overlayer of depth d of material 1. By allowing d to take negative as well as positive values, subject to the restriction $-h < d < h$, we can consider the half space formed by slicing an infinite medium by any plane within material 1, which is analogous to the problems considered in [2] and [4]. Assuming the disturbance at $x_1 = 0$ to be that corresponding to the eigenvector $\underset{\sim}{X}_1$ we may obtain the disturbance $\bar{\underset{\sim}{X}}$ at the free surface by using the propagator $\underset{\sim}{M}(d)$ to give $\underset{\sim}{X} = \underset{\sim}{M}(d)\, \underset{\sim}{X}_1$, from which the condition that the normal component of traction $\bar{T}_1$ be zero becomes

$$\overline{T}_1 = \sqrt{q_{11}q_{12}}\cosh pkd + \sqrt{q_{21}q_{22}}\, p \sinh pkd = 0 \ . \tag{17}$$

For values of ω and k corresponding to any point in the hatched regions of Figure 1, the quadratic surds in (17) are either both real or both pure imaginary, p is fixed and (17) can be regarded as an equation to determine d. Points to the right of the line $\omega = c_3k$ correspond to real values of p, for which equation (17) becomes

$$\tanh pkd = -\frac{1}{p}\sqrt{\frac{q_{11}q_{12}}{q_{21}q_{22}}} \ , \tag{18}$$

which has the solutions only for negative values of d. Equation (18) holds in particular throughout the first hatched region shown in Figure 1, bounded above by $q_{12} = 0$ and below by $\omega = 0$. Values of $\omega > c_3k$ correspond to imaginary values of p and writing $p = im$ in equation (17) leads to

$$\tan mkd = \frac{1}{m}\sqrt{\frac{q_{11}q_{12}}{q_{21}q_{22}}} \ . \tag{19}$$

All the hatched regions in which equation (19) is valid are bounded between one of the pair of curves $q_{11} = 0$, $q_{12} = 0$ and one of the pair $q_{21} = 0$, $q_{22} = 0$ with $\omega = c_3k$ forming the right hand boundary. Thus the right hand side of (19) varies between zero and infinity so that the layer depth d must be positive and satisfy $0 < d < \pi/2mk$ in order that surface waves of the given wavelength and frequency may propagate.

Consider now an overlayer of depth d of an isotropic elastic material with density $\bar{\rho}$ and Lamé constants $\bar{\lambda}$ and $\bar{\mu}$. The propagator matrix $\underset{\sim}{\overline{P}}(x_1)$ in this material is readily obtainable from equation (7) on making the appropriate substitutions. The vector $\underset{\sim}{Y}^A$ at the top of this overlayer is then given in terms of the vector $\underset{\sim}{Y}^B$ at the bottom by the relation

$$\underset{\sim}{Y}^A = \underset{\sim}{\overline{P}}(d)\, \underset{\sim}{Y}^B \ . \tag{20}$$

Continuity of T, U and W between this overlayer and the half space gives, on using the eigenvector $\underset{\sim}{X}_1$ in (14) that

$\underset{\sim}{Y}^B = (M\sqrt{q_{11}q_{12}} \quad R \quad M\sqrt{q_{21}q_{22}} \quad 0)^T$ where R and M are arbitrary.

Substituting this into (20) and using the free surface conditions $T^A = 0$, $R^A = 0$ gives two homogeneous equations for M and R which have non-trivial solutions provided

$$\Delta_{12}\sqrt{q_{11}q_{12}} + \Delta_{13}\sqrt{q_{21}q_{22}} = 0 \ , \tag{21}$$

where

$$\Delta_{12} = (\bar{p}_{11}\bar{p}_{22} - \bar{p}_{12}\bar{p}_{21}) \ , \quad \Delta_{13} = (\bar{p}_{13}\bar{p}_{22} - \bar{p}_{12}\bar{p}_{23}) \ , \tag{22}$$

and $\bar{p}_{\ell m}$ are the elements of $\underset{\sim}{\overline{P}}(d)$. For points (ω,k) in the hatched regions of Figure 1, there are solutions to (21) when Δ_{12} and Δ_{13} are of opposite sign. The equation $\Delta_{12} = 0$ is the dispersion equation for a plate of depth d of the overlayer material subject to traction-free conditions on one of the faces and the condition that the two components of displacement vanish on the other. The equation $\Delta_{13} = 0$ gives the dispersion equation in the same plate subject to traction-free conditions on one face and the conditions of vanishing normal traction and zero tangential displacement on the other face. Alternatively, we may regard $\Delta_{13} = 0$ as the dispersion equation for anti-symmetric (flexural) waves in a plate of depth 2d of the material, subject to traction free conditions on both surfaces.

Regarding Δ_{12} and Δ_{13} as functions of d for fixed values of ω and k, it may be shown that $\Delta_{12} = 1$ and $\Delta_{13} = 0$ at $d = 0$ for all ω and k and that as d increases from zero both functions initially decrease. For values of ω and k such that $v = \omega/k$ satisfies $\bar{c}_R < v < \bar{c}_2$, where $\bar{c}_R$ is the Rayleigh wave speed in the material of the overlayer, both Δ_{12} and Δ_{13} may be shown to be negative as $d \to \infty$ and hence there is some value of d for which equation (21) is satisfied. If the values of ω and k are such that

$\bar{c}_2 < v < \bar{c}_1$ than both Δ_{12} and Δ_{13} oscillate with the same period as d increases and it is possible to find values of d such that Δ_{12} is negative and Δ_{13} is positive and vice-versa, hence a solution to (21) always exists for some value of d. When ω and k correspond to $v > \bar{c}_1$ both Δ_{12} and Δ_{13} are again oscillating functions and it is again possible to find values of d which satisfy (21). Thus for this overlayer, there exists some thickness which allows the propagation of surface waves at any wavelength and frequency corresponding to the stopping bands in the laminate.

References

1. B. A. Auld, G. S. Beaupre, G. Herrmann: In Electronics Letters 13, 525 (1977)
2. B. A. Auld, G. S. Beaupre, G. Herrmann: In Modern Problems in Elastic Wave Propagation (J. Wiley & Sons, London 1978)
3. B. A. Auld, G. S. Beaupre, G. Herrmann: Mechanics Today 5, 83 (1980)
4. B. Djafari-Rouhani, L. Dobrzynkski, O. Nardouin Duparc, R. E. Camley, A. A. Maradudin: Phys. Rev. B28 1711 (1983)
5. B. Djafari-Rouhani, A. A. Maradudin, R. F. Wallis: Phys. Rev. B29 6454 (1984)
6. E. R. Baylis, W. A. Green: J. Sound and Vibration 110, 1 (1986)
7. E. R. Baylis, W. A. Green: J. Sound and Vibration 111, 181 (1986)
8. E. R. Baylis: In Mechanical Behaviour of Composites and Laminates ed. by W. A. Green and M. Mićunović (Elsevier Applied Science, London 1987)
9. E. R. Baylis: Ph.D. Thesis, University of Nottingham (1986)

Surface and Interface Waves in a Sandwich Plate with Interface Soft Layers

S.K. Datta[1], *A.H. Shah*[2], *T. Chakraborty*[1], *and R.L. Bratton*[1]

[1]Department of Mechanical Engineering and CIRES, University of Colorado, Boulder, CO 80309, USA

[2]Department of Civil Engineering, University of Manitoba, Winnipeg, Canada R3T 2N2

1. INTRODUCTION

In the past, dynamic behavior of periodically laminated medium has been studied extensively. A review of the literature on exact and approximate analyses of this problem can be found in [1,2].

The corresponding problem of a plate of finite thickness having a large number of periodic anisotropic layers has not received much attention. In [2] a stiffness method was presented for studying harmonic wave propagation in a periodically laminated infinite medium. This method is also well suited for analyzing a finite thickness plate with many layers. In this approach each lamina is divided into several sublayers and the displacement distribution through the thickness of each sublayer is approximated by polynomial interpolation functions in such a way that displacements and tractions are continuous across the interfaces between adjacent sublayers. Details of the method can be found in [2,3]. Here we summarize the pertinent equations and present numerical results showing the effect of soft interface layers between reinforced laminae on the dispersion of guided waves.

The particular systems considered are a single ply (0°) and a cross-ply (0°/90°/0°) laminated plate. Each lamina is assumed to be reinforced by continuous aligned fibers, so that it can be modeled as a transversely isotropic medium with the symmetry axis aligned with the fiber direction. The geometry of a typical cross-ply laminate is shown in Fig. 1, which also shows thin interface layers between adjacent laminae. We have also considered the case when all the fibers are in the same direction (not shown). Although it is possible to derive an exact dispersion equation for propagation of harmonic waves in the plane of the plate, the equation is rather complicated when the number of layers is large. This equation simplifies for propagation in the 0° and 90° directions when one also considers small number of layers. In this paper we have considered a five layered plate: three reinforced layers of the same thickness and two interface isotropic soft equal thickness layers. Results are presented using both an analytical technique and the stiffness method.

2. GOVERNING EQUATIONS

We consider time harmonic wave propagation along the x or y axis. Because of symmetry the problem reduces to two uncoupled ones: plane strain in which the displacement components are u_x, o, u_z, and SH when the only nonzero displacement component is u_y. In this paper we consider only the plane strain case.

2.1 Stiffness Method

Consider the ith lamina bounded by $z=z_{i-1}$ and $z=z_i$. The stress strain relation in this lamina will be given by

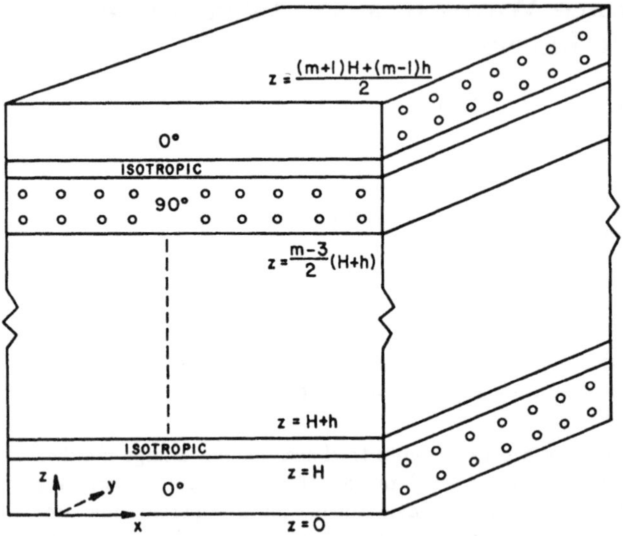

Fig. 1. Geometry of the laminated plate with interface layers. m is the total number of layers (m≥5).

$$\begin{Bmatrix} \sigma_{xx} \\ \sigma_{zz} \\ \sigma_{xz} \end{Bmatrix} = \begin{bmatrix} c_{11}^{(i)} & c_{13}^{(i)} & 0 \\ c_{13}^{(i)} & c_{33}^{(i)} & 0 \\ 0 & 0 & c_{55}^{(i)} \end{bmatrix} \begin{Bmatrix} \epsilon_{xx} \\ \epsilon_{zz} \\ \gamma_{xz} \end{Bmatrix}, \tag{1}$$

where σ_{ij} and ϵ_{ij} are the stress and strain components, respectively, and we have written $\gamma_{xz}=2\epsilon_{xz}$. Note that if the y axis is the axis of symmetry, then

$$c_{11}^{(i)} = c_{33}^{(i)} \text{ and } c_{55}^{(i)} = \frac{1}{2}\left(c_{11}^{(i)} - c_{13}^{(i)}\right).$$

Then the problem is equivalent to that of an isotropic one. In order to get good numerical results each lamina will be divided into several sublayers, p, say. Within the ith sublayer we will choose a local coordinate with the origin at the mid-plane and x_j, y_j, z_j, parallel to the global x,y,z axes, respectively. Let $2h_j$ be the thickness of this sublayer. Denoting $u^{(j)}$ to be the displacement at a point in the jth lamina we write

$$u_x^{(j)} = u_{j-1}f_1 + u_j f_2 + \left(\frac{1}{c_{55}^{(j)}}\chi_{j-1} - \frac{\partial w_{j-1}}{\partial x_j}\right) f_3 + \left(\frac{1}{c_{55}^{(j)}}\chi_j - \frac{\partial w_j}{\partial x_j}\right) f_4 \tag{2}$$

$$u_z^{(j)} = w_{j-1}f_1 + w_j f_2 + \left(\frac{1}{c_{33}^{(j)}}\sigma_{j-1} - \frac{c_{13}^{(j)}}{c_{33}^{(j)}}\frac{\partial w_{j-1}}{\partial x_j}\right) f_3 + \left(\frac{1}{c_{33}^{(j)}}\sigma_j - \frac{c_{13}^{(j)}}{c_{33}^{(j)}}\frac{\partial w_j}{\partial x_j}\right) f_4 ,$$

where f_n (n=1,...,4) are cubic polynomials in the local coordinate z_j given by

$$f_1 = \frac{1}{4}(2-3\eta_j + \eta_j^3),\quad f_2 = \frac{1}{4}(2+3\eta_j - \eta_j^3)$$

$$f_3 = \frac{h_j}{4}(1-\eta_j - \eta_j^2 + \eta_j^3),\ f_4 = \frac{h_j}{4}(-1-\eta_j + \eta_j^2 + \eta_j^3)\,.$$

Here $\eta_j = z_j/h_j$ and u_j, w_j, χ_j, σ_j are the values of u_x, u_z, σ_{xz}, and σ_{zz} at the jth node. These nodal values of the displacement and traction components are functions of x_j (=x) and t. In this paper it will be assumed that the time dependence is of the form $e^{-i\omega t}$, ω being the circular frequency. The factor $e^{-i\omega t}$ will be dropped in the sequel.

The equations governing the nodal generalized coordinates $\{u_j, \chi_j, w_j, \sigma_j\}$ are obtained using Hamilton's principle. The details can be found in [3]. It is found that, if $\{\underset{\sim}{Q}\}$ is the vector of all the generalized coordinates, and if we assume for $\{\underset{\sim}{Q}\}$ the form

$$\{\underset{\sim}{Q}\} = \{\underset{\sim o}{Q}\}e^{ikx} \tag{3}$$

then $\{\underset{\sim o}{Q}\}$ satisfies the equation

$$(k^4[\underset{\sim 1}{K}] - ik^3[\underset{\sim 2}{K}] - k^2[\underset{\sim 3}{K}] + ik[\underset{\sim 4}{K}] + [\underset{\sim 5}{K}])\,\{\underset{\sim o}{Q}\} = 0\,. \tag{4}$$

Matrices $[\underset{\sim 1}{K}]$, etc., have been defined in [3]. $[\underset{\sim 3}{K}]$, $[\underset{\sim 4}{K}]$, and $[\underset{\sim 5}{K}]$ are linear in ω^2.

For nontrivial solution the determinant of the coefficient matrix must be zero. The solution to this equation provides the dispersion relation between k and ω.

2.2 Exact Solution

Consider wave propagation in one of the layers along one of the principal directions as described above. Then we may assume that

$$u_x^{(i)} = U_1(z_i)e^{ikx-i\omega t} \tag{5}$$
$$u_z^{(i)} = U_3(z_i)e^{ikx-i\omega t}\,.$$

Further, if we let

$$\Omega_1^+ = A'e^{-is_1z_i} + A''e^{is_1z_i} \tag{6}$$
$$\Omega_1^- = A'e^{-is_1z_i} - A''e^{is_1z_i}$$
$$\Omega_2^+ = B'e^{-is_2z_i} + B''e^{is_2z_i}$$
$$\Omega_2^- = B'e^{-is_2z_i} - B''e^{is_2z_i}$$

then we may write

$$U_1 = V_1'\Omega_1^+ + V_1''\Omega_2^- \tag{7}$$
$$U_3 = V_3'\Omega_1^- + V_3''\Omega_2^+\,.$$

Here the constants V_1', V_1'', V_3', and V_3'' may be taken as

$$V_1' = ik,\quad V_3' = \frac{k_2^2 - k^2\alpha - s_1^2}{is_1\delta} \tag{8}$$
$$V_1'' = -is_2,\quad V_3'' = -\frac{k_2^2 - k^2\alpha - s_2^2}{ik\delta}$$

s_1^2 and s_2^2 are given by

$$s_{1,2}^2 = \frac{1}{2\beta}[-(k^2\gamma-k_2^2(1+\beta))\pm\{(k^2\gamma-k_2^2(1+\beta))^2-4\beta(k_2^2-k^2)(k_2^2-k^2\alpha)\}^{1/2}] . \tag{9}$$

In the above we have defined

$$\alpha = c_{11}^{(i)}/c_{55}^{(i)},\ \beta = c_{33}^{(i)}/c_{55}^{(i)},\ k_2^2=\rho\omega^2/c_{55}^{(i)}$$

$$\gamma=1+\alpha\beta-\delta^2,\ \delta=1+c_{13}^{(i)}/c_{55}^{(i)} .$$

Note that if the y-axis is the symmetry axis then

$$c_{11}^{(i)} = c_{33}^{(i)},\ c_{13}^{(i)} = c_{11}^{(i)} - 2c_{55}^{(i)} .$$

In that case the solution is the same as that for an isotropic medium.

Using the solution given by equations (7) and (6) and applying the appropriate boundary conditions of continuity of tractions and displacements at the interfaces between the layers, and the traction-free boundary conditions at the free surfaces of the plate, one arrives at the dispersion equation governing k and ω. For the sake of brevity details will be omitted here.

3. NUMERICAL RESULTS AND DISCUSSION

As an application of the techniques described above, we consider a fiber-reinforced plate when the fibers are aligned with the x-direction (0°). Properties of the plate are given in Table 1.

Table 1. Properties of 0° and 90° laminae, and the interface layer. All the stiffnesses are in the units of $10^{11}N/m^2$.

	$\rho(g/cm^3)$	c_{11}	c_{22}	c_{12}	c_{44}	c_{55}
0° lamina	1.2	1.6073	0.1392	0.0644	0.0350	0.0707
Interface	1.8	0.0865	0.0865	0.0475	0.0195	0.0195
90° lamina	1.2	0.1392	1.6073	0.0644	0.0707	0.0350

Figure 2 shows the real and imaginary branches of the frequency spectrum for propagation in the x-direction. Solid lines in this figure are the results of the numerical method and the open circles and x's are the exact solutions. Note that the two agree very well. It is seen that as $k\to\infty$ the slopes of the first symmetric and antisymmetric modes tend to the ratio $V_R/\sqrt{c_{55}/\rho}$, where V_R is the Rayleigh wave velocity in the x-direction. Results by the stiffness method were obtained using 15 sublayers.

To examine the effect of the interface soft layers we then consider a plate composed of three equal thickness unidirectional (0°) fiber-reinforced layers separated by two thin soft layers. Properties of each layer are as in Table 1. Figures 3a-3c show the dependence of the phase velocity in the x-direction on the frequency when there are no interface layers and when the ratio of the thicknesses of an interface layer and a lamina takes two values, 0.1 and 0.2. It is seen that the presence of the soft interface layers is to lower the cut-off frequencies and the phase velocities. Also it lowers the surface wave velocity. The results shown in Fig. 3 are obtained using the exact solution.

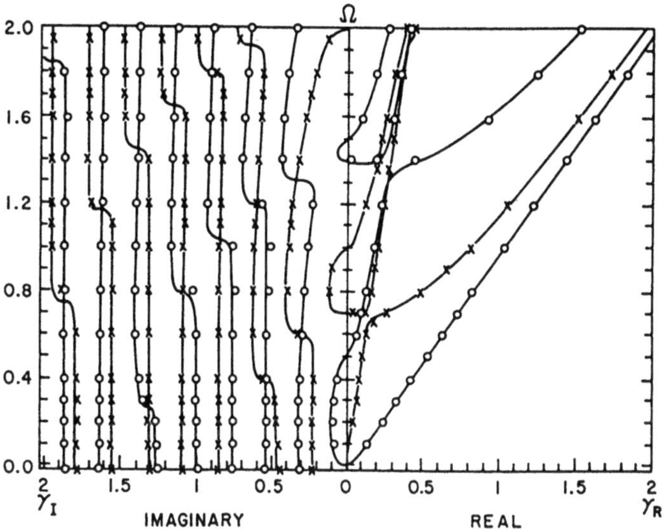

Fig. 2. Dispersion curves for the real and imaginary wave numbers. x: symmetric and o: antisymmetric modes. $\Omega = \frac{\omega H}{2\pi\sqrt{(c_{55}/\rho)_{0}}}$ and $\gamma = \frac{kH}{2\pi}$, H being the thickness of the plate.

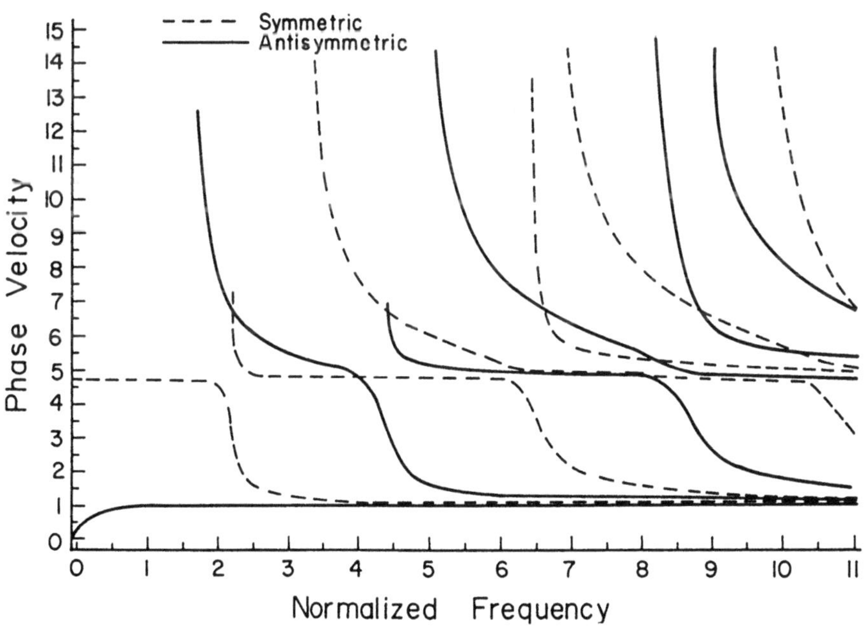

Fig. 3a. Normalized phase velocity ($c=\omega/k\sqrt{c_{55}/\rho}$) vs. normalized frequency ($\frac{1}{2}k_2H=\frac{1}{2}\omega H/\sqrt{c_{55}/\rho}$).

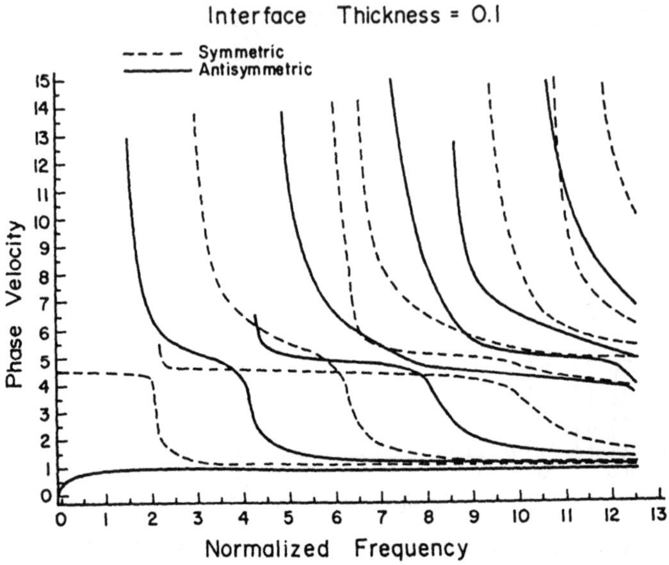

Fig. 3b. Variation of c with $\frac{1}{2}k_2H$ when the interface layer thickness ratio is 0.1.

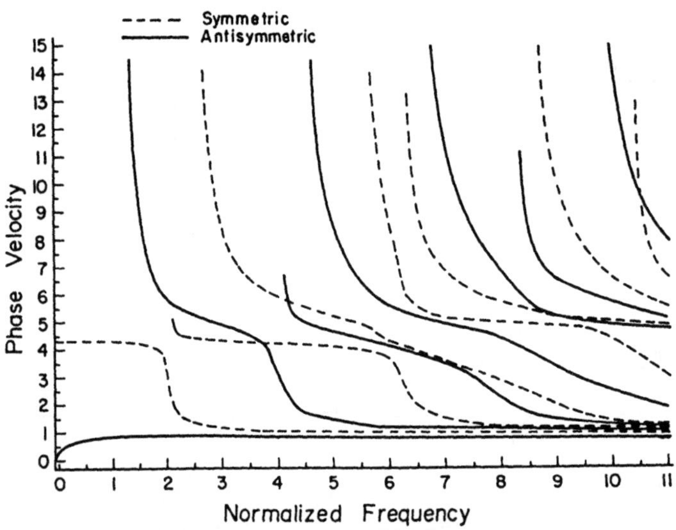

Fig. 3c. Variation of c with $\frac{1}{2}k_2H$ when the interface layer thickness ratio is 0.2.

Using the stiffness method we also studied the dispersion of guided waves in a $0°/90°/0°$ laminated plate with interface layers. Fig. 4 shows the frequency-wave number dependence when the interface layer thickness is 0.1. These curves are quite different from those shown in Fig. 2. To examine the behavior of the first three branches in more detail we show in Fig. 5 $\Omega = \frac{\omega h}{\pi\sqrt{(c_{55}/\rho)}_{\text{isotropic}}}$ vs. $\gamma = \frac{kh}{\pi}$. It is

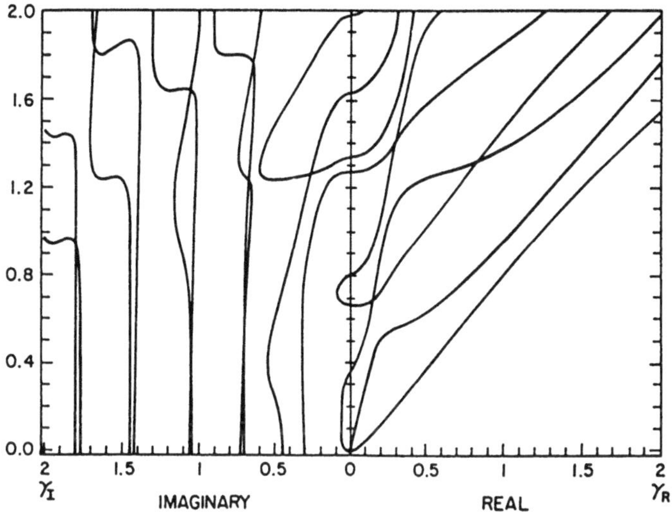

Fig. 4. Dispersion curves for a 0°/90°/0° plate with soft thin interface layers.

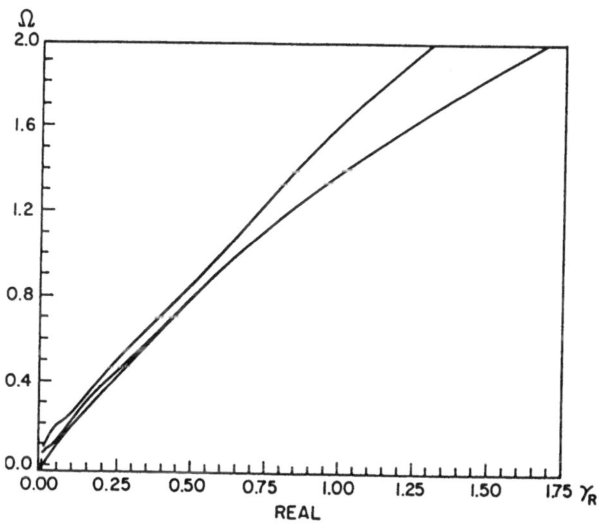

Fig. 5. First three real branches of the frequency spectrum for a 0°/90°/0° plate with soft thin interface layers.

interesting to observe the slowing down of the waves as the wavelength becomes of the order of the interface layer thickness.

4. CONCLUSION

Dispersion of guided waves in a fiber-reinforced laminated plate with or without soft interface layers between the laminae has been studied here using an exact method of solution and an approximate stiffness method. It is shown that the approximate numerical method provides a solution that agrees very well with the exact solution. The advantage of the approximate method is that it is easily used for any number of laminae with arbitrary properties. The results for a laminated plate with soft interface layers show significant lowering of cut-off frequencies and phase velocities. Velocity of surface wave is found to decrease with increasing interface layer thickness.

5. ACKNOWLEDGMENT

The work reported here has been supported by a grant from the Office of Naval Research (#N0014-86-K-0280; Program Manager: Dr. Y. Rajapakse) and grants (MSM-8609813, INT-8521422, and INT-8610487) from the National Science Foundation. Partial support was also received from the Natural Science and Engineering Research Council of Canada (A-7988). The first author gratefully acknowledges the support received from the University of Colorado in the form of a Faculty Fellowship for the academic year 1986-87. Parts of the work were done when he was visiting the University College, Galway, and the Chalmers University of Technology, Gothenburg.

6. REFERENCES

1. J.D. Achenbach: Arch. Mech. 28, 257 (1976).
2. A.H. Shah and S.K. Datta: Int. J. Solids Str. 18, 397 (1982).
3. S.K. Datta, A.H. Shah, R.L. Bratton, and T. Chakraborty: Wave Propagation in Laminated Composite Plates, Report CUMER87-4 (Department of Mechanical Engineering, University of Colorado, Boulder, 1987).

Rayleigh Waves in Materials Reinforced by Two Families of Fibres

D.I. Milosavljević

University "Svetozar Marković", Kragujevac,
Sestre Janića 6, Yugoslavia

1. INTRODUCTION

Detailed account of surface waves in anisotropic elastic materials has been given by CHADWICK and SMITH [1] and particularly in cubic elastic materials by the same authors [2]. SPENCER [3,4] has shown that anisotropic solids may be successfully modelled as fibre-reinforced solids with one or more families of fibres. Wave propagation in an infinite plate reinforced by one family of fibres has been analysed by GREEN [5] and GREEN and MILOSAVLJEVIĆ [6] and they have shown that in the short wave limit the stress-free plate behaves as a semi-infinite region, giving solutions of Rayleigh wave type. An infinite plate reinforced by two families of straight mechanically equivalent fibres has been considered by MILOSAVLJEVIĆ [7] and particularly surface waves in [8]. It has been shown that in the case of inextensible fibres Rayleigh waves can propagate only in special directions, that is in directions perpendicular to one of the families of the fibres.

The object of this paper is to examine the dependence of Rayleigh wave speed on direction of propagation in materials reinforced by two families of mechanically equivalent fibres, focusing attention on the comparison of solutions for inextensible and strongly anisotropic materials. The manner in which the wave speed depends on the direction of propagation for strongly anisotropic materials is shown in both an exact and an approximate form. The perturbation technique has been employed to display the way in which the inextensible solution should be interpreted by comparison with the solution for a strongly anisotropic material.

2. GOVERNING EQUATIONS

Here we deal with a semi-infinite body made of material reinforced by two families of straight mechanically equivalent fibres which lie in planes parallel to the stress-free boundary. Such material will have two preferred directions and since two families of fibres are mechanically equivalent, will have orthotropic symmetry. We suppose that the only anisotropic properties of the material are those which are due to the presence of the fibres which will be characterized by unit vectors $\underset{\sim}{a}$ and $\underset{\sim}{b}$. All vector and tensor components will be referred to a system of rectangular cartesian coordinates x_i $(i=1,2,3)$. Components of the infinitesimal displacement vector $\underset{\sim}{u}$ are denoted by u_i, and components of infinitesimal strain tensor $\underset{\approx}{e}$ by e_{ij} so that $e_{ij}=(u_{i,j}+u_{j,i})/2$ where commas denote partial derivatives. The Cauchy stress tensor $\underset{\sim}{t}$ has components t_{ij}, and the fibre direction vectors $\underset{\sim}{a}$ and $\underset{\sim}{b}$ have components a_i and b_i respectively. The constitutive equation is then (see SPENCER [3,4])

$$\begin{aligned} t_{ij}=&\{\lambda e_{rr}+\gamma_3(a_r a_s e_{rs}+b_r b_s e_{rs})+\gamma_4 a_r b_s e_{rs}\cos 2\phi\}\delta_{ij}+2\mu e_{ij} \\ &+\{\gamma_3 e_{rr}+2\gamma_1 a_r a_s e_{rs}+\gamma_6 b_r b_s e_{rs}+\gamma_5 a_r b_s e_{rs}\cos 2\phi\}a_i a_j \\ &+\{\gamma_3 e_{rr}+\gamma_6 a_r a_s e_{rs}+2\gamma_1 b_r b_s e_{rs}+\gamma_5 a_r b_s e_{rs}\cos 2\phi\}b_i b_j \\ &+\tfrac{1}{2}\{\gamma_4 e_{rr}\cos 2\phi+\gamma_5(a_r a_s e_{rs}+b_r b_s e_{rs})\cos 2\phi+2\gamma_2 a_r b_s e_{rs}\}(a_i b_j+a_j b_i) \\ &+\gamma_7\{a_r(e_{ri}a_j+e_{rj}a_i)+b_r(e_{ri}b_j+e_{rj}b_i)\}, \end{aligned} \tag{1}$$

where $\lambda,\mu,\gamma_1,\ldots,\gamma_7$ are even functions of $\cos 2\phi$ and 2ϕ is the angle between the two fibre directions. Indices take the values 1, 2 and 3 and summation convention is employed.

If we introduce the assumption of inextensibility in the fibre directions $\underset{\sim}{a}$ and $\underset{\sim}{b}$ then the material has two kinematic constraints given as

$$a_i a_j e_{ij} = 0 \quad \text{and} \quad b_i b_j e_{ij} = 0. \tag{2}$$

The constraints of inextensibility give rise to reaction stresses T_a and T_b whose cartesian components are $T_a a_i a_j$ and $T_b b_i b_j$ respectively. The constitutive equation is then given (see SPENCER [3,4]) in the following way:

$$\begin{aligned} t_{ij} =& (\lambda e_{rr}+\gamma_4 a_r b_s e_{rs}\cos 2\phi)\delta_{ij}+2\mu e_{ij} \\ &+\tfrac{1}{2}(\gamma_4 e_{rr}\cos 2\phi+2\gamma_2 a_r b_s e_{rs})(a_i b_j+a_j b_i) \\ &+\gamma_7\{a_r(e_{ri}a_j+e_{rj}a_i)+b_r(e_{ri}b_j+e_{rj}b_i)\}+T_a a_i a_j+T_b b_i b_j. \end{aligned} \tag{3}$$

Thus a material reinforced by two families of inextensible fibres may be regarded as the limiting case of the material reinforced by two families of extensible fibres when $\gamma_1\to\infty$ and $\gamma_6\to\infty$ with $a_i a_j e_{ij}\to 0$ and $b_i b_j e_{ij}\to 0$ in such a way that relations

$$2\gamma_1 a_r a_s e_{rs}+\gamma_6 b_r b_s e_{rs}\to T_a-\gamma_3 e_{rr}-\gamma_5 a_r b_s e_{rs}\cos 2\phi, \tag{4}$$

and

$$\gamma_6 a_r a_s e_{rs}+2\gamma_1 b_r b_s e_{rs}\to T_b-\gamma_3 e_{rr}-\gamma_5 a_r b_s e_{rs}\cos 2\phi, \tag{5}$$

are satisfied.

Choosing cartesian coordinates x_1 perpendicular to both families of fibres, x_2 and x_3 along bisectors of the two families of fibres, we assume that the semi-infinite body occupies the region $x_1 \geq 0$. The equations of motion with no body forces are

$$t_{ij,j} = \rho \ddot{u}_i \ , \quad (i=1,2,3), \tag{6}$$

where ρ is density of the material and dots denote differentiation with respect to time. Equations (6) are to be solved for displacements u_i with harmonic time dependence which leaves the boundary traction free, that is

$$t_{1i} = 0, \ (i=1,2,3), \ \text{at } x_1 = 0, \tag{7}$$

and for which $|\underset{\sim}{u}|$ tends to zero with increasing penetration into the body that is

$$u_i \to 0 \qquad \text{as} \qquad x_1 \to \infty. \tag{8}$$

Such a wave may be represented as a linear combination of plane waves of the form

$$\{u_r, T_a, T_b\} = \{U_r, T_A, T_B\} \exp\{ik(sx_2 + cx_3 - vt)\}, \tag{9}$$

where k is the wave number, v is the speed of propagation, $s \equiv \sin\alpha$, $c \equiv \cos\alpha$, and α is the angle which the wave normal makes with the x_3-axis. Components of the unit vectors $\underset{\sim}{a}$ and $\underset{\sim}{b}$ are $\{0, \sin\phi, \cos\phi\}$ and $\{0, -\sin\phi, \cos\phi\}$ respectively. By removing common factors, the equations of motion (6) take the form

$$c_{11}U_1'' + k^2[\rho v^2 - (s^2c_{66} + c^2c_{55})]\,U_1 + iks(c_{12}+c_{66})U_2' + ikc(c_{13}+c_{55})U_3' = 0,$$

$$iks(c_{12}+c_{66})U_1' + c_{66}U_2'' + k^2[\rho v^2 - (s^2c_{22} + c^2c_{44})]U_2 - ksc(c_{23}+c_{44})U_3 = 0,$$

$$ikc(c_{13}+c_{55})U_1' - k^2sc(c_{23}+c_{44})U_2 + c_{55}U_3'' + k^2[\rho v^2 - (s^2c_{44} + c^2c_{33})]U_3 = 0, \tag{10}$$

for a material with extensible fibres, and

$$a_{11}U_1'' + k^2[\rho v^2 - (s^2a_{66} + c^2a_{55})]U_1 + iks(a_{12}+a_{66})U_2' + ikc(a_{13}+a_{55})U_3' = 0,$$

$$iks(a_{12}+a_{66})U_1' + a_{66}U_2'' + k^2[\rho v^2 - (s^2a_{22} + c^2a_{44})]U_2 - k^2sc(a_{23}+a_{44})U_3$$
$$+ ik\sin\phi[(s\sin\phi + c\cos\phi)T_A + (s\sin\phi - c\cos\phi)T_B] = 0,$$

$$ikc(a_{13}+a_{55})U_1' - k^2sc(a_{23}+a_{44})U_2 + a_{55}U_3'' + k^2[\rho v^2 - (s^2a_{44} + c^2a_{33})]U_3$$
$$+ ik\cos\phi[(s\sin\phi + c\cos\phi)T_A - (s\sin\phi - c\cos\phi)T_B], \tag{11}$$

for a material with inextensible fibres, where a prime denotes differentiation with respect to x_1 and

$$c_{11}=\lambda+2\mu=a_{11},\quad c_{12}=\lambda+\sin^2\phi(2\gamma_3-\gamma_4\cos2\phi),\quad c_{13}=\lambda+\cos^2\phi(2\gamma_3+\gamma_4\cos2\phi),$$
$$c_{22}=\lambda+2\mu+2\sin^2\phi[2\gamma_3-\gamma_4\cos2\phi+2\gamma_7+\sin^2\phi(2\gamma_1+\gamma_6+\gamma_2-2\gamma_5\cos2\phi)],$$
$$c_{23}=\lambda+\cos^2\phi[2\sin^2\phi(2\gamma_1+\gamma_6-\gamma_2)+2\gamma_3+\gamma_4\cos2\phi]+\sin^2\phi(2\gamma_3-\gamma_4\cos2\phi),$$
$$c_{33}=\lambda+2\mu+2\cos^2\phi[2\gamma_3+\gamma_4\cos2\phi+2\gamma_7+\cos^2\phi(2\gamma_1+\gamma_6+\gamma_2+2\gamma_5\cos2\phi)],$$
$$c_{44}=\mu+\gamma_7+2\sin^2\phi\cos^2\phi(2\gamma_1-\gamma_6),$$
$$c_{55}=\mu+\gamma_7\cos^2\phi=a_{55},\quad c_{66}=\mu+\gamma_7\sin^2\phi=a_{66},$$
$$a_{12}=\lambda-\gamma_4\cos2\phi\sin^2\phi,\quad a_{13}=\lambda+\gamma_4\cos2\phi\cos^2\phi,$$
$$a_{22}=\lambda+2\mu-2\gamma_4\cos2\phi\sin^2\phi+2\gamma_2\sin^4\phi+4\gamma_7\sin^2\phi$$
$$a_{23}=\lambda+\gamma_4\cos^2 2\phi-\gamma_2\sin^2 2\phi/2\ ,$$
$$a_{33}=\lambda+2\mu+2\gamma_4\cos 2\phi\ \cos^2\phi+2\gamma_2\cos^4\phi+4\gamma_7\cos^2\phi,\quad a_{44}=\mu+\gamma_7. \tag{12}$$

The constraint conditions (2) become

$$k[sU_2\sin^2\phi\pm(cU_2+sU_3)+cU_3\cos^2\phi]=0, \tag{13}$$

and the boundary conditions (7) become

$$c_{11}U_1'+iksc_{12}U_2+ikc\,c_{13}U_3=0,$$
$$ikcU_1+U_3'=0,$$
$$iksU_1+U_2'=0,\qquad \text{at } x_1=0, \tag{14}$$

for material with extensible fibres whilst for material with inextensible fibres in $(14)_1$ we take a_{ij} instead of c_{ij}.

3. SURFACE-WAVE SOLUTION

3.1. Inextensible Fibres

When the fibres are inextensible the constraint conditions (13) lead to nontrivial solutions for U_2 and U_3 if and only if

$$s^2\sin^2\phi+c^2\cos^2\phi=0, \tag{15}$$

is satisfied, that is when the wave propagates perpendicularly to one of the families of fibres. To obtain nontrivial solutions in that particular direction it is necessary to postulate existence of a singular layer at the boundary, as has been done by GREEN [5], GREEN and MILOSAVLJEVIĆ [6] for plates reinforced by a single family of fibres and MILOSAVLJEVIĆ [7,8] for material reinforced by two families of fibres. In the singular layers the reaction stresses T_a and T_b become infinite, corresponding to finite loads on the boundary. Across a singular layer, the shear stresses t_{12} and t_{13} are discontinuous, being zero at $x_1=0$ but having a non-zero limit as $x_1\to0$ from the interior of the body. With the above assumptions we obtain an expression for Rayleigh wave speed in the form

$$\{e-(c^2a_{13}-s^2a_{12})^2/a_{11}-\rho v^2\}\sqrt{(d-\rho v^2)/(e-\rho v^2)}$$
$$-(\rho v^2-4s^2c^2a_{55}a_{66}/d)\sqrt{d/a_{11}}=0, \tag{16}$$

where

$$e = s^4a_{22}+c^4a_{33}-2s^2c^2a_{23}, \quad d=s^2a_{66}+c^2a_{55}. \tag{17}$$

From equation (16) we obtain, for a chosen material, a Rayleigh wave speed which is subsonic relative to the speed of propagation $(d/\rho)^{\frac{1}{2}}$ of transverse body waves.

3.2. Extensible Fibres

When the fibres are extensible, in order to satisfy condition (8), we suppose

$$U_i = D_i\, e^{-kpx}, \quad (i=1,2,3). \tag{18}$$

On substituting expressions (18) into the equations (10) and removing common factors we arrive at a system of algebraic equations which have non-zero solutions when the following relation is satisfied:

$$\begin{vmatrix} c_{11}(p^2-g_1) & sph_3 & cph_2 \\ -sph_3 & c_{66}(p^2-g_2) & -sch_1 \\ -cph_2 & -sch_1 & c_{55}(p^2-g_3) \end{vmatrix} = 0, \tag{19}$$

where

$$g_1 = (s^2c_{66}+c^2c_{55}-\rho v^2)/c_{11} \equiv (d-\rho v^2)/c_{11},$$
$$g_2 = (s^2c_{22}+c^2c_{44}-\rho v^2)/c_{66},$$
$$g_3 = (s^2c_{44}+c^2c_{33}-\rho v^2)/c_{55},$$
$$h_1 = c_{23}+c_{44},\ h_2=c_{13}+c_{55},\ h_3= c_{12}+c_{66}\,. \tag{20}$$

Equation (19) is a sextic of p with real coefficients and each of the roots p_k, $k=1,2,\dots,6$, is either real or one of a complex conjugate pair. It may be shown, for material with strong fibres, that if $g_1>0$, three solutions for p^2 are real and positive and thus all six roots p_k are real. We confine attention to surface waves which are subsonic relative to the speed of propagation $(d/\rho)^{1/2}$ of transverse body waves and, therefore, we suppose solutions of (19) in the following way

$$p_\alpha>0, \quad (\alpha=1,2,3); \quad p_\beta<0,\ (\beta=4,5,6), \tag{21}$$

and construct the general solution with $p_1<p_2<p_3$ in the form

$$U_j = \sum_{\alpha=1}^{3} D_{j\alpha}\, e^{-kp_\alpha x_1}, \quad (j=1,2,3). \tag{22}$$

We may employ the equations of motion (10) to express the nine constants $D_{j\alpha}$ in terms of three independent constants A_α and, taking the boundary conditions (14), we obtain a secular equation in the form

$$\begin{vmatrix} d_{11} & d_{12} & d_{13} \\ 1+M_1/N_1 & 1+M_2/N_2 & 1+M_3/N_3 \\ 1+M_1/L_1 & 1+M_2/L_2 & 1+M_3/L_3 \end{vmatrix} = 0, \tag{23}$$

where

$$\begin{aligned} M_\alpha &= c_{11}(p_\alpha^2-g_1)h_1-p_\alpha^2 h_2 h_3, \\ N_\alpha &= c_{66}(p_\alpha^2-g_2)h_2+s^2 h_1 h_3, \\ L_\alpha &= c_{55}(p_\alpha^2-g_3)h_3+c^2 h_1 h_2, \\ d_{1\alpha} &= (c_{11}p_\alpha^2-s^2 c_{12}M_\alpha/N_\alpha-c^2 c_{13}M_\alpha/L_\alpha)/p_\alpha. \end{aligned} \tag{24}$$

The secular equation (23) has been derived for $s\neq 0$ and $c\neq 0$. When either $s\neq 0$ or $c=0$ the governing equations are simplified and analysis is similar to the case when the material is reinforced by one family of fibres (see [5,6]). After some calculations (23) may be written in a more convenient form as follows:

$$\begin{aligned} &\rho v^2(p_2p_3+p_1p_2+p_1p_3)+\{s^2c_{66}g_3+c^2c_{55}g_2-c_{11}g_1(g_2+g_3-s^2c_{12}^2/c_{11}c_{66} \\ &-c^2c_{13}^2/c_{11}c_{55})-2s^2c^2h_1-s^2c^2(c_{12}h_2-c_{13}h_3)^2/c_{11}c_{55}c_{66}\} \\ &+g_1(s^2c_{12}^2g_3/c_{66}+c^2c_{13}^2g_2/c_{55}-2s^2c^2c_{12}c_{13}h_1/c_{55}c_{66} \\ &-c_{11}g_2g_3+s^2c^2c_{11}h_1^2/c_{55}c_{66})(p_1+p_2+p_3)/p_1p_2p_3= 0. \end{aligned} \tag{25}$$

For a given material, from (25) we obtain Rayleigh wave speed as a function of direction of propagation.

4. APPROXIMATE SOLUTION

When the material is reinforced with strong fibres the material constants γ_1 and γ_6 are dominant and, to simplify analysis, according to expressions (1), (4) and (5) we may set $\gamma_2=\gamma_3=\gamma_5=\gamma_6=0$ without losing important information in a limiting process by which a material with strong fibres changes into a material with inextensible fibres. By introducing the small parameter $\varepsilon<<1$ in the following way

$$\varepsilon^2=d/2\gamma_1, \tag{26}$$

we may deduce that c_{22}, c_{33}, c_{23}, c_{44}, g_2, g_3 and h_1 are quantities of Order $1/\varepsilon^2$, that is $O(1/\varepsilon^2)$. Furthermore by the method of dominant balance from (19) we deduce that $p_1^2=O(1)$, $p_2^2=O(1/\varepsilon^2)$ and $p_3^2=O(1/\varepsilon^2)$ for any direction of propagation except when the wave propagates perpendicular to one of the families of fibres. Equation (25) then leads to an approximate solution for squared Rayleigh wave speed in the following form

$$\rho v^2 = d - \frac{1}{c_{11}(\bar{p}_2+\bar{p}_3)^2}\{d + \frac{s^2 c_{66}\bar{g}_3 + c^2 c_{55}\bar{g}_2 - 2s^2c^2h_1}{\bar{p}_2\bar{p}_3}\}^2 + O(\varepsilon^4) , \tag{27}$$

where barred quantities are obtained when, in corresponding expressions, we use only the dominant term for the squared wave speed, that is $\rho v^2 = d$, and

$$\bar{p}^2_{2/3} = \{\bar{g}_2+\bar{g}_3 \mp \sqrt{(\bar{g}_3-\bar{g}_2)^2+4s^2c^2h^2/c_{55}c_{66}}\}/2+O(1). \tag{28}$$

When the direction of propagation is perpendicular to one of the families of fibres then $p_2^2 = O(1)$ and, therefore, has to be matched with p_1^2 leading to an expression for Rayleigh wave speed of the form

$$\begin{aligned}&\{e-(c^2c_{13}-s^2c_{12})^2/c_{11}-\rho v^2\}\{(d-\rho v^2)/(e-\rho v^2)\}^{1/2}\\ &-(\rho v^2-4s^2c^2c_{55}c_{66}/d)(d/c_{11})^{1/2} = O(\varepsilon^2),\end{aligned} \tag{29}$$

where

$$e = s^4\hat{c}_{22}+c^4\hat{c}_{33}-2s^2c^2\hat{c}_{23} , \tag{30}$$

and hatted quantities are obtained from corresponding quantities in which dominant tems depending on γ_1 are removed. It is easily seen that when $\varepsilon\to 0$ expression (29) leads to expression (16) for material with inextensible fibres.

5. NUMERICAL RESULTS AND DISCUSSION

To illustrate how Rayleigh wave speed depends on direction of propagation we construct material constants from those measured by MARKHAM [9] for carbon fibre-epoxy resin composite reinforced by one family of fibres in such a way that if two families coincide then we obtain values for one family of fibres. Thus we set constants as follows

$$\begin{aligned}&\lambda = 5.65 \text{ GPa}, \quad \mu=2.46 \text{ GPa}, \quad \gamma_4=-1.28 \text{ GPa}, \quad \gamma_7=3.20 \text{ GPa} ,\\ &2\gamma_1= 11o.45 \text{ GPa}, \quad \gamma_2=\gamma_3=\gamma_5=\gamma_6= 0.\end{aligned} \tag{31}$$

Such material may be regarded as strongly anisotropic and it is not unreasonable to seek approximations based on the assumption that $2\gamma_1$ is a large constant of order $1/\varepsilon^2$, while the remaining elastic constants are of Order one. The variation of scaled wave speed $v\sqrt{\rho/d}$ with angle of propagation α, for material with angle between two families of the fibres $2\phi=40^0$, is given in Fig.1. Solid lines correspond to exact solutions, given in eqn.(25), while broken lines correspond to approximate solutions, given in eqn.(27).

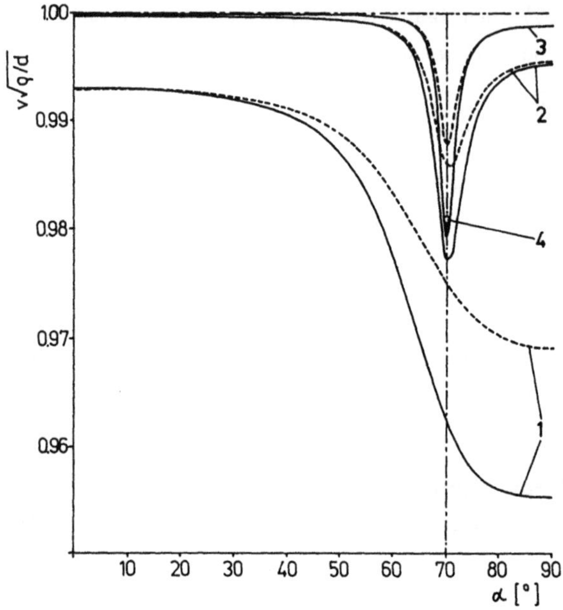

Fig.1. Variation of scaled Rayleigh wave speed with angle of propagation α ; solid lines-exact solutions, broken lines-approximate solutions

Dot (4) corresponds to the solution obtained for material with inextensible fibres. Curves (1) correspond to a material having the material constants listed in (31). To display the behaviour of solutions when the direction of propagation is almost perpendicular to one of the families of fibres for an "almost inextensible" material, we construct a material for which $\gamma_1^* = 25\gamma_1$ (curves (2)) and $\gamma_1^* = 100\gamma_1$ (curves (3)) while all other constants remain fixed. The manner in which curves (2) and (3) behave leads to the conclusion that the solution for a material with inextensible fibres should be interpreted as a solution in which the wave speed is indistinguishable from a transverse body wave speed $\sqrt{d/\rho}$ for any direction of propagation except when the wave propagates perpendicularly to one of the families of fibres. In that particular direction there is a discontinuous jump to the value given by eqn.(16).

ACKNOWLEDGEMENTS - This research was supported by the Science and Engineering Research Council of the U.K. I would like to express my thanks to Dr.W.A. Green for his invaluable comments and suggestions during preparation of this paper.

REFERENCES

1. P. Chadwick, G.D. Smith: In Advances in Applied Mechanics, ed. by C.-S.Yih, Vol. 17 (Academic Press, New York etc.1977) p.303
2. P. Chadwick, G.D.Smith: In Mechanics of Solids, eds.H.G. Hopkins and M.J. Sewell (Pergamon, Oxford 1982) p.47
3. A.J.M. Spencer: In Proceedings of the International Symposium on the Mechanical Behaviour of Structured Media, ed. by A.P.S. Selvadurai (Amsterdam: Elsevier 1981) p.3
4. A.J.M. Spencer: In Continuum Theory of the Mechanics of Fibre-Reinforced Composites, ed A.J.M. Spencer (Springer, Vienna 1984)
5. W.A. Green: Q.J. Mech. Appl. Math., Vol. 35 (1982) p.485
6. W.A.Green, D. Milosavljević: Int J. Solids Structures, Vol.21, No.4, (1985) p.343
7. D. Milosavljević: Wave Propagation in Elastic Plates reinforced by two families of fibres, Ph.D. thesis (Belgrade 1986)
8. D. Milosavljević: In Procedings of XVII Yugoslav Congress of Theoretical and Applied Mechanics (in Serbian),(Zadar 1986) p.1.
9. M.F. Markham, Composites, 1, (1970) p.145

Dispersion of Harmonic Waves in Elastic and Viscoelastic Periodic Composite Materials

Y. Chevalier

I.S.M.C.M., Laboratoire de Rhéologie et de Structures,
3, rue Fernand Hainaut, F-93407 Saint-Ouen, France

A mathematical model is developed to replace any periodic elastic or viscoelastic composite material by a homogeneous medium with the object of using it to describe propagation of plane harmonic waves. The method employed here uses the stress polarization tensor principle applied to a truncated stress and displacement expansion. Spectral lines derived from this approach are exhibited for viscoelastic laminated composites with isotropic or transversely isotropic layers.

INTRODUCTION

Composite materials are used in mechanical engineering for their high specific strength and also for their capacity to soften shocks and sound vibrations. In establishing the mathematical model, several approaches can be adopted.

The first is the exact treatment which includes the field equations of elasticity for each phase. For periodic and unbounded materials for example the solution can be expanded in Bloch expansion see N.TURBE [1], in which each term is a propagation term (Bloch waves). For special periodic structures, mostly for elastic layered media, elementary solutions can be found see S.M.RYTOV [2], C.SVE [3]. In general approximate solutions for propagation terms can be found by variational methods see W.KOHN et al. [4], S.NEMAT-NASSER [5]. As this approach makes the analysis very complicated, it is not of practical interest and thus the development of approximate theories becomes a necessity.

A great many approximate theories which construct continuum models for dynamic analysis have been proposed. In some of them the composite is replaced by a homogeneous, elastic or viscoelastic, anisotropic medium, G.DUVAUT [6]. This theory does not accomodate any dispersion and is valid only when the wave length is large. In the effective stiffness theory, equations are obtained by expanding the displacement for each constituent in power series and introducing the series into a variational functional (Hamilton's principle). This theory has rather complicated equations even for simple composites like layered composites and is also valid for large wave length, R.D.MINDLIN [7], C.T.SUN and al. [8]. Finally, composite materials can be treated by the fundamental equations of the theory of mixtures. The dispersion is accounted for by introducing interactions between the phases, see A.E.GREEN and al. [9], but the difficulty is the choice of interaction terms. McNIVEN et al. [10] suggested matching the second approximate and cut-off frequencies (when they are known).

The approximate theory we propose in this study is quite different from those described above because it needs only one computation for all periodic elastic or viscoelastic composite materials. For both the matrix and the inclusions, displacement and stress are expressed as truncated Fourier expansions on a class of characteristic functions. These functions are specific of the composite material and ensure continuity of the displacement and stress at the interface. Application of a stress polarization principle, see Y. CHEVALIER [11], [12], which is defined on a product space including displacement and stress, yields the displacement equation of motion. For longitudinal waves travelling parallel to layering of viscoelastic laminates or of composites constructed of laminae with unidirectional fibres and a viscoelastic matrix, numerical studies of the phase velocity spectra associated with the resulting theory reveal that the dispersion depends strongly upon the elasticity.

2 HOMOGENEOUS MODEL OF COMPOSITE MATERIALS

We consider a fine periodic structure $\mathcal{M}$ consisting of R linearly elastic (possibly viscoelastic) homogeneous phases Y_r. The HOOKE'S tensor and the mass density, denoted by $c(x)=(c_{ijkl}(\mathbf{x}))$ and $\rho(\mathbf{x})$ are Y-periodic, $Y=\bigcup_{r=1}^{R} Y_r$ is the basic cell. The analysis begins by assuming that all the fields depend on time t, on the coordinate $\mathbf{x}=(x_i)$, i=1,2,3, and on the local coordinate $\mathbf{y}=(y_i)$, $\mathbf{y}\in Y$:

$$\tilde{V}(\mathbf{x},t) = \tilde{\mathbf{V}}(\mathbf{x},\mathbf{y},t)\Big|_{\mathbf{y}=\mathbf{x}} . \tag{1}$$

$\tilde{\mathbf{V}}$ can be stress or displacement and we assume that $\tilde{\mathbf{V}}$ is Y periodic in y. Then $\tilde{V}$ is split into two parts ,$V^{(G)}(\mathbf{x},t)$, which is the mean value of V in Y, and $\tilde{V}^{(L)}(\mathbf{x},\mathbf{y},t)$, which is the variation

$$\tilde{V}(\mathbf{x},\mathbf{y},t) = V^{(G)}(\mathbf{x},t) + \tilde{V}^{(L)}(\mathbf{x},\mathbf{y},t). \tag{2}$$

V is expanded in infinite series in each inclusion Y_r and the truncated expansion is written in Y_r :

$$\tilde{V}^{(L)}(\mathbf{x},\mathbf{y},t) = \sum_{\mathbf{q}\in\mathcal{N}_n} V^{(r)}_{(\mathbf{q})}(\mathbf{x},t)\psi^{(r)}_{\mathbf{q}}(\mathbf{y}). \tag{3}$$

$\psi^{(r)}_{\mathbf{q}}$, $\mathbf{q}\in\mathcal{N}_n$, is a set of orthonormal functions which are specific of the composite material and ensure continuity of the displacement and stress at the interface Γ_r of each inclusion y_r.

$$\psi^{(r)}_{\mathbf{q}}(\mathbf{y}) = \frac{\partial\psi^{(r)}_{\mathbf{q}}(\mathbf{y})}{\partial y_i} = 0 \quad \text{on } \Gamma_r \tag{4}$$

$$\mathcal{N}_n=\left\{\mathbf{q}=(q_1,q_2,q_3)\ ,\ q_i \text{ is an integer} < n,\ i=1,2,3\right\}.$$

Let $\mathcal{M}^{(\circ)}$ be a comparison homogeneous medium,

$u'_i = u_i - u^{(\circ)}_i$ the difference between displacements , (5)

$P_{ij} = \sigma_{ij} - c^{(\circ)}_{ijkl} \frac{\partial u_k}{\partial x_l}$ the stress polarization tensor, (6)

where $\mathbf{c}^{(\circ)} = (c^{(\circ)}_{ijkl})$ is the elastic tensor of the static homogenized medium $\mathcal{M}^{(\circ)}$, $\mathbf{c}^{(\circ)}$ can be obtained by computation [6], or by testing the composite material ([13]). $\mathbf{u} = (u_i)$, $i=1,2,3$ is the displacement in the composite material $\mathcal{M}$ and $\mathbf{u}^{(\circ)} = (u^{(\circ)}_i)$ the displacement in $\mathcal{M}^{(\circ)}$. The stress polarization principle involving the functional $J_\sigma(\mathbf{u}',\mathbf{P})$ for independent variation of the fields $\mathbf{u}'$ and $\mathbf{P}$ expressed as a finite expansion (3) yields the equations of motion

$$c^{(\circ)}_{ijkl} \frac{\partial^2 u^c_k}{\partial x_j \partial x_l} = \rho \frac{\partial^2 u^c_i}{\partial t^2} + \sum_{r=1}^{R} \sum_{q \in \mathcal{N}_n} s_r(\rho_r - \rho) F^{(r)}_{(q)} \frac{\partial^2 u'^{(r)}_{(q)i}}{\partial t^2} \quad (7)$$

$$\sum_{q \in \mathcal{N}_n} \Bigg\{ s_r Q_{(pq)} \left[c^{(\circ)}_{ijkl} \frac{\partial^2 u'^{(r)}_{(q)k}}{\partial x_j \partial x_l} + \frac{\partial P^{(r)}_{(q)ij}}{\partial x_j} \right] + s_r c^{(\circ)}_{ijkl} \times \left[R^{(r)}_{(pq)k} \frac{\partial u'^{(r)}_{(q)l}}{\partial x_j} - R^{(r)}_{(qp)j} \frac{\partial u'^{(r)}_{(q)k}}{\partial x_l} \right] - s_r R^{(r)}_{(qp)j} P^{(r)}_{(q)ij} - s_r c^{(\circ)}_{ijkl} S^{(r)}_{(pq)jl} u'^{(r)}_{(q)k} \Bigg\} = \sum_{q \in \mathcal{N}_n} s_r \rho_r Q^{(r)}_{(pq)} \frac{\partial^2 u'^{(r)}_{(q)i}}{\partial t^2} + s_r(\rho_r - \rho) F^{(r)}_{(q)} \frac{\partial^2 u^c_i}{\partial t^2} \quad (8)$$

$$\sum_{q \in \mathcal{N}_n} \Bigg\{ s_r h_{ijkl} Q^{(r)}_{(pq)} P^{(r)}_{(q)kl} - \frac{s_r}{2} Q^{(r)}_{(pq)} \left[\frac{\partial u'^{(r)}_{(q)i}}{\partial x_j} + \frac{\partial u'^{(r)}_{(q)j}}{\partial x_i} \right] - \frac{s_r}{2} \left[R^{(r)}_{(pq)i} u'^{(r)}_{(q)j} + R^{(r)}_{(pq)j} u'^{(r)}_{(q)i} \right] \Bigg\} = 0 \quad (9)$$

and initial and boundary conditions (for elastic materials), see [12], [13]. In this model there are no body forces, ρ denotes the averaged mixture mass density, s_r, $r=1,2,..,R$ is the volume fraction of the r phase ($\sum_{r=1}^{R} s_r = 1$), $\mathbf{h} = (\mathbf{c} - \mathbf{c}^{(\circ)})^{-1}$ is a compliance tensor and the coefficients are defined by

$$F^{(r)}_{(q)} = \frac{1}{V_r} \int_{Y_r} \psi^{(r)}_{\mathbf{q}}(\mathbf{y}) dy \text{ (mean of } \psi^{(r)}_{\mathbf{q}} \text{ on } Y_r)$$

$$Q^{(r)}_{(pq)} = \frac{1}{V_r} \int_{Y_r} \psi^{(r)}_{\mathbf{p}}(y)\, \psi^{(r)}_{\mathbf{q}}(y)\, dy = \begin{cases} 0 & \text{if } \mathbf{p} \neq \mathbf{q} \\ 1 & \text{if } \mathbf{p} = \mathbf{q} \end{cases}$$

$$R^{(r)}_{(pq)} = \frac{1}{V_r} \int_{Y_r} \psi^{(r)}_p . \nabla\psi^{(r)}_q dy = (R_{(pq)i}) \quad i=1,2,3 \tag{10}$$

$$S^{(r)}_{(pq)} = \frac{1}{V_r} \int_{Y_r} \nabla\psi_q . \nabla\psi_p \quad dy = (S_{(pq)ij}) \quad i,j=1,2,3$$

where V_r is the volume of Y_r. $u_i^{(c)} = u_i^{(\circ)} + u_i^{(G)}$ is the "displacement" in the composite material. The coefficient $\rho_r - \rho$ which results from the heterogeneity of the material describes the coupling in the motion equations due to the dynamic forces. This model can be easily generalized to viscoelastic composites.

3 PROPAGATION OF PLANE HARMONIC WAVES

The stress and displacement equations of motion (7),(8), (9) may be used to study the propagation of plane harmonic waves in arbitrary direction ν. We might consider solutions of the form

$$\mathbf{X} \exp \left[i(k\nu.x - \omega t) \right], \qquad i^2 = -1 . \tag{11}$$

In equation (11), $\mathbf{X}$ is a constant amplitude, k is the wave number, ω the circular frequency and $\nu = (\nu_i)$ the unit vector defining the direction of propagation. The equation (9) is solved in stress polarization tensor $\mathbb{P}$ and equations (7) and (8) lead to the general eigenvalue problem.

$$\left\{ -k^2 \left[A(\omega)\right] + ik \left[B(\omega)\right] + \left[C(\omega)\right] + \omega^2 \left[M\right] \right\} \mathbf{X} = 0, \tag{12}$$

where $\mathbf{X} = \left(\mathbf{u}^{(c)}, \mathbf{u}^{(1)}_{(1)}, \ldots, \mathbf{u}^{(R)}_{(1)}, \ldots, \mathbf{u}^{(1)}_{(N)}, \ldots, \mathbf{u}^{(R)}_{(N)} \right)$ is the generalized displacement, [A] , [B] , [C] and [M] are (3+3RN) matrices. N is the size of the set $\mathcal{N}_n$ and [A] , [C] , [M] are symmetric matrices and [B] is skew symmetric.

-When all the inclusions in the composite material have an elastic behaviour [A] , [B] and [C] are real matrices with constant coefficients.

-When some inclusions have a viscoelastic behaviour, the FOURIER transform of the convolution equations (7),(8), (9) shows that the elastic moduli become complex moduli which depend on the circular frequency ω. The general eigenvalue problem (12), is solved by the LZ algorithm [14] and gives spectral lines $\omega(k)$ or $k(\omega)$ for elastic materials, $k(\omega)$ for viscoelastic materials. The program works on every composite material which is described by a set of orthonormal functions ψ_n and by elastic (or viscoelastic) moduli. For vanishing wave number the limiting phase velocities agree with the constant phase velocities according to the effective modulus theory and also with the limiting phase velocities obtained from exact treatment [12], [13]. The cut-off frequencies ω_c are the roots of the equation

$$\det \left\{ [C] + \omega_c^2 \quad [M] \right\} = 0 . \tag{13}$$

For very large wave number the phase velocities v_∞ satisfy the equation

$$\det \left\{ [A] - v_\infty^2 \; [M] \right\} = 0. \tag{14}$$

4 DISPERSION CURVES FOR PERIODICALLY LAYERED MATERIALS

In this section we will draw spectral lines for two visco-elastic layered materials. The unit cell of the first material consists of two isotropic layers, the second material is a 0°, ± 45°, 90°, angle-ply laminate, each lamina is transversely isotropic (Fig.1). The materials are y_2 periodic and therefore $\mathcal{N}_N=(1,2,\ldots,N)$ and the functions $\psi_n^{(r)}$ depend only on the y_2 coordinate and are defined as LEGENDRE associated functions. In (15) $P_n(z)$ is the LEGENDRE polynomial of order n.

$$\psi_{m+1}^{(r)}(z) = \sqrt{\frac{m^2(q+m+1/2)}{(2q+m)!}} \; (1-z^2)^{q/2} \; \frac{d^q P_{m+q}(z)}{d \; z^q} \tag{15}$$

$r=1,2,\ldots,R \qquad m = 0,1,\ldots, N-1$

$z \in [-1,+1]$ when the local coordinate y_2 belongs to one layer. The ability with which the model predicts the dynamic response of a layered material is assessed in comparing spectra of dilatational waves propagating parallel and perpendicular to layering and S.H. waves in a SUN material [12], [13].

The first composite material studied consists of an infinite sequence (twenty) of two alternating layers, steel and plexi-

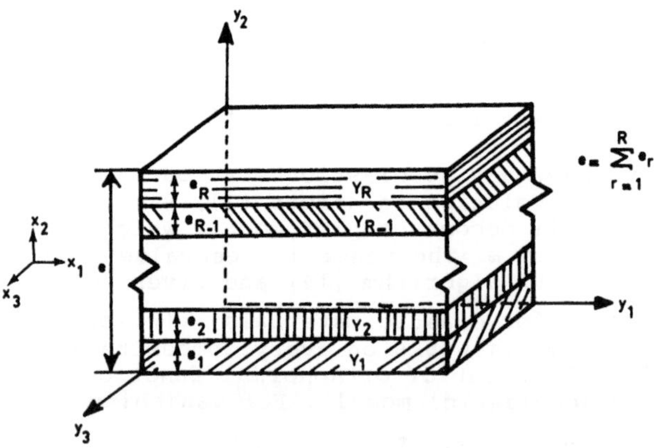

Figure 1 - A unit cell Y of a laminated medium with the local coordinate system (y_1, y_2, y_3)

glass, perfectly bonded to the adjoining layers, the properties of which are specified in Tab.1. For longitudinal waves propagating parallel to layering in the x_1 direction the only non-zero field components (5), (6) in the expression (3) are

$$u_1^{(c)}(x_1,t),\ u'^{(r)}_{1(2n+1)}(y_2,t),\ u'^{(r)}_{2(2n+2)}(y_2,t)$$

$$P^{(r)}_{11(2n+1)}(y_2,t),\ P^{(r)}_{22(2n+1)}(y_2,t),\ P^{(r)}_{33(2n+1)}(t_2,t)$$

$r=1,2$, $n=0,1,2$, $u'^{(r)}_1$ and the stress polarization tensor are symmetric with respect to the mid-planes of each layers; $u'^{(r)}_2$ is antisymmetric. The figure 2 indicates that agreement between experimental data and theoretical data is good. We note that longitudinal waves are more attenuated than the same waves in the plexiglass.

Table 1 - Properties of the laminate steel/plexiglass

r=1,2	thickness e_r (mm)	mass density ρ kg/m^3	stiffness $\lambda_r^* + 2\mu_r^*$ (MPa) ampli	stiffness Loss tg.	shear moduli μ_r^* (MPa) ampli	shear moduli Loss tg.
			static		static	
Layer (1) (plexi.)	3	1185	6500	0	1970	0
			(1MHz)		(1MHz)	
			6780	$1.37\ 10^{-2}$	2120	$2.5\ 10^{-2}$
			3MHz		(3MHz)	
			7150	$1.2\ 10^{-2}$	2260	$1.46\ 10^{-2}$
Layer (2)	2	7800	243000	0	86700	0

The second composite we studed is a 0°, ± 45°, 90° angle ply glass/epoxy composite material. Each layer is supposed to be homogeneous : the material properties used for computation of spectral lines are specified in the figure 3.

For longitudinal waves propagating parallel to the layering (x_1 direction) the non-zero field components (5), (6) in the development (3) of displacement are

$$u_1^{(c)}\ (x_1,t), u^{(r)}_{1(2n+1)}(y_2,t), u'^{(r)}_{2(2n+2)}(y_2,t)$$

$$P^{(r)}_{11(2n+1)}(y_2,t), P^{(r)}_{22(2n+1)}(y_2,t), P^{(r)}_{33(2n+1)}(y_2,t)$$

$$P^{(2)}_{23(2n+2)}(y_2,t),\ P^{(3)}_{23(2n+2)}(y_2,t)$$

$r=1,2,3,4$, $n=0,1$. $u'^{(r)}_1$, $P^{(r)}_{11}$, $P^{(r)}_{22}$, $P^{(r)}_{33}$ are symmetric with respect to the mid-planes of each layer, $u'^{(r)}_2$, $P^{(2)}_{23}$ and $P^{(3)}_{23}$

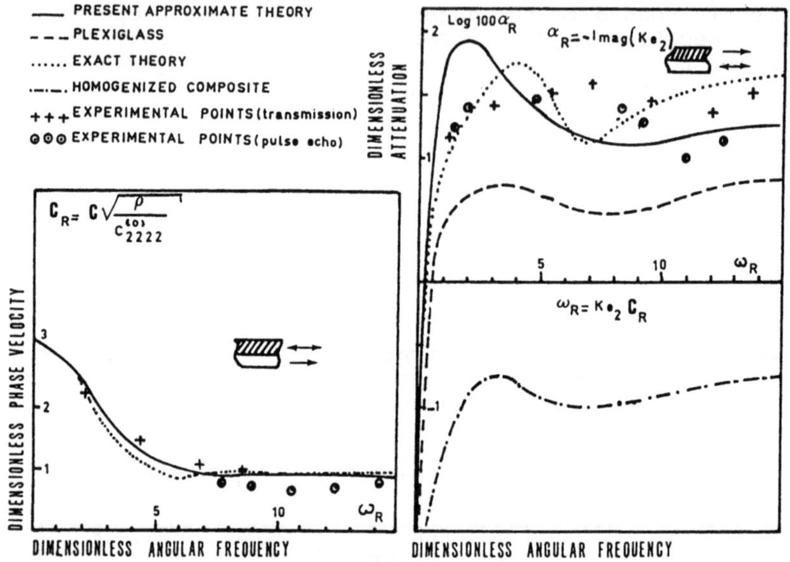

Figure 2 - Dimensionless phase velocity c_R and dimensionless attenuation α_R versus dimensionless angular frequency ω_R for dilatation waves propagating in the direction of the layering of a steel plexiglass medium.

$\rho = 3830 kg/m^3$ (average mass density) $c^{(o)}_{2222} = 10740$ MPa

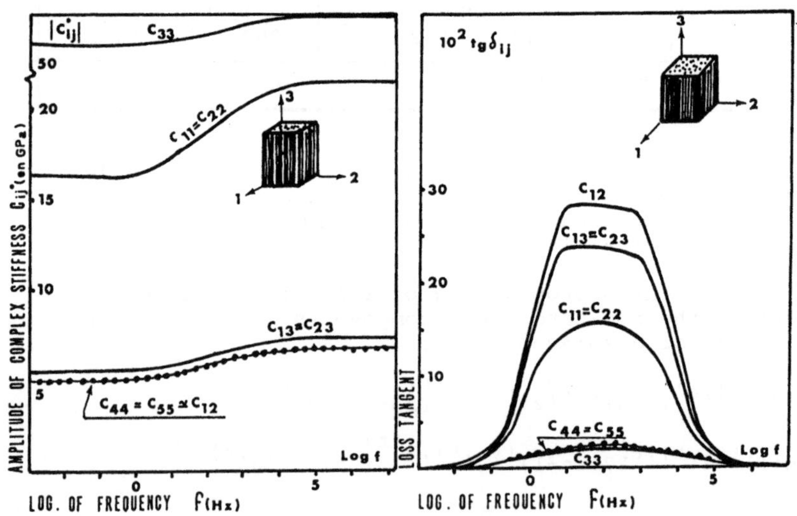

Figure 3 - Amplitude and loss tangent of the complex stiffness C^{*}_{ij} in the principal direction of orthotropy of one layer of a glass/epoxy fiber composite versus the frequency.

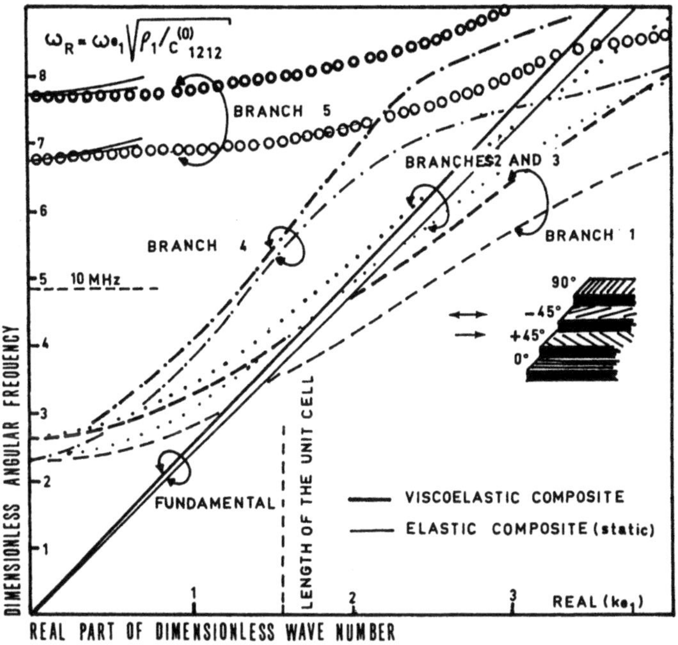

Figure 4 - Frequency spectrum of longitudinal waves for the six lowest propagating modes parallel to layerings of 0°, + 45°, 90° glass/epoxy viscoelastic luminate.

$e_1 = 0.1mm \quad \rho_1 = 2000\ kg/m^3, \quad c^{(o)}_{1212} = 5200\ MPa$

are antisymmetric. In this material the fundamental mode (branch 0) is weakly dispersive: the dispersion is due only to the viscoelastic behaviour of the layers (fundamental modes of elastic material are non-dispersive). The spectral lines (Figure 4) look like spectral lines of a plate |15| and for the higher modes (branches 1 and 2) which are local modes the influence of viscoelasticity on dispersion is very important and cannot be neglected even for weakly viscoelastic composite materials.

5 CONCLUSION

A system of displacement equations of motion has been constructed for every elastic or viscoelastic composite materials and every motion. A significant feature of this theory is its ability to account for dispersion curves of any composite by only one computational routine [12], [13]!

Time harmonic waves travelling obliquely in laminated media [3] can be treated by this approach. In special composites composed of alternating lamination of two homogeneous isotropic elastic materials, and for the lowest modes, the dispersion curves agree over a large range with the exact solution [2], [8], [12]. For infinite wave number this theory leads to the phase velocity of shear waves in the slower medium [2], [12].

R E F E R E N C E S

1.- **N.TURBE** : Application of Bloch expansion to periodic elastic media, Math.Mech. in the Appl. Sci. Vol. 4 (1982) p.433-449.

2.- **S.M.RYTOV** : Acoustical properties of a thinly laminated medium, Soviet Physics Acoustics, vol.3 n°2 (1955) p.68-80.

3.-**C.SVE** : Time harmonic waves travelling obliquely in a periodically laminated medium, J.Appl.Mechanics vol.38, n°2 (June 1971) p.477-482.

4.- **W.KOHN, J.A.KRUMHANSL** and **E.H.LEE** : Variational methods for dispersion and elastic properties of composite materials. J.Appl.Mechanics, vol.39 (June 1972) p.327-336.

5.- **S.NEMAT-NASSER** : General variational methods for waves in elastic composites, J.Elasticity, Vol. 12, n°2 (June 1972) p.73-90.

6.- **G.DUVAUT** : Analyse fonctionnelle et mécanique des milieux continus - Application à l'étude des matériaux composites à structure périodique - homogénéisation, in Theoretical and Applied Mechanics ed. by W.T.KOITER (North-Holland 1976) p.119-131.

7.- **R.D.MINDLIN** : Microsctructure in linear elasticity, Arch. for Rat.Mech. and Analysis, vol.16, (1964) p.51-78.

8.- **C.T.SUN, J.D.ACHENBACH** and **G.HERRMANN** : Continuum theory for a laminated medium, J.Appl.Mechanics, vol.35 (1968) p.467-475.

9.- **A.E.GREEN** and **P.M.NAGHDI** : A theory of mixtures.Arch.Rat. Mech. and Analysis, vol.24 (1967) p.243-263.

10.- **H.D.McNIVEN** and **Y.MENGI** : A mathematical model for the linear dynamic behaviour of two phase periodic materials (p.271-280). A mixture theory for elastic laminated composites (p.281-302). Propagation of transient waves in elastic laminated composites, Int.J.Solids and Structures, vol.15 (1979).

11.- **Y.CHEVALIER** : Application des contraintes et des déformations de polarisation à l'étude des milieux composites, I : formulation générale, J.Mécanique vol. 19 n°2 (1980) p.301-328

12.- **Y.CHEVALIER** : Application des contraintes et des déformations de polarisation à l'étude des milieux composites, II : comportement dynamique, J.Mech. Theor. et Appliquée, vol.3, N°2 (1984) p.223-254

13.- **Y.CHEVALIER** : Dispersion des ondes dans les composites élastiques et viscoélastiques périodiques. Thèse de doctorat d'état (Paris 1986)

14.- **L.KAUFMAN** : The LZ algorithm to solve the general eigenvalue problem for complex matrices [F_2], A.C.M. trans. Software, vol.1 n°3 (1975) p.271-281

15.- **J.MIKLOWITZ** : The theory of elastic waves and waveguides Applied mathematics and mechanics (North-Holland, Amsterdam 1978).

Part VI

Waves on Curved and Fluid-Loaded Surfaces, in Viscoelasticity and in Thermoelasticity

Mathematical and Computational Aspects of the Determination of Elastic Surface Waves Along Cylindrical Cavities of General Cross Section

A. Bamberger[1], *P. Joly*[2], *and M. Kern*[2]

[1]Institut Français du Pétrole, 1-4 ave. de Bois Préau, F-92500 Rueil Malmaison, France

[2]Institut National de Recherche en Informatique et Automatique, Domaine de Voluceau, F-78150 Le Chesnay Cedex, France

In this paper, we present a general mathematical theory for the analysis of guided wave problems. This theory, based on the spectral analysis of bounded self-adjoint operators in Hilbert spaces ([12], [14], [17]), leads to existence results and qualitative properties of the guided modes. It also leads to appropriate numerical methods for actually computing the guided waves (dispersion curves, displacement fields, ...). The outline of our article is as follows. In Section 1, we give the main features of the theory. In Section 2, we describe the approach in more detail for the application to elastic surface waves guided by the exterior of a cylindrical cavity of arbitrary cross section in an isotropic homogeneous medium. The results of this paper were announced in [7] and complete proofs can be found in [8].

1 - GENERAL PRESENTATION OF THE THEORY AND APPLICATIONS

Before entering more general considerations, let us briefly present, to fix some notation and definitions, the model problem we shall treat in detail in Section 2. As illustrated in Fig.1, we consider an infinite cylindrical hole in a homogeneous isotropic solid. The domain of propagation is the exterior of the hole.

The density ρ and the Lamé parameters λ and μ are positive and constant.

By definition, a guided wave is a solution of the elastodynamics equation in the propagation domain which is harmonic with respect to variables x_3 and t, where x_3 is the direction of the hole and t denotes the time, i.e. a solution in the form ($u=(u_1,u_2,u_3)$ denotes the displacement field) :

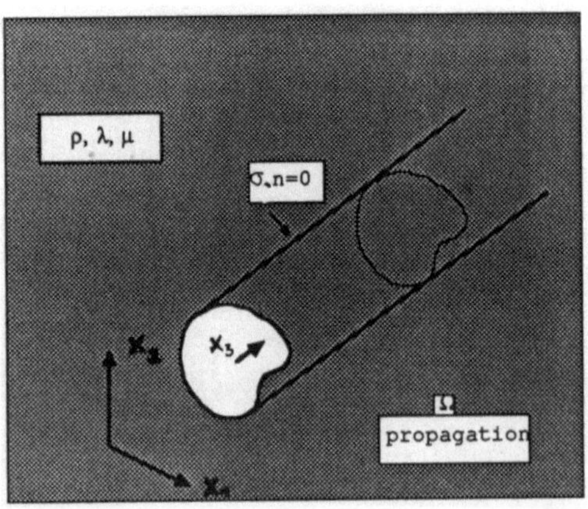

Figure 1 : The model problem.

$$V_P = \left(\frac{\lambda+2\mu}{\rho}\right)^{\frac{1}{2}}$$

pressure wave velocity

$$V_S = \left(\frac{\mu}{\rho}\right)^{\frac{1}{2}}$$

shear wave velocity.

Ω : cross section.

Γ : boundary of Ω.

$$u(x_1,x_2) \exp [i(\omega t - \beta x_3)], \qquad (x_1, x_2) \in \Omega$$

whose amplitude $|u(x_1,x_2)|$ is concentrated at the vicinity of the hole. This property is equivalent, mathematically, to the square integrability condition :

$$\int_\Omega |u(x_1,x_2)|^2 \, dx < +\infty$$

and justifies the name surface wave.

Let us emphasize the fact that the boundary condition we have chosen, namely the free boundary condition, will play a very important role as for Rayleigh waves in a half-space [1]. For instance, with the rigid boundary condition $u|_\Gamma = 0$, such surface waves cannot exist.

However, the mathematical theory we are going to present is very general and, thus, can be applied to a lot of guided wave problems. Let us quote:

- guided waves in heterogeneous fluids ([22], [26]) or solids ([5]),
- fluid-solid interface waves ([6], [24]),
- electromagnetic waves in optical fibers ([3], [4]).

Indeed, the only requirements for the theory to be valid are the following :

(i) - the equations of the problem must be self-adjoint in some functional L^2-type space,

(ii) - the propagation domain is invariant by translation in one direction,

(iii) - the cross section of the domain is a local (compact) perturbation of a reference domain for which the calculations can be made by hand. (In the problems mentioned above, the reference domain is the homogeneous whole space.)

This means that we can consider, for example, anisotropic elastic media but not viscoelastic media.

Some other applications of the theory are under study :

- Surface waves guided by a geometric perturbation of a homogeneous half-space (which is in this case the reference domain) (see [2], [16]).
- Guided waves in piezoelectric media, ...

As we said before, the relevant mathematical theory is the spectral theory of self-adjoint operators. Indeed, the existence of guided waves is shown to be equivalent to an eigenvalue problem :

$$A(\beta)\, u = \omega^2 u,$$

where $A(\beta)$ is a self-adjoint differential operator in the transverse variables (x_1, x_2) in which β appears as a parameter. Then, if there exists a guided wave, it means that the square of the frequency, ω^2, is an eigenvalue of $A(\beta)$ and the natural way to consider the problem is, mathematically,

- for a given β, find ω and u.

The technique for solving this eigenvalue problem is based, as we shall see, on a compact perturbation method (this is why *(ii)* is required) and the so called Max-Min principle (this is why *(i)* is required).

2 - APPLICATION OF THE SPECTRAL THEORY TO A SURFACE WAVE PROBLEM

We now come back to our model problem. As is well known, at least physically, surface waves of type (1.1) can only exist if and only if ω and β satisfy an equation $D(\omega,\beta) = 0$ which is, by definition, the dispersion relation of the waveguide. In general, this relation is not known explicitly and the goal of the abstract theory is, in some sense, to solve $D(\omega,\beta) = 0$ without any expression for the function D. However, in the case of a circular cross section, analytical calculations are available.

2.1 - *THE RESULTS FOR A CIRCULAR CROSS SECTION*

Many authors have been working on the problem ([9], [10], [19]); however, as far as we know, the question was solved completely only five years ago by BOSTRÖM and BURDEN [10]. Let us briefly describe the main steps of the calculations.

(i) - Following MIKLOWITZ [18], the displacement field is decomposed with the help of 3 scalar potentials (ϕ,ψ,χ) :

$$u = \text{grad}\ \phi + \text{rot}(\psi\ e_3) + \text{rot}(\text{rot}(\chi\ e_3)). \tag{2.1}$$

(ii) - One uses polar coordinates (r,θ) to separate the variables in the propagation domain $r \geq a$. This leads to the following expressions for the potentials:

$$\begin{aligned}
\phi(r,\theta) &= \phi(r)\cos n\theta \quad \text{and} \quad -\frac{d^2\phi}{dr^2} + \frac{1}{r}\frac{d\phi}{dr} + \left(\frac{n^2}{r^2} + \beta^2 - \frac{\omega^2}{V_P^2}\right)\phi = 0 \\
\psi(r,\theta) &= \psi(r)\sin n\theta \quad \text{and} \quad -\frac{d^2\psi}{dr^2} + \frac{1}{r}\frac{d\psi}{dr} + \left(\frac{n^2}{r^2} + \beta^2 - \frac{\omega^2}{V_S^2}\right)\psi = 0 \\
\chi(r,\theta) &= \chi(r)\cos n\theta \quad \text{and} \quad -\frac{d^2\chi}{dr^2} + \frac{1}{r}\frac{d\chi}{dr} + \left(\frac{n^2}{r^2} + \beta^2 - \frac{\omega^2}{V_S^2}\right)\chi = 0\ ,
\end{aligned} \tag{2.2}$$

the three unknown functions being coupled by the free boundary conditions.

(iii) - One proves that there is no square integrable solution of (2.2) when $\omega^2 \geq \beta^2 V_S^2$.

(iv) - One shows that for $\omega^2 < \beta^2 V_S^2$, the general expression for L^2-solutions of (2.2) is

$$\begin{aligned}
\phi(r) &= A\ K_n(\alpha_P \tfrac{r}{a}) \qquad & \alpha_P &= \beta a \left(1 - \frac{\omega^2}{\beta^2 V_P^2}\right)^{\frac{1}{2}} \\
\psi(r) &= B\ K_n(\alpha_S \tfrac{r}{a}) & & \\
\chi(r) &= C\ K_n(\alpha_S \tfrac{r}{a}) & \alpha_S &= \beta a \left(1 - \frac{\omega^2}{\beta^2 V_S^2}\right)^{\frac{1}{2}}\ ,
\end{aligned} \tag{2.3}$$

where $K_n(\cdot)$ is the nth order modified Bessel function [21] .

(v) - Inserting the boundary conditions leads to an explicit homogeneous linear system in (A,B,C)

$$M_n(k,\beta,\omega) \begin{pmatrix} A \\ B \\ C \end{pmatrix} = 0$$

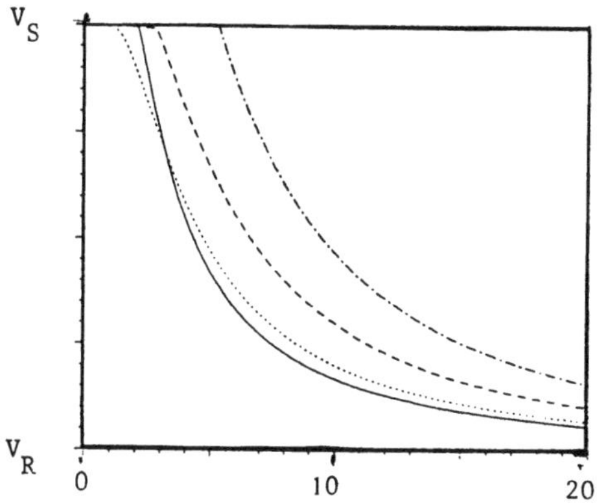

Figure 2 : The five first modes. Circular cross section of radius a. (The speeds $c_1(\beta) = c_2(\beta)$ ——— are coincident).

from which we deduce the dispersion relation

$$\det M_n(k,\beta,\omega) = 0 \; ; \tag{2.4}$$

a transcendental equation which can be solved numerically. One proves that

- for each β, there exist a finite number of guided waves

More precisely, there exists a sequence β_m of thresholds, or cut-off frequencies, such that

(1) $\beta_{m+1} > \beta_m \qquad \beta_m \sim C.m \quad \text{as} \quad m \to +\infty.$

(2) For $\beta \in (\beta_m, \beta_{m+1})$, the dispersion relation has exactly m solutions : $\omega_1(\beta) \leq \omega_2(\beta) \leq \ldots \leq \omega_m(\beta)$.

(3) For each m, the phase velocity $c_m(\beta) = \dfrac{\omega_m(\beta)}{\beta}$ decreases from V_S for $\beta = \beta_m$ to V_R, Rayleigh wave velocity, when $\beta \to +\infty$.

These results are summarized in Fig.2, depicting the dispersion curves $\beta a \to c_m(\beta)$.

One can prove that to the first curve existing for any value of β correspond two guided modes, which means that, in fact, the two first thresholds are equal to 0.

2.2 - *THE GENERAL CASE*

The goal of our work is the extension of most of the results of the circular case to arbitrary cross sections. We shall decompose the mathematical analysis into several steps.

2.2.1 - *Reduction to an eigenvalue problem*

The equations are derived from the classical elastodynamics equations by replacing the differentiation with respect to x_3 by the multiplication by $i\beta$. We get

$$\operatorname{div}^\beta(\sigma^\beta(u)) = \rho\,\omega^2 u \qquad \text{in } \Omega$$

$$\sigma^\beta(u)\cdot n = 0 \qquad \text{on } \Gamma \tag{2.5}$$

where we have set $\left(\sigma^\beta(u) = (\sigma^\beta_{ij}(u))\right)$ (the stress tensor).

$$\sigma^\beta_{ij}(u) = \lambda \operatorname{div}^\beta u\ \delta_{ij} + 2\mu\,\epsilon^\beta_{ij}(u)$$

$$\operatorname{div}^\beta u = \frac{\partial u_1}{\partial x_1} + \frac{\partial u_2}{\partial x_2} + i\beta\, u_3 \tag{2.6}$$

$$\epsilon^\beta_{ij}(u) = \frac{1}{2}\left(\frac{\partial u_i}{\partial x_j} + \frac{\partial u_j}{\partial x_i}\right) \qquad 1\le i,j\le 2$$

$$\epsilon^\beta_{i3}(u) = \epsilon^\beta_{3i}(u) = \frac{1}{2}\left(\frac{\partial u_3}{\partial x_i} + i\beta\, u_i\right) \qquad 1\le i\le 2 \tag{2.7}$$

$$\epsilon^\beta_{33}(u) = i\beta\, u_3.$$

To give the abstract formulation of (2.5), we need a functional framework, so we introduce the Hilbert spaces :

$$H = L^2(\Omega)^3 \qquad V = H^1(\Omega)^3 \text{ (Sobolev space [11], [13])}$$

and consider in H the following scalar product :

$$(u,v) = \int_\Omega \rho\ u\cdot v\ dx \qquad \forall\ (u,v) \in H^2. \tag{2.8}$$

We introduce the unbounded self-adjoint operator ([22]) defined in H by

$$D(A(\beta)u) = \left\{u \in V/\ \operatorname{div}^\beta(\sigma^\beta(u)) \in H;\ \sigma^\beta(u)\cdot n = 0\right\}$$

$$A(\beta)\ u = \operatorname{div}^\beta(\sigma^\beta(u))\ . \tag{2.9}$$

Then, (2.5) is equivalent to writing

$$A(\beta)\ u = \omega^2\ u\,, \qquad u \in D(A(\beta))\,, \tag{2.10}$$

which means that ω^2 is an eigenvalue of $A(\beta)$ and that u is the corresponding eigenvector. We are thus reduced to looking for the eigenvalues and eigenvectors of $A(\beta)$, β playing the role of a parameter. Note that as $A(\beta)$ is unbounded, the question of the existence of eigenvalues is not a trivial problem. For instance, if we replace the free boundary condition by the null displacement condition $u_{|\Gamma} = 0$, one easily sees that there is no eigenvalue.

2.2.2 - *Some classical results in abstract spectral theory*

We give in this section the main mathematical tools we shall use in the next sections. It is useful to introduce the bilinear form associated with $A(\beta)$:

$$\forall\ (u,v) \quad a(\beta;u,v) = \int_\Omega \sigma^\beta_{ij}(u)\ \overline{\epsilon^\beta_{ij}(v)}\ dx\ ;$$

$a(\beta,u,v)$ is symmetric and positive. By Green's formula, we get :

$$a(\beta;u,v) = (A(\beta)u,v), \qquad \forall\ (u,v) \in D(A(\beta)) \times V.$$

The quantity $R(\beta;v) = \frac{a(\beta;v,v)}{\|v\|^2}$ is the Rayleigh quotient associated with $A(\beta)$. Before stating the Max-Min principle we need to recall some definitions of spectral theory.

- *The spectrum of* $A(\beta)$

$$\sigma(A(\beta)) = \{\ \lambda \in \mathbb{R}\ /\ (A-\lambda) \text{ has no bounded inverse } \}$$

- *The point-spectrum of* $A(\beta)$ *(eigenvalues)*

$$\sigma_p(A(\beta)) = \{\ \lambda \in \mathbb{R}\ /\ \exists\, u \neq 0 \in D(A(\beta)),\quad A(\beta)\, u = \lambda u\ \}$$

- *The discrete spectrum of* $A(\beta)$

$$\sigma_d(A(\beta)) = \{\ \lambda \in \sigma_p(A(\beta))\ /\ \lambda \text{ is isolated in } \sigma(A(\beta)) \text{ and } \lambda \text{ is of finite multiplicity}\}$$

- *The essential spectrum of* $A(\beta)$

$$\sigma_{ess}(A(\beta)) = \sigma(A(\beta)) \setminus \sigma_d(A(\beta))$$

One number plays a very important role in the theory, namely the lower bound of the essential spectrum :

$$\sigma_*(\beta) = \inf \sigma_{ess}(A(\beta)) \qquad (\geq 0 \text{ as } A(\beta) \text{ is positive}).$$

The Max-Min principle characterizes all the eigenvalues of $A(\beta)$ which are strictly smaller than $\sigma_*(\beta)$. More precisely, consider the sequence $s_m(\beta) \in \mathbb{R}$ defined by the two equivalent formulas ([14]) :

$$s_m(\beta) = \underset{V_{m-1} \subset H}{\text{Max}} \quad \underset{v \in V_{m-1}^{\perp}}{\text{Min}} \quad R(\beta;v)$$

$$= \underset{V_m \subset V}{\text{Min}} \quad \underset{v \in V_m}{\text{Max}} \quad R(\beta;v)$$

- V_m denotes any m-dimensional subspace (of V or H) and
- $V_m^{\perp} = \{\ u \in V\ /\ (u,v) = 0,\ \ \forall\ v \in V_m\} \subset V.$

In finite dimension or more generally for compact operators, the numbers $s_m(\beta)$ are known as being the eigenvalues of $A(\beta)$. In our case, the situation is slightly different and the precise result is the following [22]:

THEOREM (Max-Min principle)

$$s_1(\beta) \leq s_2(\beta) \leq \ldots \leq s_m(\beta) \leq s_{m+1}(\beta) \leq \ldots \leq \sigma_*(\beta)$$

$$s_m(\beta) \uparrow \sigma_*(\beta) \qquad \text{when } m \to +\infty.$$

For each m, the following alternative holds:

(i) $s_m(\beta) = \sigma_*(\beta)$ $(\Rightarrow s_{m+j}(\beta) = \sigma^*(\beta) \ \forall j \geq 1)$.

Then $A(\beta)$ has at most (m-1) eigenvalues in $]-\infty, \sigma_*(\beta)[$.

(ii) $s_m(\beta) < \sigma_*(\beta)$.

Then $A(\beta)$ has at least m eigenvalues in $]-\infty, \sigma_*(\beta)[$ which are $\{s_1(\beta), s_2(\beta), s_m(\beta)\}$.

Now, the program we shall follow in the remainder of the article is the following :

(1) Find the number $\sigma_*(\beta)$.
(2) Prove that there is no eigenvalue in $]\sigma_*(\beta), +\infty[$.
(3) Prove the existence of eigenvalues in $]-\infty, \sigma_*(\beta)[$, with the help of the Max-Min principle.

2.2.3 – *Determination of the essential spectrum of* $A(\beta)$

Most of the results are obtained thanks to a transformation of the bilinear form $a(\beta,u,v)$ which leads to the following identity (see [8] for details) :

$$a(\beta;v,v) - \beta^2 V_S^2 \|v\|^2 = d(\beta;v,v) + p(\beta;v,v), \tag{2.11}$$

where $d(\beta,v,v)$ is positive :

$$d(\beta;v,v) = \lambda \int_\Omega |\mathrm{div}^\beta v|^2 \, dx + \mu \int_\Omega |\nabla v_3|^2 \, dx + \mu \int_\Omega \left| \frac{\partial v_1}{\partial x_2} + \frac{\partial v_2}{\partial x_1} \right|^2 dx$$
$$+ 2\mu \int_\Omega \left(\left|\frac{\partial v_1}{\partial x_1} + i \frac{\beta}{2} v_3\right|^2 + \left|\frac{\partial v_2}{\partial x_2} + i \frac{\beta}{2} v_3\right|^2 \right) dx \tag{2.12}$$

and $p(\beta;v,v)$ is a compact perturbation of $d(\beta;v,v)$ with variable sign :

$$p(\beta;v,v) = 2\beta \, \mathrm{Im} \left(\int_\Gamma (v_1 n_1 + v_2 n_2) \, \overline{v}_3 \, d\sigma \right). \tag{2.13}$$

Note that the compactness of $p(\beta;v,v)$ is a consequence of the trace theorem from $H^1(\Omega)$ in $H^{\frac{1}{2}}(\Gamma)$ and the fact that, as the boundary (Γ) is bounded, the injection from $H^{\frac{1}{2}}(\Gamma)$ in $L^2(\Gamma)$ is compact.

From (2.11), one can show that the essential spectrum of $A(\beta)$ is the same as the one of the non-perturbed operator $A_0(\beta)$ (without hole) :

$$\sigma_{ess}(A(\beta)) = [\beta^2 V_S^2, +\infty[\ \Rightarrow \sigma_*(\beta) = \beta^2 V_S^2. \tag{2.14}$$

Moreover one proves that if ω^2 is an eigenvalue of $A(\beta)$, the following inequalities hold :

$$\gamma_0 \beta^2 V_S^2 \leq \omega^2 < \beta^2 V_S^2 \qquad (\gamma_0 = \gamma_0(\Omega) < 1). \tag{2.15}$$

- The inequality $\omega^2 \geq \gamma_0 \beta^2 V_S^2$ is a consequence of a generalized Korn's inequality for the bilinear form $a(\beta;v,v)$. ([20]).

- The inequality $\omega^2<\beta^2 V_S^2$ makes use of Rellich's theorem, namely [23] ,

$$\Delta\phi + \lambda\phi = 0 \qquad \text{for } |x| \geq R$$

$$\lambda > 0, \quad \phi \in L^2(|x| \geq R) \quad \Rightarrow \quad \phi(x) = 0 \quad \text{for } |x| > R.$$

In particular this last inequality proves that all the eigenvalues of $A(\beta)$, i.e. all the eigenmodes, can be characterized by the Max-Min principle.

2.2.4 - *The two main existence theorems.*

One can show (see [8] for details) the two following results :

THEOREM 1

For each $m\geq 1$, there exists $\beta_m\geq 0$ such that for $\beta\geq\beta_m$, $A(\beta)$ has at least m eigenvalues:

$$\omega_1(\beta)^2 \leq \omega_2(\beta)^2 \leq \;....\; \leq \omega_m(\beta)^2 \qquad \text{(with } \omega_j(\beta)^2 = s_j(\beta)).$$

THEOREM 2

The first two eigenvalues $\omega_1(\beta)^2$ and $\omega_2(\beta)^2$ exist for any value of β, i.e. $\beta_1=\beta_2 = 0$. Conversely $\beta_4>0$.

The existence proofs use the Min-Max characterization of $s_{m(\beta)}$:

$$s_m(\beta) = \min_{V_m \subset V} \; \max_{v \in V_m} \; \frac{a(\beta;v,v)}{\|v\|^2} ,$$

so, if one can find n test functions v_1, v_2, ..., v_m in V such that, setting $V_m = \text{span}\,[v_1,v_2, ..., v_m]$,

$$\forall\; v \in V_m \qquad a(\beta,v,v) - \beta^2\, V_S^2\, \|v\|^2 < 0,$$

then, from the Max-Min principle, one knows that $s_m(\beta)$ is the mth eigenvalue of $A(\beta)$.

From (2.11), we see that we must find V_m in order to make the sum $d(\beta,v,v) + p(\beta,v,v)$ negative on the subspace V_m. As $d(\beta,v,v)$ is positive, we have to choose v such that $p(\beta,v,v) <0$. Note that, if the boundary term $p(\beta,v,v)$ did not exist - for example when $\Omega = \mathbb{R}^2$ or with the boundary condition $u_{|\Gamma}=0$ - then one would have $a(\beta,v,v) \geq \beta^2\, V_S^2\, \|v\|^2$ for any v in V which would mean that $A(\beta)$ has no eigenvalue.
Of course, the main points of the proof are the construction of the test functions $[v_1,v_2, ..., v_m]$ which is guided by the physical intuition (see [8]).

2.2.5 - *Further theoretical results*

Concerning the cut-off frequencies β_m, one proves the

THEOREM 3

The sequence β_m tends to $+\infty$ when $n \to +\infty$.

This result uses explicitly the Max-Min characterization of the numbers $s_m(\beta)$. It is obtained thanks to a comparison with the case of the circular cross section, for which the result is known (Section 2.1).

Concerning the asymptotic behaviour at high frequency ($\beta \to +\infty$) of the phase velocities $c_m(\beta) = \dfrac{\omega_m(\beta)}{\beta}$, one has

THEOREM 4

For any $m \geq 1$, $\limsup_{\beta \to +\infty} c_m(\beta) \leq V_R$.

This result is obtained, in some sense, via comparison with the case of the half-space.

In fact, the high frequency analysis remains, at least from a purely mathematical point of view, an open question. Our conjecture is :

(i) - If the boundary (Γ) is smooth, the limit velocity of $c_m(\beta)$ is exactly equal to V_R.

(ii) - If the open set Ω presents an acute angle, this limit velocity is, at least for smaller values of m, strictly smaller than V_S.

The point (i) is based on calculations made by WILSON and MORRISON [25] showing that $\lim_{\beta \to 0} c_m(\beta) = V_R$ and that the energy of the corresponding guided wave concentrates at the vicinity of the point of (Γ) of minimal algebraic curvature. The second point (ii) is a consequence of finite element calculations performed by LAGASSE [15], for the case of a wedge.

3 - CONCLUSION AND NUMERICAL ASPECTS

The mathematical analysis appears as a very elegant and powerful tool to prove theoretical results, as existence theorems or asymptotic behaviours, in very general situations, when analytical calculations are not possible. Nevertheless, the theory does not provide all the desired information. That is why, in order to complete the theoretical contribution, it is necessary to use numerical methods to actually compute, for example, the cut-off frequencies, the dispersion curves, the displacement fields, ... etc.

Two numerical approaches seem rather natural. The first one is directly derived from the mathematical approach and consists in approximating the partial differential equation $A(\beta)u = \omega^2 u$ in the unbounded domain Ω. This leads to two successive approximations :

1) Reduce the domain of calculation to a bounded domain Ω_R, "*sufficiently large*" (for instance $\Omega_R = \{x \in \Omega / |x| < R\}$) and impose a priori boundary conditions on the artificial boundary Γ_R (for example, u=0 or $\sigma(u).n=0$).

2) Discretize the domain Ω_R and approximate the differential operator $A(\beta)$ by a discrete operator $A_h(\beta)$ with the finite element method (h denotes the approximation parameter).

Thus, one is led to a classical algebraic eigenvalue problem :

$$A_h(\beta)\, u_h = \omega^2\, u_h \qquad u_h \in V_h \qquad \text{(finite dimension space).}$$

Note that the "*spectral*" consequences of the approximations 1) and 2) are the following :

- the spectrum of $A_h(\beta)$ is purely punctual (because of the domain truncation).
- the spectrum of $A_h(\beta)$ is bounded (because of the discretization of the equations).

This observation is very important in interpreting the numerical results and in discriminating between eigenvalues corresponding to the continuous spectrum of $A(\beta)$ and eigenvalues actually approximating eigenvalues of $A(\beta)$.

A second approach, that we expect to be more accurate and efficient than the first one, consists in reducing the problem to the boundary (Γ) with the help of the integral equation method. One proves that ω^2 is an eigenvalue of $A(\beta)$, if and only if there exists $(\omega,\beta) \in \mathbb{R}^2$ and a function ϕ defined on (Γ) such that

$$T(\omega,\beta)\ \phi = 0, \qquad\qquad \phi \neq 0,$$

where $T(\omega,\beta)$ is a singular integral operator whose kernel can be deduced from the fundamental solution $G(\omega,\beta)$ of the operator $A(\beta)-\omega^2$ in the whole space. Moreover, one shows that the function ϕ is the trace on (Γ) of the displacement field associated with the guided mode.

As the specturm of $T(\omega,\beta)$ is made purely of eigenvalues we deduce that

$$\left| \begin{array}{c} \omega^2 \text{ is an eigenvalue of } A(\beta) \\ \Leftrightarrow \\ 0 \text{ is eigenvalue of } T(\beta,\omega) \end{array} \right.$$

The principle of the numerical methods is then very natural. It contains two steps :

1) Find the eigenvalues $\{\lambda_1(\omega,\beta),\ \lambda_2(\omega,\beta),\ \ldots\}$ of $T(\beta,\omega)$.

2) Solve the equations $\lambda_k(\omega,\beta) = 0, \qquad k=1,2,3\ \ldots$.

One obtains a finite dimensional problem with the help of the boundary element method.

Indeed, in this approach, the problem to be solved becomes non-linear but, on the other hand, its dimension is reduced from 2 to 1 and any truncation problem is avoided.

The implementation of the second approach is now under study at INRIA.

4 - REFERENCES

1 ACHENBACH J.D. : "*Wave propagation in elastic solids*", North-Holland, Amsterdam (1973).

2 AULD B.A. : "*Acoustic fields and waves in solids*", I and II, Wiley-InterScience, New York (1973).

3 BAMBERGER A., BONNET A.S. : *"Calcul des modes guidés d'une fibre optique"* Analyse mathématique, Ecole Polytechnique, CMAP. Rapport interne n°143 (1986).

4 BAMBERGER A., BONNET A.S., DJELLOULI R. *"Calcul des modes guidés d'une fibre optique. Différentes formulations mathématiques du problème"*. Ecole Polytechnique. CMAP. Rapport interne n°142 (1986).

5 BAMBERGER A., DERMENJIAN Y., JOLY P. : *"Ondes guidées en milieu élastique tridimensionnel hétérogène : analyse mathématique"* Rapport INRIA à paraître.

6 BAMBERGER A., JOLY P., KERN M., NICOLETIS L., QUIBLIER J. : *"Hole geometry and anisotropy effects on tube wave propagation : a quasi-static study"*. IFP Report 35-529. Submitted to Geophysics (1987).

7 BAMBERGER A., JOLY P., KERN M., *"Propagation d'ondes élastiques guidées par la surface libre d'une cavité de section arbitraire"* Note aux C.R.A.S., t.304. Série I n°3 (1987).

8 BAMBERGER A., JOLY P., KERN M. :*"Etude mathématique des modes guidés par l'extérieur d'une cavité"* Rapport INRIA n° 650 (1987).

9 BIOT M., : *"Propagation of elastic waves in a liquid-filled cylindrical bore, surrounded by an elastic solid"*, J. Appl. Physics 24, 515-521 (1953).

10 BOSTRÖM A., BURDEN A.D. : *"Propagation of elastic surface waves and their excitation by a point force"*, J. Acoust. Soc. Amer. 72, 998-1004 (1982).

11 BREZIS H. : *"Analyse fonctionnelle : Théorie et Applications"*. Masson (1983).

12 COURANT R., HILBERT D. : *"Methods of Mathematical Physics"*, Vol. I, Wiley, New York (1962).

13 DAUTRAY R., J.L LIONS. : *"Analyse mathématique et calcul numérique pour les sciences et les techniques"*, Vol. 2, Masson, Paris (1985).

14 DUNFORD N., SCHWARTZ J.T. : *"Linear operators"*, Part II, Wiley, New York (1963).

15 LAGASSE P.E. : *"Higher order finite elements analysis of topographic guides supporting elastic surface waves"*, J. Acoust. Soc. Amer. 53,pp. 1116-1122 (1973).

16 MARADUDIN A, A. : *"Surface waves on rough surfaces"*, Proceedings of the Second International Conference in Plasmas and Solids, Ohrid, Yugoslavia, Sept. 5-11 (1985).

17 KATO T. *"Perturbations theory for linear operators"*, Springer, Berlin (1966)

18 MIKLOWITZ J. : *"The theory of elastic waves and waveguides"*, North-Holland, Amsterdam (1980).

19 MINDLIN I.A. : *"Free elastic waves on the surface of a tube of infinite thickness"*, J. Appl. Math. Mech., 27, pp.823-828 (1963).

20 NITSCHE J.A. : *"On Korn's second inequality"*, RAIRO Analyse Numérique 15, pp. 237-248 (1981)

21 OLVER F. : *"Asymptotics and special functions"*, Academic Press, New York (1974).

22 REED M., SIMON B. : "*Methods of modern mathematical physics*", Vols. I-IV, Academic Press, New York (1981).

23 RELLICH F. : "*Über das asymptotische Verhalten der Lösungen von Δu+ku=0 in unendlichen Gebieten*", Jber. Deutsche Math. Verein. 53, pp. 57-64 (1943).

24 WHITE J.E.: "*Underground Sound*" . Elsevier, Amsterdam (1983).

25 WILSON L.O., MORRISON J.A. : "*Propagation of high frequency elastic surface waves along cylinders of general cross section*", J. Math. Phys. 16, pp. 1695-1805 (1975).

26 WILCOX C.H. : "*Sound propagation in stratified fluids*" Applied Mathematical Sciences, Vol. 50, Springer-Verlag, Berlin (1984).

Low Frequency Propagation of Elastic Waves in a Fluid-Filled Borehole

A. Bamberger[1], P. Joly[2], M. Kern[2], L. Nicoletis[1], and J. Quiblier[1]

[1]Institut Français du Pétrole, 1-4 ave. de Bois Préau, F-92500 Rueil Malmaison, France

[2]Institut National de Recherche en Informatique et Automatique, Domaine de Voluceau, F-78150 Le Chesnay Cedex, France

1 - INTRODUCTION

A tube wave is the lowest order mode propagating in a fluid-filled borehole. It is the only one without a cut-off frequency and is similar to the Stoneley interface wave. It has been studied, for instance by WHITE [1] in the case of a circular cylinder.

In this paper, we use abstract variational methods to study the low frequency velocity of this wave, in the case of an arbitrary hole geometry. The most notable result is that this limit velocity is always smaller than the velocity of the fluid. Its computation involves the solution of a static exterior problem in the elastic medium.

An outline of the paper is as follows : in Section 2 we give the equations of the problem, and state the various hypotheses underlying our work. We give the main result in Section 3, together with various comments and the proof. We conclude in Section 4 with some applications.
More complete results, and several applications, may be found in NICOLETIS et al. [2], to which we refer the reader.

2 - STATEMENT OF THE PROBLEM

We consider a fluid-filled borehole embedded in a solid medium. The hole is assumed to be invariant along the x_3 direction, but may have an arbitrarily shaped cross-section in the (x_1,x_2) plane.

The inner medium in Ω^I is an isotropic fluid, described by its density ρ_F, and its velocity c. The outer medium is perfectly elastic, and in this work will be assumed to be homogeneous and isotropic. It is characterized by its density ρ_S and its Lamé constants λ and μ. An extension to anisotropic and locally heterogeneous media is considered in NICOLETIS et al. [2].

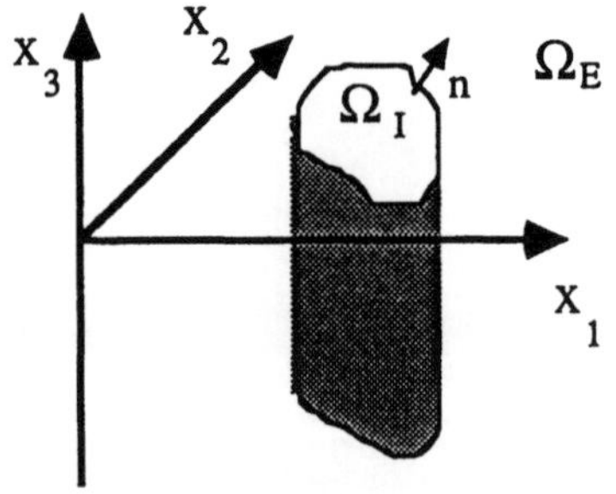

Figure 1 : The geometry of the problem.

Ω : cross section ; Γ : boundary of Ω ; S : area of cross section ; n : unit normal.

The unknown field in the fluid will be the pressure p and in the solid we shall be looking for the displacement field u. All quantities will be assumed to propagate in the x_3 direction with a propagation constant (axial wave number) β, and to be harmonic in time with frequency ω. The factor $\exp[i(\omega t-\beta x_3)]$ will be suppressed in all that follows, so that u and p are only functions of (x_1,x_2). The velocity of such a wave is ω/β.

With this notation, the equations in the fluid in Ω^I are simply given by

$$-\Delta p + \beta^2 p = \frac{\omega^2}{c^2} p \qquad \text{in } \Omega^I, \tag{1}$$

where Δ denotes the Laplace operator in 2 dimensions.

The equations of motion in the solid are

$$\frac{\partial \sigma^\beta_{ij}(u)}{\partial x_j} - i\beta\, \sigma^\beta_{i3}(u) = -\rho_s \omega^2 u_i, \qquad i = 1,2,3, \quad \text{in } \Omega^E, \tag{2}$$

where the stress tensor σ^β is linked to the strain tensor ϵ^β by Hooke's law :

$$\sigma^\beta_{ij}(u) = \lambda\, \epsilon^\beta_{kk}(u)\, \delta_{ij} + 2\mu\, \epsilon^\beta_{ij}(u) \tag{3}$$

and ϵ^β is given by

$$\left|\begin{array}{ll} \epsilon^\beta_{ij}(u) = \frac{1}{2}\left(\frac{\partial u_i}{\partial x_j} + \frac{\partial u_j}{\partial x_i}\right) & i,j = 1,2 \\ \epsilon^\beta_{i3}(u) = \frac{1}{2}\left(\frac{\partial u_3}{\partial x_i} - i\,\beta u_i\right) & i = 1,2 \\ \epsilon^\beta_{33}(u) = -i\beta\, u_3. & \end{array}\right. \tag{4}$$

The superscript β indicates that in both σ and ϵ, derivatives with respect to β have been replaced by multiplications by $(-i\beta)$.

At the interface Γ, the boundary conditions are continuity of normal displacement :

$$\frac{1}{\rho_F}\frac{\partial p}{\partial n} - \omega^2(u\cdot n) = 0, \tag{5}$$

and continuity of the normal stress :

$$\sigma^\beta(u)\cdot n + p\cdot n = 0. \tag{6}$$

We shall say that (p,u) is a guided mode if it is a (suitably smooth) solution of (1)-(6), such that

$$\int_{\Omega^E} |u|^2\, dx < \infty.$$

This condition means that the energy of the field remains confined to the vicinity of the borehole and the smoothness requirement on (p,u) simply means that the fields are sought in a Sobolev space.

The existence of guided modes is only possible for certain values of ω and β satisfying a *"dispersion relation"*. It is not possible, in general, to give its analytical form. Following BAMBERGER et al. [3], [7], we use abstract methods to derive its properties. This problem is studied by JOLY [4], where it is shown that the dispersion curves $\beta \to \omega(\beta)$ consist of a countable hierarchy of modes (see also PAILLET and WHITE [5]). All but the first among them have a cut-off frequency above which they do not propagate (they are called *"normal modes"* in [5], and correspond to *"Pseudo-Rayleigh"* waves), whereas the first one may propagate for low frequencies, and resembles the Stoneley wave that propagates at the interface of two elastic half-spaces. It is called the *"tube-wave"*. Its existence depends on the values of c and V_s (see below).

We denote this tube-wave by $(u(\beta),\ p(\beta))$, and its frequency by $\omega(\beta)$. Following Joly, we know that u,p and ω are continuous functions as $\beta \to 0$ (or equivalently as $\omega \to 0$, since we have $\omega^2 \le \beta^2 V_s^2$). We denote the limits by (u^0, p^0) :

$$\lim_{\beta \to 0} u(\beta) = u^0 \qquad \lim_{\beta \to 0} p(\beta) = p^0.$$

In addition, we assume that these functions are differentiable, for β sufficiently small. The question we wish to address is the following :

What is the limit of the velocity of the tube wave as the frequency goes to 0 ? (Assuming the limit exists.)

Let us note : $V_{st} = \lim_{\beta \to 0} \dfrac{\omega(\beta)}{\beta}$.

3 - THE MAIN RESULT

Theorem 1

$$V_{st}^2 = \frac{c^2}{1 + \dfrac{\rho_F c^2}{S} \displaystyle\int_\Gamma \frac{1}{p^0} (u^0 \cdot n)\, d\sigma} , \tag{7}$$

where p^0 is the constant pressure in the fluid, and u^0 is the solution of the static problem :

$$\left| \begin{array}{lll} \dfrac{\partial}{\partial x_j} \sigma^0_{ij}(u^0) = 0 & i = 1,2 & \text{in } \Omega^E \\ \sigma(u^0) \cdot n + p^0 n = 0 & & \text{on } \Gamma. \end{array} \right. \tag{8}$$

A few comments are in order :

i) The expression (7) is independent of p^0, since the displacement u^0 is proportional to p^0 (we may take $p^0 = 1$).

ii) The value of the Stoneley velocity is always smaller than the velocity of the fluid. This results from Green's formula :

$$p^0 \int_\Gamma (u^0 \cdot n)\, d\sigma = \int_{\Omega^E} \sigma_{ij}(u^0)\, \epsilon_{ij}(u^0)\, dx > 0.$$

iii) This result was obtained under the assumption that the tube wave exists at low frequency. The result is thus only valid as long as we have

$$V_{st} < V_s,$$

i.e. the tube wave is a true guided wave.

It remains to be proved that if this inequality is valid, then the tube wave does propagate at low frequency. We now turn to the proof of theorem 1.

Proof of Theorem 1

We shall divide it into several steps, for each of the quantities involved :

α) **Limit pressure.**

We begin by letting $\beta \to 0$ in (1). As we have already noted, we also have $\omega \to 0$, so that $p^0 = \lim_{\beta \to 0} p(\beta)$ satisfies

$$-\Delta p^0 = 0.$$

Next, we use the transmission condition (5) to find that

$$\frac{1}{\rho_F} \frac{\partial p^0}{\partial n} = 0.$$

Thus, p^0 is solution of a homogeneous Neuman problem :

$$p^0 \text{ is a constant in } \Omega^I. \tag{9}$$

β) **Behavior of the strain and stress tensors**

We first go to the limit in (4). We have

- For i,j = 1,2 : $\epsilon^{\beta}_{ij}(u) \to \epsilon_{ij}(\overline{u^0}) = \frac{1}{2} \left(\frac{\partial u_i^0}{\partial x_j} + \frac{\partial u_j^0}{\partial x_i} \right)$
- For i = 1,2 : $\epsilon^{\beta}_{i3}(u) \to \frac{1}{2} \frac{\partial u_3}{\partial x_i}$

  $\epsilon^{\beta}_{33}(u) \to 0,$

where $\overline{u^0}$ denotes the transverse part of u_0.

This result gives the limiting stress :

$$\left| \begin{array}{lll} - \ \sigma^{\beta}_{ij}(u) & \to \lambda[\epsilon_{11}(\overline{u^0}) + \epsilon_{22}(\overline{u^0})]\ \delta_{ij} + 2\mu\,\epsilon_{ij}(\overline{u^0}) & i,j = 1,2, \\ - \ \sigma^{\beta}_{i3}(u) & \to \mu \dfrac{\partial u_3^0}{\partial x_i} & i = 1,2, \\ - \ \sigma^{\beta}_{33}(u) & \to 0. & \end{array} \right.$$

γ) Limit equation for $\overline{u}$

Using the previous results, we let $\beta \to 0$ in (2), for i,j = 1,2 we get

$$\frac{\partial}{\partial x_j} \sigma_{ij}(\overline{u^0}) = 0 \qquad i = 1,2 \tag{10}$$

and using only the first two components of (6) we obtain boundary conditions for $\overline{u}$:

$$\sigma(\overline{u^0})\cdot n + p^0 n = 0. \tag{11}$$

$\overline{u^0}$ is solution of (10)-(11), which is a static 2D problem. One should note that $\overline{u^0}$ is decoupled from u_3^0, which does not enter the low frequency result at all. This is not surprising, since the same technique shows that u_3^0 is solution of

$$\left| \begin{array}{ll} -\Delta u_3^0 = 0 & \text{in } \Omega^E \\ \dfrac{\partial u_3^0}{\partial n} = 0. & \end{array} \right. \tag{12}$$

It is possible to show that $u_3^0 \in L^2(\Omega^E)$, and since we have an exterior problem, one must have $u_3^0=0$.

δ) Derivation of the limiting velocity

This is the bulk of the proof. It results from a variational formulation of the inner problem. Multiplv eq.(1) by a test function q, and use Green's formula to obtain

$$\int_{\Omega^I} \nabla p\ \nabla q\cdot dx - \int_{\Gamma} \frac{\partial p}{\partial n} q\ d\sigma + \beta^2 \int_{\Omega^I} pq\ dx = \frac{\omega^2}{c^2} \int_{\Omega^I} pq\ dx, \qquad \forall\ q,$$

where $\nabla p = (\frac{\partial p}{\partial x_1}, \frac{\partial p}{\partial x_2})$ is the gradient of p.

Using the boundary condition (5), we transform the integral over Γ :

$$\int_{\Omega^I} \nabla p\ \nabla q\ dx + \beta^2 \int_{\Omega^I} pq\ dx = \rho_F\ \omega^2 \int_{\Gamma} (u\cdot n)\ q d\sigma + \frac{\omega^2}{c^2} \int_{\Omega^I} (pq)\ dx, \qquad \forall\ q. \tag{13}$$

The next step is to differentiate (13) with respect to β^2. For brevity we note $\dot{p} = \frac{\partial p}{\partial \beta^2}$, $\dot{u} = \frac{\partial u}{\partial \beta^2}$. Then

$$\int_{\Omega^I} \nabla\dot{p}\nabla q \, dx + \beta^2 \int_{\Omega^I} \dot{p}q \, dx + \int_{\Omega^I} pq \, dx \tag{14}$$

$$= \frac{d\omega^2}{d\beta^2} \left[\rho_F \int_\Gamma (u{\cdot}n)q \, d\sigma + \frac{1}{c^2}\int_{\Omega^I} pq \, dx\right] + \omega^2 \left[\rho_F \int_\Gamma (\dot{u}{\cdot}n)q \, d\sigma + \frac{1}{c^2}\int_{\Omega^I} \dot{p}q \, dx\right].$$

In the low frequency limit, β and ω^2 go to zero, then (14) reduces to

$$\int_{\Omega^I} \nabla\dot{p}(0) \nabla q \, dx + \int_{\Omega^I} p^0 q \, dx = \frac{d\omega^2}{d\beta^2}\left(\rho_F \int_\Gamma (\overline{u}^0 n) \, q \, d\sigma + \frac{1}{c^2}\int_{\Omega^I} p^0 q \, dx\right).$$

The choice of $q=p^0$ (constant after (9)) eliminates the first term in this equation. Solving for $\frac{d\omega^2}{d\beta^2}$ gives

$$\frac{d\omega^2}{d\beta^2}(0) = \frac{\int_{\Omega^I} |p^0|^2 \, dx}{\rho_F \int_\Gamma (\overline{u}^0{\cdot}n) \, p^0 \, d\gamma + \frac{1}{c^2}\int_{\Omega^I} |p^0|^2 \, dx} . \tag{15}$$

Since p^0 is constant $\int_{\Omega^I} |p^0|^2 \, dx = |p^0|^2 \times S$, where S is the cross section of the hole. The right hand side of (15) is the same one as in (7).

Assuming a Taylor expansion for $\omega^2(\beta^2)$ (an even function) in the form $(\omega^2(0) = 0)$

$$\omega^2(\beta^2) = \beta^2 \times \frac{d\omega^2}{d\beta^2}(0) + O(\beta^4)$$

shows that $\frac{d\omega^2}{d\beta^2}$ (0) is the desired Stoneley velocity.

□

4 - NUMERICAL RESULTS

We present a simple application of Theorem 1 : the propagation of a tube-wave in an elliptical hole. In NICOLETIS et al. [2], this is compared with the propagation in a circular, but anisotropic, hole. The importance of the effects of ellipticity versus anisotropy are discussed.

The geometry of the ellipse is presented in Fig. 2.

It is also useful to introduce the static coefficients of the medium : Young's modulus E and Poisson's ratio ν.

The integral in (7) has been computed by QUIBLIER [6], giving

$$\int_\Gamma (\overline{u}^0{\cdot}n) \, d\sigma = 2\pi \, R^2 \, \frac{1+\nu}{E} \, \left[1+(3-4\nu) \, m^2\right] \, p^0.$$

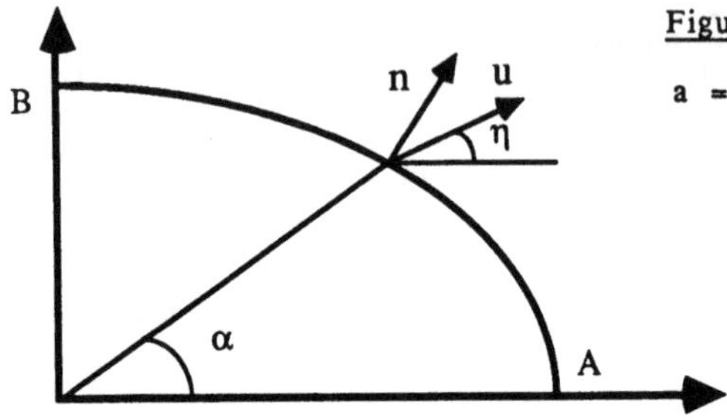

Figure 2 : The geometry of the ellipse.

$a = OA,\ b = OB,\ R = \frac{a+b}{2},\ m = \frac{a-b}{a+b}$

Then (7) becomes (in this case $S=\pi ab$)

$$V_{st}^2 = c^2 \left[1 + 2\rho_F c^2 \frac{(1+\nu)}{E} \frac{1+(3-4\nu)m^2}{1-m^2}\right]^{-1}.$$

As a first check, we note that for m=0 (a circle), we recover the well-known result (WHITE [1])

$$V_{st}^2 = c^2 \left(1 + \rho_F \frac{2c^2}{E}(1+\nu)\right)^{-1}.$$

Figure 3 shows the variations of the wave velocity normalized by the velocity of the circular case, for 3 different media. The parameters are given in Table 1. Notice that the ellipticity effect is stronger for "*soft*" media than for "*hard*" media, and that the tube wave velocity decreases with increasing velocity.

QUIBLIER [6] has also computed the deviation of the displacement vector from the radial direction. (In the circular case, the displacement is radial.) It is the angle $\eta-\theta$ in Fig. 2. He found that

$$\tan \eta = \frac{(1+m)\ [1+(3-4\nu)m]}{(1-m)\ [1-(3-4\nu)m]} \tan \alpha \qquad (16)$$

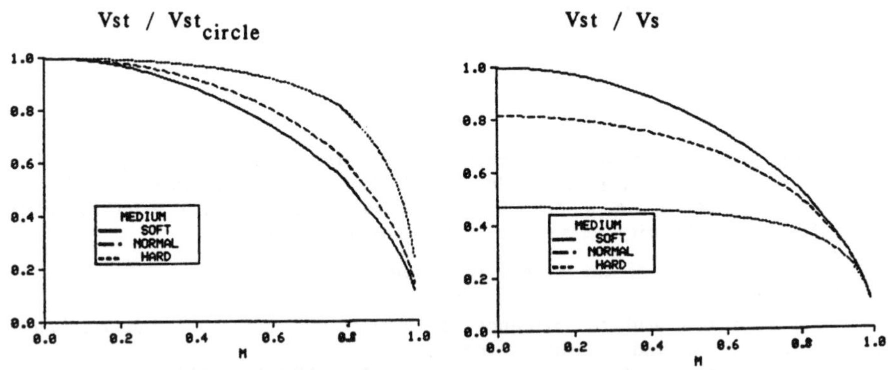

Figure 3 : Variation of the tube-wave velocity as a function of m.

a) Normalized by the tube wave velocity in a circular hole.

b) Normalized by the shear wave velocity.

Table 1 : Parameters of the three different media.

Medium	$c(ms^{-1})$	ρ_F/ρ_s	ν	$\rho_F\, c^2/\mu$
Soft	1650	0.5	0.25	1
Normal	1650	0.5	0.25	0.5
Hard	1650	0.5	0.25	0.125

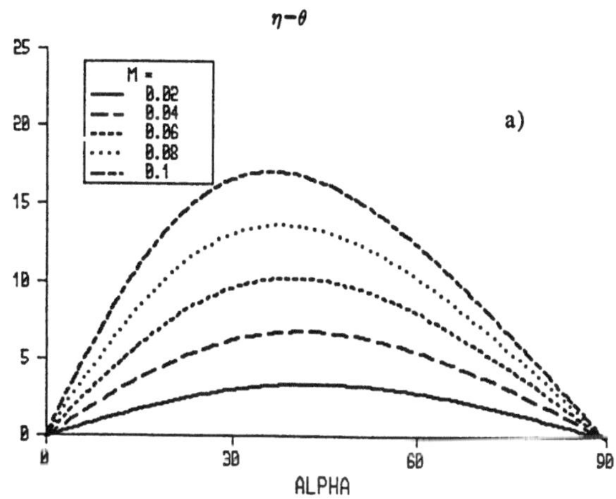

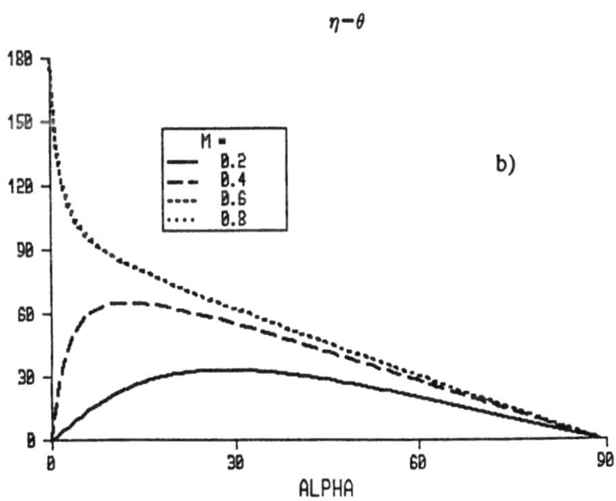

Figure 4 : Deviation between the tube-wave polarization plane and the radial plane along the azimuth of the hole :

a) small ellipticities b) large ellipticities

A plot of η-θ as a function of θ is presented in Fig.4. We see that, even though for small values of m the deviation remains small, it can reach 90° or even 180° for larger ellipticities. This means that a dilatation inside the hole produces a shortening of the larger axis of the ellipse.

REFERENCES

1. WHITE J.E. *"Underground Sound"*. Elsevier, Amsterdam (1983).

2. NICOLETIS L., QUIBLIER J., BAMBERGER A., JOLY P., KERN M. *"Hole geometry and anisotropy effects on tube wave propagation : a quasi-static study"*. IFP Report 35-529. Submitted to Geophysics (1987).

3. BAMBERGER A., JOLY P., KERN M. *"Etude mathématique des modes élastiques guidés par l'extérieur d'une cavité cylindrique de section arbitraire"*. INRIA Report n°650 (1987).

4. JOLY P. *"Low frequency behavior of elastic waves in a fluid filled borehole"*. INRIA Report, to appear (1988).

5. PAILLET F., WHITE J.E. *"Acoustic modes of propagation in the borehole and their relationship to rock properties"*. Geophysics, Vol. 47 pp. 1215-1228 (1982).

6. QUIBLIER J. *"Comportement d'un trou dans une plaque chargée"*. IFP report 35-213 (1987).

7. BAMBERGER A., JOLY P., KERN M. *"Mathematical methods for the analysis of elastic surface waves"*. These Proceedings (1988).

Surface Vibration of a Semi-infinite Viscoelastic Medium

P.M. O'Leary

Department of Mathematical Physics, University College, Galway, Ireland

The work treats the forced vibration of a semi-infinite viscoelastic medium due to a point load applied at the free surface . It has been previously shown [CURRIE, HAYES, and O'LEARY, 1977] that for certain materials two Rayleigh waves may be expected. In this work the displacement amplitudes corresponding to both Rayleigh waves due to the point source are compared for a wide range of materials.

1 Introduction

The problem of the displacement of a viscoelastic half-space due to a periodic force is considered in this work. The elastic case was first considered by LAMB [1] and the viscoelastic case has been considered by many authors since then, including NEWLANDS [2], TSAI and KOLSKY [3]. The analysis used here follows closely that used by TSAI and KOLSKY [3]. In their work, and in all previous work, it is assumed that the Rayleigh function has only one pole in the region of contour integration used for the inversion of the Hankel transform. This assumption is valid in the elastic case [4]. However, in the viscoelastic case two poles may exist for a wide range of materials [5, 6, 7]. In this work the effect of the second pole is considered for an oscillating source on the surface of a viscoelastic half-space. It is found that the amplitude of the displacement corresponding to the second pole is of the same order of magnitude as that which occurs in the elastic case.

2 Analysis

Consider a viscoelastic half-space subjected to a periodic force of the type $p(r)e^{i\omega t}$. The displacement vector **u** can be written in the form $\mathbf{u} = Ue^{i\omega t}$ and the equation of motion is

$$[\lambda(\omega) + 2\,\mu(\omega)]\,\nabla\,(\nabla.\mathbf{U}) - \mu(\omega)\,\nabla\times(\nabla\times\mathbf{U}) = -\rho\,\omega^2\mathbf{U}, \qquad (1)$$

where λ and μ are the complex Lamé constants which are functions of frequency ω, ρ is the density of the material and **u** is the displacement vector. We are going to consider the axisymmetric problem where

$$\mathbf{u} = \mathbf{u}\,(r, z, t) = u\mathbf{e}_r + w\mathbf{e}_z\,. \qquad (2)$$

We shall consider the case where the boundary conditions may be expressed in the form

$$\sigma_{zz} = -\,p(r)e^{i\omega t} \qquad \text{on } z = 0 \qquad (3)$$

and

$$\sigma_{rz} = 0 \qquad \text{on } z = 0 . \tag{4}$$

If the vector **U** has components u and w in the radial and z directions, we define

$$\Delta = \operatorname{div} \mathbf{U} = \frac{1}{r}\frac{\partial}{\partial r}(ru) + \frac{\partial w}{\partial z} \tag{5}$$

and

$$\Omega = [\frac{\partial u}{\partial z} - \frac{\partial w}{\partial r}]. \tag{6}$$

Taking the curl and the divergence of (1), we obtain

$$\frac{\partial}{\partial r}[\frac{1}{r}\frac{\partial}{\partial r}(r\Omega)] + \frac{\partial^2\Omega}{\partial z^2} + k_2^2\Omega=0 \tag{7}$$

and

$$\frac{1}{r}\frac{\partial}{\partial r}[r\frac{\partial\Delta}{\partial r}] + \frac{\partial^2\Delta}{\partial z^2} + k_1^2\Delta = 0, \tag{8}$$

where

$$k_1^2 = \frac{\rho\omega^2}{(\lambda+2\mu)} \quad \text{and} \quad k_2^2 = \frac{\rho\omega^2}{\mu} . \tag{9}$$

We now take the Hankel transforms, using the following notation:

$$D(k, z) = \int_0^\infty rJ_0(kr)\Delta(r,z)dr \tag{10}$$

and

$$\omega_1(k, z, w) = \int_0^\infty rJ_1(kr)\Omega(r,z,w)dr. \tag{11}$$

Using a similar notation for the other quantities, we obtain

$$D = kU_1 + \frac{\partial W}{\partial z} \tag{12}$$

and

$$\omega_1 = \frac{\partial U_1}{\partial z} + kW. \tag{13}$$

Taking the Hankel transforms of (7), (8) respectively,

$$\frac{\partial^2\omega_1}{\partial z^2} = s^2\omega_1 \tag{14}$$

and

$$\frac{\partial^2 D}{\partial z^2} = q^2 D, \tag{15}$$

where

$$s^2 = k^2 - k_2^2 , \qquad q^2 = k^2 - k_1^2 \quad . \tag{16}$$

Thus we find that

$$D(k,z) = A(k)\, e^{-qz} \tag{17}$$

and

$$\omega_1 = B(k)\, e^{-sz} \quad , \tag{18}$$

where A(k) and B(k) are to be determined from the boundary conditions (3,4) of the problem, P(k) the transformed load function, and the branches of q, s chosen in order that we do not obtain infinite displacements as $z \to \infty$. Hence we require that

$$\mathrm{Re}(s) > 0 \quad \text{and } \mathrm{Re}(q) > 0. \tag{19}$$

Transforming the boundary conditions , after some algebra, we obtain

$$U_1 = \frac{P(k)k}{\mu F(k)} \{[2k^2 - k_2^2]e^{-qz} - 2qse^{-sz}\} \tag{20}$$

and

$$W = \frac{P(k)q}{\mu F(k)} \{[2k^2 - k_2^2]e^{-qz} - 2k^2e^{-sz}\} \, . \tag{21}$$

The function F(k) is the well known Rayleigh function defined by

$$F(k) = [2k^2 - k_2^2]^2 - 4qsk^2 \, . \tag{22}$$

Using the inversion formulae for the Hankel transforms, we obtain the following forms for the displacement components:

$$u = \int_0^\infty \frac{\{[2k^2 - k_2^2]e^{-qz} - 2qse^{-sz}\}}{F(k)} \frac{P(k)}{\mu} J_1(kr)k^2dk \tag{23}$$

and

$$w = \int_0^\infty \frac{\{[2k^2 - k_2^2(w)]qe^{-qz} - 2qk^2e^{-sz}\}}{F(k,w)} \frac{P(k)}{\mu} J_0(kr)kdk \, . \tag{24}$$

Let us now consider the case of a concentrated periodic force applied on the surface of the medium. If we choose the origin of the r-z axis to coincide with the point load, we may write

$$p(r,t) = P_0 \;\; \delta(r)\, e^{\,i\omega t} \tag{25}$$

with P_0 being the amplitude of the oscillation, ω the oscillating frequency and $\delta(r)$ the Dirac delta function. The Hankel transform of (25) is

$$P(k) = P_0\, e^{i\omega t} \tag{26}$$

and so (23) and (24) take the forms

$$u = \frac{P_0 e^{i\omega t}}{2\pi\mu(\zeta)} \int_0^\infty \left\{\frac{[2k^2-k_2^2]e^{-qz} - 2qse^{-sz}}{F(k)}\right\} J_1(kr)k^2dk \tag{27}$$

and

$$w = \frac{P_0 e^{i\omega t}}{2\pi\mu} \int_0^\infty \left\{\frac{[2k^2-k_2^2]qe^{-qz} - 2qk^2e^{-sz}}{F(k,z)}\right\} J_0(kr)kdk. \tag{28}$$

Since our interest is the surface measurement, we let $z = 0$ and obtain the surface motions

$$u(0) = \frac{P_0 e^{i\omega t}}{2\pi\mu(\omega)} \int_0^\infty \left\{\frac{[2k^2-k_2^2] - 2qs}{F(k)}\right\} J_1(kr)k^2dk \tag{29}$$

and

$$w(0) = \frac{-P_0 e^{i\omega t}}{2\pi\mu(\omega)} \int_0^\infty \frac{k_2^2 q}{F(k)} J_0(kr)kdk\,. \tag{30}$$

Since we are using the positive roots for s and q the quantity $kk_2^2q/F(k)$ of the integrand in (30) is an odd function of k, and if we use the relationship

$$J_0 = \frac{1}{2}[\,H_0^{(1)}(kr) + H_0^{(2)}(kr)\,] \tag{31}$$

in the integrand, then the integral in (30) can be written as

$$I = \frac{1}{2} \int_{-\infty}^{\infty} \frac{kk_2^2 q}{F(k)} H_0^{(1)}(kr)dk\,, \tag{32}$$

where $H_0^{(1)}$ denotes the Hankel function of the first kind of order zero and $H_0^{(2)}$ denotes the Hankel function of the second kind of order zero.

Changing the variable from k to a complex variable $\xi = k + i\tau$, we may follow the contour indicated in Fig. 1 for contour integration in the ξ-plane.

Noting that for large argument

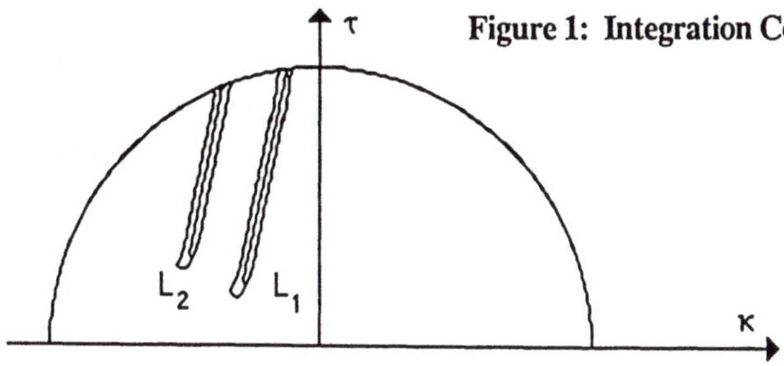

Figure 1: Integration Contour in Complex ξ-Plane

$$H_0^{(1)}(\xi r) \sim \sqrt{\frac{2}{\pi \xi r}} \exp\left[i\left(\xi r - \frac{\pi}{4}\right) \right] \tag{33}$$

$$H_0^{(2)}(\xi r) \sim \sqrt{\frac{2}{\pi \xi r}} \exp\left[-i\left(\xi r - \frac{\pi}{4}\right) \right] \tag{34}$$

the integrand vanishes on the infinite arc and we get

$$I = -\pi i \sum \frac{\kappa_i k_2^2 q(\kappa_i)}{F'(\kappa_i)} H_0^{(2)}(\kappa_i r) - \int_{L_1} - \int_{L_2} \tag{35}$$

where $\int_{L_1}$ and $\int_{L_2}$ are the integrals along the cuts L_1 and L_2 respectively and where κ_i are the Rayleigh pole(s). We note that for elastic materials there is only one such root, but for certain viscoelastic materials, there may be two roots of the Rayleigh equation [5]. The branch line integrals are the dilatational waves and shear waves, respectively. In the far field it is seen that the first term predominates, and we shall omit the effects of the second and third terms.

Hence in the far field, we write

$$w(0) = \sum \frac{-P_0 e^{i\omega t}}{2\pi\mu(\omega)} \frac{\kappa_i k_2^2 q(\kappa_i)}{F'(\kappa_i)} H_0^{(2)}(\kappa_i r) . \tag{36}$$

Using the asymptotic form of $H_0^{(2)}$ for large $\kappa_i r$ we get

$$w(0) = \sum \sqrt{\frac{1}{2\pi\kappa_i r}} \frac{-P_0}{2\pi\mu(\omega)} \frac{\kappa_i k_2^2 q(\kappa_i)}{F'(\kappa_i)} \exp\left[i\left(\omega t - k_i r - \frac{\pi}{4}\right)\right] . \tag{37}$$

Similarly it can be shown that

$$u(0) = \sum \sqrt{\frac{1}{2\pi k_i r}} \frac{iP_0}{2\pi\mu(\omega)} \frac{\kappa_i^2[2\kappa_i^2 - k_2^2 - 2q(\kappa_i)s(\kappa_i)]}{F'(\kappa_i)} \times \exp\left[i\left(\omega t - \kappa_i r - \frac{\pi}{4}\right)\right]. \tag{38}$$

3.Numerical Results

The displacements corresponding to the two roots were plotted for different material constants at various distances from the source, and it was found that for both cases the displacement ratios remain constant for a wide range of r. If we denote the root which occurs in the elastic case as the quasielastic root, and the second root as the viscoelastic root, it was found that in all cases the amplitude of the vertical displacement was greater for the quasielastic root than for the viscoelastic root. However, the horizontal displacement amplitude was found to be greater in the case of the viscoelastic root. For the case where $\lambda = \mu'(1 + .02i)$, $\mu = \mu'(1 + .04i)$ it was found that the vertical displacement for the quasielastic case was 1.3 times that for the viscoelastic case, and the ratio of quasielastic horizontal displacement amplitude to that of the viscoelastic case was .5. For the case where $\lambda = 2\mu'$, $\mu' = (1 + .1i)$ it was found that the corresponding ratios were 2.5 and .95 respectively. In order to test the dependence of these ratios on the material constants the following cases were considered:

(i) If we write $\lambda = \lambda' + ia\mu'$, $\mu = \mu' + ib\mu'$ (39)

then the case of varying the ratio a/b while fixing the Poisson's ratio for small values of a and b was considered.

In Fig. 2 , the displacement ratios for various ratios a/b are shown for the case when the Poisson's ratio is 0.3, and we see that when a/b increases, the ratios increase, but that they do not change significantly.

(ii) If we fix a and b and vary the Poisson's ratio, we see the dependence of the ratios on the Poisson's ratio.

In Fig. 3, we see the displacement ratio decreases for values of Poissons ratio between .21 and .26, and increases subsequently. At no stage does the ratio of the vertical displacements decrease below 1.3 and at Poisson's ratio of .33, it reaches 2.5.

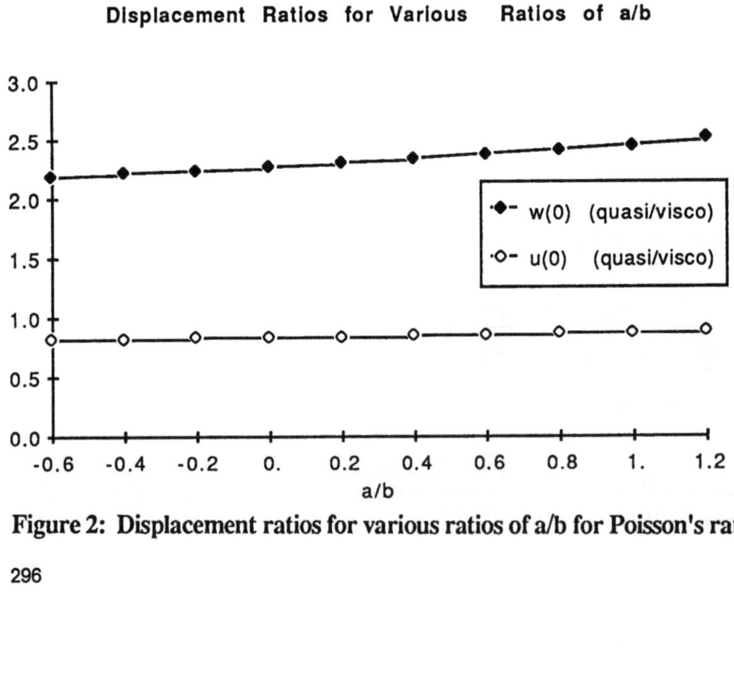

Figure 2: Displacement ratios for various ratios of a/b for Poisson's ratio = 0.3

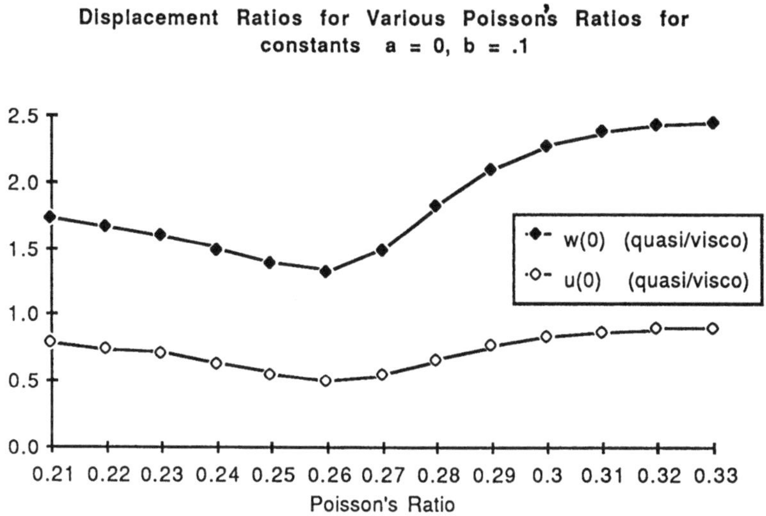

Figure 3: Displacement ratios at a point away from the source for the case where a = 0 , b = 0.1.

(iii) A common assumption made in the solution of boundary value problems in viscoelasticity is that the Poisson's ratio is constant. Previously it has been shown [7] that for a wide range of loss tangents two roots are admissible. In Fig. 4, we consider the displacement ratios for the case where $\tan\delta = .16$ ($\tan\delta = \mathrm{Im}(\mu)/\mathrm{Real}(\mu)$), since this spans the range of values of Poisson's ratios. It is found that the displacement ratio decreases with Poisson's ratio, but that the ratio of vertical displacements is always greater than 1.32.

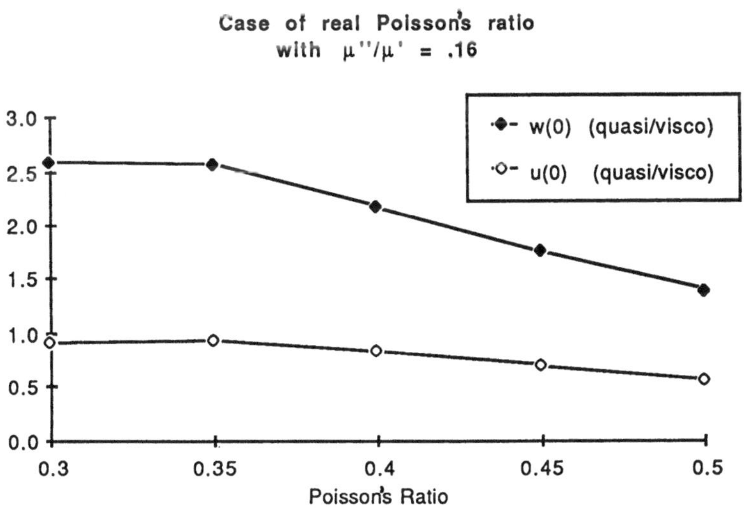

Figure 4: The displacement ratios for real values of Poisson's ratio

4. Conclusions

It has been shown that the predicted displacement due to the second root of the Rayleigh equation, which occurs in certain viscoelastic materials, is a significant ratio of the quasielastic wave, and that the radial displacement is predicted to be greater for this root. The relative sizes of the displacements depend on the material constants arising.

References

1. H. Lamb: Phil. Trans. Roy. Soc. London Ser. A, 203: 1- 42 (1904)

2. M. Newlands: J. of Acoust. Soc. Amer., 26 : 3, 434- (1954)

3. Y. M. Tsai , H. Kolsky: J. Mech. Phys. Solids 16: 99-109 (1968)

4. M. Hayes, R. S. Rivlin: ZAMP 13: 80-83 (1962)

5. P. K. Currie, M. Hayes, P. O'Leary: Quart. App. Math. 35 : 35-53 (1977)

6. P. K. Currie, P. O'Leary: Quart. App. Math. 35 : 445-454 (1978)

7. P. O'Leary: Proc. R. Ir. Acad. 81A : 147-155 (1981)

Surface Waves on an Isotropic Viscoelastic Half-Space: The Method of Generalized Rays

P. Borejko and F. Ziegler

Institut für Allgemeine Mechanik, Technische Universität Wien, Karlsplatz 13 – 201, A-1040 Wien, Austria

1. INTRODUCTION

The surface motions of an elastic and isotropic half-space produced by a localized source were studied by LAMB /1/ in a classic paper. The comprehensive survey of the literature on the Lamb problem can be found in EWING et al. /2/ and MIKLOWITZ /3/.

Recently PAO and GAJEWSKI /4/ developed the technique of generalized rays providing exact representations for transient responses due to the localized source perturbing a layered elastic medium. A review is given in /5/.

The first purpose of this paper is to apply the method of generalized rays to the study of surface motions generated by the line source buried within an elastic isotropic half-space. In section 2 we set down the solution to the problem and discuss its numerical evaluation. We obtain the graphs showing the horizontal and vertical displacements as a function of time at three different receivers located at the free surface of the half-space. We also obtain the curves representing the orbits of the particles at these receivers. Finally, we examine the development of the Rayleigh wave and conclude that the Rayleigh wave is elliptical retrograde.

The surface motions of a viscoelastic isotropic half-space generated by the localized source were investigated by NEWLANDS /6/. Viscoelastic Rayleigh waves were studied by CURRIE, HAYES and O'LEARY in a series of papers /7-11/. They examined particular models of viscoelastic behaviour and found that (in contrast to elastic materials) (i) more than one elliptically polarised Rayleigh wave may be possible; (ii) the waves may be either prograde or retrograde at the free surface.

The second aim of this work is to study surface motions generated by the line source buried within a viscoelastic isotropic half-space. This is accomplished in section 3 by applying the correspondence principle formulated by DASGUPTA and SACKMAN /12/ to the numerical solution of the corresponding elastic problem provided by the technique of generalized rays. For different viscoelastic behaviour of the half-space, we obtain the graphs showing the horizontal and vertical displacements as a function of time at the surface receivers. We also obtain the curves representing the particle orbit at one surface receiver. In particular, for the Maxwell model, it is found that two retrograde elliptically polarised Rayleigh waves are possible.

2. SURFACE MOTIONS OF AN ELASTIC HALF-SPACE

We are concerned with an elastic, isotropic and homogeneous material of equal Lamé constants λ and μ and density ρ occupying the half-space defined by $-\infty < x < \infty$, $-\infty < y < \infty$ and $z \geq 0$. The surface of the half-space, defined by the xy-plane, is traction free. The medium is perturbed at time $t=0$ by the line source of explosion represented by the distribution of the body force $\underline{F}$, measured per unit mass, of the form: $\underline{F} = \mathrm{grad}\,[f(t)\delta(x)\delta(z-z_0)]$ (cf. /4, eqs. (2.4) and (2.24)/) and located at the depth $z=z_0$. Here grad denotes the gradient operation, $f(t)$ is the time function of the source, and $\delta(x)$ and $\delta(z-z_0)$ are the Dirac delta functions. The source radiates a compressional cylindrical pulse /13, § 2.5/ and, since the distribution of $\underline{F}$ is uniform with respect to y, the problem is two-dimensional and satisfies the conditions of plane strain /4, § II B.1, 13, § 2.6/; i.e., all field variables are independent of y and the non-vanishing components of the displacement $\underline{u}$ are $u_x = u_x(x,z,t)$ and $u_z = u_z(x,z,t)$, respectively.

The exact expressions for the horizontal u_x and vertical u_z displacements at the free surface $z=0$ are provided by the method of generalized rays in the form of convolution integrals

$$u_i(x,0,t) = A\,H(t-t_M) \int_{t_M}^{t} \ddot{f}(t-\tau) I_i(x,0,\tau)\,d\tau, \quad i=x,z. \qquad (2.1)$$

It has been assumed that $f(0^+)=0=\dot{f}(0^+)$ and $\ddot{f}(0^+)=0$ (cf. /4, eqs. (4.9), (7.1)/). Here $A=(4\pi c^2)^{-1}$ and $c=[(\lambda+2\mu)/\rho]$ is the speed of the compressional wave. $H(t)$ is the Heaviside unit step function, t_M is the arrival time of the disturbance at the receiver located at the free surface $z=0$, t is the actual observation time, and a dot denotes the time differentiation. The ray integrals $I_x(x,0,\tau)$ and $I_z(x,0,\tau)$ are given by

$$I_x(x,0,\tau) = 2\,\mathrm{Im} \int_A^{\xi_1(x,0,\tau)} \xi E_x(\xi)\,d\xi, \quad I_z(x,0,\tau) = 2\,\mathrm{Re} \int_A^{\xi_1(x,0,\tau)} E_z(\xi)\,d\xi, \qquad (2.2)$$

where Re and Im stand for the real and imaginary parts, respectively (cf. /4, eq. (7.2)/). The path of integration is along the contour $AM\xi_1$ in the complex ξ-plane (Fig. 1) and the integrands are defined by the relations $(2.3)_{1,2}$ given in /14/.

As a preliminary to the evaluation of $I_x(x,0,\tau)$ and $I_z(x,0,\tau)$, we first note that the equation

$$-\,ix + z_0\,\xi\,(\xi^2 + c^{-2})^{-1/2} = 0 \qquad (2.3)$$

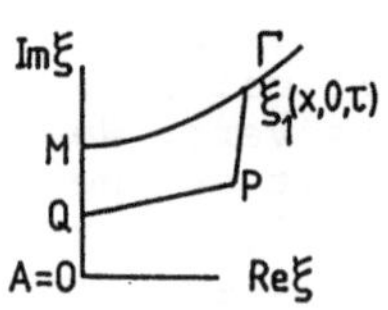

Fig.1. Integration in the complex ξ-plane

has only one root $\xi_M = ib_M$, b_M being a positive real number less than $1/c$ for all positive values of x, representing the stationary point, labelled M on the imaginary ξ-axis (Fig. 1), of the function

$$t = -i\xi x + z_o(\xi^2 + c^{-2})^{1/2}, \tag{2.4}$$

which, for fixed x and z_o, defines in the complex ξ-plane the Cagniard path Γ (Fig. 1) with time t ($t=\tau$) as a parameter /4, pp. 207-210/. In (2.3) and (2.4) x ($x \geq 0$) is the horizontal distance of the surface receiver from the epicentre.

Next, we fix the surface receiver (i.e., we specify the ratio x/z_o), calculate the root ξ_M of equation (2.3), compute from (2.4) the arrival time t_M of the disturbance at the receiver /4, pp. 212-213/ by setting $\xi = \xi_M$ and finally, for a prescribed value of τ ($\tau > t_M$), determine from (2.4), by putting $t = \tau$ and $\xi = \xi_1$, the upper limit of integration $\xi_1(x,0,\tau)$ in (2.2): the location of $\xi_1(x,0,\tau)$ on Γ /4, pp. 247-248/. The values of ξ_M and ξ_1 are calculated numerically by using Müller's method /4, p. 248/.

Since the ray integrals $I_x(x,0,\tau)$ and $I_z(x,0,\tau)$ vanish along the path AM /4, pp. 215-216/ and both integrands in (2.2) have no singular points to the right of the path $AM\xi_1$ /4, pp. 210, 250/, we change, by choosing $\xi_Q = 0.8\,\xi_M$ and $\xi_P = \xi_M + 0.2|\xi_M|$, the original path $AM\xi_1$ to a new path $QP\xi_1$ consisting of two straight paths QP and $P\xi_1$; P is the pivotal point (Fig. 1) /4, p.250/. The changes of the integration variable $\xi = \frac{1}{2}[(\xi_P-\xi_Q)p + \xi_P + \xi_Q]$ and $\xi = \frac{1}{2}[(\xi_1-\xi_P)p + \xi_1 + \xi_P]$ map the intervals [Q,P] and [P,ξ_1] in ξ onto the intervals [-1,1] in p, p being real, so that the integration in the complex ξ-plane is replaced by an integration with respect to the real variable p, which, in turn, is accomplished numerically by using the Gaussian quadrature /4, pp. 250-251/. The desired accuracy in the numerical integration of (2.2) can be achieved by a suitable increase of the number of Gaussian points.

The procedure of evaluation of the ray integrals $I_x(x,0,\tau)$ and $I_z(x,0,\tau)$ provides their values at a number of equidistant points τ of the time interval $[t_M,t]$ and hence, on specifying f(t), the values of the integrands in (2.1) are known at these points. Thus, the convolution integrals in (2.1) are computed numerically using Simpson's rule /4, p. 251/. We choose f(t) to be a pulse with rounded shoulders: f(t)=0 for $t<0$, $f(t)=t^2/2$ for $0<t<\Delta$, $f(t) = t^2/2 - (t-\Delta)^2$ for $\Delta<t<2\Delta$, $f(t) = \Delta^2$ for $2\Delta<t<4\Delta$, $f(t) = (t-2\Delta)^2/2 - (t-3\Delta)^2$ for $4\Delta<t<5\Delta$, $f(t) = (t-6\Delta)^2/2$ for $5\Delta<t<6\Delta$, $f(t) = 0$ for $t>6\Delta$, where 6Δ is the width of f(t), the sum of rise- and fall-time of the initial disturbance. Note that f(t) satisfies the assumptions following (2.1).

The results of the numerical evaluation of the solution (2.1) for $C = 1.0$, $C = (\mu/\rho)^{1/2}$ being the speed of the shear wave, $c=(3)^{1/2}C$, $\Delta = 0.1$ and $z_o = 0.5$ are presented in Fig. 2 and 3. The normalized time $\tau = t/t_o$, $t_o = 2z_o/C$ being the characteristic time, is plotted on the abscissae in Fig. 2; Δ is also normalized by t_o.

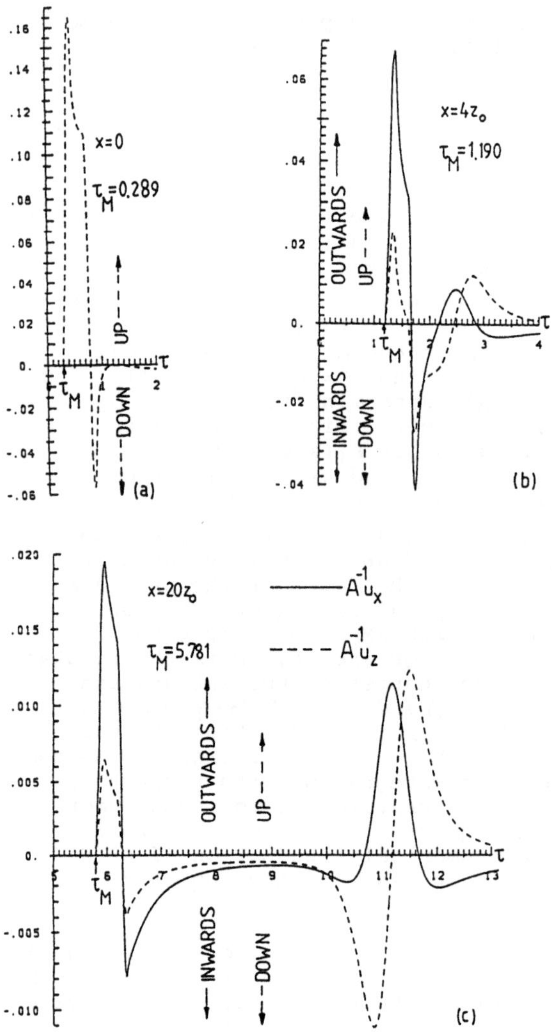

Fig. 2. Displacements on the surface of an elastic half-space. A compressional line source is situated below the surface at a depth $z_0 = 0.5$. The observation points are located on the surface at the epicentre and at distances of four and twenty source depths measured from the epicentre ($C=1.0$, $c=(3)^{1/2}C$ and $\Delta=0.1$).

Figure 2(a) shows the surface response at the epicentre; i.e., for $x = 0$. In this special situation, $\xi_M = 0$, $t_M = z_0/c$, Γ coincides with the real ξ-axis in Fig. 1 and hence the path of integration $QP\xi_1$ is now along the real ξ-axis in Fig. 1 with $\xi_Q = 0$ and $\xi_P = 0.25$. In the numerical integration of (2.2) 20 Gaussian points were used. The u_x-component, as expected, is absent at the epicentre and the u_z consists of two sudden and opposite jerks following in rapid succession and the protracted "tail" representing a typical property of two-dimensional transient waves /1, p.28, 3, p.107/. In Fig. 2(b) and 2(c) are shown the responses at the surface receivers located at the distances $x = 4z_0$ and $x = 20z_0$ from the epicentre; the orbits of the surface particles at these receivers are given in Fig. 3(a) and 3(b). In the numerical integration of (2.2) 96 Gaussian points were used. The u_x and u_z consist both of two sudden and opposite jerks following in rapid succession, one smooth pulse and the protracted "tail". The motion of the surface particle is initially prograde (at $x = 20z_0$ nearly linearly polarised) and then it becomes retrograde (at $x = 20z_0$ nearly elliptically polarised). Since in the elastic plane Rayleigh wave the particle motion is elliptical retrograde /2, p.33/ and at the large distance from the epicentre $x = 20z_0$ the Rayleigh wave appears at about the time required to traverse this distance with its characteristic velocity $C_R = 0.9194$ C /2, pp. 66, 33/, we conclude that the smooth pulse shown in Fig. 2(b) and 2(c) can be identified with the Rayleigh wave. The development of the Rayleigh wave is clearly seen with the increase of the epicentral distance; at $x = 20z_0$, the major part of the surface motion of the elastic half-space is the Rayleigh wave, as originally supposed by LORD RAYLEIGH /15/ and subsequently confirmed by LAMB /1/. The particle motion in the elastic Rayleigh wave is elliptical retrograde, the major axes of the ellipse is at the right angle to the free surface.

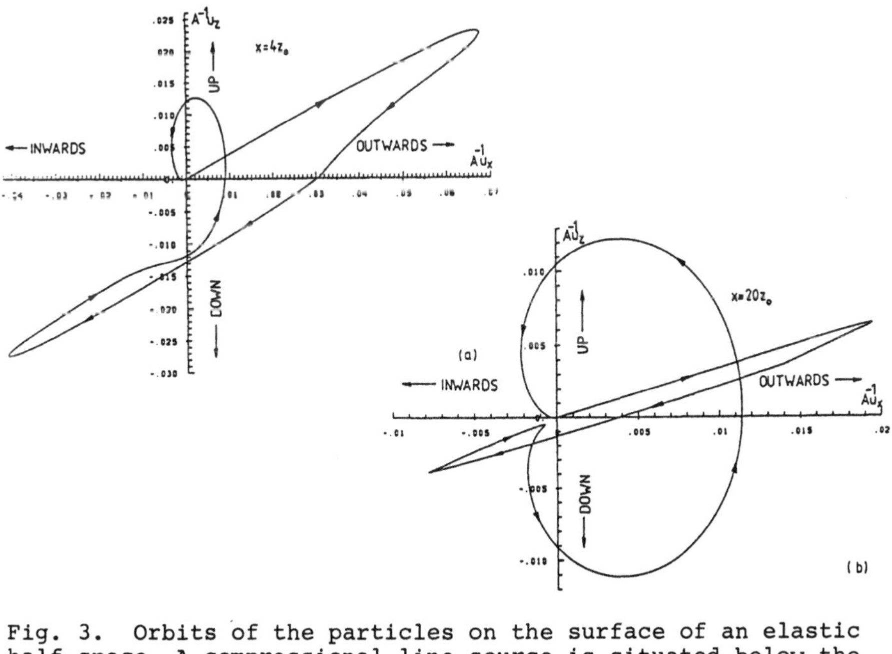

Fig. 3. Orbits of the particles on the surface of an elastic half-space. A compressional line source is situated below the surface at a depth $z_0=0.5$. The observation points are located on the surface at distances of four and twenty source depths measured from the epicentre ($C=1.0$, $c=(3)^{1/2}C$ and $\Delta=0.1$).

3. SURFACE MOTIONS OF A VISCOELASTIC HALF-SPACE

In this section a viscoelastic, isotropic and homogeneous half-space is assumed to have constant Poisson's ratio. Thus, the material behaviour is characterized by a single modulus, the complex shear modulus: $\mu^*(\Omega) = \mu[\alpha(\Omega) + i\beta(\Omega)]$, where $\Omega = \omega t_0$ is the non-dimensional frequency, ω being the circular frequency (cf. /12, eq. (12)/). In order that the material is stable and the energy dissipated over the cycle is non-negative it is necessary that $\alpha(\Omega)>0$ for $\Omega\varepsilon(-\infty,\infty)$ and sign $[\beta(\Omega)]$ = sign (Ω). Since the free surface of the elastic and viscoelastic half-space is traction free, and the location and the time behaviour of the source for the two problems is identical, the solution of the viscoelastic problem is obtained from the solution of the elastic problem by the correspondence principle /16, p.96/. However, the classical form of this principle cannot be applied here; the solution (2.1) is not given in closed form. In this situation the alternative form of the correspondence principle is used to give

$$U_i^*(x,0,\Omega) = \int_{-\infty}^{\infty} G(\Omega,\Omega')\, U_i(x,0,\Omega')\, d\Omega', \qquad i = x,z , \qquad (3.1)$$

(cf. /12, eq. (20)/). Here the non-dimensional frequency Ω' is taken as the integration variable, $U_i(x,0,\Omega')$ and $U_i^*(x,0,\Omega')$ are in turn the Fourier transforms of the elastic, $u_i(x,0,t)$, and viscoelastic, $u_i^*(x,0,t)$, displacements, $G(\Omega,\Omega') = b(\Omega)/\pi\, \{b^2(\Omega) + [a(\Omega) - \Omega']^2\}$ is the Green's function for the half-plane b>0 of the ab-plane, $a(\Omega)=\Omega\ [\alpha^2(\Omega) + \beta^2(\Omega)]^{-1/4}$ $\cos[\Phi(\Omega)/2]$ and $b(\Omega)=\Omega[\alpha^2(\Omega)+ \beta^2(\Omega)]^{-1/4} \sin[\Phi(\Omega)/2]$, where $\Phi(\Omega)$ = arcsin $\{\beta(\Omega)\ [\alpha^2(\Omega) + \beta^2(\Omega)]^{-1/2}\}$. Invoking that Re $[U_i(x,0,\Omega')]$ and Im $[U_i(x,0,\Omega')]$ are even and odd functions in Ω' respectively, the relation (3.1) takes the form

$$U_j^*(x,0,\Omega) = \int_0^{\infty} [G(\Omega,\Omega') + G(\Omega,-\Omega')]\ \mathrm{Re}[U_j(x,0,\Omega')]\ d\Omega'$$
$$+ i\int_0^{\infty} [G(\Omega,\Omega')-G(\Omega,-\Omega')]\,\mathrm{Im}\ [U_j(x,0,\Omega')]d\Omega', \quad j=x,z, \quad i^2 =-1. \qquad (3.2)$$

Since $G(\Omega,\Omega') + G(\Omega,-\Omega')$ and $G(\Omega,\Omega')-G(\Omega,-\Omega')$ are even and odd functions in Ω' respectively, the first and second integrals in (3.2) are even and odd functions in Ω respectively. Thus, on applying the inverse Fourier transform to $U_i^*(x,0,\Omega)$, we obtain the solution to the viscoelastic problem: The surface displacements $u_i^*(x,0,t)$ which are real-valued functions in time.

The numerical evaluation of (2.1) yields the values of $u_i(x,0,t)$ for the finite sequence of t. Hence, the approximation to $U_i(x,0,\Omega')$ over the finite range $[0,\Omega'_{max}]$ is obtained by the use of the discrete Fourier transform; its evaluation is effected numerically with the aid of the FFT algorithm. When $\mu^*(\Omega)$ is specified, the values of the integrands in (3.2) are known at a number of equidistant points Ω' of the finite frequency interval $[0,\Omega'_{max}]$ and thus the integrals in (3.2) are evaluated numerically using Simpson's rule for the finite sequence of $\Omega\varepsilon[0,\Omega'_{max}]$. The approximation to the solution $u_i^*(x,0,t)$ of the viscoelastic Lamb problem is obtained by taking the real part of the inverse discrete Fourier transform of $U_i^*(x,0,\Omega)$; this is accomplished numerically with the use of the FFT algorithm. We observe that the numerically computed $u_i^*(x,0,t)$ are not real-valued as they should be: this is mainly due to the fact that we are limited in (3.2) to a finite range of integration. The minimum arrival time of the disturbance is $t_M^* = r/c^*$, where r is the distance between the receiver and the source and $c^* \doteq (3)^{1/2}\, R_e\,\{[\mu^*(\Omega\,\max)/\rho]^{1/2}\}$ is the speed of the compressional viscoelastic wave (cf. /16, p. 60/).

The results are shown in Fig. 4-9 for different models of viscoelastic behaviour. It is assumed that $\mu=1.0$, $\rho=1.0$, $\Delta=0.1$ and $z_0=0.5$. The normalized time $\tau-\tau_M = (t-t_M^*)/t_0$ is plotted on the abscissae in Fig. 4,7 and 9; Δ is also normalized by t_0.

(i) For the Maxwell model, the displacements at the receiver located at the distance $x = 4z_0$ from the epicentre and the orbit of the particle at this receiver are shown in Figs.4-6.

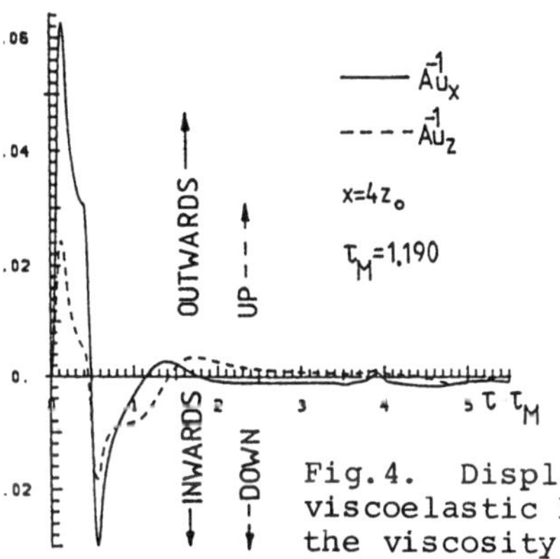

Fig.4. Displacements on the surface of a viscoelastic half-space (Maxwell model with the viscosity $\eta=1.0$). A compressional line source is situated at depth $z_0=0.5$. The observation point is as in Fig.5.

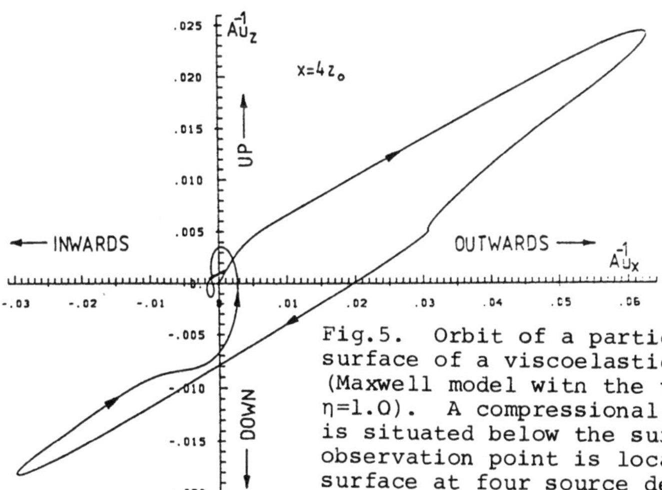

Fig.5. Orbit of a particle on the surface of a viscoelastic half-space (Maxwell model witn the viscosity $\eta=1.0$). A compressional line source is situated below the surface. The observation point is located on the surface at four source depths from the epicentre.

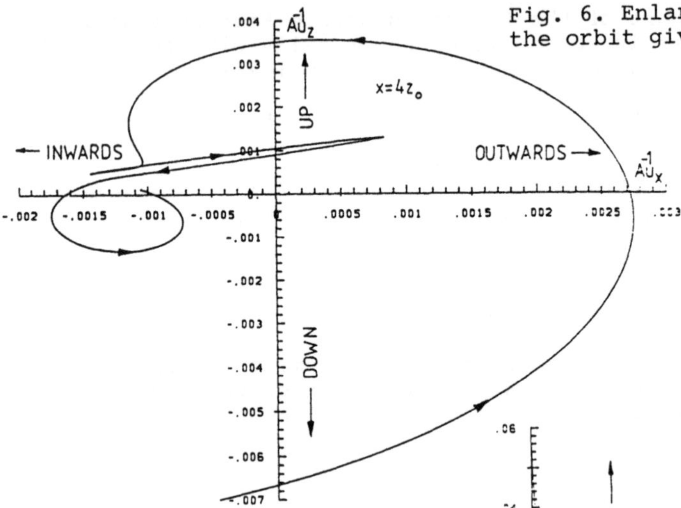

Fig. 6. Enlarged part of the orbit given in Fig. 5.

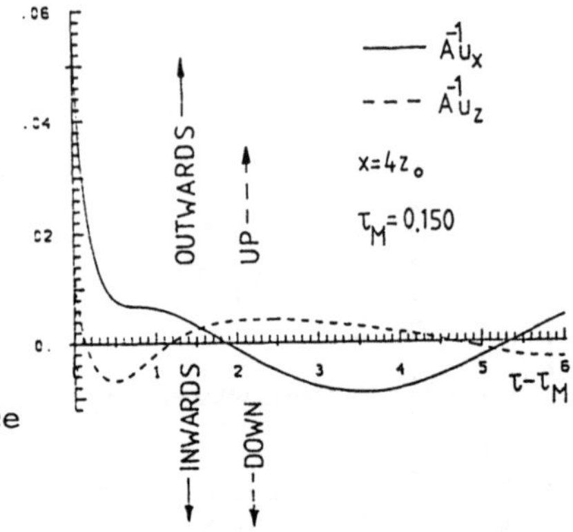

Fig.7. Displacements on the surface of a viscoelastic half-space (Kelvin-Voigt model with the viscosity η=0.2). Line source and observation point are as in Fig.5.

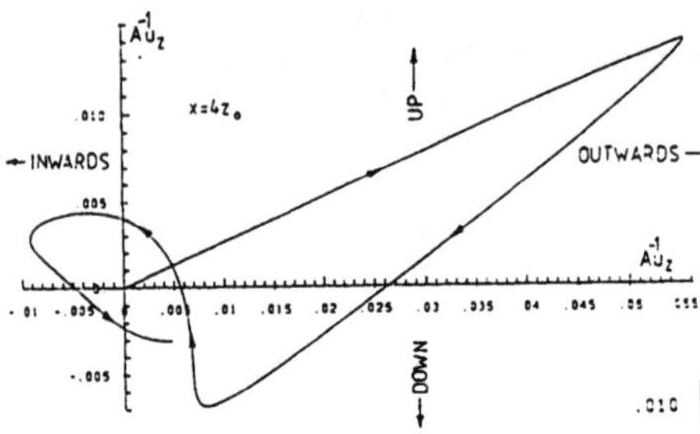

Fig.8. Orbit of a particle on the surface of a viscoelastic half-space (Kelvin-Voigt model with the viscosity η=0.2). The line source and observation point are as in Fig.5.

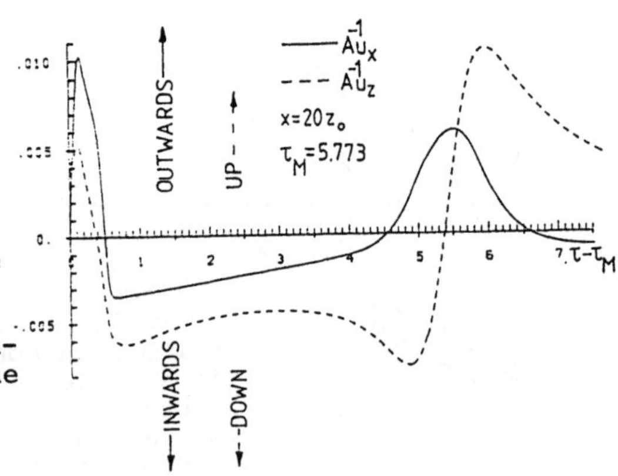

Fig.9. Displacements on the surface of a half-space with constant damping $\mu^*=\mu(1+i0.1)$. The line source is as in Fig.5. The observation point is located on the surface at a distance of twenty source depths measured from the epicentre.

The u_x^* and u_z^* consist both of two sudden and opposite jerks following in rapid succession, the smooth pulse, the jerk (very small compared with the initial jerks), the second smooth pulse (much smaller than the first one) and the protracted "tail" (not shown in Fig. 4). The motion of the surface particle is initially prograde then elliptical retrograde, again prograde (almost linearly polarised) and finally elliptical retrograde. These two elliptical retrograde pulses are presumably the viscoelastic Rayleigh waves (cf. /8/).

(ii) For the Kelvin-Voigt model, the displacements and the orbit of the particle at the same receiver as before are shown in Fig. 7 and 8. The u_x^* and u_z^* consist both of one sudden jerk, one smooth pulse and the protracted "tail" (not shown in Fig. 7). The motion of the surface particle is initially prograde and then almost elliptical retrograde. This elliptical retrograde pulse represents presumably the viscoelastic Rayleigh wave (cf. /8/).

(iii) In Fig. 9 the displacements are shown at the receiver located at the distance $x = 20z_0$ from the epicentre. The damping characteristic of the half-space is assumed to be constant within the range of frequencies $[0, \Omega max]$. The viscoelastic signal is damped and distorted in comparison with the elastic signal.

Acknowledgement

The support of the "Österreichischer Fonds zur Förderung der wissenschaftlichen Forschung", Project S 30/01, 1985/87 is gratefully acknowledged.

REFERENCES

1. H. Lamb: Philos. Trans. R. Soc. London A 203, 1 (1904)

2. W.M. Ewing, W.S. Jardetzky, F. Press: Elastic Waves in Layered Media (McGraw-Hill, New York 1957)

3. J. Miklowitz: The Theory of Elastic Waves and Waveguides (North-Holland, Amsterdam 1978)

4. Y.H. Pao, R.R. Gajewski: In Physical Acoustics, ed. W.P. Mason, R.N. Thurston, Vol. 13 (Academic Press, New York 1977) p. 183

5. F. Ziegler: Z. Angew. Math. u. Mech. 65, T15 (1985)

6. M. Newlands: J. Acoust. Soc. Am. 26, 434 (1954)

7. P.K. Currie, M. Hayes, P.M. O'Leary: Q. appl. Math. 35, 35 (1977)

8. P.K. Currie, P.M. O'Leary: Q. appl. Math. 35, 445 (1978)

9. P.K. Currie, Q. appl. Math. 37, 332 (1979)

10. P.M. O'Leary: Proc. R. Ir. Acad. 81A, 147 (1981)

11. P.M. O'Leary: In Recent Developments in Surface Acoustic Waves, ed. by D.F.Parker, G.A.Maugin, Springer Ser. Wave Phen. (Springer, Berlin, Heidelberg 1988)

12. G. Dasgupta, J.L. Sackman: ASME J. Appl. Mechanics 44, 57 (1977)

13. J.A. Hudson: The Excitation and Propagation of Elastic Waves (Cambridge University Press, Cambridge 1980)

14. P. Borejko: Acta Mechanica 67, 79 (1987)

15. Lord Rayleigh (J.W. Strutt): Proc. Lond. Math. Soc. 17, 4 (1885)

16. D.R. Bland: The Theory of Linear Viscoelasticity (Pergamon Press, Oxford 1960)

Acoustical Scattering by a Shallow Surface-Breaking Crack in an Elastic Solid Under Light Fluid Loading

R.H. Tew[1], J.R. Ockendon[2], and G.A.D. Briggs[1]

[1]Department of Metallurgy and Science of Materials, University of Oxford, Parks Road, Oxford OX1 3PH, Oxford

[2]Mathematical Institute, 24-29 St. Giles, Oxford OX1 3LB, U.K.

A mathematical framework is developed for the interpretation of data from acoustic microscopes. The main part of this work is devoted to the simplest case of a localised traction in the surface of the specimen when the acoustic field is incident normally; oblique incidence is described briefly in the appendix.

1. INTRODUCTION

A non-destructive evaluation of the surface and near sub-surface structure of a material can be obtained using an acoustic microscope [1]. A focused wave propagates from an acoustic lens through a liquid couplant towards the surface of the specimen. An acoustic image is formed from the reflected and scattered waves received at the microscope transducer. To understand such image formation, we must analyse the scattering of an acoustic wave in a fluid by a surface-breaking inhomogeneity, in particular a crack, in an adjacent solid. To this end, we suppose that a time-harmonic plane wave is incident from a fluid upon such an elastic solid and we take the plane of incidence to be normal to the crack direction. The crack depth is assumed small compared to an acoustic wavelength whilst the length of the crack is taken to be infinite. We also assume that the effect of the crack under plane wave incidence is to generate a distribution of surface traction along the crack direction, the z-axis, which is represented by a multipole expansion. This is similar to the approach taken by HOWE [2].

Subtracting the incident and geometrically reflected and transmitted waves from the fluid velocity potential and elastic displacement vector and assuming a time dependence $e^{-i\omega t}$, we arrive at the following problem to solve for the scattered fields ϕ, $\underline{u}$:

$$(\nabla^2 + k_F{}^2)\phi = 0, \quad y > 0 \tag{1}$$

$$-\frac{1}{k_p{}^2}\nabla(\nabla\cdot\underline{u}) + \frac{1}{k_s{}^2}\nabla\wedge(\nabla\wedge\underline{u}) = \underline{u}, \quad y < 0 \tag{2}$$

$$2\mu\frac{\partial u_y}{\partial y} + \lambda\nabla\cdot\underline{u} = -p + F_0\delta(x), \qquad y = 0 \tag{3}$$

$$\frac{\partial u_x}{\partial y} + \frac{\partial u_y}{\partial x} = 0, \qquad y = 0 \tag{4}$$

$$\frac{\partial\phi}{\partial y} = -i\omega u_y, \qquad y = 0 \tag{5}$$

along with suitable radiation/decay conditions at infinity. Equation (3) represents the first term of a particular multipole expansion; higher order terms may be obtained by differentiation and other localised tractions with different orientation are amenable to the analysis we now present. This model should still give a qualitative picture of the scattering phenomena to be expected in the full problem. An exact solution to this problem can be obtained by transform methods and asymptotic results derived by considering deformations of the integration path in the complex plane, taking due account of singularities being in close proximity or even coalescing. To avoid this intricate analysis, we solve this problem by ray methods; ray methods are also applicable to more complicated geometries and more realistic crack models where transforms cannot be used easily. We scale

$$\underset{\sim}{x} = L\underset{\sim}{x}', \quad |\underset{\sim}{x}'| = O(1) \tag{6}$$

for some large, but as yet undetermined, length scale L and assume ray series solutions of the form

$$\phi(\underset{\sim}{x}') \sim e^{ik_F L\phi^{(F)}(\underset{\sim}{x}')} \sum_{n=0}^{\infty} \frac{A_n^{(F)}(\underset{\sim}{x}')}{(ik_F L)^n}, \quad y' > 0 \tag{7}$$

$$\underset{\sim}{u}(\underset{\sim}{x}') \sim e^{ik_p L\phi^{(P)}(\underset{\sim}{x}')} \sum_{n=0}^{\infty} \frac{\underset{\sim}{u}_n^P(\underset{\sim}{x}')}{(ik_p L)^n} + e^{ik_s L\phi^{(S)}(\underset{\sim}{x}')} \sum_{n=0}^{\infty} \frac{\underset{\sim}{u}_n^S(\underset{\sim}{x}')}{(ik_s L)^n}, y'<0. \tag{8}$$

Henceforth, all primes are omitted. The irrotational and divergence-free components of $\underset{\sim}{u}$ can be written as

$$\underset{\sim}{u}_n^P = \underset{\sim}{\nabla} A_{n-1}^{(P)} + A_n^{(P)} \underset{\sim}{\nabla}\phi^{(P)}, \; n = 0, 1, 2, \ldots, \; A_{-1}^{(P)} \equiv 0 \tag{9}$$

$$\underset{\sim}{u}_n^S = (\underset{\sim}{\nabla} A_{n-1}^{(S)} + A_n^{(S)} \underset{\sim}{\nabla}\phi^{(S)}) \wedge \underset{\sim}{k}, \; n = 0, 1, 2, \ldots, \; A_{-1}^{(S)} \equiv 0 \tag{10}$$

respectively. Substituting (7-10) into (1,2) gives the eikonal equations

$$|\underset{\sim}{\nabla}\phi^{(\alpha)}|^2 = 1, \quad \alpha = F, P, S \tag{11}$$

whilst the amplitudes satisfy the recursive transport equations

$$A_n^{(\alpha)}\nabla^2\phi^{(\alpha)} + 2\underset{\sim}{\nabla} A_n^{(\alpha)} \cdot \underset{\sim}{\nabla}\phi^{(\alpha)} = -\nabla^2 A_{n-1}^{(\alpha)} \; (\alpha = F, P, S) . \tag{12}$$

Solving the eikonal equations by Charpit's method yields three families of rays emanating from the inhomogeneity itself - one family propagating into the fluid and the other two propagating into the solid, all three having cylindrical wavefronts. Also, any rays propagating away from the interface at points other than the origin correspond to plane wave disturbances, of which there are two types. They are (i) interfacial waves, corresponding to free modes of the coupled fluid-solid configuration and (ii) head waves, induced by the interaction of the cylindrical waves with the interface. A brief discussion of each type of wave now follows. Further details may be found in [3].

(i) Interface Waves

Substituting the expansions (7,8) into equations (3-5) and considering the O(1) terms we obtain, using Snell's law, the following algebraic equations for $A_0^{(\alpha)}(x,0)$ (α = F, P S):

$$\frac{2}{C_P}\frac{\partial\phi^{(P)}}{\partial x}\frac{\partial\phi^{(P)}}{\partial y}A_0^{(P)} + \frac{1}{C_S}\{1-2(\frac{\partial\phi^{(S)}}{\partial x})^2\}A_0^{(S)} = 0; \quad y = 0 \tag{13}$$

$$\frac{\partial\phi^{(P)}}{\partial y}A_0^{(P)} - \frac{\partial\phi^{(S)}}{\partial x}A_0^{(S)} + \frac{1}{C_F}\frac{\partial\phi^{(F)}}{\partial y}A_0^{(F)} = 0; \quad y = 0 \tag{14}$$

$$\frac{1}{C_P}\{C_P^2 - 2C_S^2(\frac{\partial\phi^{(P)}}{\partial x})^2\}A_0^{(P)} - 2C_S\frac{\partial\phi^{(S)}}{\partial x}\frac{\partial\phi^{(S)}}{\partial y}A_0^{(S)} + \frac{\varepsilon C_P}{C_F}A_0^{(F)}=0; \quad y=0, \tag{15}$$

where $\varepsilon=(\rho_F C_F)/(\rho_S C_P)<<1$ with ρ_F being the density of the fluid and ρ_S that of the solid. The assumption $\varepsilon << 1$ corresponds to light fluid loading. The terms omitted in deriving (13-15) are $O(k_\alpha^{-1}L^{-1})$ and so whether or not the $O(\varepsilon)$ term should be retained in (15) depends upon the relative sizes of ε and $k_\alpha^{-1}L^{-1}$. In the cases $\varepsilon \lesssim k_\alpha^{-1}L^{-1}$, this term should be removed from (15) to reappear in some higher order equation. For a non-trivial determination of $A_0^{(\alpha)}(x,0)$ (α = F, P, S), we obtain a determinantal equation for the wavespeeds of the interface waves; these are the leaky Rayleigh wave and the Scholte wave, which we consider in turn.

The Leaky Rayleigh Wave

From consideration of the $O(k_\alpha^{-1}L^{-1})$ terms from the boundary conditions, we obtain the radiation into the fluid due to the leaky Rayleigh wave (LRW) in the form

$$\phi_L \sim \beta_0 \exp[ik_F L(\frac{C_F}{C_R}x + \gamma_F y) - \frac{C_F\gamma_P\varepsilon k_F L}{C_R d\gamma_F}(x - \frac{C_F y}{C_R\gamma_F})]H(x - y\cot\theta_M), \tag{16}$$

where C_R is the Rayleigh wavespeed ($C_F < C_R$), $\gamma_F = \surd(1-C_F^2/C_R^2)$, $\gamma_P = \surd(C_P^2/C_R^2-1)$, β_0, d are known constants, H(x) is the Heaviside step function and $\theta_M = \cos^{-1}(C_F/C_R)$. In deriving this, the regions $k_F^{-1}L^{-1} >>$, ~, << ε were treated separately and a matching procedure used [3]. We see that ϕ_L is a plane wave with a weak exponential decay (since we assume $\varepsilon << 1$) and is 'cut-off' across the polar direction $\theta = \theta_M$, this being the direction along which the wavefronts to the cylindrical wave in the fluid, ϕ_C, and ϕ_L are tangential. To remove this discontinuity, we scale

$$\theta = \theta_M + \varepsilon\bar{\theta}, \quad r = \varepsilon^{-2}\bar{r}; \quad \bar{\theta}, \bar{r} = O(1) \tag{17}$$

and seek a transition solution of the form

$$\phi = A(\bar{r},\bar{\theta})e^{ik_F\bar{r}/\varepsilon^2}. \tag{18}$$

This solution must match onto ϕ_C as $\bar{\theta} \to +\infty$ and onto $\phi_C + \phi_L$ as $\bar{\theta} \to -\infty$. To leading order, A satisfies the equation

$$\frac{1}{\bar{r}^2}\frac{\partial^2 A}{\partial\bar{\theta}^2} + 2ik_F\frac{\partial A}{\partial\bar{r}} + ik_F\frac{A}{\bar{r}} = 0 \tag{19}$$

and the appropriate solution is found to be

$$A(\bar{r},\bar{\theta}) = \beta_1 e^{-i\eta^2/2}\int_\eta^\infty e^{is^2/2}\,ds; \qquad \eta = (\bar{\theta} + \chi)\sqrt{k_F\bar{r}} \tag{20}$$

for known constants β_1, χ. The cut-off phenomenon in ϕ_L is provided by the Stokes discontinuity in the Fresnel integral in (20). Asymptotic analysis of (20) indicates that the cut-off angle is corrected from $\theta = \theta_M$ to

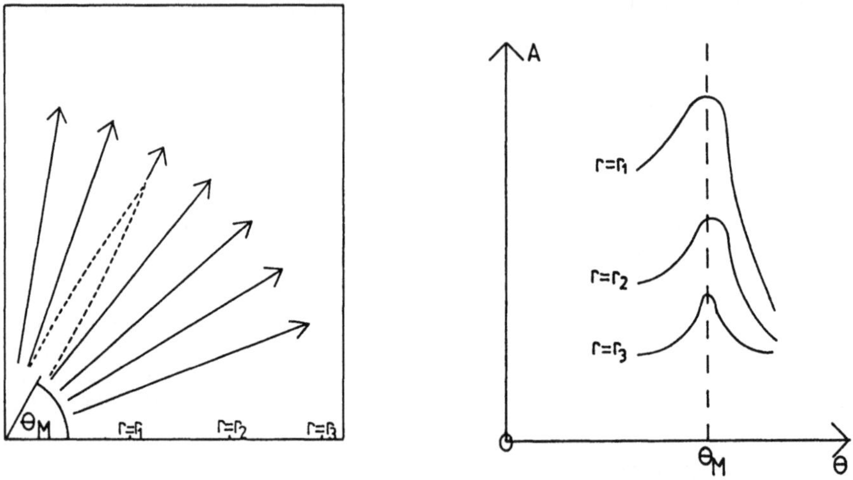

Fig.1: (a) Ray pattern with approximate beam boundary (dotted) (b) Radiated amplitude around the 'Mach' angle θ_M

$\theta = \theta_M' = \theta_M - O(\varepsilon)$, the implication being that the LRW does not propagate any unattenuated energy to infinity and that ϕ_C dominates except in the region

$$|\theta - \theta_M| = O(\varepsilon), \; 1 \ll k_F r \ll \varepsilon^{-2} . \tag{21}$$

In this region, an unattenuated beam structure of precisely one half the amplitude of the LRW is generated. This beam is intense since it does not have the $O(k_F^{-1/2} r^{-1/2})$ decay of the surrounding field. A similar beam formation was also obtained by CRIGHTON [4] when considering the forced motion of a plate under light fluid loading.

The Scholte wave

This wave propagates unattenuated along the interface and is evanescent into both media. The form of the Scholte wave is found to be

$$\phi_{Sch} \sim A_{Sch} \exp[ik_F L(1 + \frac{\alpha^2\varepsilon^2}{2}) x - k_F L\alpha\varepsilon y] \tag{22}$$

for a known $O(1)$ constant α, showing that the decay into the fluid is weak if $\varepsilon \ll 1$. The wavefronts to the Scholte and cylindrical waves are tangential in the vicinity of the interface, again suggesting the need for a transition analysis. Hence, we need an approximate solution to equation (1) which satisfies the boundary conditions and which matches $\phi_{Sch} + \phi_C$ near the interface and ϕ_C alone at large distances. Proceeding as for the LRW, the appropriate solution is found to be

$$\phi \sim \gamma_1 e^{ik_F\bar{r}/\varepsilon^2} \{e^{-i\eta^2/2} \int_\eta^\infty e^{is^2/2} ds + \frac{1}{\alpha\sqrt{(k_F\bar{r})}}\}; \quad \eta = (\bar{\theta} - i\alpha)\sqrt{k_F\bar{r}}, \tag{23}$$

where, unfortunately, the constant γ_1 cannot be determined from this analysis. A more detailed asymptotic analysis [3] does show, however, that the Scholte and cylindrical waves can be decomposed into separate,

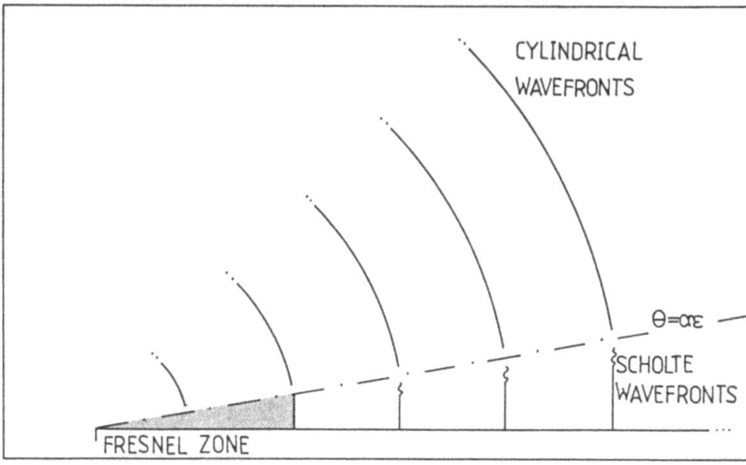

Fig.2: Leading order wavefronts in the fluid close to the interface.

identifiable contributions to the acoustic field only at large distances $\gtrsim O(k_F^{-1}\varepsilon^{-2})$ from the inhomogeneity. Even then, the Scholte wave is only felt for polar angles $\theta \lesssim \alpha\varepsilon$, across which it is cut-off. For $\theta > O(\varepsilon)$, it is found that ϕ_C dominates. It can also be shown from this analysis that the cylindrical wave vanishes to leading order on the interface.

(ii) Head Waves

Head waves are induced in the fluid by the interfacial disturbances caused by the cylindrical waves in the solid. Ray theory predicts the head waves to be of the form

$$\phi_H^{(\alpha)} \sim A_\alpha \frac{e^{ik_F r\cos(\theta-\theta_\alpha)}}{[k_F r\sin(\theta_\alpha-\theta)]^{\frac{3}{2}}} \quad \theta < \theta_\alpha, \quad \alpha = P,\ S \tag{24}$$

for constant A_α. We see that $\phi_H^{(\alpha)}$ is singular across $\theta = \theta_\alpha$, which is the direction of tangency between the head wavefronts and the cylindrical wavefronts in the fluid. Hence, we must find a transition solution $\phi_T^{(\alpha)}$ valid in the vicinity of $\theta = \theta_\alpha$, which is finite along $\theta = \theta_\alpha$, matches onto ϕ_C alone for $\theta > \theta_\alpha$ and matches onto $\phi_C + \phi_H^{(\alpha)}$ for $\theta < \theta_\alpha$. Details of the mathematical formulation of these matching conditions are in [3] and the result is that if we scale

$$\theta = \theta_\alpha + L^{-\frac{1}{2}}\bar{\theta}, \quad r = L\bar{r}, \quad \bar{\theta},\ \bar{r} = O(1), \tag{25}$$

then the transition solution is

$$\phi_T^{(\alpha)} = e^{ik_F L\bar{r}}\left\{\frac{F(\theta_\alpha)}{\sqrt{(k_F\bar{r}L)}} + B_\alpha \frac{e^{-\frac{1}{2}ik_F L\bar{r}(\theta-\theta_\alpha)^2}}{(k_F L\bar{r})^{\frac{3}{4}}} D_{\frac{1}{2}}[(\theta-\theta_\alpha)e^{-i\pi/4}\sqrt{k_F L\bar{r}}]\right\} \tag{26}$$

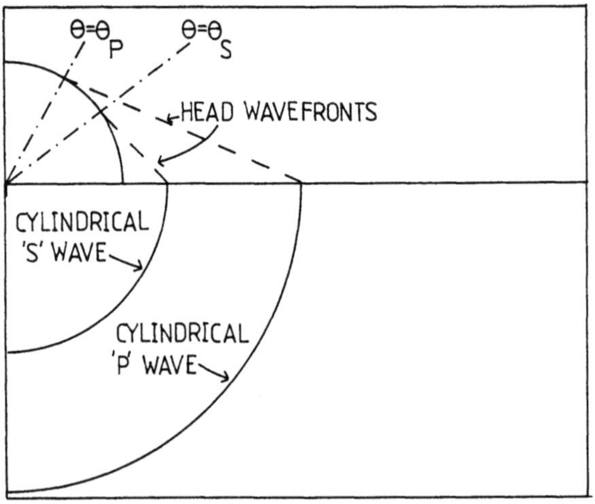

Fig.3: Geometrical construction of the head wavefronts

for constant B_α, where $F(\theta)e^{ik_F r}/\sqrt{k_F r}$ is the leading order estimate of ϕ_C and $D_{\frac{1}{2}}$ is the parabolic cylinder function [5]. Head waves are therefore a second order effect and are normally dominated by the cylindrical wave. Away from $\theta = \theta_\alpha$, they have an algebraic decay $O\left(k_F^{-3/2} r^{-3/2}\right)$, but along $\theta = \theta_\alpha$, analysis of (26) shows that the second order term is $O(k_F^{-\frac{3}{4}} r^{-\frac{3}{4}})$, a factor $O(k_F^{\frac{3}{4}} r^{\frac{3}{4}})$ stronger than that elsewhere. We have therefore predicted another beam-like structure which, this time, is not a leading order effect.

Our main conclusion is that a non-uniformity arises in the asymptotic solution for the acoustic field in the fluid whenever wavefronts to different waves are tangential. The removal of such a non-uniformity generally results in the existence of a beam-like structure in the fluid. Work is currently in progress to test these results experimentally. We note that the LRW is likely to be a dominant feature in image formation in the acoustic microscope whilst the head wave phenomenon may also be responsible for some interesting features on such images. However, the exponential decay into the fluid of the Scholte wave may prevent this wave from having any effect upon acoustical micrographs.

APPENDIX

Oblique Plane Wave Incidence Upon the Inhomogeneity

The structure of the scattered field in the fluid changes when the incoming wave is obliquely incident upon the crack. A brief discussion of the qualitative features of this case now follows.

All scattered waves have equal components of wavevector along the crack direction, which is given by that of the incident wave, m, say. This is because there is no mechanism to affect the wave motion along the direction of the crack. This distribution in phase implies that the waves emanating directly from the crack now have conical wavefronts, and are not cylindrical as before. The axis of the cones is along the crack direction. If $m < k_p$, there are two conical waves excited in the solid which, in turn, induce plane head waves in the fluid.

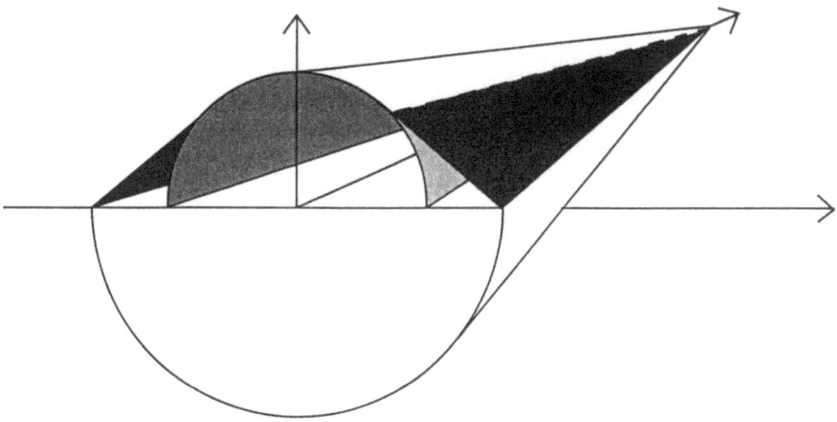

Fig.4: Conical wavefronts emanating from the inhomogeneity inducing a head wave

As m increases, the generator of tangency of each head wavefront with the conical wavefronts slides around the surface of the cone until it lies in the vertical plane through the crack as m approaches k_p. For $m > k_p$, the longitudinal conical wave in the solid is replaced by a disturbance which decays exponentially with distance from the crack, as is the corresponding head wave in the fluid. These phenomena also occur to the shear conical and head waves as m increases past k_s.

The LRW excited still generates a finite, intense beam into the fluid with its wavefronts lying tangential to generators of the conical wavefronts in the fluid. As m is increased, this generator of tangency also slides around the surface of the cone until, as m increases past k_R the beam approaches the vertical plane through the crack and becomes both shorter and narrower, eventually to become an exponentially small disturbance.

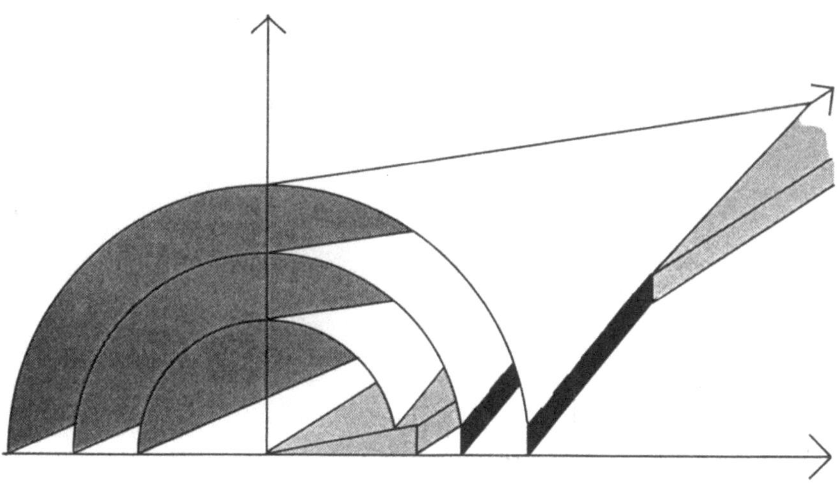

Fig.5: Rayleigh beam formation in the case of oblique incidence

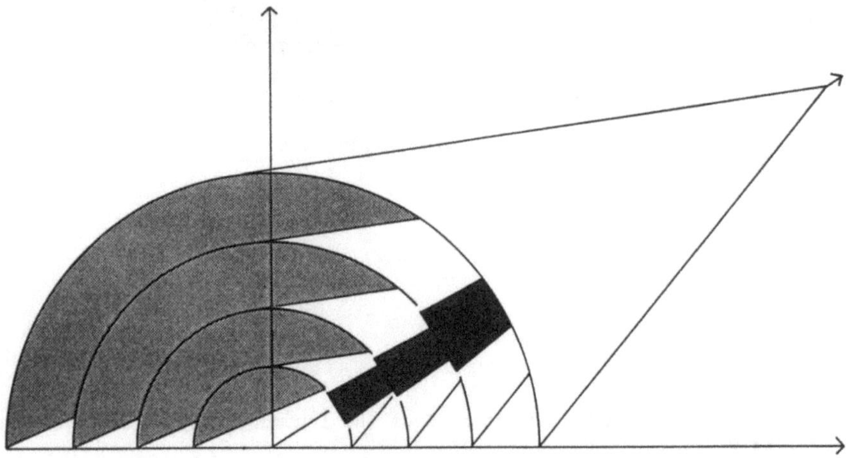

Fig.6: Interaction between Scholte wavefronts and conical wavefronts

Figures 5 and 6 show the beam formation associated with the LRW and also the Scholte wave, which has wavefronts tangential to the conical wavefronts in the vicinity of the interface. There is still the weak exponential decay into the fluid and so the conical wave eventually dominates. The region in which the Scholte and conical waves interfere, so that they cannot be decomposed into separate, identifiable contributions, is shown as a wedge running parallel to the crack with the conical and Scholte waves 'slicing' into it.

REFERENCES

1. G.A.D. Briggs: An Introduction to Scanning Acoustic Microscopy, (Oxford University Press/Royal Microscopical Society 1985)
2. M.S. Howe: I.M.A. Journal of Applied Mathematics, 36 (3) pp247-262 (1986)
3. R.H. Tew: D Phil Thesis University of Oxford (1987)
4. D.G. Crighton: J. Sound Vibration, 63 (2) pp225-235 (1979)
5. E.T. Whittaker and G.N. Watson: A Course of Modern Analysis, (Cambridge University Press 1927)

Determination of the Rheological Parameters of Viscoelastic Liquids Using Shear Surface Waves

P. Kiełczyński and W. Pajewksi

Section of Acoustoelectronics, Institute of Fundamental Technological Research, Polish Academy of Sciences, PL-00-049 Warsaw, ul. Świętokrzyska 21, Poland

1. Introduction

Measurements of the viscosity of liquids are of great importance in determining the internal structure of these liquids as well as in the control of industrial processes, e.g. during the fabrication of oils, resins or biological liquids. To this end, only shear horizontal (SH) bulk ultrasonic waves have been employed. For example, in the method proposed by *Mason* [1] one can determine the shear impedance Z_l of a liquid from the measured modulus and phase of the reflection coefficient of the volume wave at the solid-liquid interface. Due to a big difference between the impedances of the solid and liquid, the volume wave method is of low sensitivity.

In order to increase the sensitivity of the measurements the authors have proposed the use of shear *surface* waves propagating in solids, i.e. Love waves and Bleustein-Gulyaev (B-G) waves. The appropriate analytical formulae have been derived using Auld's perturbation method. From the obtained results it follows that the proposed method can be a few orders more sensitive than the volume wave one.

2. Description of the Proposed Method

An ultrasonic unit consists of a solid rectangular rod, which can support on one of its surfaces shear horizontal (SH) surface waves. For Love waves this waveguide should have a layered subsurface structure, e.g. a copper layer on a steel substrate [2]. For B-G surface waves [3] the waveguide must be made of a piezoelectric material of a given symmetry (6 mm, for instance).

Similarly to Mason's method, the proposed method is comparative, i.e. the results of two measurements are taken into account; the first for a free rod and the second for one loaded with a measured liquid. The length of interaction L between the surface wave and the investigated liquid can, of course, be simply regulated. This possibility does not exist for the classical reflectance method of Mason.

In the measurements one tries to determine the real part R and the imaginary part X of the shear impedance $Z_l = R + jX$ of a liquid, where $j^2 = -1$. Subsequently, from R and X one can determine the other parameters of the liquid, such as dynamic viscosity, shear modulus, etc. [4].

3. Perturbation Method

All the media considered in this paper are linear, i.e. the elastic body is described by linear elasticity and a piezoelectric material by the quasi-static approximation. Moreover, except for the viscoelastic liquid, all the media are lossless.

It is assumed that the shear surface wave propagates on the surface $x_3 = 0$ of the half space $x_3 > 0$, in the direction x_1 and is harmonic in time $[\exp(j\omega t)]$. Moreover, one assumes that the considered surface wave is uniform in the direction x_2, i.e. the considered problem is two-dimensional (x_1, x_3).

In determining the difference $\Delta\beta$ between the wave number of the surface wave propagating on the surface loaded with a viscoelastic liquid and the free surface, a perturbation formula is of essential significance. This formula is derived by [Ref. 5, Eq. (12.7)], namely,

$$\Delta\beta = \frac{j[v_2^*(x_3) Z_A v_2(x_3)]|_{x_3=0}}{4P} \quad , \tag{1}$$

where P, given by

$$4P = 2\,\mathrm{Re}\left\{\int_0^\infty \{-v_2^*(x_3)T_{21}(x_3) + \varphi(x_3)[j\omega D_1(x_3)]^*\}dx_3\right\} \quad , \tag{2}$$

is the average power flow per unit width along x_2 for the unperturbed surface. The asterisk (*) denotes complex conjugation. Moreover, v_2, T_{21} and φ, D_1 correspond, respectively, to particle velocity, stress, electric potential and displacement components. The acoustic surface impedance for the perturbed surface, i.e. loaded with the measured liquid, is given by

$$Z_A = -\left.\frac{T_{23}(x_3)}{v_2(x_3)}\right|_{x_3=0} \quad , \tag{3}$$

where the stress component T_{23} and particle velocity v_2 correspond to the perturbed surface of the waveguide.

Due to the fact that the velocity of the SH volume wave in the viscoelastic liquid is many times lower than in the solid, one can demonstrate that

$$Z_A \approx -Z_l \quad . \tag{4}$$

Finally, one obtains that

$$\Delta\beta = -\mathrm{j}CZ_l \quad , \tag{5}$$

where the coefficient of proportionality C is given by

$$C = \frac{[v_2(x_3)]^2|_{x_3=0}}{4P} \tag{6}$$

and depends on the type and frequency ω of the surface wave and on the material and geometrical parameters of the surface waveguide. In (6) $v_2(x_3)$ is taken for the unperturbed surface.

Equation (5) confirms the intuitively obvious statement that the real part R of the shear impedance Z_l of the liquid corresponds to the additional attenuation of the surface wave $\Delta\alpha$, and the imaginary part X causes an additional phase shift of the wave $\Delta\theta$, namely

$$\Delta\alpha = 20 \ln CLR \tag{7}$$

$$\Delta\theta = CLX \quad , \tag{8}$$

where L is the actual length of interaction between the surface wave and the viscoelastic liquid.

The equations (7) and (8) are fundamental for the determination of R and X from direct measurements of $\Delta\alpha$ and $\Delta\theta$.

4. The Coefficient of Proportionality C for Love Waves

Employing the appropriate constitutive relations [2], one obtains from (6) that for the SH surface waves of the Love type one has

$$C = \frac{\omega}{\frac{c_{44}^s \cos^2(k_L h b^1)}{b^a} + k_L h c_{44}^h \left[1 + \frac{\sin(2k_L h)}{2k_L h b^1}\right]} \quad , \tag{9}$$

where

$$b^1 = \sqrt{(v_L/v_h)^2 - 1} \quad , \quad b^a = \sqrt{1 - (v_L/v_s)^2} \tag{10}$$

and

$$v_h = \sqrt{c_{44}^h/\varrho_h} \quad , \quad v_s = \sqrt{c_{44}^s/\varrho_s} \quad . \tag{11}$$

Moreover, $k_L = \omega/v_L$ denotes the wave number of the Love wave, and c_{44}^s, ϱ_s and c_{44}^h, ϱ_h correspond, respectively, to the elastic constant and mass density in the substrate $x_3 > h$ and in the surface layer of thickness h.

It is interesting to note that the coefficient C for the high frequency region of the Love wave ($k_L h \gtrsim 2\pi$) is independent of frequency and tends to $1/(Z_h h)$.

5. The Coefficient of Proportionality for B-G Surface Waves

Taking into account the appropriate constitutive relations [3] for the B-G surface wave propagating on a metallized surface of the piezoelectric material of symmetry 6 mm, one has

$$C = \frac{\omega k_{15}^2}{c_{44}(1 + k_{15}^2)} \quad , \tag{12}$$

where k_{15}^2 is the electromechanical coupling coefficient of the piezoelectric material and c_{44} is its elastic constant.

Unlike the case of Love waves, the coefficient C for the B-G wave is always a linear function of the frequency, as in the case of Rayleigh surface waves [5].

6. Sensitivity of the Proposed and the Classical Reflectance Methods

For the classical reflectance method employing volume SH waves, one has [4]

$$R = Z_q(1 - r)/2 \quad , \tag{13}$$

$$X = Z_q \Delta\theta/2 \quad , \tag{14}$$

where Z_q is the shear impedance of the solid probe, e.g. quartz, and r and $\Delta\theta$ denote, respectively, the modulus and phase of the reflection coefficient at the solid-liquid interface (for SH volume waves).

The appropriate quantities corresponding to the method proposed by the authors employing shear surface waves will be further denoted by the upper index "saw". Analogously, those for the volume reflectance method by "vol".

From (7) and (8), and (13) and (14), one can derive that

$$\frac{\Delta\alpha^{\mathrm{saw}}}{\Delta\alpha^{\mathrm{vol}}} = \frac{\Delta\theta^{\mathrm{saw}}}{\Delta\theta^{\mathrm{vol}}} = CLZ_q \quad . \tag{15}$$

In the case of Love surface waves, in the high frequency region ($k_{\mathrm{L}}h \gtrsim 2\pi$), this ratio tends to

$$C \underset{k_{\mathrm{L}}h \to \infty}{\longrightarrow} \frac{L}{h}\frac{Z_q}{Z_h} \quad . \tag{16}$$

Since, in practice, $Z_q/Z_h \approx 1$ and $L/h \cong 10^2 - 10^4$, the proposed method can be significantly more sensitive than the classical one which employs volume waves [6].

7. Conclusion

The application of SH surface waves to the measurement of the rheological parameters of viscoelastic liquids leads to a method which can be a few orders of magnitude more sensitive than the classical Mason method employing volume waves. As a consequence, the proposed surface wave method can be used for measurements of liquids having much lower viscosity than has been possible previously.

References

1. W.P. Mason, W.O. Baker, H.J. McSkimin, J.H. Heiss: Phys. Rev. **75**, 936–946 (1949)
2. G.W. Farnell, E.L. Adler: In *Physical Acoustics*, Vol. IX, ed. by W.P. Mason (Academic, New York 1972)
3. E. Dieulesaint, D. Royer: *Ondes Elastiques dans les Solides* (Masson, Paris 1974)
4. R.S. Moore, H.J. McSkimin: In *Physical Acoustics*, Vol. VI, ed. by W.P. Mason (Academic, New York 1970)
5. B.A. Auld: *Fields and Waves in Solids*, Vol. II (Wiley, New York 1973)
6. P. Kielczyński, W. Pajewski: J. Acoust. Soc. Am. **81**, 3 (1987)

The Influence of Off-Centre Defects on SAW Characteristics

J. Kapelewski and W. Przybylski

Institute of Technical Physics, Kaliski st.
PL-01-908 Warsaw 49, Poland

There is a growing interest recently in studying the effects of impurity defects of various kinds on SAW characteristics. One stimulus for such investigations is to provide SAW substrates, the mechanical and electrical parameters of which would be most appropriate for such or the other sort of applications. Especially interesting in this respect is the possibility to employ peculiar properties of many off-centre impurity defects to induce or enhance piezoelectric features in the, appropriately doped, near-surface region of some crystals [1].

The objective of this paper is to outline a consistent semi-phenomenological treatment of the parameters determining the surface acoustic impedance changes induced by defects of this sort. The proposed approach is formulated in quite general terms and can easily accommodate any type of off-centre defects, apart from its physical nature. Its important advantage lies in its inherent analytical simplicity as well as in a capability to treat even complex crystal structures. For the sake of brevity, we focus here on a description of the mechanical side of the problem, relating all the phenomena responsible for inducing local (average) piezoelectric features to the piezoelastic ones. The latter, having common source with the former for any crystal with ionic components, consists in the coupling between the local distortion of a dipole (mechanical) moment density with the corresponding strain field, considered, in the last resort, to be induced by a propagating SAW.

1. The Formulation of the Mechanical Model

Let us start from the general formula for the potential energy of a harmonic lattice with point defects

$$U = -\sum_{s} w_{\alpha}(s)\, u_{\alpha}(s) + \frac{1}{2}\sum_{ss'} K_{\alpha\beta}(s,s')\, u_{\alpha}(s)\, u_{\beta}(s') \qquad (1.1)$$

with $s=(\underline{m}, \varkappa)$ denoting lattice sites, $\underline{m}$ - the position of the unit cell, $\varkappa$ is to distinguish different atoms in the same unit cell, $\underline{u}$ denotes the displacement vector, α, β - the Cartesian coordinates.

In constructing a physically reasonable model treatment of such a defect system one should take into account the qualitative distinctiveness of a force and displacement distribution in microscopic regions around defects. Consequently, these regions should be separately treated in a way, which would be capable of describing the effect of a near-defect microstructure on the elastic behaviour of the crystal, expressed by the strain tensor. In view of the inherently macroscopic nature of the latter quantity, it is advisable to treat volume elements of such a size, in the crystal, globally, in an averaged way. It effectively leads to a description in terms of a uniform lattice, composed either of unit cells or extended unit cells of a fixed volume with the same symmetry, chosen appropriately for the size of the foregoing "defected" regions (for simplicity we will assume a principal identity of all the defects to be treated here). Although the latter construction becomes apparently ill defined, when one considers defects as located at the centre of the corresponding elements, it is however quite acceptable as a contrivance in a modelling procedure. The reason for this is a slowly varying (macroscopic) character of the strain tensor as compared to the inherently microscopic thickness of an "unfitness" shell , effectively contributing to a set of measure zero when passing to a continuum picture. For the volume elements, one can define the dipole moment and the elasticity tensor by means of the well known formulae

$$P_{\alpha\beta} = (2\Omega)^{-1} \sum_{\Omega} [X_\alpha(s) F_\beta(s) + X_\beta(s) F_\alpha(s)] \tag{1.2}$$

$$C_{\alpha\beta\gamma\delta} = \Omega^{-1} \sum_{\Omega} X_\alpha(s)\, \Phi_{\beta\gamma}(s,s')\, X_\gamma(s') \tag{1.3}$$

with s denoting here atomic sites within Ω , and X(s) - the corresponding position vector in the local frame of coordinates.

For describing the fields of relative displacements near to the defect sites by means of quantities relating to the particular volume elements treated as a whole, one can use the method described in [2] , or simply perform averaging over symmetry-adapted components of the displacement vectors taken for all the coordination spheres (the position of the symmetry centre is to be assigned to the site of the substituted atoms).

The resulting quantities $Q_{\Gamma\gamma}(L)$ being characterized by the irreducible representation of the crystal point group (the subscript γ is to label various linear combinations of displacement vector transforming itself according to Γ) provide an important, for many applications, symmetry characteristic of the near-defect microdisplacement distribution.

Such a representation greatly facilitates the analytical formulation of the proposed model description. After doing the respective transformation and separating out the contribution to the displacements related to a local deformation, the elastic energy given in (1.1) can be modelled according to the above mentioned principles with the help of the formula

$$U=-\sum P_{\Gamma\gamma}(\underline{L})\varepsilon_{\Gamma\gamma}(\underline{L})-\sum w_{\Gamma\gamma}(\underline{L})Q_{\Gamma\gamma}(\underline{L})c(L)+\frac{1}{2}\sum C_{\Gamma}\varepsilon^{2}_{\Gamma\gamma}(\underline{L})-$$
$$-\sum g_{\Gamma}Q_{\Gamma\gamma}(\underline{L})\varepsilon_{\Gamma\gamma}(\underline{L})c(L)+\frac{1}{2}\sum Q_{\Gamma\gamma}(\underline{L})K_{\Gamma\Gamma'}(\underline{L},\underline{L}')Q_{\Gamma'\gamma'}(\underline{L}')c(\underline{L})c(\underline{L}'), \tag{1.4}$$

where $c(\underline{L})=0$ or 1, - denotes the number indicating the occupation of the $\underline{L}$-th "cell" by the defect.

In the adiabatic approximation it leads to the coupled equations of motion

$$-M\omega^2\underline{u}(\underline{L})+\underline{\underline{C}}\,\underline{\varepsilon}(\underline{L})-\underline{\underline{g}}\,\underline{Q}(\underline{L})\,c(\underline{L})=0 \qquad \text{(a)}$$
$$\sum_{L'}\underline{\underline{K}}(\underline{L},\underline{L}')\,\underline{Q}(\underline{L})\,c(\underline{L}')-\underline{\underline{g}}\,\underline{\varepsilon}(\underline{L})=0. \qquad \text{(b)} \tag{1.5}$$

For the sake of brevity, in (1.5) a shorthand operator notation has been used. By eliminating $Q(\underline{L})$ in (1.5a) with the help of (1.5b) one obtains an effective nonlocal elasticity tensor in the form

$$\underline{\underline{C}}_{eff}(\underline{L},\underline{L}')=\underline{\underline{C}}\,\delta_{\underline{L},\underline{L}'}-\underline{\underline{D}}(\underline{L},\underline{L}')\,\underline{\underline{g}}^2\langle c(\underline{L})\,c(\underline{L}')\rangle\equiv\underline{\underline{C}}\,\delta_{\underline{L},\underline{L}'}-\delta\underline{\underline{C}}(\underline{L},\underline{L}') \tag{1.6}$$

with $\underline{\underline{D}}=\underline{\underline{K}}^{-1}$.

Regarding the statistical side of the model we assume as in [4] a random (Gaussian) character of the defect distribution along the lattice planes of the surface cut, and with distance from the surface a deterministic one, having exponential decay. This is embodied in the formulae

$$c(\underline{L})=c(\underline{l})\exp(-\alpha n)\quad,\quad\langle c(\underline{l})\rangle=c$$
$$\langle c(\underline{L})\,c(\underline{L}')\rangle=\delta W(\underline{l}-\underline{l}',n,n')=\frac{\Omega_s\delta}{\pi R^2}\exp\left[\frac{-(\underline{l}-\underline{l}')^2}{R^2}-\alpha(n+n')\right] \tag{1.7}$$

with $\underline{L}=(\underline{l},n)$, Ω_s- the elementary area and R denoting the correlation radius, $\delta=c^2$.

An important advantage of the foregoing description lies in its spatial uniformity, which allows a direct passage to the continuous one in the framework of the conventional pseudocontinuum theory. Keeping the same notation, the strain field contribution to the density of the elastic energy can then be written as

$$U_\varepsilon(\underline{r})=\frac{1}{2}\underline{\varepsilon}(r)\int d\underline{r}'\,\underline{\underline{C}}_{eff}(\underline{r},\underline{r}')\,\underline{\varepsilon}(\underline{r}')-c(z)\,\underline{\underline{P}}(\underline{r}) \tag{1.8}$$

with $c(z)=c\exp(-gz)$.

Consequently, for the stress tensor we obtain

$$\sigma_{\alpha\beta} = C_{\alpha\beta\gamma\delta}\,\varepsilon_{\gamma\delta}(\underline{r}) - \int \delta C_{\alpha\beta\gamma\delta}(\underline{r},\underline{r}')\,\varepsilon_{\gamma\delta}(\underline{r}')\,d\underline{r}' - c(z)\,P_{\alpha\beta}(\underline{r}) \qquad (1.9)$$

and the boundary conditions at the free surface take (in the Cartesian coordinates) the nonlocal form

$$C_{3\beta\gamma\delta}\,\varepsilon_{\gamma\delta}(\underline{\rho},0) - \int \delta C_{3\beta\gamma\delta}(\underline{\rho}-\underline{\rho}',0,z')\,\varepsilon_{\gamma\delta}(\underline{\rho}',z')\,d\underline{r}' - c\,P_{3\beta}(\underline{\rho},0) = 0. \qquad (1.10)$$

These relations, together with the conventional equilibrium conditions within the body, fully determine the elastic distortion of the crystal in the presence of the near-surface noncentral defects.

2. The Role of the Temperature

Temperature effects originating from the defects can be treated within the thermal equilibrium model, as follows. The effective force density induced by an external distortion $\underline{Q}$ at $\underline{r}$, after taking into account the polarising effect of the other defects, is given by

$$\underline{f}(\underline{r},\underline{r}_o) = \underline{\underline{K}}(\underline{r},\underline{r}_o)\,\underline{Q}(\underline{r}_o)\,c(z_o) + Z^{-1}\int d\underline{r}'\,\underline{\underline{K}}(\underline{r}',\underline{r}_o)\int d\underline{\xi}(\underline{r}')\,c(z')\,\underline{\xi}(\underline{r}')\times$$
$$\times\exp\left\{-\beta U_o(\underline{\xi}(\underline{r}')) + \beta\underline{\xi}(\underline{r}')\,\underline{f}(\underline{r}',\underline{r}_o)\right\} \qquad (2.1)$$

with $\beta = (k_B T)^{-1}$, U - denoting the corresponding local ("diagonal") portion of (1.4), and

$$Z = \int d\underline{\xi}\,\exp(-\beta U_o(\underline{\xi})).$$

The second term in (2.1) expresses the contribution from the neighbouring defects due to the thermal phonon field by means of the familiar Boltzmann form of the density matrix.

Expanding this to the linear terms in $\beta\,\underline{\xi}\,\underline{f}$ yields

$$\underline{f}(\underline{r},\underline{r}_o) = \underline{\underline{K}}(\underline{r},\underline{r}_o)\,\underline{Q}(\underline{r}_o)\,c(z_o) + \beta\,\underline{\underline{\Delta}}\int d\underline{r}'\,\underline{\underline{K}}(\underline{r},\underline{r}_o)\,c(z')\,\underline{f}(\underline{r}',\underline{r}_o) \qquad (2.2)$$

where

$$\Delta_{\alpha\beta} = Z^{-1}\int d\underline{\xi}\,\xi_\alpha\xi_\beta\,\exp\left\{-\beta U_o(\underline{\xi})\right\}. \qquad (2.3)$$

In the 2D Fourier transforms (corresponding to the wave vector of the propagating SAW), we get

$$\underline{f}(\underline{k},z,z_0)=\underline{\underline{K}}(\underline{k},z,z_0)\underline{Q}(\underline{k},z_0)c(z_0)+ \\ +\beta\underline{\underline{\Delta}}\int dz'\underline{\underline{K}}(\underline{k},z,z')c(z')\underline{f}(\underline{k},z',z_0). \tag{2.4}$$

Hence, (in the simplified continuous matrix notation with respect to z,z´), we read

$$\underline{\tilde{f}}(\underline{k})=[\underline{1}-\beta\underline{\underline{\tilde{K}}}(\underline{k})\underline{c}\underline{\underline{\Delta}}]^{-1}\underline{\underline{\tilde{K}}}(\underline{k})c\underline{Q}(\underline{k}). \tag{2.5}$$

Consequently, for the temperature-dependent matrix of the effective defect--defect interactions one finally obtains

$$\underline{\underline{\tilde{K}}}_T(\underline{k})=[\underline{1}-\beta\underline{\underline{\tilde{K}}}(\underline{k})\underline{c}\underline{\underline{\Delta}}]^{-1}\underline{\underline{\tilde{K}}}(\underline{k}). \tag{2.6}$$

The respective temperature - dependent contribution to the P tensor can be obtained by inserting into the density matrix $Z=\exp(-\beta U)$, the externally (SAW) induced component of $\underline{Q}$ [3].
In this way one gets

$$P_T(\underline{r})=P(\underline{r})[1-\beta\int d\underline{r}'\underline{w}(\underline{r}')\underline{Q}(\underline{r}')c(z')]. \tag{2.7}$$

Thus the foregoing contribution is

$$\delta\underline{\underline{P}}(\underline{r})=-\beta\underline{\underline{P}}(\underline{r})\int d\underline{r}'\underline{w}(\underline{r}')\underline{Q}(\underline{r}')c(z'). \tag{2.8}$$

Comparing (2.8) with (1.4), we arrive at the following formula for the temperature-dependent contribution to the piezoelastic tensor:

$$\delta\underline{\underline{g}}(\underline{r},\underline{r}')=-\beta\underline{\underline{P}}(\underline{r})\underline{w}(\underline{r}')c(z'). \tag{2.9}$$

A set of the defect dipole moments produces displacements of the local mass centres given by

$$u_\alpha(\underline{r})=\int d\underline{r}'G_{\alpha\beta,\gamma}(\underline{r},\underline{r}')c(z')P_{\beta\gamma} \tag{2.10}$$

with $\underline{G}$ being the Green´s function for the host semi-infinite solid. It leads to the near-surface strain tensor of the form

$$\underline{\underline{\varepsilon}}_P(\underline{r})=\int d\underline{r}'[\underline{\underline{G}}^{(2)}(\underline{r},\underline{r}')c(z')\underline{\underline{P}}(\underline{r}')\underline{n}-\underline{\underline{G}}^{(1)}(\underline{r},\underline{r}')c'(z')\underline{\underline{P}}(\underline{r}')\underline{n}]+ \\ +\int_S d\underline{\rho}'\underline{\underline{G}}^{(1)}(\underline{\rho}-\underline{\rho}';z,0)c\underline{\underline{P}}(\underline{r}')\equiv\int d\underline{r}'\underline{\underline{H}}^{(s)}(\underline{r},\underline{r}')\underline{\underline{P}}(r') \tag{2.11}$$

where

$$G^{(2)}_{\alpha\beta,\gamma'\gamma} = \frac{1}{2}\left[G_{\alpha\beta,\gamma'\gamma} + G_{\gamma\beta,\gamma'\alpha}\right]$$

$$G^{(1)}_{\alpha\beta,\gamma} = \frac{1}{2}\left[G_{\alpha\beta,\gamma} + G_{\gamma\beta,\alpha}\right] \tag{2.12}$$

$$c'(z') = \frac{dc(z')}{dz'}\ , \quad \underline{n} = (0.0.1)\ .$$

Inserting (2.11) into (1.9) and (1.10) and making use, in the bulk region, of the set of equilibrium conditions $\delta_{\alpha\beta,\beta} = 0$, results in the closed set of homogeneous equations for the macroscopic dipole-force tensor which can easily be treated, with the help of the surface fluctuation waves method [4].

From the harmonic contribution to the respective energy density term $P_{\gamma\gamma}(r)\cdot \varepsilon_{\gamma\gamma}(r)$ one obtains the additional perturbation to $\underline{K}$, as given by the following simple formula

$$\delta\hat{\underline{K}}(\underline{r},\underline{r}') = \beta^2 \underline{P}^2 \underline{H}\,\delta w(\underline{r})\, w(\underline{r}')\, W(\underline{\rho}-\underline{\rho}', z, z') \tag{2.13}$$

$$\text{with} \quad \underline{H} = \iint d\underline{r}_1\, d\underline{r}_2\, \underline{H}^{(s)}(\underline{r}_1, \underline{r}_2)\ . \tag{2.14}$$

The above outline of the way in which the effect of the temperature on mechanical and coupling parameters of a defect near-surface layer can be consistently treated, clearly indicates a distinct generally nonlocal nature of the corresponding contributions. The extension of the approach presented here to more specific types of noncentral defects, as applied to a controlled modification of SAW parameters, will be reported later.

3. Literature

1. V.S. Vihnin: In Fiz. tverd. tela, 26, p.906 (1984).
2. J. Kapełewski: In Electromechanical Interactions in Deformable Solids and Structures, North-Holland 1987.
3. The use of the equilibrium form of the density matrix is justified here by small values of the frequency in the acoustic limit.
4. J. Kapelewski: In J. Tech. Phys., 25, p.147 (1984).

On the Propagation of Surface Waves

T.P. Ivanov

Faculty of Mathematics and Mechanics, University of Sofia,
5, Anton Ivanov Blvd., BG-1126 Sofia, Bulgaria

1. Introduction

Rayleigh waves in classical, isotropic elasticity /1/ may be regarded as a superposition of two nondispersive inhomogeneous plane waves with the same direction of propagation along the stress-free surface and direction of attenuation orthogonal to the surface, equal phase speeds but with different values of the attenuation coefficients. The displacement field decays to zero uniformly with increasing distance from the stress-free surface. These characteristic features of Rayleigh waves do not remain valid when thermoelastic interactions /2-6/, viscoelastic properties /7-8/, constrained materials /9/, etc., are considered. Then the surface waves may be interpreted as a superposition of dispersive inhomogeneous plane waves. The superposed waves have different directions of propagation and attenuation and different phase speeds and attenuation coefficient values. Their directions of propagation are not parallel to the stress -free surface. For each of them the plane of constant phase and the plane of constant amplitude are not orthogonal. The displacement /or displacement-temperature/ field decays again to zero with increasing distance from the surface but not uniformly. These results were obtained for the thermoelastic Rayleigh waves in /10/. Appropriate criteria for behaviour at infinity were also discussed there in order to preserve some of the most characteristic features of the Rayleigh waves known from classical elasticity.

In the present paper a general treatment of the problem is proposed. It concerns propagation of surface waves in linear media and their behaviour at infinity. Some numerical results for the speeds and the attenuation coefficients of the superposed waves and the angles between the normals of the planes of constant phase (or amplitude) and the surface of the half-space are also presented when generalized thermoelastic media are considered.

2. Solution for Surface Waves in Linear Media

Let the semi-infinite body occupy the region $z \geq 0$, where Oxyz is a rectangular Cartesian coordinate system. The displacement components are denoted by u_i, $i=1,2,3$. When thermal effects are also included, the temperature change with respect to constant surrounding temperature is denoted by u_4. The motion is governed by the linearized equations of elasticity (isotropic or anisotropic, without or with constraints), thermoelasticity (classical or generalized), viscoelasticity, etc., where external body forces and heat sources are assumed to be absent. It is a straightforward task to construct a surface-wave solution when it is assumed that, throughout the motion, the boundary of the body is stress-free and that heat is exchanged freely with the surroundings according to the linearized radiative condition. Then the solution which does not depend on y has the form /1,2,6, 7,9,11/

$$u_m = \sum_{n=1}^{N} B_{mn} e^{F_n}, \tag{1}$$

where the complex constants B_{mn} are determined up to a multiplier; m=1,2,3 and N=2 if thermal effects are not included; m=1,2,3,4 and N=3 if thermomechanical models are considered. The exponents $F_n = F_n(x,z,t)$ are given by

$$F_n = -(q_x x + q_z^{(n)} z) - i\omega\left(t - \frac{x}{v_x} - \frac{z}{v_z^{(n)}}\right), \tag{2}$$

where t is the time,

$$v_x = v_S(\mathrm{Re}\, s^{-1})^{-1}, \qquad v_z^{(n)} = -v_S(\mathrm{Im}(b_n s^{-1}))^{-1},$$
$$q_x = q_o \Omega \,\mathrm{Im}\, s^{-1}, \qquad q_z^{(n)} = q_o \Omega \,\mathrm{Re}(b_n s^{-1}). \tag{3}$$

Here ω is the angular frequency, taken to be real and positive and p is the complex wave number. The following constants are also used: the rotational wave speed v_S, the dimensionless complex wave speed $s=\omega/pv_S$, the dimensionless frequency $\Omega=\omega/\omega_0$ (ω_0 is a properly chosen characteristic frequency), the characteristic attenuation $q_o=\omega_0/v_S$. The complex constants b_n depend on the material constants and s, where s satisfies a transcendental secular equation. The explicit form of this equation and the constants b_n are presented for different cases in /1,2,6,7,9,11/ respectively.

Let us concentrate now our attention on the nth superposed wave in the surface wave. Planes of constant phase are given by

$$\frac{x}{v_x} + \frac{z}{v_z^{(n)}} = t + C_1 \tag{4}$$

at any particular time t and constant C_1. The phase speed is defined as

$$v_n = |v_x v_z^{(n)}| (v_x^2 + v_z^{(n)2})^{-\frac{1}{2}}. \tag{5}$$

The quantities v_x and $v_z^{(n)}$ represent the speeds of the points at which the plane of constant phase intersects the axis x and z respectively and

$$\alpha_n = \arctan(v_x / v_z^{(n)}) \tag{6}$$

is the angle between the normal to the planes of constant phase and the x-axis (Fig.1). Since v_x is one and the same for each of the superposed waves, their planes of constant phase for any fixed t and C_1 intersect the surface of the half-space in one and the same line that moves in the direction of the x-axis with a speed v_x, usually called the speed of the surface wave.

Planes of constant amplitude are given by

$$q_x x + q_z^{(n)} z = C_2 , \tag{7}$$

where C_2 is a constant. Here q_x and $q_z^{(n)}$ may be considered as attenuation coefficients of the points along the axes x and z respectively. The angle β_n between the normal to the planes of constant amplitude and the x-axis is equal to

$$\beta_n = \arctan(q_z^{(n)}/q_x) . \tag{8}$$

Since q_x is one and the same for each of the superposed waves, their planes of constant amplitude for any fixed C_2 intersect the surface z=0 in one and the same line for which the attenuation are equal (Fig.1).

The criteria for the behaviour of the superposed waves at infinity adopted in /10/ are:

(i) the amplitudes of the superposed waves do not increase for the points of intersection of the propagating planes of constant phase with any fixed line parallel to the x-axis,

(ii) the amplitudes of the superposed waves decay exponentially with distance from the surface on any fixed line parallel to the z-axis,

(iii) the amplitudes of the superposed waves do not increase for any fixed point on a plane of constant phase in the direction of its propagation,

(iv) the amplitudes of the superposed waves tend to zero with distance from the surface for the points of the lines in which the planes of constant phase at any fixed time t intersect the coordinate plane Oxz.

If we assume for definiteness hereafter that $v_x>0$, we have

$$\text{(i)} \quad q_x \geqslant 0 , \qquad \text{(ii)} \quad q_z^{(n)} > 0 \tag{9}$$

and also

$$\text{(iii)} \ v_z^{(n)} > 0 \ (\alpha_n > 0), \quad \text{(iv)} \ v_z^{(n)} q_z^{(n)} > v_x q_x \ (\beta_n > \alpha_n) \tag{10}$$

or

$$\text{(iii)} \ v_x q_z^{(n)} + v_z^{(n)} q_x \leqslant 0 \ (\beta_n \leqslant \alpha_n + \tfrac{\pi}{2}), \quad \text{(iv)} \ v_z^{(n)} < 0 \ (\alpha_n < 0). \tag{11}$$

The case $\alpha_n=0$ is trivial since then the criteria (iii) and (iv) coincide with the criteria (i) and (ii) respectively and do not add extra restrictions. From (8) and (9) it follows that $0<\beta_n \leqslant \frac{\pi}{2}$. The conditions (10) and (11) show that the criterion (iii) does not add additional restrictions on the behaviour at infinity of the nth wave if $\alpha_n>0$ and the criterion (iv) does not add restrictions if $\alpha_n<0$. In classical elasticity we have $B_{2n}=0$, $\alpha_n=0$, $\beta_n=\frac{\pi}{2}$, n=1,2.

The field functions $u_-(x,z,t)$ decay to zero as $z\to\infty$ uniformly in x if and only if $q_x=0$. This statement follows from (1) and (2) since the amplitudes of the superposed waves increase infi-

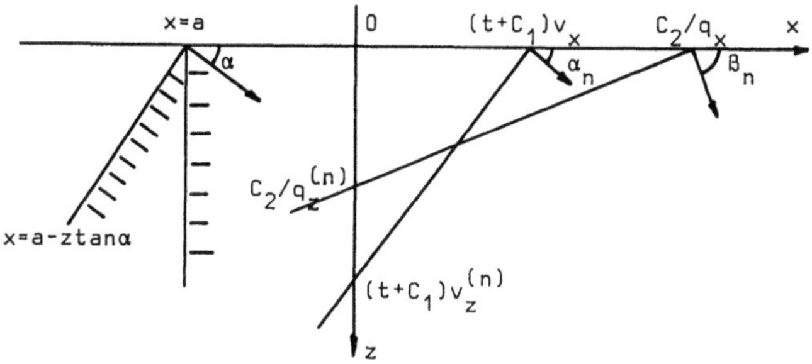

Fig.1

nitely for any fixed $z>0$ if $q_x>0$ and $x\to-\infty$. If we like to preserve the property that solution (1) decays to zero as $z\to\infty$ uniformly in x, y and t for $q_x>0$, the variable x has to be bounded below. From the criteria (i) and (ii) it follows that the solution (1) tends to zero as $z\to\infty$ uniformly in x, y and t in the portion of the half-space for which $x\geqslant a$, where a is an arbitrarily fixed constant. This region could be extended if $\min\alpha_n=\alpha>0$ and criterion (iv) is applied. Then the solution tends to zero as $z\to\infty$ uniformly in x, y and t in the region $z\geqslant 0$ and $x\geqslant a-z\tan\alpha$ shown in Fig.1.

The speeds and the attenuation coefficients of the superposed waves along the z-axis and the phase speeds v_n could be determined from (6), (8) and (5) respectively when v_x, q_x, α_n and β_n are known.

The basic open problem in these considerations is the determination of the number of surface waves that satisfy the secular equation and desired regularity properties at infinity (i)-(iv). This problem is solved completely only for classical elasticity where a unique surface wave exists /12/.

The analytical and numerical results in /7,8/ show that a unique surface wave exists also in the cases of viscoelastic media considered in these papers. It is perhaps worth noting here that the conclusion of possible existence of more than one surface wave refers to surface waves that satisfy the secular equation and only the criteria (i) and (ii) at infinity, i.e. at least one of the criteria (iii) or (iv) is violated. The results in this section give also an explanation of the fact why the speed of the surface wave may be greater than the body-wave speeds /7,8/. The reason is that the surface wave speed v_x is not a phase speed and it is greater than the phase speeds of the superposed waves when $\alpha_n\neq 0$.

It appears likely that in classical thermoelasticity there always is a unique quasi-elastic wave (E-mode), which resembles the classical Rayleigh wave but is subject to damping and dispersion, and a unique quasi-thermal wave (T-mode) in which the disturbance is dominated by thermal diffusion /4,10/. From the numerical results in the next section the same conclusion follows in the case of generalized thermoelasticity when the modified Fourier's law /13/ or the generalized entropy production inequality /14/ is used.

3. Numerical Results for Surface Waves in Generalized Thermoelasticity

In generalized thermoelasticity when the half-space is thermally insulated the secular equation has the form given in (12) viz.

$$(2-s^2)^2(b_1^2+b_1b_2+b_2^2-1+v_o^2s^2)-4b_1b_2b_3(b_1+b_2)=0 \ . \qquad (12)$$

Here $v_o=v_S/v_T$ and v_T is the dilatational wave speed. The constants b_n are given by

$$b_1^2,b_2^2=1-\frac{1}{2}v_o^2s^2(d\pm(d^2-4f)^{\frac{1}{2}}) \ , \qquad b_3^2=1-s^2 \ . \qquad (13)$$

The square root in $(13)_1$ is chosen to have a negative imaginary part,

$$d=1+h+eg+i(1+e)\Omega^{-1} \ , \qquad f=h+i\Omega^{-1} \qquad (14)$$

and e is the thermoelastic coupling constant /2/. The constants g and h are the dimensionless constants connected with the finite speed of heat propagation. When the modified Fourier's law is used /13/, $g=h(>0)$ is equal to the relaxation time multiplied by ω_0. In the model based on the generalized entropy production inequality /14/, the constants connected with the finite speed (again multiplied by ω_0) satisfy the inequality $g\geqslant h(>0)$. The case of classical thermoelasticity corresponds to $g=h=0$ /2,10/.

The numerical computations refer to media of polyvinylbutyral ($e=0.432$, $\nu=0.4$) in the frequency range $10^{-3}\leqslant\Omega\leqslant10^{2}$ at different values of the constants g and h. Here ν is the Poisson's ratio. Some of the results obtained for the E-mode are presented in Figs.2-6. They show the dimensionless speed v_x/v_S and the dimensionless attenuation coefficient q_x/q_o along the x-axis, the angles α_n and β_n, n=1,2,3, between the normals of the planes of constant phase and amplitude respectively and the x-axis, all of them versus the dimensionless frequency Ω. The superposed waves corresponding to subscripts 1 and 3 have elastic behaviour whereas the wave corresponding to subscript 2 represents the

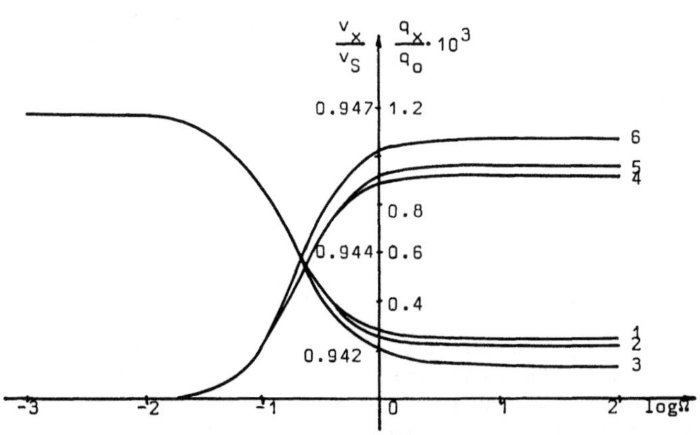

Fig.2. Dimensionless speeds and attenuation coefficients along the x-axis as functions of the dimensionless frequency: curve 1 corresponds to v_x/v_S and g=h=0; 2 - v_x/v_S and g=h=0.1; 3 - v_x/v_S and g=0.5, h=0.1; 4 - q_x/q_o and g=h=0; 5 - q_x/q_o and g=h=0.1; 6 - q_x/q_o and g=0.5, h=0.1

thermal influence. From the numerical results it follows that models with finite speed of heat propagation do not lead to considerable quantitative changes of the surface wave characteristics in comparison with classical thermoelasticity.

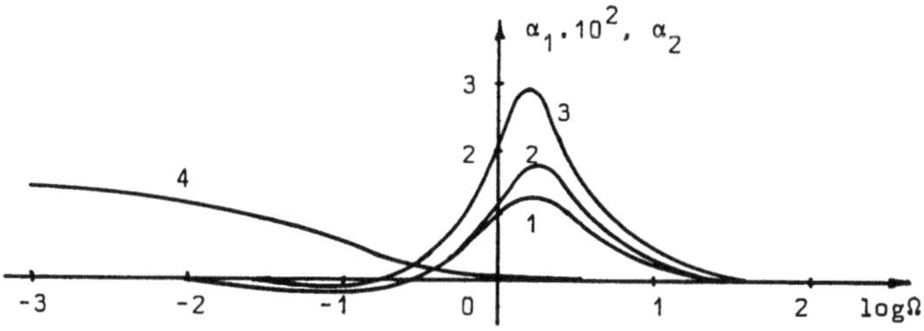

Fig.3. Dependence of the angles between the normals of the planes of constant phase of the 1st and 2nd superposed waves and the x-axis versus the dimensionless frequency: curve 1 corresponds to α_1 and g=h=0, 2 - α_1 and g=h=0.1, 3 - α_1 and g=0.5, h=0.1, 4 - α_2 and g=h=0, g=h=0.1 or g=0.5, h=0.1

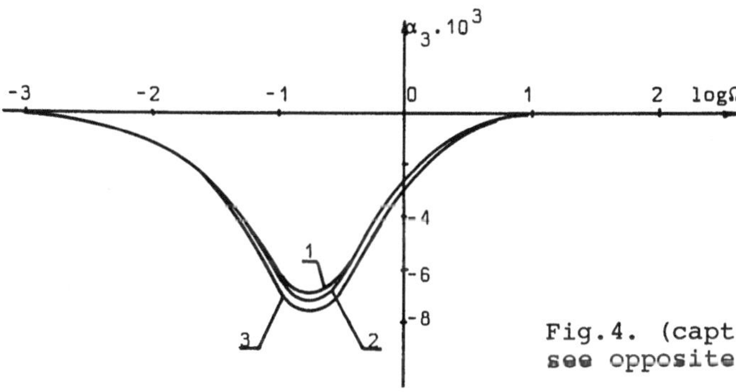

Fig.4. (caption see opposite page)

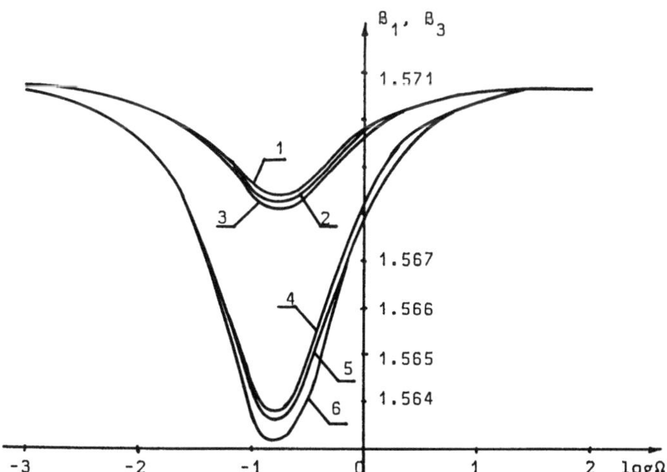

Fig.5. (Caption see opposite page)

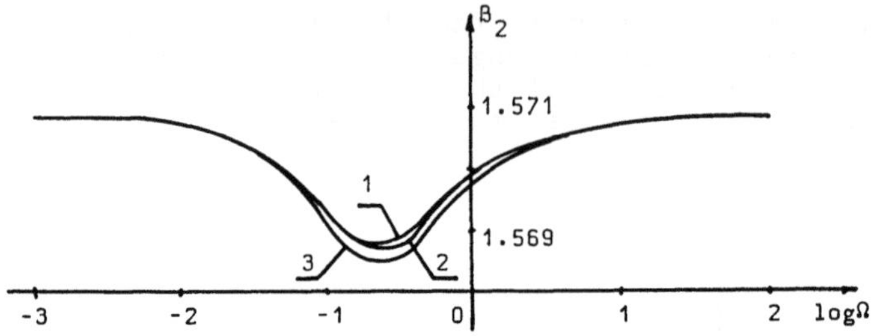

Fig.6. Dependence of the angle between the normal of the planes of constant amplitude of the 2nd superposed wave and the x-axis versus the dimensionless frequency: curve 1 corresponds to g=h=0; 2 - g=h=0.1; 3 - g=0.5, h=0.1

References

1. J.W. Rayleigh: Proc. Lond. Math. Soc. 17, 1 (1885)
2. P. Chadwick: In Progress in Solid Mechanics, ed. by I.N.Sneddon and R. Hill, Vol.1 (North-Holland, Amsterdam 1960)
3. P. Chadwick and D. W. Windle: Proc. R. Soc. London A280, 1 (1964)
4. R. J. Atkin and P. Chadwick: J. Therm. Stresses 4, 3-4 (1981)
5. A. Nayfeh and S. Nemat-Nasser: Acta Mech. 12, 1 (1971)
6. V. K. Agarwal: J. Elast. 8, 2 (1978)
7. P. K. Currie, M. A. Hayes and P. M. O'Leary: Quart. Appl. Math. 35, 1 (1977)
8. P. K. Currie and P. M. O'Leary: Quart. Appl. Math. 35, 4 (1978)
9. V. S. Captain and P. Chadwick: Quart. J. Mech. Appl. Math. 39, 3 (1986)
10. Ts. P. Ivanov: Wave Motion 10 (1988) (in press)
11. P. K. Currie: Quart. J. Mech. Appl. Math. 27, 4 (1974)
12. M. Hayes and R. Rivlin: ZAMP 13, 1 (1962)
13. H. W. Lord and Y. Shulman: J. Mech. Phys. Solids 15, 3 (1967)
14. A. E. Green and K. L. Lindsay: J. Elast. 2, 1 (1972)

Fig.4. Dependence of the angle between the normal of the planes of constant phase of the 3rd superposed wave and the x-axis versus the dimensionless frequency: curve 1 corresponds to g=h=0; 2 - g=h=0.1; 3 - g=0.5, h=0.1

Fig.5. Dependence of the angles between the normals of the planes of constant amplitude of the 1st and 3rd superposed waves and the x-axis versus the dimensionless frequency: curve 1 corresponds to $ß_1$ and g=h=0; 2 - $ß_1$ and g=h=0.1; 3 - $ß_1$ and g=0.5, h=0.1; 4 - $ß_3$ and g=h=0; 5 - $ß_3$ and g=h=0.1; 6 - $ß_3$ and g=0.5, h=0.1

Surface Waves in Thermoelasticity with Relaxation Times

R. Wojnar

IPPT-PAN, ul. Swiętokrzyska 21, PL-00-049 Warsaw, Poland

1. INTRODUCTION

In this study we deal with a stationary thermoelastic surface wave within a linear generalized thermoelasticity in which thermoelastic pulses propagate with finite wave speeds, [1].

There are at least two different theories that generalize classical linear theormoelasticity and describe thermoelastic disturbances in a wave like manner. One is called a thermoelasticity with one relaxation time, [2,3] and another is referred to as a thermoelasticity with two relaxation times, [1,4]. A study of thermoelastic surface waves based on the one relaxation time theory was given in [5] where a dispersion relation involving a frequency and a relaxation time t^0 was obtained. Also, an analysis of thermoelastic surface waves within the two relaxation times theory was presented by several authors [6,7]. The analysis of these authors ends up however with a formal dispersion relation.

In the present study we go a step further and give a qualitative and quantitative analysis of the dispersion relation for a thermoelastic surface wave in the two relaxation times theory. The analysis is based on a concept of the complex valued relaxation time and on a concept of the complex valued thermoelastic coupling parameter, and it follows up the results of the author given in [8]. We show that there is a correspondence between the dispersion relations in the one relaxation time and two relaxation times that allows us to analyse the dispersion relation of the two relaxation times theory similarly to that of the one relaxation time theory.

2. BASIC RELATIONS

We consider a thermoelastic surface wave in a half-space $x_2 \geq 0$ that is free of body forces and heat sources. The boundary of the semi-space is assumed to be free of tractions and thermally insulated.

Let u_i and Θ represent the Cartesian components of the displacement and the temperature above a reference state, respectively.

Moreover, let ρ, λ and μ, α, C_E and K, in this order, denote the density of medium, Lamé moduli, coefficient of linear thermal expansion, specific heat at constant deformation and coefficient of heat conductivity. Let $\gamma = (3\lambda + 2\mu)\alpha$ and let t_0, t_1 be the relaxation times of the two relaxation times theory, [1,4].

Starting from the displacement - temperature equations of linear thermoelasticity, i.e. from the equations of motion

$$\mu\, u_{i,jj} + (\lambda+\mu) u_{j,ji} - (\Theta + t_1\dot{\Theta})_{,i} = \rho\, \ddot{u}_i$$

$$K\, \Theta_{,ii} = C_E(\Theta + t_0\ddot{\Theta}) + \gamma\, \Theta_0\, \dot{u}_{i,i}\ , \qquad (x_2>0)$$

and from the boundary conditions

$$u_{1,2} + u_{2,1} = 0, \quad \lambda\, u_{1,1} + (\lambda+2\mu) u_{2,2} - \gamma(\Theta + t_1\dot{\Theta}) = 0$$

$$\Theta_{,2} = 0\ , \qquad (x_2=0)$$

we look for $(\underline{u},\Theta)$ in the form

$$(u_1, u_2, \Theta) = (U,V,\theta)\ \exp[-\tilde{a}x_2 + i(\tilde{q}x_1 - \tilde{\omega}t)].$$

As a result we arrive at the dispersion relation, [8],

$$G^2[q^2 - \omega^2(\tau_0 + \varepsilon\tau_1) - i\omega(1+\varepsilon) + a_2a_3]^2$$
$$= (a_2^2\, a_3^2/q^2)[2q^2 - \omega(1+\tau_0+\varepsilon\tau_1) - i\omega(1+\varepsilon) + 2a_2a_3]. \tag{1}$$

Here

$$G = \frac{q}{a_1}\ \left(\frac{a_1^2 + q^2}{2q^2}\right)^2$$

$$a_1^2 = q^2 - b^2\omega^2, \qquad a_2^2 + a_3^2 = - P - Q - R, \qquad a_2^2a_3^2 = P\,Q - q^2R$$

and

$$P = \omega^2 - q^2, \qquad Q = - q^2 + \omega^2\tau_0 + i\omega, \qquad R = \varepsilon\omega(i + \omega\tau_1).$$

Also ω and q are frequency and length of the wave vector in the units of Chadwick-Sneddon, $\omega = \tilde{\omega}/\omega^*$, $q = \tilde{q}\ c_1/\omega^*$, $\omega^* = c_1^2\ C_E/K$, τ_0 and τ_1 are relaxation times in these units. Moreover $b^2 = 2 + \lambda/\mu = c_1^2/c_2^2$ where c_1, c_2 are the velocities of the longitudinal and shear waves, respectively, and ε is the coupling constant

$$\varepsilon = \frac{\gamma^2}{\lambda + 2\mu}\ \frac{\Theta_0}{C_E}\ .$$

3. COMPLEX VALUED RELAXATION TIMES

We introduce the notation

$$T_0 = \tau_0 + \frac{i}{\omega}\ , \qquad T_1 = \tau_1 + \frac{i}{\omega} \tag{2}$$

and

$$X = \frac{\omega^2}{q^2}\ . \tag{3}$$

Then eq.(1) can be written in the form

$$G[1 - X(T_0+\varepsilon T_1) + A]^2 = A^2[2 - X - X(T_0+\varepsilon T_1) + 2A], \tag{4}$$

where

$$G \equiv G^2 = (1 - \frac{b^2}{2} X)^4/(1 - b^2 X)$$
$$A^2 = (a_2 a_3/q^2)^2 = (1-X)(1-XT_0) - \varepsilon XT_1. \qquad (5)$$

Letting $\omega \to 0$, $q \to 0$ and $X \to$ constant $< \infty$ in eq.(4) we arrive at Lockett's result, [9],

$$G = \frac{1 + \varepsilon - X}{1 + \varepsilon}.$$

If the relaxation times, τ_0 and τ_1, coincide, $T_1 = T_0 = T^0$, and we have

$$G[1 - (1+\varepsilon)XT^0 + A]^2 = A^2[2 - X - (1+\varepsilon)XT^0 + 2A] \qquad (6)$$

with

$$A^2 = (1 - X)(1 - XT^0) - \varepsilon XT^0.$$

If the thermoelastic coupling constant vanishes, $\varepsilon=0$, eq.(4) becomes

$$G[1 - XT_0 + A]^2 = A^2[2 - X - XT_0 + 2A] \qquad (7)$$

with

$$A^2 = (1 - X)(1 - XT_0). \qquad (8)$$

Since

$$2 - X - XT_0 + 2A = (\sqrt{1-X} + \sqrt{1-XT_0})^2$$
$$[1 - XT_0 + A]^2 = (1-XT_0)(2 - X - XT_0 + 2A)$$

and

$$1 + A^2 = 2 - XT_0 - X + X^2T_0$$

therefore eq.(7) can be written as

$$[G - (1-X)](1-XT_0)[\sqrt{1-X} + \sqrt{1-XT_0}]^2 = 0. \qquad (9)$$

Because the last bracket of L.H.S. cannot vanish, we have the relations

$$G = 1 - X, \qquad X = \frac{1}{T_0} \qquad (10)$$

the first of which corresponds to a classical mechanical Rayleigh wave, while the second corresponds to a thermal wave.

4. COMPLEX VALUED THERMOELASTIC COUPLING PARAMETER

Let us note that the coupling constant ε goes along with T_1 in (4). We introduce the parameter

$$E = \varepsilon \frac{T_1}{T_0}, \qquad (11)$$

which can be called a complex valued thermoelastic coupling parameter.

Clearly, E is no longer a constant since T_1 and T_0 depend on frequency ω. In terms of E eq.(4) takes the form

$$G[1 - XT_0(1+E) + A]^2 = A^2[2 - X - XT_0(1+E) + 2A], \tag{12}$$

where

$$A^2 = (1 - X)(1 - XT_0) - E\,XT_0. \tag{13}$$

If E in (12) is identified with a real valued thermoelastic coupling constant ε, then eq.(12) is identical with the dispersion relation of the one relaxation time theory, cf.[5].

5. NUMERICAL ANALYSIS

The solution of the dispersion equation (12) is now obtained numerically by a Newton method. We let

$$X_{n+1} = X_n - \frac{f(X_n)}{f'(X_n)}, \qquad n = 0,1,2,\ldots$$

where

$$f' = \frac{df}{dX}$$

and

$$f(X) = G(1 - X\,T + A)^2 - A^2(2 - X - X\,T + 2A)$$

with G and A given by (5) and $T = T_0 + \varepsilon T_1$.
For

$$\frac{1}{\tau_0} < c_R^2 = 0.2817$$

we put $X_0 = (1/T_0, 0)$ (thermal mode) while for

$$\frac{1}{\tau_0} > c_R^2$$

we let $X_0 = (c_R, 0)$ (mechanical Rayleigh mode).

The convergence of the iteration process was shown to be good. Generally, about ten iterations (less than 20) were needed to satisfy the estimate

$$|X_{n+1} - X_n| < 10^{-6}, \qquad n < 20.$$

However, for some values of the pairs (τ_0, τ_1), e.g. (10/3, 10/3), (10/3, 20/3), (3,6), (2.5, 12.5) the iteration process had to be modified by choosing another initial value from a graphical interpolation of the known result, (see Fig.1).

Graphs of the squared velocity

$$c^2 = \left[\frac{1}{\mathrm{Re}(1/\sqrt{X})}\right]^2$$

and the attenuation coefficient

$$\eta = \omega \ \mathrm{Im}(1/\sqrt{X})$$

as functions of $(1/\tau_0)$ obtained by the numerical method are shown in Fig.1.

6. ASYMPTOTICAL ANALYSIS

The behaviour of X for small values of $1/\tau_0$, ($\lesssim 0.2$) can also be studied by solving the dispersion equation (12) in an asymptotic way.

To this end we seek X in the form

$$\frac{1}{X} = T_0(1 + E\,\xi_1 + E^2\,\xi_2 + \ldots), \tag{14}$$

where ξ_1, ξ_2, ... are unknown coefficients.

Proceeding in a way similar to that of [5], we find (see Appendix)

$$\xi_1 = \frac{T_0}{T_0 - 1} \tag{15}$$

Retaining only the term linear in E we have

$$\frac{1}{X} = T_0\left(1 + E\,\frac{T_0}{T_0 - 1}\right)$$

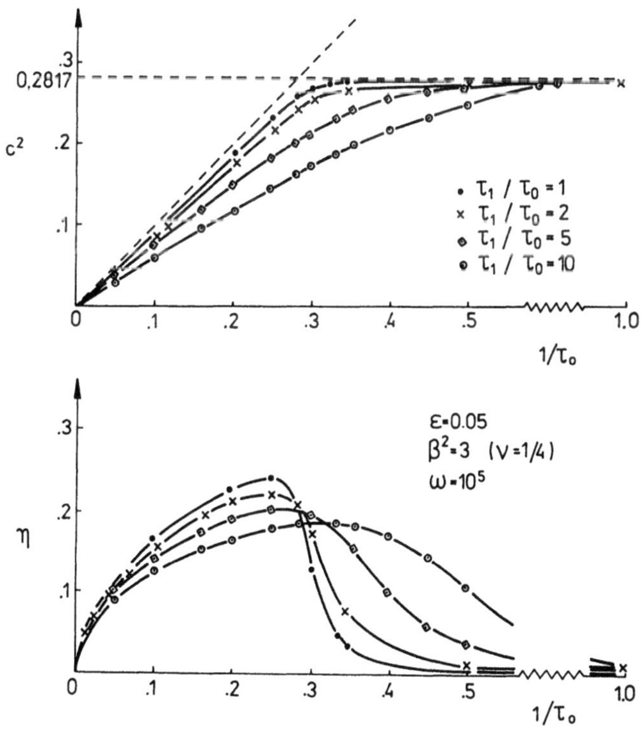

Fig.1. The squared velocity c^2 and the attenuation coefficient η as functions of $(1/\tau_0)$ for different ratios (τ_1/τ_0)

or, by the definition of E, (11), we obtain

$$\frac{1}{X} = T_0(1 + E\frac{T_1}{T_0 - 1}).$$

If, additionally, $\omega >> 1$, then

$$c = \frac{1}{\mathrm{Re}(1/\sqrt{X})} = \frac{1}{\sqrt{\tau_0}}(1 - \frac{1}{2}\varepsilon\frac{\tau_1}{\tau_0})$$

$$\eta = \omega\,\mathrm{Im}(1/\sqrt{X}) = \frac{1}{2\sqrt{\tau_0}}[1 + \varepsilon(1 - \frac{1}{2}\frac{\tau_1}{\tau_0})].$$

Thus, $c^2 \sim 1/\tau_0$, (see the Fig.1) and $\eta \sim 1/\sqrt{\tau_0}$. Therefore, for small values of τ_0^{-1} the phase velocity c and the damping coefficient η are the same as for a plane harmonic wave propagating in an infinite thermoelastic body with two relaxation times.

7. REFERENCES

1. J.Ignaczak, Linear Dynamic Thermoelasticity - a Survey, The Shock and Vibration Digest 13, 3-8, (1981)
2. S.Kaliski, Wave Equations in Thermoelasticity, Bull.Acad.Pol.Sci. techn. 13, 253-260, (1965)
3. H.W.Lord, Y.Shulman, A Generalized Dynamical Theory of Thermoelasticity, J.Mech.Phys.Solids 15, 299-309, (1967)
4. A.E.Green, K.A.Lindsay, Thermoelasticity, J.Elasticity 2, 1-7, (1972)
5. A.Nayfeh, S.Nemet-Nasser, Thermoelastic Waves with Thermal Relaxation, Acta Mech. 12, 53-69, (1971)
6. V.K.Agarval, On Surface Waves in Generalized Thermoelasticity, J.Elasticity 8, 171-177, (1978)
7. D.S.Chandrasekharaiah, K.R.Srikantaiah, On Temperature - Rate Dependent Thermoelastic Rayleigh Waves in Halfspace with Material Boundary, Gerlands Beitr.Geophys. 93, 133-141, (1984)
8. R.Wojnar, Rayleigh Waves in Thermoelasticity with Relaxation Times, 2nd Intern.Conference on Surface Waves in Plasmas and Solids, Ohrid, Yugoslavia, Sept.5-11, 1985, (World Sci.Publ.,Singapore 1986)
9. F.J.Lockett, Effect of Thermal Properties of a Solid on the Velocity of Rayleigh Waves, J.Mech.Phys.Solids 7, 71-75, (1958)
10. D.Tao, J.H.Prevost, Relaxation Effects on Generalized Thermoelastic Waves, J.Therm.Stresses 7, 79-89, (1984)

APPENDIX

We divide both sides of Eq.(12) by X^2, and obtain

$$G[\frac{1}{X} - T_0(1 + E) + A]^2 = A^2[\frac{2}{X} - 1 - T_0(1 + E) + 2A]X, \tag{A1}$$

where

$$A^2 = \frac{1}{X^2}A^2 = (\frac{1}{X} - 1)(\frac{1}{X} - T_0) - E\,T_0\frac{1}{X}. \tag{A2}$$

Next, we transform (A1) to the form involving A^2 only

$$[G\frac{1}{X}\left\{[\frac{1}{X} - T_0(1 + E)]^2 + A^2\right\} - A^2[\frac{2}{X} - 1 - T_0(1 + E)]]^2$$

$$= \left\{ A^2 - G \frac{1}{X} \left[\frac{1}{X} - T (1 + E)\right]\right\}^2 4A^2. \tag{A3}$$

If we let

$$W = 1 + E W_1, \tag{A4}$$

where

$$W_1 = \xi_1 + E \xi_2 + \dots, \tag{A5}$$

the expansion (14) can be written as

$$\frac{1}{X} = T (1 + W_1) \tag{A6}$$

or

$$X = \frac{1}{T_0(1 + E W_1)} . \tag{A7}$$

Now, if we insert (A6) into (A2) we get

$$A^2 = E T_0[T_0 W(W_1 - 1) - W_1]$$

or after using (A5) we obtain

$$A^2 = E T_0 B, \tag{A8}$$

where

$$B = Z_0 + E T_0 M^2 + E^2 T_0 N$$

with

$$Z_0 = (T_0 - 1) \xi_1 - T_0 \tag{A9}$$

and

$$T_0M^2 = (T_0 - 1) \xi_2 + T_0\xi_1(\xi_1 - 1)$$

$$N = \xi_2(2\xi_1 - 1 + E \xi_2) .$$

From (A6) we find

$$\frac{1}{X} - T_0(1 + E) = E T_0(\xi_1 - 1 + E \xi_2) \tag{A10}$$

and

$$\frac{2}{X} - 1 - T_0(1 + E) = T_0 - 1 + E T_0(2\xi_1 - 1) + 2E^2 T_0\xi_2 . \tag{A11}$$

Moreover, to within a linear term in E, we have

$$G = g_0 + E g_1,$$

where

$$g_0 = G\Big|_{E=0} = (1 - \frac{\beta^2}{2T_0})^4/(1 - \frac{\beta^2}{T_0}) \quad \text{and} \quad g_1 = \frac{\partial G}{\partial E}\Big|_{E=0} .$$

Thus

$$G \frac{1}{X} = g_0 T_0 + E T_0(g_1 + g_0 W_1 + E g_1 W_1 . \tag{A12}$$

Now, if (A8) - (A12) is inserted into (A3) we get an equation in which the coefficient of $0(E^0)$ must vanish. As a result we obtain

$$Z_0 = 0,$$

which by (A9) leads to the desired result (15).

Surface Waves in a Nonlocal Elastic Medium

C. Rymarz

Military Academy of Technology, PL-00-908 Warsaw, Poland

1 Introduction

Spatial nonlocality, as developed in the literature |1| - |4|, is involved in modelling such physical phenomena as:
- existence of long-range interactions in some types of solid state media (ionic crystals),
- micro-inhomogeneous material behaviour (polycrystalline media) and some others.

In all these cases a nonlocality parameter of the dimensions of length is introduced, being interpreted as a range of interatomic or intermolecular interactions, or as a correlation length in a stochastic description of polycrystalline media. For investigation of dynamic processes in such media in a majority of cases the nonlocal theory of elasticity is suitable.

The aim of this paper is to consider the impact of spatial nonlocality on the behaviour partly of bulk, but mainly of surface, elastic waves.

2 Nonlocal Theory Construction for a Crystalline Body

We are going to show how the nonlocal theory of elasticity relates to the crystalline medium. Starting from crystal lattice theory we obtain the equations of motion

$$m\ddot{u}_i = \sum_{k'} \Phi_{ij}(\underline{k}-\underline{k}')\, u_j(\underline{k}') . \tag{1}$$

After an interpolation process is performed, the relation (1) is transformed into the continuous one

$$\varrho \ddot{u}_i = \int_{\Omega} \Phi_{ij}(\underline{x}-\underline{x}')\, u_j(\underline{x}')\, d\vartheta', \tag{2}$$

where: ϱ-is a mass density, Φ_{ij} - dynamic matrix of the crystalline lattice, u_j - displacement field, and Ω - nonlocality area determined by the nonlocality parameter. The next step in constructing the model of nonlocal elasticity is the transition from the dynamic continuous matrix $C_{ijkl}(\underline{x}-\underline{x}')$ to the nonlocal "Hooke tensor" $\Phi_{ij}(\underline{x}-\underline{x}')$

$$\Phi_{ij}(\underline{x}-\underline{x}') \longrightarrow C_{ijkl}(\underline{x}-\underline{x}') . \tag{3}$$

Depending on the nature of the physical interactions in a crystal, there are two different possible forms for the tensor C_{ijkl}.

When interactions may be strictly divided into long-range and short-range ones, the tensor C_{ijkl} takes the form

$$C_{ijkl} = C^1_{ijkl}\,\delta(\underline{x}-\underline{x}') + C^2_{ijkl}(\underline{x}-\underline{x}'). \qquad (4)$$

The tensor C^1_{ijkl} is determined according to the known procedure presented by BORN and HUANG in [5] . For the determination of C^2_{ijkl} we use the equality of elastic energies in the interpolated crystalline medium and in the nonlocal elastic one

$$\iint_{V\Omega} d\vartheta\, d\vartheta'\, u_i(\underline{x})\,\Phi_{ij}(\underline{x}-\underline{x}')\,u_j(\underline{x}') = \iint_{V\Omega} d\vartheta\, d\vartheta'\, u_{ij}(\underline{x})\,C^2_{ijkl}(\underline{x}-\underline{x}')\,u_{kl}(\underline{x}'). \qquad (5)$$

From this relation we get the following differential equation for C^2_{ijkl}:

$$\partial_j \partial_{l'} C^2_{ijkl}(\underline{x}-\underline{x}') = \Phi_{ik}(\underline{x}-\underline{x}') \qquad (6)$$

and the set of boundary conditions omitted here for simplicity. For bounded bodies an inhomogeneity and additional anisotropy of $C_{ijkl}(x,x')$ appear in some layers close to the boundary (Boundary Layer).

When the division of physical interactions into long and short range parts is impossible, then

$$C_{ijkl} = C^2_{ijkl}\,. \qquad (7)$$

After the tensor C_{ijkl} is determined we use Hamilton's principle to get the equations of motion for the nonlocal theory of elasticity:

$$C^1_{ijkl}\,u_{k,lj} + \int_\Omega C^2_{ijkl}(\underline{x}-\underline{x}')\,u_{k,lj}(\underline{x}')\,d\vartheta' - \rho\, u_{i,tt} = 0 \qquad (8)$$

or for an isotropic medium

$$\mu_1 \nabla^2 u_i + (\lambda_1+\mu_1)u_{j,ji} + \int_\Omega d\vartheta'[\mu_2(r)\nabla^2 u_i(\underline{x}') + (\lambda_2(r)+\mu_2(r))\,u_{j,ji}(\underline{x}')] - \rho\, u_{i,tt} = 0, \qquad (9)$$

where λ_1, μ_1 are Lamé constants and $\lambda_2(r), \mu_2(r)$ nonlocal "Lamé" functions, $r = |\underline{x}-\underline{x}'|$. The integral transformations being performed, we get the functional equation

$$(\mu_1+\mu_2(k))\,k^2\underline{u}+[\lambda_1+\mu_1+\lambda_2(k)+\mu_2(k)]\,\underline{k}(\underline{k}\cdot\underline{u})-\varrho\omega^2\underline{u}=0, \qquad (10)$$

from which the dispersion relation for longitudinal and transverse waves in the nonlocal medium may be obtained:

$$\text{for } \underline{k}\parallel\underline{u} \quad [\lambda_1+2\mu_1+\lambda_2(k)+2\mu_2(k)]\,k^2-\varrho\omega^2=0, \qquad (11)$$

$$\text{for } \underline{k}\perp\underline{u} \quad [\mu_1+\mu_2(k)]\,k^2-\varrho\omega^2=0. \qquad (12)$$

Before considering boundary - value problems for surface waves (SW) in the nonlocal elastic medium some analysis of the dispersion properties of bulk waves should be performed. They depend on the behaviour of the functions $\lambda_2(k)$, $\mu_2(k)$.

3 Dispersion Properties of Bulk Waves

We assume that the nonlocality functions $\lambda_2(r)$, $\mu_2(r)$ are of finite support and of the following form:

$$\lambda_2(r),\,\mu_2(r)=\begin{cases}(A_1,\,A_2)\,G(r) & \text{for } r\leqslant R\\ 0 & \text{for } r>R\,.\end{cases} \qquad (13)$$

This means that we take into account finite range nonlocality. Generally this should be treated as some asymptotic approach. The Fourier transform of (13) is an entire function with an infinite number of zeros, which means that in a nonlocal elastic medium may exist generally an infinite number of modes.

In a nonlocal medium the following stability conditions should be satisfied:

$$[\lambda_1+2\mu_2+\lambda_2(k)+2\mu_2(k)]>0 \qquad (14)$$

$$[\mu_1+\mu_2(k)]>0 \qquad (15)$$

for $k\neq 0$, since otherwise $\omega^2<0$, which implies occurrence of instability (infinite growth of wave amplitudes). To obtain the asymptotic transition to the local theory of elasticity when $k\to 0$, the classical acoustic waves should exist, which is satisfied when

$$\lambda_2(k),\,\mu_2(k)\longrightarrow 0 \quad \text{for} \quad k\longrightarrow 0, \text{ if } \lambda_1\neq 0,\ \mu_1\neq 0$$

$$\lambda_2(k),\,\mu_2(k)\longrightarrow 1 \quad \text{for} \quad k\longrightarrow 0, \text{ if } \lambda_1=\mu_1=0.$$

Moreover there are two infinite sets of complex zeros

$k^1_n=\alpha^1_n+i\delta^1_n$ L (longitudinal modes)

$k^2_n=\alpha^2_n+i\delta^2_n$ T (transverse modes).

They are so-called bound modes having the form of spatially attenuated waves and thus being observable only in some vicinity of an excitation source. Violating the above-formulated stability conditions we can get more acoustic modes than classical ones, but in this case some regions of the spectrum are prohibited.

4 Surface Waves in a Nonlocal Elastic Medium (SW)

For the frame reference as in Fig.1 boundary conditions have the classical form $\sigma_{31} = \sigma_{33} = 0$, or taking into account non-local constitutive relations,

$$\lambda_1(u_{1,1}+u_{3,3}) + 2\mu_1 u_{3,3} + \int_{\Omega_p} \lambda_2(r)(u_{1,1}+u_{3,3})\,d\vartheta' + \int_{\Omega_p} \mu_2(r)\,u_{3,3}\,d\vartheta' = 0, \tag{16}$$

$$\mu_1(u_{1,3}+u_{3,1}) + \int_{\Omega_p} \mu_2(r)\,(u_{1,3}+u_{3,1})\,d\vartheta' = 0, \tag{17}$$

where Ω_p is presented in Fig.1

We assume that solutions for surface waves (SW) have such a form

$$u_1 = \sum_{n=1}^{N} [A_n^1 e^{ik_{3n}^1 x_3} + A_n^2 e^{ik_{3n}^2 x_3}]\, e^{i(k_1 x_1 - \omega t)} \tag{18}$$

$$u_3 = \sum_{n=1}^{N} [B_n^1 e^{ik_{3n}^1 x_3} + B_n^2 e^{ik_{3n}^2 x_3}]\, e^{i(k_1 x_1 - \omega t)} . \tag{19}$$

For SW, k_{3n}^i $(i=1,2)$ should satisfy the following conditions:

if $k_n^1(\omega)$, $k_n^2(\omega)$ are real, then $(k_{3n}^i)^2 = (k_n^i)^2 - k_1^2$,

if $k_n^1(\omega)$, $k_n^2(\omega)$ are complex, then $|k_1| \leqslant |k_n^i|$.

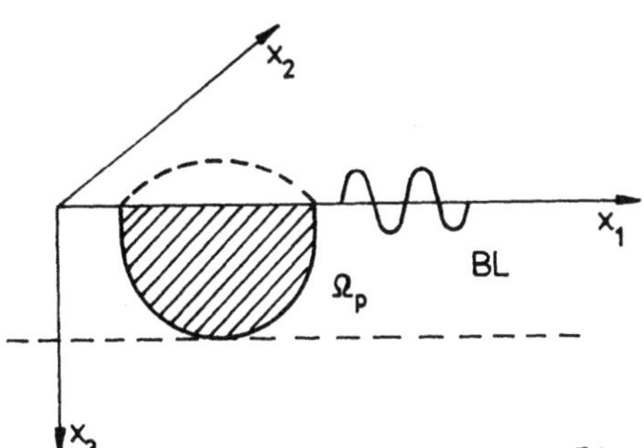

Fig.1

These relations determine the number of modes that should be taken in (18), (19). Substituting (18), (19) into the equations of motion, after some transformations we obtain

$$\sum_{n=2}^{N} L_{11}(\underline{x})\, A_n^1 + L_{12}(\underline{x}) A_n^2 = 0 \qquad (20)$$

$$\sum_{n=2}^{N} L_{21}(\underline{x})\, A_n^1 + L_{22}(\underline{x}) A_n^2 = 0. \qquad (21)$$

These functional equations are satisfied identically inside the body, beyond some boundary layer (BL) determined by the range of nonlocality. The meaning of BL is shown in Fig.2.

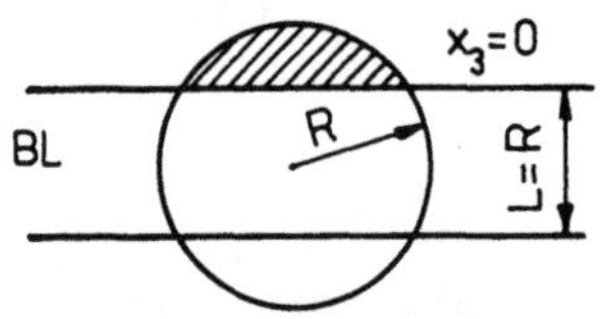

Fig.2

When $x \in$ BL, then we observe the lack of symmetry (shadowed area) and the internal forces remain uncompensated. Using the projection procedure with the nonclassical modes as base functions we obtain 2(n-1) algebraic homogeneous equations. From boundary conditions (16), (17) we get two additional algebraic equations for A_n^1, A_n^2. From the requirement that the set of 2n algebraic equations should have nontrivial solutions we get the relation

$$\det[B_{ij}] = 0 \qquad (22)$$

and from this the dispersion relation for SW.

The procedure for construction of (22) is here much more complicated than in local elasticity.

5 Examples of Nonlocal SW

In the first example we assume that $\lambda_1 = \mu_1 = 0$, and

$$\lambda_2(r),\ \mu_2(r) = \begin{cases} (A_1, A_2)\exp[-\dfrac{r}{hR(h)}] & r \leqslant R \\ 0 & r > R. \end{cases} \qquad (23)$$

The constants A_1, A_2 are determined from the conditions that

$$\lambda_2(r) \rightarrow \lambda_2 \delta(r),\quad \mu_2(r) \rightarrow \mu_2 \delta(r) \quad \text{as } h \rightarrow 0.$$

Then

$$A_1, A_2 = \frac{\lambda_2, \mu_2}{6V_\Omega} \frac{1}{h^3} \tag{24}$$

and

$$\lambda_2(k), \mu_2(k) = (\lambda, \mu) H(k) k^{-1} [1 + (hk)^2]^{-2}, \tag{25}$$

where

$$H(k) = k - 0{,}5 e^{-\frac{1}{h}} A(k) \sin(k - \varphi(k)) . \tag{26}$$

The function H(k) has the following properties:

$$H(k) \longrightarrow k \quad \text{for} \quad k \longrightarrow 0, \quad H(k) \longrightarrow 0 \quad \text{for} \quad k \longrightarrow k_0$$

and for $k > k_0$, $H(k) < 0$. This means an instability of wave motion for $k > k_0$. The dispersion curves for bulk waves (L,T) and SW of Rayleigh type are presented in Fig.3

The most characteristic phenomenon here is the existence of ω_{max} for longitudinal (L) and transverse (T) waves, above which there is no wave motion (filtration property). Necessary conditions for SW existence are satisfied in regions I and II. The performed calculations showed that nonlocal SW exists only in region I. In region II SW should be constructed from four modes having rather different phase velocities. These differences are perhaps the main reason for the nonexistence of SW, since the boundary "is no longer able to coordinate" four different modes.

A transversly polarized SW does not exist in an elastic medium. According to a calculation undertaken, such SW exists in nonlocal elasticity (Fig.4).

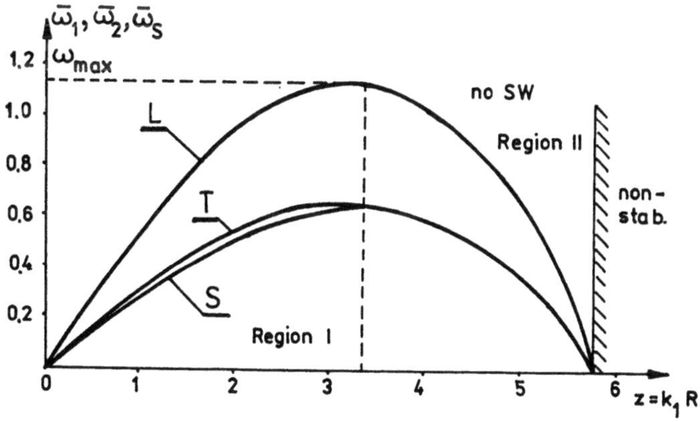

Fig.3. Dispersion curves for bulk (L,T) and surface (S) waves of Rayleigh type

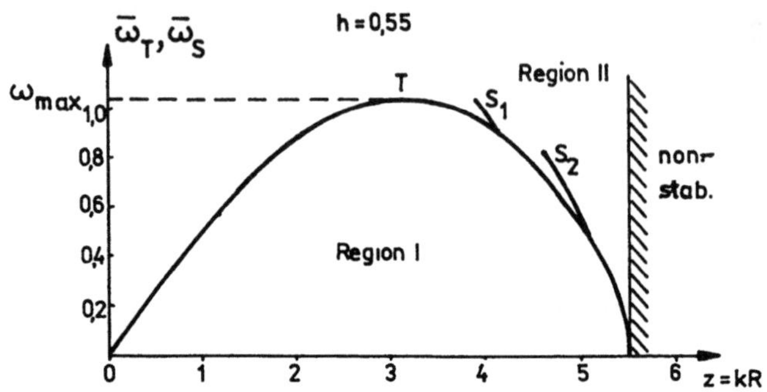

Fig.4. The dispersion curves for bulk (T) and transverse surface (S_1, S_2) waves

Such a phenomenon can be explained by the existence of two bulk modes, which being coordinated by the surface form surface waves. Therefore SW appear only for sufficiently great ω, where the phase velocities of two modes are not so different. It comes out from calculation that in this case two branches of SW are possible. This fact is easily explained.

In the second example $\lambda_1 \neq 0, \mu_1 \neq 0$ and

$$\lambda_2(r), \mu_2(r) = \begin{cases} (\lambda_2, \mu_2)\, G(r) & r \leqslant R \\ 0 & r > R \end{cases} \qquad (27)$$

where

$$G(r) = \exp\left[-\frac{hR^2}{R^2 - r^2}\right]. \qquad (28)$$

Three material nondimensional constants are introduced

$$\gamma_1 = \frac{\lambda_1}{\mu_1}, \qquad \gamma_2 = \frac{\lambda_2}{\mu_2}, \qquad \beta = \frac{\mu_1}{\mu_2} \qquad (29)$$

and stability conditions formulated

$$[1 + \beta G(k)] > 0 \qquad (30)$$

$$[1 + G(k)\beta \frac{2 + \gamma_2}{2 + \gamma_1}] > 0. \qquad (31)$$

The results of calculation are presented in Fig.5.

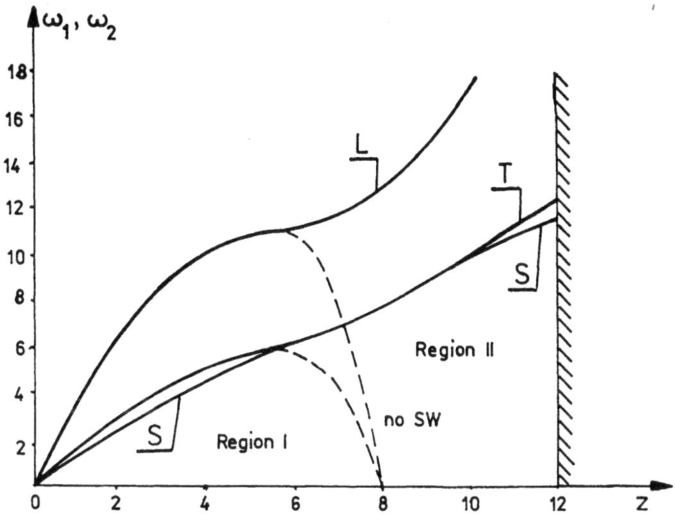

Fig.5. Dispersion curves for bulk and SW waves

Here the cooperation between local and nonlocal parts of the elasticity tensor (constants) is well observable. For sufficiently long waves (small z=kR) the nonlocality prevails, which is reflected in the curving of the dispersion curves. When the wavelength decreases (z growing) the influence of nonlocality diminishes and in the limit of very short waves we pass to the local theory of elasticity. (S,T curves tend to straight lines).

The SW of Rayleigh wave type exists in regions I,II separated by a narrow gap. This gap is here explicable on the grounds of the first example.

6 Conclusions

The presented considerations showed that waves in a nonlocal medium exhibit such nonclassical properties as: existence of ω_{max}, a possible instability region of wave motion, existence of transverse surface waves and some others. Since, as has been shown, the nonlocal elasticity has a good background in the theory of crystal lattices, these properties are in the majority interpretable.

Readers wanting more information on these subjects are referred to the author's papers [6], [7].

7. Literature

1. D.Rogula, On Nonlocal Continuum Theory of Elasticity, AMS, 25, 2, 1973
2. E.Kröner, Elasticity Theory of Materials with Long - range Cohesive Forces, Int.J.Sol. Str.3, 3, 1967.

3. M.J. Beran, Application of Statistical Theories to Heterogeneous Materials, Phys.Stat.Solid.(a), 6, 365, 1971.
4. Cz.Rymarz, Continuous Nonlocal Models of Bounded Elastic Medium, Proc.Vibr.Probl., 15, 4, 1974
5. M.Born, K.Huang, Dynamical Theory of Crystal Lattices, Oxford, Clarendon Press, 1954.
6. S.Kaliski, Cz.Rymarz, Surface Waves in Continuous Medium with Non-local Interactions, Proc.Vibr.Prob., 12, 2, 1971
7. S.Kaliski, Cz.Rymarz, Transverse Surface Waves in a Continuous Body with Non-local Interaction, AMS, 25, 5, 1971.

Index of Contributors

MIX
Papier aus verantwortungsvollen Quellen
Paper from responsible sources
FSC® C105338

If you have any concerns about our products,
you can contact us on
ProductSafety@springernature.com

In case Publisher is established outside the EU,
the EU authorized representative is:
Springer Nature Customer Service Center GmbH
Europaplatz 3, 69115 Heidelberg, Germany

Printed by Libri Plureos GmbH
in Hamburg, Germany